Targeted Drug Delivery

Methods and Principles in Medicinal Chemistry

Edited by
R. Mannhold, H. Buschmann, J. Holenz

Previous Volumes of the Series

Alza, E. (Ed.)

Flow and Microreactor Technology in Medicinal Chemistry

2022
ISBN: 978-3-527-34689-9
Vol. 81

Rübsamen-Schaeff, H., and Buschmann, H. (Eds.)

New Drug Development for Known and Emerging Viruses

2022
ISBN: 978-3-527-34337-9
Vol. 80

Gruss, M. (Ed.)

Solid State Development and Processing of Pharmaceutical Molecules
Salts, Cocrystals, and Polymorphism

2021
ISBN: 978-3-527-34635-6
Vol. 79

Plowright, A.T. (Ed.)

Target Discovery and Validation Methods and Strategies for Drug Discovery

2020
ISBN: 978-3-527-34529-8
Vol. 78

Swinney, D., Pollastri, M. (Eds.)

Neglected Tropical Diseases Drug Discovery and Development

2019
ISBN: 978-3-527-34304-1
Vol. 77

Bachhav, Y. (Ed.)

Innovative Dosage Forms Design and Development at Early Stage

2019
ISBN: 978-3-527-34396-6
Vol. 76

Gervasio, F. L., Spiwok, V. (Eds.)

Biomolecular Simulations in Structure-based Drug Discovery

2018
ISBN: 978-3-527-34265-5
Vol. 75

Sippl, W., Jung, M. (Eds.)

Epigenetic Drug Discovery

2018
ISBN: 978-3-527-34314-0
Vol. 74

Giordanetto, F. (Ed.)

Early Drug Development

2018
ISBN: 978-3-527-34149-8
Vol. 73

Handler, N., Buschmann, H. (Eds.)

Drug Selectivity

2017
ISBN: 978-3-527-33538-1
Vol. 72

Targeted Drug Delivery

Edited by
Yogeshwar Bachhav

Series Editors
Raimund Mannhold
Helmut Buschmann
Jörg Holenz

WILEY-VCH

Editor

Dr. Yogeshwar Bachhav
B 401, No 126
Florencio
Tilak Nagar
400089 Mumbai
India

Series Editors

Prof. Dr. Raimund Mannhold
Rosenweg 7
40489 Düsseldorf
Germany

Dr. Helmut Buschmann
Sperberweg 15
52076 Aachen
Germany

Dr. Jörg Holenz
BIAL - PORTELA & CA., S.A.
São Mamede Coronado
Portugal

Cover Image:
© SciWhaleDesign/Shutterstock

Library of Congress Card No.: applied for

British Library Cataloguing-in-Publication Data
A catalogue record for this book is available from the British Library.

Bibliographic information published by the Deutsche Nationalbibliothek
The Deutsche Nationalbibliothek lists this publication in the Deutsche Nationalbibliografie; detailed bibliographic data are available on the Internet at <http://dnb.d-nb.de>.

Print ISBN: 978-3-527-34781-0
ePDF ISBN: 978-3-527-82786-2
ePub ISBN: 978-3-527-82787-9
oBook ISBN: 978-3-527-82785-5

Cover Design: SCHULZ Grafik-Design
Typesetting: Straive, Chennai, India
Printing and Binding: CPI Group (UK) Ltd, Croydon, CR0 4YY

C106471_051022

Contents

A Personal Foreword

Paul Ehrlich, the German scientist who was awarded Nobel Prize in Physiology or Medicine in the year 1908, formulated the "lock-and-key theory" of drug and receptor interaction. This was a significant milestone to understand the interaction of the drug molecules with the target receptor to induce any physiological activity, either desired or undesired. All the adverse events or side effects of any drug are linked to the undesired physiological activity induced by these therapeutic agents. And that's where the concept of the targeted drug delivery gains prime importance in the field of medicine. The overall and foremost objective of the drug-delivery scientist is to deliver the drug at the right site in a desired amount to achieve the cure without inducing any adverse effects.

The present edition focuses especially on the development of drug-delivery systems on the interface discovery and of clinical development of investigational drugs which covers both approved and marketed drugs, where targeting concept should allow precise delivery of a moiety at the site of action. Most importantly, the issue focuses on the targeted drug delivery of not only small molecules but also complex biologicals such as peptides, proteins, and antibodies.

I would like to thank all the authors for their contribution to this book; without their support, this compilation was not possible. You guys really went out of way to collect all the novel findings in the designated area which significantly increases the overall value of this edition.

My sincere thanks to Dr. Frank Weinreich for his timely guidance, fruitful discussion, and feedback to all the urgent queries. I would also thank to Ms. Stefanie Volk for all her help during the publication work.

Finally, I would like to thank Dr. Helmut Buschmann for all his encouragement; without that, it was simply not possible to edit two books for Wiley-VCH. Furthermore, I would like to dedicate this book to my Mother (a school drop-out from rural India) whose teachings always helped to fight against all the odds and to excel further.

Mumbai
June 2022

Dr. Yogeshwar Bachhav
Founder and Director
Adex Pharma Consultancy Services

Preface

The latest volume in our book series *Methods and Principles in Medicinal Chemistry* with the book title *Targeted Drug Delivery*, edited by Yogeshwar Bachhav, highlights the critical role of targeted drug delivery for unmet medical needs, by describing a wide range of different approaches for targeting small-molecule as well as peptide and macromolecular drugs. In the 15 chapters written by world-renowned experts in their field, a broad range of specific formulation aspects for different types of drug classes are provided. The art and science to craft good drugs require an ever-growing plethora of skills from a medicinal chemist. The overall process of drug discovery however requires significant changes to become a more sustainable endeavor. There are many aspects in the evolving role of medicinal chemistry to consider – from potentially encompassing different molecule types to moving closer to biological and pharmaceutical development to chemically navigate efficiently in biological and drug delivery space to better designing effective molecules and addressing the biological target within a living system. The best-designed molecule for a specific disease is successful only if it can be introduced to a patient safely and efficiently. Formulation is becoming a fast-growing success factor in drug development, and matrix interactions of drug-delivery systems are becoming a key role in early development stages.

The editor has intentionally included chapters which put an emphasis on the development on the interface of drug discovery and drug development and not just on approved or marketed drugs. Above all, the book covers the targeting options not only for small molecules but also for complex biologicals such as nucleic acids, peptides/proteins, and antibodies. Key aspects of this field are presented in the following chapters:

- Basics of targeted drug delivery
- Addressing Unmet Medical Needs using Targeted Drug-Delivery Systems: Emphasis on Nanomedicine-Based Applications
- Nanocarriers-Based Targeted Drug-Delivery Systems: Small and Macromolecules
- Liposomes as Targeted Drug-Delivery Systems
- Antibody–Drug Conjugates: Development and Applications
- Gene-Directed Enzyme Prodrug Therapy (GDEPT) as a Suicide Gene Therapy Modality for Cancer Treatment

- Targeted Prodrugs in Oral Drug Delivery
- Exosomes for Drug-Delivery Applications in Cancer and Cardiac Indications
- Delivery of Nucleic Acids such as siRNA and mRNA using Complex Formulations
- Application of PROTAC Technology in Drug Development
- Metal Complexes as the Means or the End of Targeted Delivery for Unmet Needs
- Formulation of peptides for targeted delivery
- Antibody-Based Targeted T-Cell Therapies
- Devices for Active Targeted Delivery: A Way to Control the Rate and Extent of Drug Administration
- Drug Delivery to the Brain: Targeting Technologies to Delivery Therapeutics to Brain Lesions

The present book provides exhaustive review of the advanced technologies evolving in the field of the targeted drug-delivery systems. The authors especially tried to cover the approaches used to develop fast-track vaccines in the covid pandemic. Also, there is particular emphasis on the different routes where the targeted delivery concept can be used.

The book editor Yogeshwar Bachhav is a pharmacist by training and has a PhD in advanced drug-delivery systems from ICT, Mumbai (India). He has around 16 years of post-PhD experience in Europe in the field of pharmaceutical development of investigational drugs. He has contributed to the success of the clinical candidates ranging from preclinical to phase 1, 2, and 3 trials followed by commercial launch.

Yogesh has worked as a Research Scientist for around four years on a collaborative project between Pantec Biosolutions AG (Lichtenstein) and University of Geneva, Switzerland. After this, he has worked as a Formulation Manager at Debiopharm Group, Lausanne, Switzerland, for around four years in the capacity of a lab head, where he successfully developed preclinical and clinical formulations for oncology indication.

Currently Yogesh is working as a Senior Director at AiCuris Anti-infective Cures AG Germany and responsible for pharmaceutical development of investigational drugs in the domain of innovative anti-viral and anti-bacterial drugs.

Yogesh has also started a consultancy firm called Adex Pharma which deals with solving complex issues in the pharmaceutical development of new and approved drugs since 2016.

Yogesh's expertise in the field of advanced drug-delivery system comprises pre-formulation, formulation development of small molecules and/or peptides for oral, dermal, and parenteral applications. Also, he has exposure to in-house development and outsourcing these novel dosage forms.

Besides several publications in the targeted formulation field, Yogesh is a well-known expert with over 30 conference proceedings and has been named as inventor in several patent applications. He has already edited a book for Wiley-VCH in the same book series titled *Innovative Dosage Forms: Design and Development at Early Stage*.

In summary, the present book will be a very good source of the advanced knowledge in the field of targeted drug-delivery systems for many medicinal chemists facing the interdisciplinary drug discovery and development interfaces.

With this, we – the series editors – sincerely believe that readers would be highly benefited from the contents of this book.

We, as series editors, would like to thank Yogesh for putting together the brilliant contributions of the authors, all authors for their brilliant contributions, and Frank Weinreich, Stefanie Volk, and their co-workers for their great support to make this book possible.

May 2022
Aachen, Porto, and Frankfurt

Helmut Buschmann
Jörg Holenz
Raimund Mannhold

1

Basics of Targeted Drug Delivery

Kshama A. Doshi

Sutro Biopharma, 111 Oyster Point Blvd, South San Francisco CA, 94080, United States

1.1 Introduction

Biological effects conferred by drugs are associated with drug mechanism of action, and drug pharmacological and physicochemical properties. To elicit pharmacological response, drugs are commonly designed to bind to a target and activate or inhibit them, for example, chemotherapy drug belonging to the class of topoisomerase-2 inhibitors binds to and stabilizes enzyme topoisomerase-2 in cells to induce cell death, antidiabetic medication exenatide binds to and activates good lab practices (GLP)-1 to increase insulin secretion. Further, depending upon the route of drug administration, drugs undergo four main processes – absorption (absorption of drug from site of administration into blood), distribution (distribution of drug to different tissues from bloodstream), metabolism (breakdown of drug), and excretion (elimination out of the body) which are predominantly affected by the physicochemical properties of the drug. These factors largely account for the rate and extent of drug efficacy and overall potency.

In addition to the above-mentioned processes, pharmacological response and efficacy induced by the drug are also governed by its delivery to the site of action, the selective delivery to the target, and associated safety. To facilitate safe and effective drug transport, various drug-delivery systems (formulations, dosage forms, drug-device combinations, etc.) have been developed thus far. During the last several decades, multiple technologies and formulations, including controlled-release drug-delivery technology, oral and transdermal drug-delivery systems, nanotechnology-based products, have significantly improved patient outcomes [1]. While significant improvements have been made in multiple disease indications, there continue to remain areas that require attention to fulfill the unmet need in terms of increasing drug efficacy by improving patient compliance, reducing side effects, and reducing dosing frequency. Targeted drug-delivery systems have gained wide attention in recent years to selectively target the drug at the site of action and thereby facilitate site-specific delivery to ensure high safety, efficacy, and patient compliance. This chapter introduces some basic concepts

Targeted Drug Delivery, First Edition. Edited by Yogeshwar Bachhav.

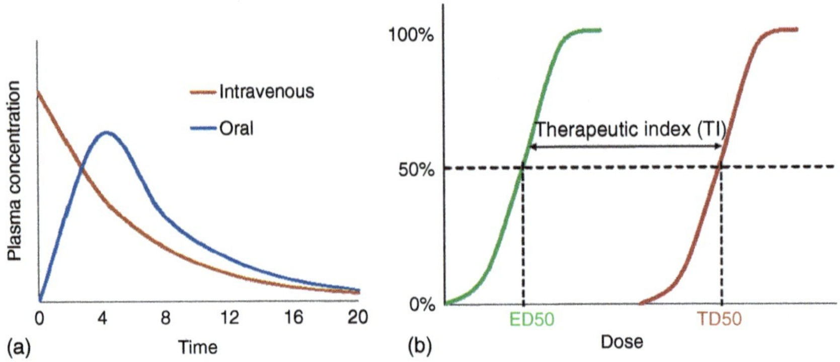

Figure 1.1 (a) Bioavailability of an agent administered intravenously (in red) and orally (in blue). (b). Therapeutic index (TI) of an agent as defined by the ratio of ED50 to TD50. ED50: Effective dose for 50% response points, TD50: Toxic dose for 50% response points.

followed by the rationale for development of targeted drug-delivery approaches, different approaches to achieve this, commercial success to date, and challenges associated with this approach.

1.1.1 Concept of Bioavailability and Therapeutic Index

Bioavailability (BA) is the rate and extent to which the drug is absorbed from the drug product and becomes available at the site of action [2]. BA of an agent administered intravenously is high as compared to oral administration. This is a result of instant entry of the agent in the systemic blood circulation following intravenous dosing as compared to absorption from the Gastrointestinal (GI) tract followed by entry into systemic circulation with oral dosing (Figure 1.1a). Therapeutic index (TI) is an indicator of relative safety of a drug. TI is defined as the ratio of maximally tolerated toxic dose to minimum effective dose. A common method used to calculate TI of an agent is to calculate ratio of dose that induces toxic effects in 50% response points (TD50) to the dose that induces therapeutic effects in 50% response points (ED50) (Figure 1.1b).

1.2 Targeted Drug Delivery

The terms "targeted drug delivery" and "targeted drug therapy" are frequently used in drug discovery research; however, both these terms are distinct from one another and cannot be used interchangeably. Targeted drug therapy refers to specific interaction between drug and a certain protein or moiety on target/disease cells [3]. Targeted drug delivery, on the other hand, refers to predominant accumulation of the drug/drug formulation in the target/disease zone [4]. Effective drug-delivery system design, for all kinds of formulation, requires four key requirements – retain, evade, target, and release.

Retain: The delivery system should remain intact in its original form throughout the course of formulation development, processing, and administration.

Evade: Upon administration, it should be retained in the form such that it evades body defense mechanisms, stays protected from the body's immune system attacks, and reaches desired target zone in an optimal time frame.

Target: Drug-delivery system should be designed to result in exclusive drug accumulation at the intended site of action, i.e. disease area, while avoiding healthy tissues and drug-associated toxicity.

Release: Once at the desired site of action, the system should be capable of releasing drug from the formulation for the agent to confer its therapeutic effect.

The goal of targeted drug-delivery system is to increase TI of a drug over a nonspecific drug-delivery system. A delivery system that results in preferential accumulation of drug at the disease site while sparing nondisease sites in the body and limiting overall toxicity is considered to have a higher TI as compared to a system that results in equal accumulation of the drug in both disease and nondisease sites [5]. A general rule is delivery system that confers higher drug TI is clinically safer as compared to lower TI.

1.3 Strategies for Drug Targeting

Over the last few decades, multiple ideas have evolved ranging from identification of different materials to invention of novel concepts to potentiate and improve delivery of drugs to intended target region. Strategies for drug targeting are often classified into three main categories – passive targeting, active targeting, and physical targeting (Figure 1.2).

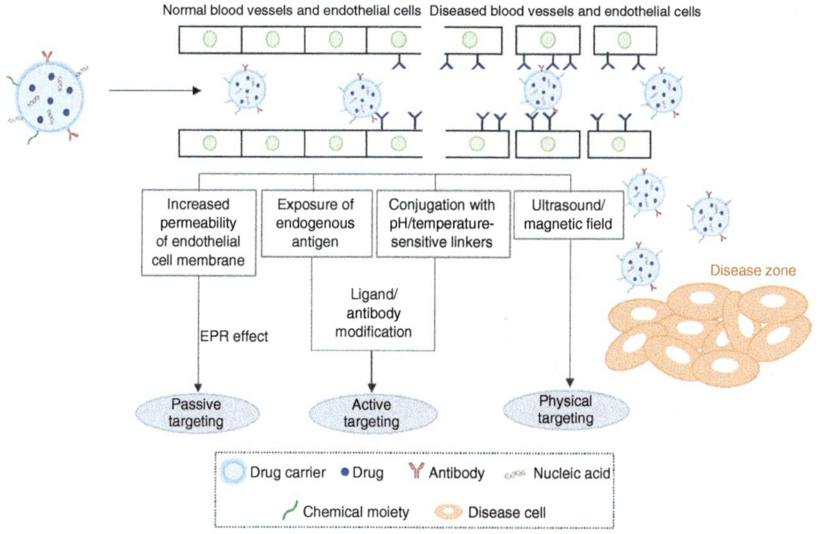

Figure 1.2 Schematic representing different directed drug-delivery-targeting techniques.

1.3.1 Passive Targeting

Often referred to as "no targeting," passive targeting utilizes the principle to accumulate drugs into specific regions of the body due to inherent features and characteristics of the said tissue. Passive targeting makes use of differences in anatomical features between target tissue and nontarget tissue to ensure preferential accumulation of drug. Common examples of passive targeting include accumulation of drugs via the reticuloendothelial system (RES), increased accumulation of drugs due to enhanced permeability and retention (EPR) effect, and localized delivery.

1.3.1.1 Reticuloendothelial System (RES) System

RES is an essential part of the immune system that lines organs, including liver and spleen. RES consists of phagocytic immune cells, including monocytes and macrophages, that can recognize and uptake foreign moieties. Biological function of monocytes and macrophages includes opsonization or capturing foreign substances that reach the systemic circulation. Thus, the RES system enables preferential uptake of nanoparticles by organs, including liver and spleen. For example, nanoparticles with strong hydrophobic surfaces are preferentially taken up by the liver followed by spleen and lungs.

1.3.1.2 Enhanced Permeability and Retention (EPR) Effect

Tumor vasculature is highly leaky and discontinuous as compared to normal tissue vasculature. Unlike normal vasculature, which is lined with endothelial cells tightly held together, tumor vasculature is more heterogeneous in size and permeability. Depending on the stage of tumor progression and anatomical location, gaps between endothelium range in size from 100 to 780 nm [6, 7]. Additionally, elevated expression of proteins, including vascular epithelial growth factor (VEGF), epithelial growth factor (EGF), and basic fibroblast growth factor (bFGF), enhances vasodilation and extravasation of drugs from the leaky vasculature in tumors [8]. These characteristics of tumor vasculature enable enhanced delivery and retention of high-molecular-weight drugs in the target region. Augmented therapeutic effect achieved as a result of this phenomenon is associated with EPR effect. EPR effect is commonly used for passive targeting of agents >40 kDa in molecular weight. Additionally, low-molecular-weight agents that are administered in drug carriers, including conjugates, nanoparticles, and liposomes, can also be delivered preferentially to the tumor by leveraging the EPR effect.

Examples of commercially available formulations that target drug to tumor region leveraging the EPR effect include Daunosome™ and Doxil™, clinically used anticancer agents. Both Daunsosome and Doxil are liposomal formulations that efficiently accumulate in the tumor cells minimizing the frequency of drug-induced adverse effects [9].

1.3.1.3 Localized Delivery

As the name suggests, localized delivery emphasizes direct delivery of the drug to the disease site or organ, thus limiting systemic exposure of drug to blood circulation and minimizing adverse drug toxicities. Localized delivery is often amenable to certain tumor types, including some forms of prostate and breast cancer, but not all tumor types or all diseases, thus limiting its use. Preclinical work has shown intratumoral delivery of paclitaxel nanoparticles conjugated to transferring ligand was effective in inducing tumor regression in mice models of prostate cancer. This treatment was significantly more effective as compared to systemic administration of paclitaxel [10]. Corticosteroids, a class of drug commonly used in asthma maintenance, are administered locally by using metered-dose inhalers. Other examples of drugs administered via local delivery systems include corticosteroids, used in metered-dose inhalers for asthma management and metronidazole, an antibiotic used in a gel formulation for treatment of periodontal diseases.

1.3.2 Active Targeting

Active targeting is by far the most well-recognized and implemented form of targeted drug delivery. This approach confers targeting properties to the drug that enables accumulation and consecutively pharmacological action toward specific molecule or region. Commonly used strategy to enable active targeting includes techniques that impose targeting properties on the drug, i.e. combining drug with other components that possess targeting features. This can be done in one of two ways. Firstly, by coupling drug with components that do not display affinity or binding toward a specific target but enable release of drug under a unique environment, e.g. sensitive to diseased (impacted) tissue pH, temperature, or enzymes. Many pharmaceutical and biotechnology companies are undertaking development of prodrugs – where drugs are conjugated and masked by enzyme-sensitive linkers to maintain them in an inactive state. On reaching the target site, these linkers are cleaved by enzymes specifically known to be upregulated in tumor microenvironment, thus making the active drug moiety selectively available for tumor region and limiting off-target adverse events.

The other technique, and which is often used, includes coupling drugs to components that display potent affinity and binding to a particular receptor expressed in the pathological tissues. This form of active targeting is also called ligand-mediated targeting. Ligand-based active targeting is commonly used in the development of many therapeutic and diagnostic modalities. Active cellular-targeting strategies involve use of affinity ligands on the surface of nanocarriers or developing antibodies against a certain ligand that can induce specific homing along with increased retention and uptake by the target cells. Antibody–drug conjugates (ADCs) utilize the principle of conjugating a drug to an antibody directed against antigens with increased expression on disease cells using cleavable linkers, thus ensuring selective binding of the ADC to the target cell over other tissues to minimize adverse drug reaction (ADR).

1.3.3 Physical Targeting

Physical targeting refers to a technique that utilizes external stimuli to induce release of the drug at a specific target site in the body. Common indications that utilize physical targeting to achieve targeted drug delivery include cancer treatment, chronic lung diseases, including chronic obstructive pulmonary disease (COPD) and cystic fibrosis (CF). Commonly used techniques for physical targeting of drug to pathological area include using ultrasound and magnetic field to target the pathological tissue.

1.3.3.1 Ultrasound for Targeting

Research focused on utilizing ultrasound waves to target tissue to release drug from polymeric micelles and enable uptake by disease cells is underway for over a decade now. Ultrasound waves can induce delivery of anticancer agents by either degradation of micelles to release drug at target site or partition of drug out of the micelles at the target tissue [11]. One of the main advantages of this technique is its noninvasive nature, leading to increased patient compliance. This technique also offers the unique advantage of deep penetration into the body along with extensive control to cater the waves to specific target sites. Despite the advantages, there are concerns associated with use of ultrasound radiations, including their effect on cell plasma membrane. Preclinical studies addressing the effect of lower-energy ultrasound radiations on the efficacy of drug release from micelles and damage to cellular membranes are underway [11].

1.3.3.2 Magnetic Field for Targeting

Magnetic targeting utilizes an external magnetic field to induce preferential localization of an intravenously injected therapeutic agent bound to or encapsulated in a magnetic drug carrier. Such drug carriers include magnetic liposomes, nanospheres, and magnetic ferrofluids and incorporate materials, such as iron, nickel, and magnetite [12]. Preclinical investigation for many magnetic drug carriers or various anticancer agents, including mitoxantrone, etoposide, and epirubicin, is currently underway [13, 14].

1.4 Therapeutic Applications of Targeted Drug Delivery

Nanocarriers are the most commonly used drug carrier system to mediate targeted drug delivery. These employ nanosized materials, including nanoparticles, liposomes, micelles, and dendrimers for targeted and controlled drug-delivery systems [15]. These delivery systems are commonly used for a wide range of purposes, ranging from disease diagnosis to management. Different disease indications that can be detected and treated with targeted drug-delivery systems are discussed below.

1.4.1 Diabetes Management

Diabetes mellitus (DM) is a chronic metabolic disorder that has significantly impacted lifestyle due to increased frequency of occurrence over the last decade. DM can be classified into Type 1 (T1DM) and Type 2 (T2DM), where T1DM results due to absolute deficiency of insulin and T2DM is a result of insulin resistance, increased glucose production, or impaired insulin secretion. Liposomes, composed of phospholipids and cholesterol, can entrap and deliver both hydrophobic and hydrophilic agents to site-specific regions. Many reports have used liposome-based delivery systems to improve site-specific delivery of insulin. Zhang et al. have shown liposomes composed of 3 : 1 ratio of lipid – cholesterol show increased entrapping of insulin, optimal membrane fluidity along with minimal insulin leakage [16]. Additional reports have shown enhanced target-specific delivery when the liposomes are coated with folic acid [17]. Nanoparticle-based targeted therapy has also been developed and tested for targeted delivery of insulin in DM management. Nanoparticles encapsulating DNA-encoding interleukins, including IL-10 and IL-14, have been designed and tested in prediabetic animal models. Results from these studies showed nanoparticles encapsulating these interleukins were potent in inhibiting response of T-cells against native islet cells and significantly inhibited development of DM [18]. Overall, treatment with nanoparticle and liposomal-based approaches has significantly improved DM management as compared to conventional treatment.

1.4.2 Neurological Diseases

Incidence of neurological diseases, including Alzheimer's and Parkinson's, has significantly risen over the last few years. While Alzheimer's is associated with extracellular deposition of amyloid beta-peptide and tau proteins, Parkinson's is associated with degeneration of dopaminergic neurons in the brain. Effective targeting of neurological disorders is often complex due to the inability or limited ability of treatment modalities to cross the blood–brain barrier (BBB). However, nanomedicines have evolved with positive outcomes in overcoming the BBB and increasing BA of therapeutic agents in neurological disorders. Acetylcholinesterase inhibitors (AChEs) inhibitors, including donepezil, rivastigmine, and galantamine, are commonly used therapeutic agents for Alzheimer's management [19]. Preclinical studies with rivastigmine-loaded poly(lactide-*co*-glycolide) (PLGA) and polysorbate 80 (PBCA-80)-coated poly(*n*-butylcyanoacrylate) nanoparticle formulation have demonstrated improved memory in mice behavioral studies as compared to rivastigmine-in solution [20]. Furthermore, nanoformulations for donepezil encapsulated in PLGA particles demonstrated higher penetration and accumulation in the brain compared to drug in solution formulation [21]. Nerve growth factor (NGF), an essential protein in survival of neurons, is currently being investigated for its therapeutic potential for neurological diseases. While NGF has limited ability to penetrate the BBB, NGF adsorbed on PBCA-80-coated poly(*n*-butylcyanoacrylate) nanoparticles have shown beneficial effects in slowing neurodegeneration and reversing amnesia in rat models [22]. Furthermore, encapsulation of curcumin and NGF in nanoformulation induced synergy and enhanced therapeutic effect in preclinical studies [23].

Treatment with dopaminergic agents, including levodopa and carbidopa, is the first-line therapy for management of patients with Parkinson's. However limited permeability across the BBB and BA of dopamine agonists necessitates increased dosing frequency of these agents. This has, however, resulted in lower patient compliance given the systemic side effects induced by increased dosing frequency. Nanodrug-delivery strategy has shown promising outcomes in management of Parkinson's. Dopamine-loaded chitosan nanoparticles demonstrated dose-dependent increase in dopamine levels and increased BA in preclinical settings [24]. Continuous stimulation of dopaminergic neurons is beneficial in the treatment of Parkinsons disease. While dopamine receptor agonist rotigotine is a potent stimulator of dopaminergic neurons in in vitro systems, its utility is limited due to poor penetration across the BBB in animal models. However, chronic administration of rotigotine loaded in PLGA-MS demonstrated sustained exposure of drug in the brain over an extended period along with improved safety and tolerability in monkeys and rats [25, 26]. In addition to nanoformulations, ADCs administered subcutaneously or systemically are being studied for management of neurological diseases. SER-241 is an investigational once-a-week ADC from Serina Therapeutics that utilizes apomorphine conjugated to an antibody for treatment of Parkinson's. SER-214 is currently in Phase 2 clinical testing in patients with advanced Parkinson's disease.

1.4.3 Cardiovascular Diseases

Cardiovascular diseases (CVDs) are the leading cause of death in the United States. Targeted drug delivery offers the potential of fulfilling unmet needs in treatment of CVDs by minimizing renal excretion of the drug, which in turn elongates residence time of the drug in systemic circulation.

Atherosclerosis is a CVD characterized by hardening and narrowing of arteries due to excessive plaque formation that eventually decreases blood flow to the heart and brain ultimately leading to conditions, such as stroke and coronary heart disease. Targeted drug delivery not only offers therapeutic options in treatment of CVD, but has also shown significant improvement in diagnosis and imaging of plaques. N1177, an iodinated aroyloxy ester, has successfully been used to identify macrophage accumulation in arterial walls in animal models of atherosclerosis [27]. This approach has shown promising results and is currently undergoing clinical testing in human patients. Targeted therapy combining physical and active targeting showed increased internalization of nanoparticles in atherosclerotic macrophages when super-paramagnetic iron oxide nanoparticles were used [28].

Myocardial ischemia–reperfusion (IR) injury is a cardiovascular condition characterized by apoptosis of cardiomyocytes due to mitochondrial disturbances and generation of reactive oxygen species. Multiple promising therapeutic agents tested for treatment of myocardial IR have failed clinical testing due to inefficient delivery of drug within a critical time frame. Nanodrug-delivery vehicles, including PLGA nanoparticles as well as PEGylated liposomes, have shown significant promise in targeting inflammatory cells due to increased inflammation-induced permeability of myocardium [29]. ONO-1301, a synthetic prostacyclin IP receptor agonist, is currently under development for myocardial IR. Preclinical work has demonstrated selective accumulation of the drug in the ischemic myocardial tissue when administered intravenously as a nanoparticle formulation as compared to ONO-1301 solution. Furthermore, ONO-1301 NPs also led to increased secretion of cytokines and tumor necrosis factor-alpha in turn increasing myocardial blood flow and reduction in infarct size [30].

1.4.4 Respiratory Diseases

Targeted drug-delivery systems administered intranasally are known to be highly effective in management of respiratory diseases, including asthma and chronic obstructive pulmonary disorder. Advantage of intranasal formulation includes minimizing drug resistance, increasing lung deposition of the drug, and minimizing toxic effects to nonpulmonary tissue. Targeted drug delivery in the form of nanoformulations, including liposomes and nanoparticles, is the new paradigm for the treatment of respiratory diseases.

Asthma is a common chronic condition characterized by shortness of breath, coughing, and wheezing. Corticosteroids and bronchodilators are commonly used in management of asthma. Preclinical studies showed nanoparticles containing salbutamol resulted in long-term relief due to sustained accumulation in the lungs as compared to solution formulation. Liposomal formulation of salbutamol sulfate also resulted in extended retention of the drug in lungs, ~10 hours, thus prolonging therapeutic effect [31]. *Mycobacterium tuberculosis* (MTB), commonly known to cause tuberculosis (TB), is one of the leading causes of fatalities worldwide. MTB reaches lung alveoli and resists macrophage-mediated destruction by preventing formation of phagolysosome. Standard-of-care drugs for the treatment of TB include rifampicin, isoniazid, and ethambutol used either alone or in combination with injectable agents (streptomycin and viomycin), fluoroquinolones, or few oral agents (ethionamide and para-aminosalicylic acid). Targeted drug delivery using the platform of mesoporous silica nanoparticles (MSNPs) has shown promising outcomes for the delivery of anti-TB drugs. Surface functionalization with poly(ethylene imine) (PEI) yielded higher loading and controlled drug delivery of rifampicin MSNPs. Furthermore, MSNPs-containing pH-sensitive pores have been shown to release isoniazid directly to MTB-infected macrophages following endocytosis [32].

1.4.5 Cancer Indications

Cancer, also referred to as malignant tumors, is characterized by a condition where genetic or acquired mutation in DNA leads to uncontrolled proliferation of cells that also has the potential of migrating from primary site of origin and invading into a secondary site. Heterogeneous nature of tumors along with dense tumor microenvironment makes treatment of cancers much more complex. Multiple technologies, including nanoformulations, radiation therapy, immunotherapy, and chemotherapy, have shown improvement in cancer management; however, toxicities associated with systemic delivery, poor drug accumulation at tumor site, and nonspecific drug effects limit the benefits offered by current drug-delivery technologies.

Antibody-mediated target engagement, a commonly used form of active targeting, has shown promising success in oncology treatment. Antibodies are commonly raised against tumor-associated antigens (TAA) that provide critical downstream signaling for cancer cell survival, thus providing therapeutic option for targeting them. Many such antigens show increased expression on cancer cells as compared to nonmalignant tissue, thus making this a targeted therapy approach. Examples include Trastuzumab developed by Genentech against Her-2 receptor and is upregulated in breast cancer cells. Another FDA-approved monoclonal antibody is bevacizumab which targets VEGF and inhibits angiogenesis in tumors. Both Trastuzumab and bevacizumab have shown improved patient survival in cancer management [33]. Drugs conjugated to TAA antibody using cleavable linkers, i.e. antibody–drug conjugates, are extensively being evaluated in preclinical and clinical studies to achieve tumor-specific targeted delivery of cytotoxic drugs. Liposomal formulations of anticancer agents have demonstrated a promising strategy for many chemotherapeutic agents, including doxorubicin and paclitaxel. PEGylated liposomal doxorubicin (Doxil) showed potent anticancer activity and reduced cardiotoxicity for first-line treatment of metastatic breast cancer [34]. DaunoXome (daunorubicin liposomes) has shown significant improvement in therapeutic efficacy and survival in patients with Kaposi's sarcoma [35]. In addition to antibodies and nanoparticles, dendrimers have also shown promise in delivering anticancer agents to specific targets. Doxorubicin-conjugated dendrimers using polyamidoamine significantly reduced tumor burden through enhanced drug accumulation in B16F10 melanoma tumors in mice [36]. Another group also showed pH-sensitive dendrimers increased tumor penetration and release of drugs into tumor microenvironment [37].

1.5 Targeted Dug-Delivery Products

Over the past few decades, multiple targeted drug-delivery products have received Food and Drug Administration (FDA) approval. Currently, the market has more than 50 products based on this technology (Table 1.1) [38, 39]. Notably, targeted delivery systems are extensively developed for drugs, which have low aqueous solubility and high toxicity, such that when administered as nanoformulations, these drugs show enhanced BA, better accumulation, pharmacokinetic properties, and reduced toxicity.

Table 1.1 Nanomedicines approved by FDA classified by type of carrier/material used in preparation of the formulation.

Drug name	Active agent	Carrier	Company	Indication
Doxil®	Doxorubicin	Liposomes	Janssen	Ovarian Cancer; Myeloma
Marqibo kit®	Vincristine	Liposomes	Onco TCS	Acute lymphoblastic leukemia
Onivyde®	Irinotecan	Liposomes	Merrimack	Pancreatic cancer
DaunoXome®	Daunorubicin	Liposomes	Galen	Kaposi's sarcoma
DepoCyt©	Cytarabine	Liposomes	(Sigma-Tau)	Lymphomatous meningitis
AmBisome®	Amphotericin B	Liposomes	Gilead Sciences	Fungal and/or protozoal infections
Adagen®	Pegademase bovine	PEGylated adenosine deaminase enzyme	Sigma-Tau Pharmaceuticals)	Immunodeficiency disease
Oncaspar®	L-Asparaginase	PEGylated L-asparaginase	Enzon Pharmaceuticals	Acute lymphoblastic leukemia
Copaxone®	Glatopa	L-Glutamate, L-alanine, L-lysine, and L-tyrosine random copolymer	Teva	Multiple sclerosis
Bydureon®	Exenatide synthetic	PLGA	AstraZeneca AB	Type 2 diabetes
Atridox®	Doxycycline hyclate	PLA	Tolmar	Chronic adult periodontitis
Abraxane	Paclitaxel	Albumin-based particles	Celgene	Metastatic Breast Cancer; NSCLC
Zyprexa Relprevv®	Olanzapine pamoate	Microcrystal	Eli Lilly	Schizophrenia
Invega Sustenna®	Paliperidone palmitate	Nanocrystal	Janssen	Schizophrenia

Source: Adapted from Patra et al. [38] and Zhong et al. [39].

1.6 Challenges

Despite the preclinical promise illustrated by targeted drug delivery in mediating disease effects, there has been limited clinical success for the therapeutic potential of this strategy in many disease indications, including cancer. Key challenges associated with active and passive drug-delivery strategies are discussed below.

1.6.1 Passive Targeting and EPR Effect

Multiple physiological barriers are involved in delivery of drug systems that leverage the EPR effect. Nanocarriers are often cleared by the mononuclear phagocytic system (MPS) in the leaky blood vessels. Many drug carriers get trapped in the sinusoids of the liver, while others are taken up by the hepatocytes and macrophages of liver (Kupffer cells) [40]. Drug delivery of passively targeted systems is also governed by the heterogeneity of the EPR effect in the disease area. Indications, such as cancer, are characterized by highly heterogeneous disease environment. Several factors, including spatial changes within the target zone, variable endothelial gaps (ranging from 1 to 100 nm) as well as temporal heterogeneity, contribute to variable permeability and perfusion of drug carriers [41, 42]. Furthermore, there are limited clinical data surrounding the potency of EPR effect in different disease conditions. To date understanding of the effectiveness of the EPR effect is primarily based on preclinical model of the disease; however, these animal models do not accurately recapitulate human anatomy or progression of disease in human settings. Limited clinical data on the effectiveness of the EPR effect in inducing accumulation of drug at the disease site and associated therapeutic benefits make translation of preclinical results more challenging [43].

A recently conducted meta-analysis on preclinical studies using nanocarriers suggested about 0.7% of injected dose of drug reaches the tumor site. Additional efforts are underway to increase the drug-delivery efficiency of nanocarriers. Preclinical studies have shown angiotensin II-induced vasodilation can enhance the EPR effect. Furthermore, cell-mediated delivery of drug carriers can overcome areas of low EPR and still offer increased drug accumulation at the disease site. This approach exploits the ability of certain cell types, specifically immune cells, to penetrate target area due to disease pathology. For example, preclinical studies have shown targeting of chemotherapy drugs to tumors using T-cell [44]. Tumors are often penetrated by immune cells, including T-cells. This phenomenon can be leveraged by administering nanoparticle-carrying T-cells that can target chemotherapy drugs to the tumor microenvironment.

1.6.2 Active Targeting

Ligand-based targeting is the most commonly used form of active-targeted drug-delivery system. Ligand conjugation of drug carriers facilitates uptake of the carrier by target cells, thus offering a platform to enhance delivery of macromolecules, including proteins and nucleic acids. However, targeted carriers carrying macromolecules often undergo endocytosis in the target cell, resulting in degradation of the macromolecule. Preclinical studies with transferrin-targeted nanocarriers have shown that they undergo clathrin-mediated endocytosis and degradation in the lysosome [43, 45]. Many ongoing efforts are addressing ways to facilitate endosomal escape of these drug carriers, e.g. pore formation proteins and pH-buffering substances [1]. Ligands chosen for actively targeted drug carriers are most commonly selected on the basis of classical disease markers, e.g. CD19 for B-cell malignancies and HER2 for breast cancer. However, given the heterogeneous nature of many diseases, cell-specific drug carriers have a high probability of promoting selection toward survival of resistant cells, since cells that do not express the classical disease marker escape being targeted by drug carriers. Furthermore, despite increased surface expression on cells, not all ligands are suitable for internalization of drug carriers into the cell thus limiting drug uptake. Therefore, ligand selection for targeted drug delivery is an important consideration, and ligands should be screened and selected not only based on their expression profile but also on their ability to be internalized by target cells.

Compared to manufacture of passively targeted drug carriers, conjugation of ligands to drug carriers for active delivery involves a complex manufacturing process. Multiple designs and engineering steps, including ligand synthesis, purification, and stability of drug carrier–ligand conjugate, make active drug delivery significantly more challenging with longer timelines and increased cost. Additionally, active-targeting strategies are also associated with complex pharmaceutical development and scale-up under good manufacturing practice (GMP) laws that further add to the cost of this therapy.

1.7 Scale-up and Challenges

Several methods have been developed and reported for the manufacture of targeted drug-delivery products. The process of manufacturing depends on whether the nanocarrier is composed of polymer, lipids, or is metal based. Table 1.2 lists different manufacturing processes that are commonly used for each of the nanocarrier types [46].

Table 1.2 Methods of nanocarrier production with various materials.

Nanocarrier type	Manufacturing processes
Polymeric nanocarriers	Nanocrystallization
	Extrusion
	Supercritical fluid technology
	Sonication method
	Salting out
Lipid nanocarriers	High-pressure homogenization
	Solvent emulsification evaporation
	Solvent emulsification diffusion
	Ultrasonication
Metallic nanocarriers	
Carbon	Chemical vapor deposition, laser ablation, combustion process
Gold	Chemical reduction, UV irradiation
Silica	Etching, deposition, photolithography
Iron	Co-precipitation, thermal decomposition, hypodermal synthesis

All the methods listed above can be classified as bottom-up or top-down processes. Bottom-up processes include processes where the final product is produced as a result of precipitation whereas top-down process starts with a macro-size drug power that further undergoes size reduction. Multiple factors need to be accounted for while choosing the scale-up method for nanocarriers. These include toxicological features, size and shape, nature of the material, generally regarded as safe (GRAS) status, and biodegradable nature of the material [47]. Hence, it is essential to ensure that the key features of the drug-delivery carrier are retained, and not lost, during the process of scale-up.

Given the engineering and chemical complexity of nanocarriers, commercialization and regulatory approval constitute the most time-limiting factors in commercial success of nanocarriers to date. One of the most common obstacles is presented by the lack of GLP compliance during preclinical studies in academic setting which, in turn, limits their collaboration with pharmaceutical sector. While incorporating GLP is not crucial for proof of concept (PoC) aimed at preclinical studies, it is critical that studies be conducted in a GLP setting when they are aimed at demonstrating the promise of the technology and its translational application. GLP compliance is also associated with significant increase in overall costs and time, and careful assessment should be conducted with respect to the objective of preclinical studies before embarking on the GLP route. Design of the clinical trial also significantly influences success rate of a nanocarrier. Recent advances in clinical trials for nanocarriers have highlighted the importance of factors, including companion diagnostics, patient selection criteria (extent of EPR effect), disease heterogeneity, presence of target receptor, and the ability of drug carrier, to bind the target. All these factors are responsible to govern the success of targeted drug-delivery systems. Merrimack Pharmaceuticals achieved significant success in clinical testing of their nanoliposomal irinotecan (nal-IRI), using a companion diagnostic tool of ferumoxytol (FMX) iron nanoparticles. Their studies demonstrated positive correlation between accumulation of ferumoxytol (FMX) iron nanoparticles and response to nal-IRI, such that tumors that accumulated more FMX were more responsive to nal-IRI [48].

1.8 Current Status

Continued research and preclinical success in optimizing targeted drug-delivery systems have resulted in ongoing multiple clinical trials by using targeted nanocarriers. Table 1.3 lists some of the currently ongoing trials that are testing targeted nanocarriers for different therapeutic indications. Table below summarizes the ongoing clinical trials using targeted nanocarriers.

Table 1.3 List of ongoing clinical trials that utilize targeted nanocarriers.

Nanocarrier type	Drug	Therapeutic indication	Clinical trial identifier #
Polymeric nanoparticles	Cetuximab	Colon cancer	NCT03774680
Silver nanoparticles	Antimicrobial drugs	Bacterial and fungal infection	NCT03752424
Albumin-stabilized nanoparticles	Paclitaxel	Breast cancer (Stage III, Stage IV)	NCT00785291
Ultrasmall silica nanoparticles	(64Cu)-labeled PSMA-targeting particle tracer	Diagnostic tool for prostate cancer	NCT04167969
Topical fluorescent nanoparticles	Quantum dots coated with veldoreotide	Breast cancer, skin cancer	NCT04138342
Cholesterol-rich nonprotein nanoparticle	Paclitaxel	Coronary artery disease	NCT04148833
Targeting-enhancing Nanoparticle	Paclitaxel	Solid cancer	NCT02979392
Targeted silica nanoparticle	Fluorescent-dye labeled particles cRGDY-PEG-Cy5.5-C	Head and neck cancer	NCT02106598
Magnetic nanoparticle	Chemotherapy	Prostate cancer	NCT02033447

1.9 Conclusion and Prospects

Research focused on identifying, improving, and applying targeted drug-delivery systems has seen unprecedented advances in the last few decades. The rationale supporting this strategy includes improving therapeutic efficacy, minimizing drug-induced adverse effects, developing improved versions of current drugs as well as better patient compliance. An ideal drug-delivery system should deliver maximum drug at the disease site; however, this is often not the case in diseases, such as cancer, where less than 5% of administered drug reaches the tumor site even when delivered using targeted delivery systems. While nanodrug carriers have made extensive contributions to increase the circulation time to better leverage the EPR effect to reach target site, additional efforts need to be made on improving the delivery of these nanocarriers to the disease site. This requires better understanding of multiple factors, including disease physiology, regulation of blood vessels and blood flow, heterogeneity of disease region as well as physiological barriers. Furthermore, improving models, laboratory practices, and techniques used in conducting preclinical research can assist in achieving successful bench to bedside translation. A modified regulatory framework focused on evaluating safety and quality of targeted drug-delivery systems will further enable clinical success of emerging technologies. While efforts aimed at improving targeting specificity of delivery systems are underway, many products, including Abraxane®, an albumin-bound paclitaxel formulation for the treatment of cancer; liposome-based drugs Caelyx®, Myocet® (doxorubicin), and Mepact® (mifamurtide); and nanoparticle-based therapeutic agents Emend® (aprepitant) for nausea and Rapamune® (sirolimus) for graft rejection have been marketed for human use and are widely improving patient outcomes.

References

1 Sahay, G., Querbes, W., Alabi, C. et al. (2013). Efficiency of si RNA delivery by lipid nanoparticles is limited by endocytic recycling. *Nat. Biotechnol.* 31 (7): 653–658.

2 Millar, S.A., Stone, N.L., Yates, A.S., and O'Sullivan, S.E. (2018). A systematic review on the pharmacokinetics of cannabidiol in humans. *Front. Pharmacol.* 9: 1365.

3 Gerber, D.E. (2008). Targeted therapies: a new generation of cancer treatments. *Am. Fam. Physician* 77 (3): 311–319.

4 Torchilin, V.P. (2000). Drug targeting. *Eur. J. Pharm. Sci.* 11 (Suppl. 2): S81–S91.

5 Bodor, N. (1987). Redox drug delivery systems for targeting drugs to the brain. *Ann. N. Y. Acad. Sci.* 507: 289–306.

6 Hobbs, S.K., Monsky, W.L., Yuan, F. et al. (1998). Regulation of transport pathways in tumor vessels: role of tumor type and microenvironment. *Proc. Natl. Acad. Sci. U.S.A.* 95 (8): 4607–4612.

7 Yuan, F., Salehi, H.A., Boucher, Y. et al. (1994). Vascular permeability and microcirculation of gliomas and mammary carcinomas transplanted in rat and mouse cranial windows. *Cancer Res.* 54 (17): 4564–4568.

8 Dellian, M., Witwer, B.P., Salehi, H.A. et al. (1996). Quantitation and physiological characterization of angiogenic vessels in mice: effect of basic fibroblast growth factor, vascular endothelial growth factor/vascular permeability factor, and host microenvironment. *Am. J. Pathol.* 149 (1): 59–71.

9 Gregoriadis, G. (1995). Engineering liposomes for drug delivery: progress and problems. *Trends Biotechnol.* 13 (12): 527–537.

10 Sahoo, S.K., Ma, W., and Labhasetwar, V. (2004). Efficacy of transferrin-conjugated paclitaxel-loaded nanoparticles in a murine model of prostate cancer. *Int. J. Cancer* 112 (2): 335–340.

11 Rapoport, N. (2004). Combined cancer therapy by micellar-encapsulated drug and ultrasound. *Int. J. Pharm.* 277 (1-2): 155–162.

12 Hafeli, U.O. (2004). Magnetically modulated therapeutic systems. *Int. J. Pharm.* 277 (1–2): 19–24.

13 Lubbe, A.S., Alexiou, C., and Bergemann, C. (2001). Clinical applications of magnetic drug targeting. *J. Surg. Res.* 95 (2): 200–206.

14 Alexiou, C., Arnold, W., Klein, R.J. et al. (2000). Locoregional cancer treatment with magnetic drug targeting. *Cancer Res.* 60 (23): 6641–6648.

15 Mahapatro, A. and Singh, D.K. (2011). Biodegradable nanoparticles are excellent vehicle for site directed in-vivo delivery of drugs and vaccines. *J. Nanobiotechnol.* 9: 55.

16 Zhang, X., Qi, J., Lu, Y. et al. (2014). Enhanced hypoglycemic effect of biotin-modified liposomes loading insulin: effect of formulation variables, intracellular trafficking, and cytotoxicity. *Nanoscale Res. Lett.* 9 (1): 185.

17 Agrawal, A.K., Harde, H., Thanki, K., and Jain, S. (2014). Improved stability and antidiabetic potential of insulin containing folic acid functionalized polymer stabilized multilayered liposomes following oral administration. *Biomacromolecules* 15 (1): 350–360.

18 Ko, K.S., Lee, M., Koh, J.J., and Kim, S.W. (2001). Combined administration of plasmids encoding IL-4 and IL-10 prevents the development of autoimmune diabetes in nonobese diabetic mice. *Mol. Ther.* 4 (4): 313–316.

19 Hernando, S., Gartziandia, O., Herran, E. et al. (2016). Advances in nanomedicine for the treatment of Alzheimer's and Parkinson's diseases. *Nanomedicine (London)* 11 (10): 1267–1285.

20 Joshi, S.A., Chavhan, S.S., and Sawant, K.K. (2010). Rivastigmine-loaded PLGA and PBCA nanoparticles: preparation, optimization, characterization, in vitro and pharmacodynamic studies. *Eur. J. Pharm. Biopharm.* 76 (2): 189–199.

21 Md, S., Ali, M., Baboota, S. et al. (2014). Preparation, characterization, in vivo biodistribution and pharmacokinetic studies of donepezil-loaded PLGA nanoparticles for brain targeting. *Drug Dev. Ind. Pharm.* 40 (2): 278–287.

22 Kurakhmaeva, K.B., Djindjikhashvili, I.A., Petrov, V.E. et al. (2009). Brain targeting of nerve growth factor using poly(butyl cyanoacrylate) nanoparticles. *J. Drug Targeting* 17 (8): 564–574.

23 Kuo, Y.C. and Lin, C.C. (2015). Rescuing apoptotic neurons in Alzheimer's disease using wheat germ agglutinin-conjugated and cardiolipin-conjugated liposomes with encapsulated nerve growth factor and curcumin. *Int. J. Nanomed.* 10: 2653–2672.

24 Trapani, A., De Giglio, E., Cafagna, D. et al. (2011). Characterization and evaluation of chitosan nanoparticles for dopamine brain delivery. *Int. J. Pharm.* 419 (1–2): 296–307.

25 Ye, L., Guan, X., Tian, J. et al. (2013). Three-month subchronic intramuscular toxicity study of rotigotine-loaded microspheres in SD rats. *Food Chem. Toxicol.* 56: 81–92.

26 Tian, J., Du, G., Ye, L. et al. (2013). Three-month subchronic intramuscular toxicity study of rotigotine-loaded microspheres in Cynomolgus monkeys. *Food Chem. Toxicol.* 52: 143–152.

27 Weissleder, R., Nahrendorf, M., and Pittet, M.J. (2014). Imaging macrophages with nanoparticles. *Nat. Mater.* 13 (2): 125–138.
28 Morishige, K., Kacher, D.F., Libby, P. et al. (2010). High-resolution magnetic resonance imaging enhanced with superparamagnetic nanoparticles measures macrophage burden in atherosclerosis. *Circulation* 122 (17): 1707–1715.
29 Hausenloy, D.J. and Yellon, D.M. (2013). Myocardial ischemia-reperfusion injury: a neglected therapeutic target. *J. Clin. Invest.* 123 (1): 92–100.
30 Yajima, S., Miyagawa, S., Fukushima, S. et al. (2019). Prostacyclin analogue-loaded nanoparticles attenuate myocardial ischemia/reperfusion injury in rats. *JACC Basic Transl. Sci.* 4 (3): 318–331.
31 Chen, X., Huang, W., Wong, B.C. et al. (2012). Liposomes prolong the therapeutic effect of anti-asthmatic medication via pulmonary delivery. *Int. J. Nanomed.* 7: 1139–1148.
32 Clemens, D.L., Lee, B.Y., Xue, M. et al. (2012). Targeted intracellular delivery of antituberculosis drugs to *Mycobacterium* tuberculosis-infected macrophages via functionalized mesoporous silica nanoparticles. *Antimicrob. Agents Chemother.* 56 (5): 2535–2545.
33 Kabbinavar, F., Hurwitz, H.I., Fehrenbacher, L. et al. (2003). Phase II, randomized trial comparing bevacizumab plus fluorouracil (FU)/leucovorin (LV) with FU/LV alone in patients with metastatic colorectal cancer. *J. Clin. Oncol.* 21 (1): 60–65.
34 O'Brien, M.E.R., Wigler, N., Inbar, M. et al. (2004). Reduced cardiotoxicity and comparable efficacy in a phase III trial of pegylated liposomal doxorubicin HCl (CAELYX/Doxil) versus conventional doxorubicin for first-line treatment of metastatic breast cancer. *Ann. Oncol.* 15 (3): 440–449.
35 Sadava, D., Coleman, A., and Kane, S.E. (2002). Liposomal daunorubicin overcomes drug resistance in human breast, ovarian and lung carcinoma cells. *J. Liposome Res.* 12 (4): 301–309.
36 Zhong, Q., Bielski, E.R., Rodrigues, L.S. et al. (2016). Conjugation to poly(amidoamine) dendrimers and pulmonary delivery reduce cardiac accumulation and enhance antitumor activity of doxorubicin in lung metastasis. *Mol. Pharmaceutics* 13 (7): 2363–2375.
37 Li, H.J., Du, J.Z., Liu, J. et al. (2016). Smart superstructures with ultrahigh pH-sensitivity for targeting acidic tumor microenvironment: instantaneous size switching and improved tumor penetration. *ACS Nano.* 10 (7): 6753–6761.
38 Patra, J.K., Das, G., Fraceto, L.F. et al. (2018). Nano based drug delivery systems: recent developments and future prospects. *J. Nanobiotechnol.* 16 (1): 71.
39 Zhong, H., Chan, G., Hu, Y. et al. (2018). A comprehensive map of FDA-approved pharmaceutical products. *Pharmaceutics* 10 (4): 263.
40 Sadauskas, E., Wallin, H., Stoltenberg, M. et al. (2007). Kupffer cells are central in the removal of nanoparticles from the organism. *Part. Fibre Toxicol.* 4: 10.
41 Chauhan, V.P. and Jain, R.K. (2013). Strategies for advancing cancer nanomedicine. *Nat. Mater.* 12 (11): 958–962.

42 Ernsting, M.J., Murakami, M., Roy, A., and Li, S.D. (2013). Factors controlling the pharmacokinetics, biodistribution and intratumoral penetration of nanoparticles. *J. Controlled Release* 172 (3): 782–794.

43 Bogart, L.K., Pourroy, G., Murphy, C.J. et al. (2014). Nanoparticles for imaging, sensing, and therapeutic intervention. *ACS Nano.* 8 (4): 3107–3122.

44 Huang, B., Abraham, W.D., Zheng, Y. et al. (2015). Active targeting of chemotherapy to disseminated tumors using nanoparticle-carrying T cells. *Sci. Transl. Med.* 7 (291): 291ra94.

45 Rosenblum, D. and Peer, D. (2014). Omics-based nanomedicine: the future of personalized oncology. *Cancer Lett.* 352 (1): 126–136.

46 Paliwal, R., Babu, R.J., and Palakurthi, S. (2014). Nanomedicine scale-up technologies: feasibilities and challenges. *AAPS Pharm. Sci. Tech.* 15 (6): 1527–1534.

47 Colombo, A.P., Briancon, S., Lieto, J., and Fessi, H. (2001). Project, design, and use of a pilot plant for nanocapsule production. *Drug Dev. Ind. Pharm.* 27 (10): 1063–1072.

48 Ramanathan, R.K., Korn, R.L., Raghunand, N. et al. (2017). Correlation between ferumoxytol uptake in tumor lesions by MRI and response to nanoliposomal irinotecan in patients with advanced solid tumors: a pilot study. *Clin. Cancer Res.* 23 (14): 3638–3648.

2

Addressing Unmet Medical Needs Using Targeted Drug-Delivery Systems: Emphasis on Nanomedicine-Based Applications

Chandrakantsing Pardeshi[1], *Raju Sonawane*[1], *and Yogeshwar Bachhav*[2]

[1] *R. C. Patel Institute of Pharmaceutical Education and Research, Department of Pharmaceutics, Shirpur 425 405, India*
[2] *Adex Pharmaceutical Consultancy Services (OPC) Pvt Ltd., Mumbai, India*

2.1 Introduction

The prime goal of drug-delivery research is to improve patient compliance by formulating clinically efficient drug products. Current scenario aims at developing targeted drug-delivery systems that deliver therapeutics at specific target sites in the body in therapeutically acceptable doses, after administration [1]. Here, while delivering the drug at desired sites (cell, tissue, or organ), the severity of the adverse effects could be reduced significantly. Thus, ideal drug delivery should be capable of finding a potential, highly specific target. In addition, the targeted drug delivery would reduce the effective dose and dosing frequencies required for disease treatment [2].

An ideal targeted drug-delivery system should avoid the off-target toxicity in normal cells [3, 4]. A targeted drug-delivery system additionally protects the therapeutic agent from degradation by the pH of the tissue environment where it is supposed to exert the therapeutic effect or due to the enzymatic activity.

A successful targeting can be achieved by chemically attaching a receptor-specific ligand to the surface of the drug carrier or by the use of external stimuli, such as magnetic field, pH change, temperature change, or ultrasound [5, 6].

During the course of targeting, the targeted drug-delivery systems are expected to reach the intended target with minimal nonspecific accumulation. The physical adsorption, chemical conjugation with the targeting ligand, or encapsulation of drug within the carrier should neither alter the drug's action at target site nor the function of ligand or carrier to reach the intended site of action [7, 8]. Thus, effective targeted drug-delivery systems require four key elements – retainment, evasion, targeting, and release. The targeted drug-delivery systems are highly complex systems and involve integration of various disciplines viz. biology, chemistry,

Targeted Drug Delivery, First Edition. Edited by Yogeshwar Bachhav.

physics, engineering, and medicine [9]. The current unmet needs and challenges in this area were summarized by Alexander Florence who is one of the few who raised awareness on the exaggerated claims of nanoparticle-based drug targeting [1, 10].

The major strategies of drug targeting have been categorized into two types viz. active and passive targeting. The term *active targeting* represents the interactions between drug or drug-loaded carriers with the specific receptors present on the surface of the target cells, usually mediated through specific ligands [11]. *Passive targeting* refers to the accumulation of a drug-carrier system or drug targeting at a precise site (particular regions of the body, due to the natural features and physiological role of said tissues); it may be attributed to chemical, physical, pharmacological, and biological aspects of the disease. The carrier size and surface properties of the drug-targeted systems must be specially controlled to prevent the uptake by the reticuloendothelial system (RES) to maximize the targeting capability, and increase the systemic circulation of the drug-containing carrier [12, 13].

The present idea of developing targeted drug-delivery systems was originated in early twentieth century from the observations of Paul Ehrlich (a German physicist and Nobel laureate), who proposed a hypothesis of *magic bullet* concept. Strategic efforts have been taken over the past few decades to investigate several options for achieving selective drug targeting [13, 14]. While developing targeted drug-delivery systems, pharmaceutical nanoparticulate carriers (nanomedicines) offer several benefits, *as listed below to minimize the potential systemic side effects.*

- protection of encapsulated drug from biological and/or chemical degradants
- increased apparent aqueous solubility of drug
- enhanced residence time at the site of absorption
- enhanced cellular internalization
- controlled release kinetics of the encapsulated drug
- targeted drug delivery through surface modification of the therapeutic nanocarriers with specific ligand
- In addition, targeted drug-delivery systems also suffer few limitations viz rapid clearance of the targeted systems, sometimes immune reactions against such targeted systems, requirement of highly skilled manufacturing and sophisticated technology for formulation development, and difficulty in maintaining stability of formulation at the target site [13–15].

Though the present chapter preliminarily focuses on the basic concepts, types, and strategies of targeted drug-delivery systems, the major emphasis is also to explore the nanomedicine-based systems to address the unmet medical needs.

In view of this, Figure 2.1 illustrates the diagrammatic representation of the various targeting approaches, types of targeting, ligands, and targeted nanocarriers. We believe that this compilation would be helpful for the readers and researchers who engaged in developing novel nanomedicines intended for various targeted drug-delivery applications.

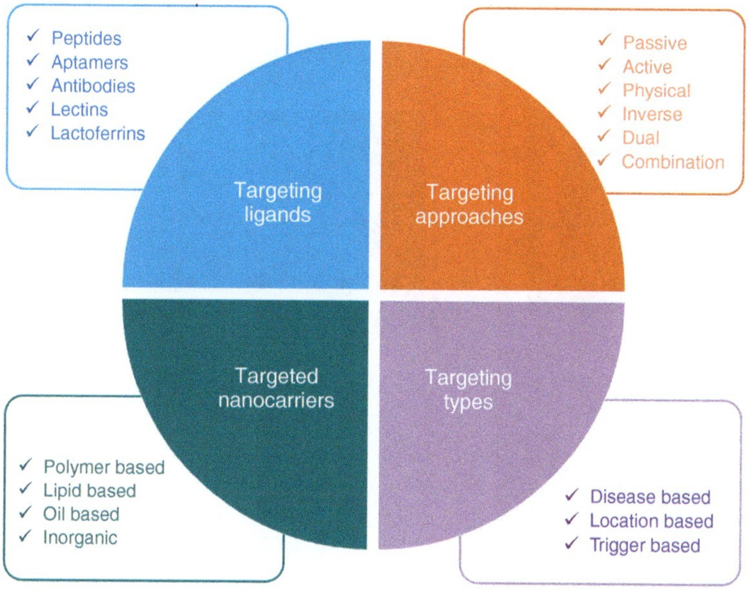

Figure 2.1 Schematic representation of the various targeting approaches, types of targeting, ligands, and targeted nanocarriers.

2.2 Targeted Drug-Delivery Systems for Unmet Medical Needs

Target refers to the specific tissue, organ, or a cell or group of cells, which in chronic or acute pathophysiological condition needs medical treatment. *Carrier* is one of the special molecules or systems essentially required for effective transportation of laden drugs to the preselected target site. They have engineered vectors, which retain the drug inside or onto them either via encapsulation or via spacer moiety and transport or deliver it either inside or into the vicinity of the target cell [16].

While adopting different drug-targeting strategies, the complex nature of biological systems, from macroscale level (tissues and organs) to the molecular or atomic level (cell or cell constituent) need to be considered (Figure 2.2). It is expected that the drug binds specifically and effectively to the molecular determinant involved in a disease. Moreover, safe and efficient transport of the therapeutic cargo from the tissue or organ to the cell, or cell constituent could be achieved by utilizing a drug-targeting strategy that overcomes the most common challenges posed by the variety of barriers in the body, such as low diffusion of ligand-gated carrier into tissues and tissue heterogeneity [17].

Figure 2.3 illustrates the design of an ideal ligand-targeted drug-delivery system. It is made of a targeting ligand (lectin, lactoferrin, peptide, aptamer, or antibody) linked to the therapeutic or diagnostic payload via a cleavable linker and a spacer [19].

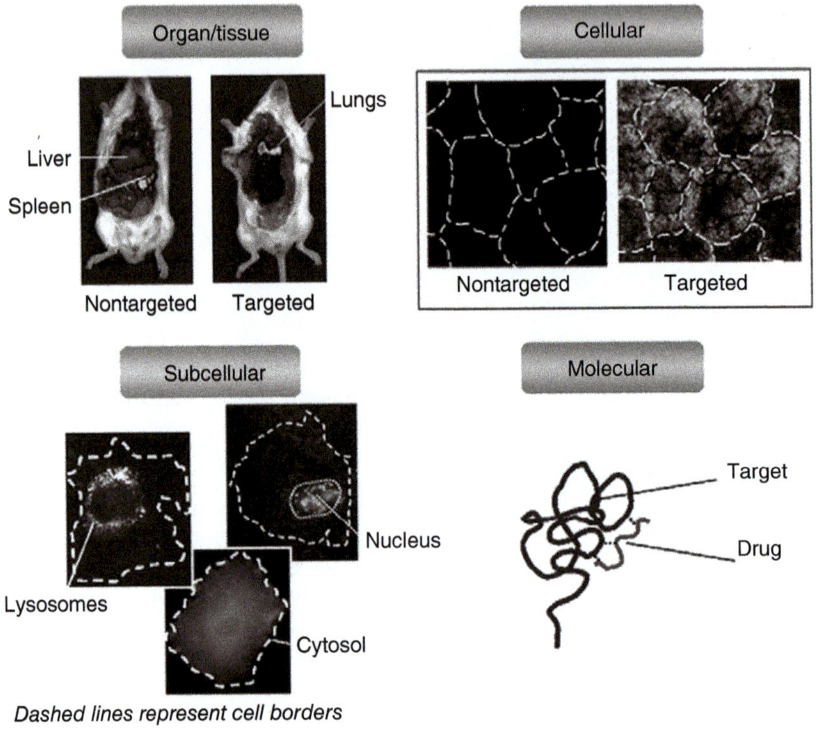

Figure 2.2 Levels of drug targeting. Source: Reproduced from Muro [17] with kind permission of Elsevier.

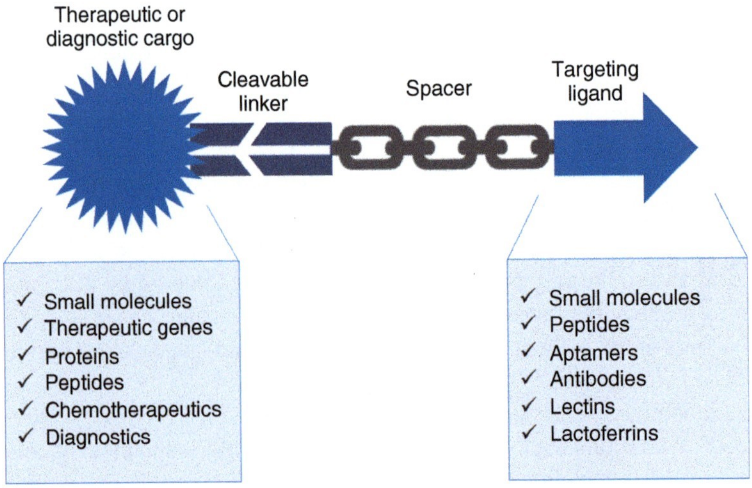

Figure 2.3 A schematic illustration of an ideal ligand-targeted drug-delivery system. Source: Modified after Vyas and Sihorkar [18].

2.2.1 Targeting Ligands

The term *ligand* refers to a component that is coupled to the drug molecule to achieve active targeting and that possess affinity toward a specific receptor present at the disease site (on the surface of a cell, tissue, or organ). A drug can be coupled to the ligand either directly by chemical conjugation or indirectly through encapsulation into a carrier, most preferably nanocarrier [17, 18].

Basically, there are two main types of ligands viz intracellular ligands (ligands that bind to receptors inside the cell), and extracellular ligands (ligands that bind to receptors outside the cell). Several strategies have been explored by researchers around the globe for the design and investigation of the installation of ligands on the surface of nanocarriers intended for therapeutic, diagnostic, and theranostics applications. The attachment of ligands on the nanocarrier surface is a complex task involving knowledge of the properties of both the ligand and the nanocarrier to maintain the structures of both the participants as well as the stability of the formed ligand–nanocarrier construct [20].

The targeting ligands can be installed onto the surface of nanocarrier either through preconjugation of the ligands to the materials prior to formulating nanocarriers (Figure 2.4a) or by physically attaching the ligands to the surface after formulating nanocarriers (Figure 2.4b). In the preconjugation strategy, ligands can be applied to initiate polymerization of the polymers that will assemble into nanocarriers. In general, preconjugation of ligands is ideal for chemistry-based conjugation, which is suitable for most of the ligands, including small molecules, peptides, and aptamers, whereas the postattachment route may benefit the ligands that are sensitive to chemical environments, such as antibodies. Conjugation of ligands can also be done by covalent bonding (Figure 2.4c) through amide bond formation between the amine groups present in ligand and carboxyl groups on the surface of nanocarriers, mediated using cross-linkers, such as 3-(3-dimethylaminopropyl)-1-ethylcarbodiimide hydrochloride (EDC), and *N*-hydroxysuccinimide (NHS). In addition, targeting ligands can be attached to the surface of nanocarriers through physical adsorption by utilizing affinity complexes (Figure 2.4d) [20].

In execution of successful targeting, the ligand-installed nanocarriers suffer many challenges viz enzymatic degradation, potential immune reaction, and uptake by RES. Therefore, various ligand presenting strategies have been investigated by the researchers to maximize the ligand efficacy in the diseased region while minimizing the potential risks during administration [20].

These ligand presentation strategies include (i) directly presenting ligands on the surface of nanocarriers, (ii) uncaging of the ligand through the light-triggered activation, (iii) ligand exposure through pH, enzyme-triggered degradation of the surface polymeric coating, (iv) conformation change of ligand-supporting materials, and (v) reversible shielding between dual ligands [20–26].

2.2.1.1 Small Molecules as Targeting Ligands

Li et al. [27] have recently developed a novel glucose-functionalized (glucose transporter-1 [GLUT-1]) targeting, tumor microenvironment responsive, and

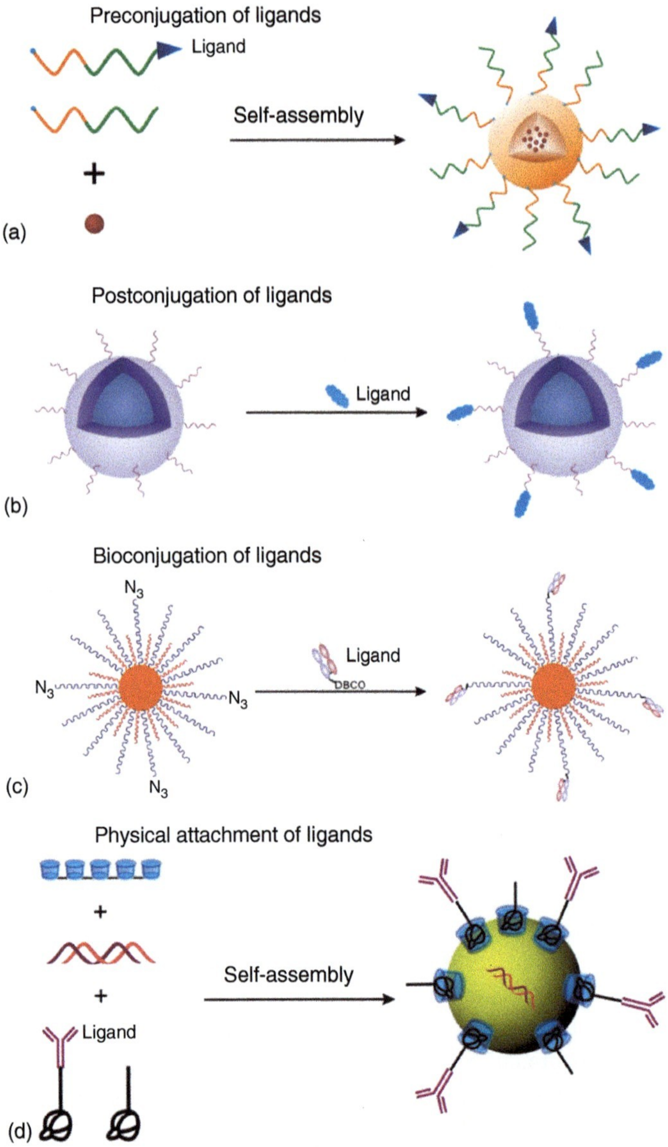

Figure 2.4 Strategies of installation of targeting ligands on the surface of nanocarriers. (a) Strategies for preconjugation and (b) postconjugation of ligands during nanocarrier formulation. Strategies of (c) bioconjugation and (d) physical attachment of ligands on formulated nanocarrier surface. Source: From Mi et al. [20]. Republished with permission of John Wiley and Sons.

drug-delivery system made of polydopamine nanoparticles (PDA NPs). Photothermally triggered cytosolic drug-delivery system was designed by conjugation of bortezomib (BTZ, an anticancer agent) to the catechol groups of PDA NPs in a pH-dependent manner so as to efficiently accumulate in tumor site and localize in subcellular endo/lysosomes of tumor cells. In addition, they could respond to tumor microenvironment and endo/lysosomal pH as well as near infrared irradiation (NIR) to promote the robust release of BTZ. Authors concluded that the GLUT1-targeting, pH, and photothermal-responsive drug-delivery NPs showed great potential for widely used chemophotothermal tumor therapy [27].

2.2.1.2 Aptamers as Targeting Ligands

The word *aptamer* is derived from the Latin word *aptus*, which means, "to fit." Aptamers (first reported by Ellington and Gold in 1990) are single-stranded DNA or RNA sequences that can specifically bind to targets by folding into well-defined three-dimensional structures [28, 29].

Xing et al. [30] developed a doxorubicin (DOX) doped liposome for the real drug delivery in vitro and in vivo cancer models.

As shown in Figure 2.5, the chemotherapeutic drug DOX was doped in the aqueous core of the liposome, and AS1411aptamer (targeted to nucleolin) was attached to liposome through the hydrophobic interaction between lipid bilayer membranes

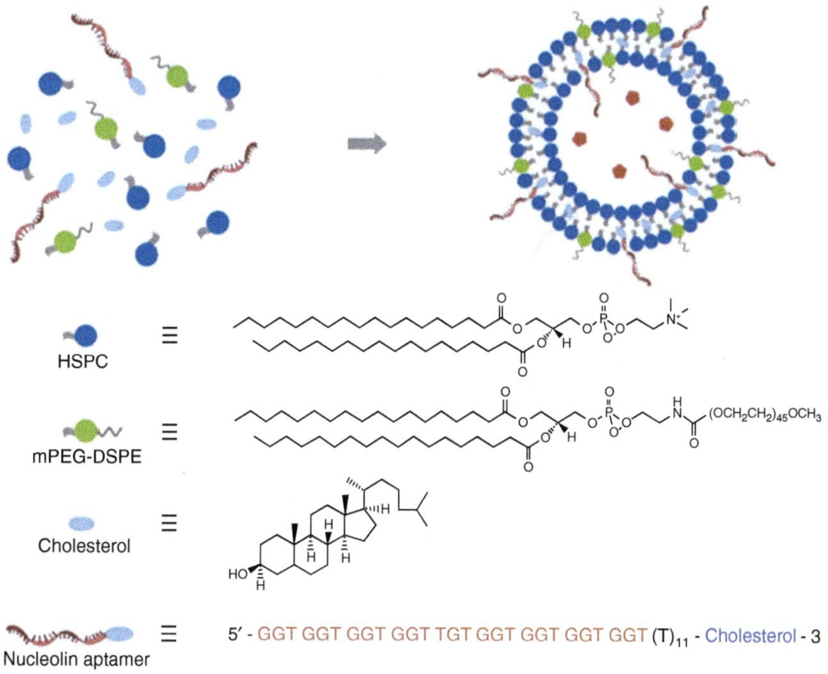

Figure 2.5 Schematic illustration of nucleolin aptamer-functionalized liposome for in vitro and in vivo cancer therapy. Cholesterol-modified DNA strands were immobilized onto the surface of liposome by intercalating the 30-cholesterol modification into the lipid bilayer. Source: Xing et al. [30] with kind permission of the, The Royal Society of Chemistry, London.

and the cholesterol end on the aptamer. With this functionalized liposome, the drug delivery and therapeutic efficiency to target cancer cells and tissues were investigated, which showed improved anticancer efficacy targeting the MCF-7 breast cancer cells and xenograft MCF-7 breast tumors in nude mice. The modification of polyethylene glycol (PEG) on the surface of liposome not only increased its biocompatibility but also prolonged the circulating half-life, which enhanced the targeting efficiency to the tumor tissues in mice [30].

2.2.1.3 Antibodies as Targeting Ligands

Antibodies are typically 200–300 nm-sized proteins and part of immune system. Following conjugation to nanoparticles, these antibodies can deliver a therapeutic or theranostic agent to a specific diseased site, maybe a tissue or organ [31].

Scindia et al. [32] specifically targeted the alpha 8 (α8) integrin receptors overexpressed on the glomerular mesangial cells in lupus glomerulonephritis (GN; inflammation of the glomeruli of kidneys) by preparing anti-α8 integrin antibody conjugated immunoliposomes (ILs). These immunoliposomes (ILs) werę loaded with DiI, a red fluorescent dye, to allow tracking in vivo and injected into the tail vein of female mice at different ages. Specificity of targeting was studied by fluorescence microscopy and flow cytometry. Delivery by anti-α8 integrin ILs was specific to the kidney even in the severely nephritic mice. Thus, targeted delivery by anti-α8 integrin ILs opens a novel method for treatment of glomerular diseases in humans. Authors stated that this method can be used for the treatment of a diverse range of chronic glomerular diseases, in addition to lupus GN, and will facilitate the concept of end-organ-targeted therapies in renal disease [32].

2.2.1.4 Lectins as Targeting Ligands

Lectins are plant-derived proteins or glycoproteins with high specificity for sugar (carbohydrate) moieties and are able to recognize and bind specifically to the glycan arrays of glycosylated lipids and proteins. In the last two decades, several investigations have reported the conjugation of lectins on colloidal carrier systems. Among them, wheat germ agglutinin (WGA), odorranalectin, *Solanum tuberosum* lectin, and *Ulex europaeus* agglutinin I (UEA-I) have been investigated widely for targeted drug delivery [15].

Chen et al. [33] constructed the *S. tuberosum* lectin (STL)-conjugated poly (DL-lactic-*co*-glycolic acid) nanoparticles (PLGA NPs) as novel biodegradable and targeted nose to brain drug-delivery system. The aim of this study was to investigate the in vitro cellular uptake and brain-targeting efficiency of STL–PLGA NPs compared to unmodified NPs. The in vitro uptake study showed markedly enhanced endocytosis of STL–PLGA NPs compared to unmodified PLGA NPs in Calu-3 cells. In addition, STL–PLGA NPs demonstrated 1.89–2.45 times ($p < 0.01$) higher brain-targeting efficiency in different brain tissues of Sprague–Dawley rat model than unmodified NPs. These findings suggested that the conjugation of STL on the surface of PLGA NPs could enhance the targetability of prepared nanoformulation and could serve as a promising brain drug-delivery system [33].

2.2.1.5 Lactoferrins as Targeting Ligands

Lactoferrin (Lf) is a natural cationic iron-binding glycoprotein belonging to the transferrin (Tf) family. Extensive histological investigations reported that the Lf receptors (LfR) are highly expressed on the apical surface respiratory epithelial cells and the brain cells, such as brain endothelial cells and neurons. Lf also possesses the ability to cross the BBB, and therefore, it has also been widely investigated for the targeted delivery of neurotherapeutics [15]. Furthermore, it has been proved that the Lf is safer than lectins [34].

Meng et al. [35] developed a Huperzine A (HupA)-loaded PLGA NPs surface modified with Lf-conjugated *N*-trimethylated chitosan (TMC) (HupA Lf-TMC NPs) for safe and efficient intranasal delivery of HupA to the brain for the treatment of Alzheimer's disease (AD). The HupA Lf-TMC-PLGA NPs were prepared by emulsion–solvent evaporation method and optimized using Box–Behnken experimental design. MTT assay was used to evaluate the cytotoxicity of the NPs. In vivo imaging system was used to investigate brain-targeting effect of NPs after intranasal administration. The biodistribution of HupA-loaded NPs after intranasal administration was determined by liquid chromatography–tandem mass spectrometry. HupA Lf-TMC NPs showed lower toxicity in the 16HBE cell line compared with HupA solution. Qualitative and quantitative cellular uptake experiments indicated that accumulation of Lf-TMC NPs was higher than nontargeted analogs in 16HBE and SH-SY5Y cells. In vivo imaging results showed that Lf-TMC NPs exhibited a higher fluorescence intensity in the brain and a longer residence time than nontargeted NPs.

2.2.2 Targeting Approaches

2.2.2.1 Disease-Based Targeting

This section highlights the disease-based targeting approaches for the delivery of nanomedicines as an effective strategy against therapy of various diseases (cancer and infectious diseases viz. tuberculosis, malaria, human immunodeficiency virus [HIV], and so on).

2.2.2.1.1 Tumor Targeting

Nanoparticle-mediated targeted drug-delivery systems have been designed to promote the therapeutic delivery to the tumor sites, minimizing the side effects associated with the use of free drug and main aim should be to have accumulation of NPs in tumors by the enhanced permeation and retention (EPR) effect. Tumor treatment necessitates optimum dose of therapeutic agent to be delivered to the tumors. The intravenously (IV) administered drug or drug-NPs need to circulate in the bloodstream, extravasate (cross the vascular walls) into the interstitium, and penetrate the tumors [36, 37]. Several nanoformulations have recently been approved for the treatment of cancer clinically viz. Myocet® (liposomal doxorubicin), Daunoxome® (liposomal daunorubicin), Doxil® (liposomal doxorubicin), Depocyt® (liposomal cytarabine), Genexol-PM® (paclitaxel-loaded polymeric micelle), Abraxane® (albumin-bound paclitaxel particles), and additionally, other products are in the various stages of the clinical trials viz. Rituxan® (rituximab), Herceptin® (trastuzumab), Campath® (alemtuzumab) and so on [3].

Tumors can be targeted both passively as well as actively by using NPs. Passively, NPs of size up to 500 nm can extravagate the leaky tumor vasculature via EPR. To avoid RES uptake, often these NPs are PEGlyated [38]. NPs can also be targeted actively by conjugating targeting ligands (aptamers, lectins, transferrin, folate, and antibodies) on their surface [39].

2.2.2.1.2 Targeting Infectious Diseases

Bacteria, viruses, fungi, and other such microorganisms are the major sources of infectious diseases (malaria, tuberculosis, HIV/AIDS, cholera, and several other diseases). Essentially, the antimicrobial compounds inhibit or interrupt important cell cycle processes of the microorganisms and kill them. However, there is increased risk of developing resistance and many new antimicrobials are currently in the early stages of clinical trials to address this multidrug resistance. This situation requires the use of targeted drug-delivery approach for treating infectious diseases [3, 40].

Nanoparticles-mediated targeted drug-delivery systems have attracted several research groups worldwide to focus potentially on the advantages of these systems, such as:

- ability to deliver drugs/genes
- sustained action, ability to deliver drug combinations
- fewer incidence of systemic side effects
- overcome drug resistance [40].

Ligand-conjugated nanocarriers have been found to be highly effective against infectious diseases. Lectin-conjugated gliadin NPs were investigated for targeting *Helicobacter pylori* infections [41]. Also, a DC3-9dR-mediated delivery of siRNA targeting the tumor necrosis factor-alpha (TNF-α) was investigated for targeting the dendritic cells and to suppress the viral replication in dengue [42].

Li et al. [43] have recently developed PLGA-based NPs of dapsone and clofazimine for the treatment of tuberculosis. The aim of the investigation was to overcome poor solubility of clofazimine and to avoid production of toxic metabolites due to first-pass metabolism of dapsone following oral administration. The NPs were prepared by emulsion–solvent evaporation method. It was observed that the PLGA-based NPs of dapsone and clofazimine targeted the deep lungs and exhibited higher therapeutic efficiency in the in vivo experiments compared to dapsone–clofazimine solution which might be attributed to the enhanced distribution and sustained release of the drugs from the nanoparticulate system. In conclusion, authors stated that the PLGA NPs of dapsone–clofazimine combination would be a cost-effective, safe, efficacious, and clinically pertinent novel dosage form for the delivery of these anti-TB drugs systemically [43]. A similar study has been reported by Pandey & Khuller [44] who developed nebulized solid lipid particles (SLNs) incorporating a combination of rifampicin, isoniazid, and pyrazinamide for bronchoalveolar drug delivery in *Mycobacterium tuberculosis*-infected guinea pigs. The nebulized formulation achieved complete removal of the tubercle bacilli from the lungs and spleen after just seven doses of pulmonary administration compared to forty-six oral doses of bulk drug solutions [44].

Malaria is another widespread infectious disease caused by the four species of the parasitic protozoans of the genus *Plasmodium*: *Plasmodium falciparum*, *Plasmodium vivax*, *Plasmodium malariae*, and *Plasmodium ovale*. Owing to serious side effects and drug resistance associated with current therapy, there is an urgent need for targeted drug therapy against malaria. In milieu of this, Marques et al. [45] developed primaquine-loaded liposomes functionalized with covalently bound heparin for targeted delivery to *Plasmodium*-infected RBCs (pRBCs) [45]. Mosqueira et al. [46] formulated parenteral formulation of poly(D,L-lactide) (PLA) or PEGylated PLA nanocapsules loaded with halofantrine. The passively targeting PEGylated nanocapsules were observed to be both long-circulating and cytotoxic to the parasites. The halofantrine-loaded nanocapsules showed activity that was similar to or better than that of the solution-based formulation in the 4-day test and as a single dose in severely infected mice. Nanocapsules increased the area under the curve for halofantrine in plasma more than sixfold compared with the solution throughout the experimental period of 70 hours. In addition, nanocapsules induced a significantly faster control of parasite development than the solution in the first 48 hours posttreatment. The parasitemia has fallen more rapidly using PLA nanocapsules, and the effect was more sustained in case of PEGylated PLA nanocapsules [46].

Antiretroviral drugs are the major means of effective management of HIV and prevent its progression toward AIDS. Theoretically, all stages in the life cycle of HIV are potential targets for antiretroviral therapy [47, 48]. Along with the treatment based on different categories of anti-HIV/AIDS drugs, combination therapy has been a major revolution in the treatment of HIV/AIDS. This therapy is specifically designated as highly active antiretroviral therapy (HAART) [48]. Today, the most severe problem with HIV/AIDS therapy is the emergence of multidrug resistance (MDR) against antiretroviral agents during the course of therapy, which leads to poor clinical outcomes [48]. NPs-based targeted delivery of antiretroviral therapeutics has widely been investigated to target the mononuclear phagocytic system (MPS) cells viz. monocytes/macrophages (Mo/Mac), common reservoirs for HIV virus [49]. Wu et al. [50] reported the galactosylated liposomes of azidothymidine palmitate (AZTP) for active targeting of AZT to the hepatocytes which otherwise act as reservoirs for HIV [50]. Garg and Jain [51] studied the azidothymidine (AZT)-loaded galactosylated liposomes targeting the lectin receptors present on the macrophages.

A single IV dose of galactosylated liposomal encapsulated AZT was investigated in Sprague–Dawley rats to get insight into bone marrow toxicity, plasma and tissue distribution of free drug.

The maximum cellular uptake and no significant hematological toxicity were observed with AZT-loaded galactosylated liposomes. Present formulation has maintained a significant level of AZT in the tissues rich with galactose-specific receptors and had a prolonged residence in the body resulting in enhanced half-life of AZT, thereby enhancing the anti-HIV efficiency in the selected animal model [51].

2.2.2.2 Location-Based Targeting

Present section highlights the targeted drug-delivery approaches for the organ-specific (brain, lungs, liver, and kidney) delivery of nanotherapeutics.

2.2.2.2.1 Brain Targeting

A range of polymers and/or polymer derivatives has been explored for surface modification of the fabricated nanocarriers to enhance the residence time of the nanoformulations on the nasal mucosal membrane, which will further aid in improved targetability; following the direct nose-to-brain delivery of neurotherapeutics [15].

Recently, Pardeshi and Belgamwar [52] have investigated the fabrication of flaxseed oil-based neuronanoemulsions (NNEs), which were further surface-modified with TMC to prepare mucoadhesive neuronanoemulsion (mNNE) intended for the direct nose-to-brain drug delivery. The NNEs were loaded with high-partitioning ropinirole-dextran sulfate (ROPI-DS) nanoplex and fabricated using the hot high-pressure homogenization (HPH) technique. The NNEs were optimized using central composite experimental design. The primary objective was to provide a controlled drug release with prolonged residence on the nasal mucosa for the treatment of Parkinson's disease (PD). Enhanced brain targeting through direct nose-to-brain drug delivery and improved therapeutic efficacy through enhanced retention of mNNE formulation over the nasal mucosal membrane led to the reduction of the dose and frequency of administration and also the higher safety.

The mNNE formulation was administered to the Swiss albino mice model via the intranasal route, and both the plasma and brain pharmacokinetics were estimated in comparison to intravenously administered mNNE formulation and intranasally administered aqueous ROPI suspension. The *in vivo* studies performed on mice exhibited the high brain-targeting efficiency of mNNE formulation through the nose-to-brain delivery via the olfactory pathway. Following the intravenous administration, at each time point, the concentration of ROPI from mNNE formulation was higher in the plasma compared with the brain (Figure 2.6a). Following the intranasal administration, at each time point studied, the concentration of ROPI was higher in the brain from mNNE formulation compared with aqueous ROPI suspension (Figure 2.6b). The mNNEs formulation also showed the highest brain–blood ratio (Figure 2.6c), thereby suggesting the high brain-targeting efficiency as compared to aqueous ROPI PDS suspension and NNEs formulation, after the intranasal administration [52].

2.2.2.2.2 Lung Targeting

The lungs provide a large surface area and a thin epithelial layer perfused with continuous blood flow. Targeting the lungs assists in avoiding the first-pass effect faced by oral drugs and provides quick systemic drug administration for the treatment of respiratory diseases, *such as* asthma, TB, cystic fibrosis, chronic obstructive pulmonary disease (COPD), and lung cancer. Targeted drug delivery *via* pulmonary administration overcomes the major barriers viz. mucus layer and enzymatic degradation, thereby making lungs favorable site for drug and/or gene delivery [3, 53].

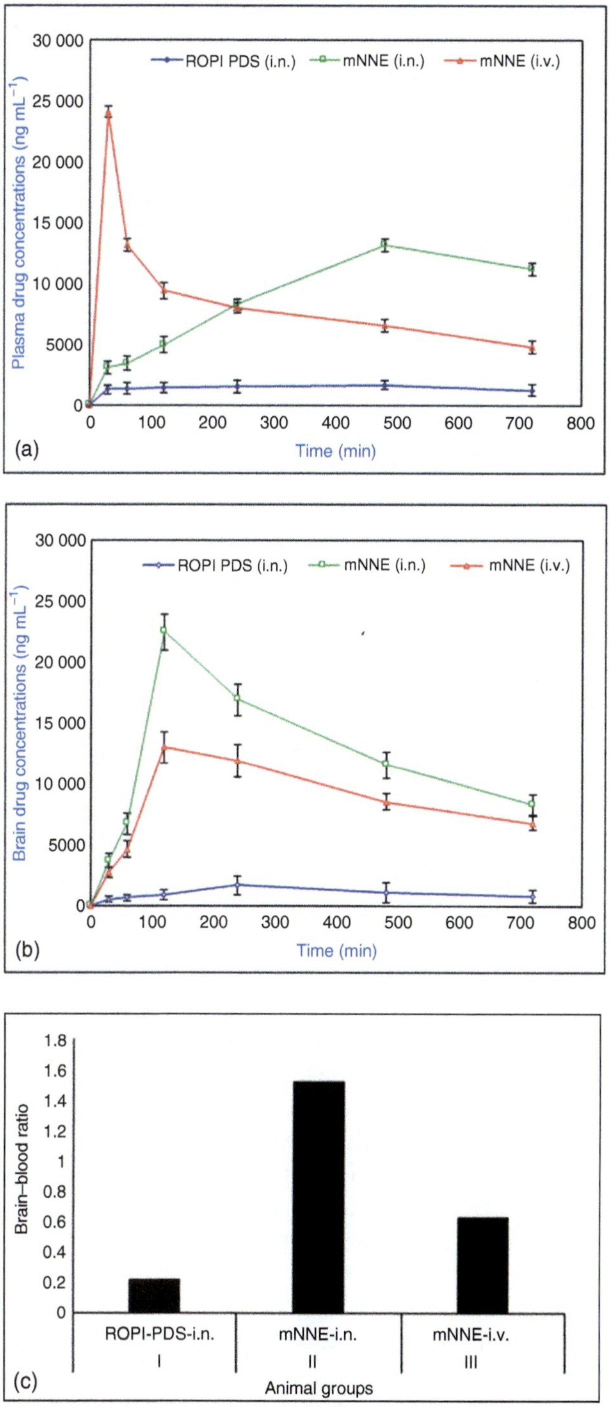

Figure 2.6 (a) Plasma concentrations vs. time profile of ROPI PDS and mNNE formulations after intranasal and intravenous administration. (b) Brain concentrations vs. time profile of ROPI PDS and mNNE formulations after intranasal and intravenous administration, and (c) the brain–blood ratio of ROPI PDS (i.n.), mNNEs (i.n.), and mNNEs (i.v.) formulations in Swiss albino mice model. Source: from Pardeshi and Belgamwar [52] with kind permission of the, Elsevier, Amsterdam.

Rosiere et al. [54] developed temozolomide (TMZ)-loaded nanomicelles using the folate-grafter self-assembling copolymer folate-polyethylene glycol-hydrophobically modified dextran (F-PEG-HMD) and investigated as dry powder for inhalation (DPI) formulation for lung cancer therapy. In the presence of F-PEG-HMD, the solubility of TMZ was improved significantly led to enhanced accumulation of TMZ at the tumor site. The TMZ-loaded nanomicelles were then spray dried to produce dry powders intended for inhalation purposes. The developed DPI formulation showed wide pulmonary deposition in the deep lungs (lower respiratory tract), the prime location of adenocarcinomas that overexpress folate receptors (FRs). The in vitro antiproliferative activity of TMZ–DPI formulations was compared to that of the TMZ solution in murine M109-HiFR lung carcinoma Human A549 non-small cell lung cancer (NSCLC) cell line which was assayed by colorimetric MTT assay. It was observed that TMZ was more effective when given as DPI formulation [54].

2.2.2.2.3 Liver Targeting

The liver cancer is most prevalent among the various types of cancers, and chemotherapy has been the main treatment option against this condition. However, high systemic toxicity and lack of specificity limit the use of traditional chemotherapy approach to treat liver cancer and hence, there is a need for novel therapeutic strategy to address the unmet medical need. Targeted drug delivery that can carry the drug to the specific organ or tissue is highly desirable strategy, in this case, to improve the therapeutic efficiency and reduce the off-target toxicity [55, 56].

Tian et al. [57] developed a liver-targeted drug-delivery system composed of chitosan/poly(ethylene glycol)-glycyrrhetinic acid (CTS/PEG-GA) NPs. The NPs were prepared by ionotropic gelation technique. Glycyrrhetinic acid was used as targeting ligand whereas doxorubicin hydrochloride was employed as an antineoplastic agent. The biodistribution of the prepared NPs was assessed by single-photon emission computed tomography (SPECT), while cellular uptake was evaluated using human hepatic carcinoma cells (QGY-7703 cells) cell line. Real-time SPECT images of the NPs in Balb/c nude mice model revealed that the DOX-loaded CTS/PEG-GA NPs were accumulated extensively in the liver compared to kidney and bladder, owing to conjugation of GA on the NPs surface. Also, the antitumor efficacy of the DOX-loaded CTS/PEG-GA NPs was studied in H22 cell-bearing nude mice. The tumor growth was observed for 10 days in mice injected with physiological saline, free DOX, and DOX-loaded CTS/PEG-GA NPs. All mice were sacrificed, and tumors were taken out after 10 days of treatment. It was found that the tumor size was high for saline group compared to free DOX or DOX-loaded CTS/PEG-GA NPs group, demonstrating the efficacy of prepared NPs in targeting liver cancer, actively [57].

2.2.2.2.4 Kidney Targeting

The renal disorders require high renal drug concentrations, which may not translate in reaching the target cells. Also, renal dysfunctions may affect the distribution of drugs to the kidneys. Drug-delivery systems targeted to the kidneys are of great importance to overcome the above-mentioned problems and to improve the therapeutic efficiency in treating renal diseases at the same time by reducing renal toxicity

[58]. Kidney-targeted drug-delivery systems have been evaluated at different key targets viz. glomerular endothelial cells, glomerular basement membrane, podocytes, glomerular mesangial cells, and proximal tubules [59].

Asgeirsdottir et al. [60] designed dexamethasone liposomes which were modified by anti-E-selectin antibody (AbEsel-Liposomes) and used this formulation for targeting glomerular endothelial cells (GECs) in glomerulonephritis. In glomerulonephritis model rats, E-selectin was expressed on glomerular endothelial cells, which resulted in homing of Ab(Esel) liposomes to glomeruli after intravenous administration. Accumulation of Ab(Esel) liposomes in the kidney was 3.6 times higher than nontargeted IgG liposomes. In addition, the glomerular endothelial activation and albuminuria were reduced after 7 days of postadministration. Moreover, the surface of activated GECs also expresses other selectin and integrin receptors, which can serve as potential targets for delivering drugs to GECs [60].

Tuffin et al. [61] prepared OX7-coupled immunoliposomes (OX7-IL) by coupling liposomes with F_{ab} fragments of OX7 mAb directed against Thy1.1 antigen. Because the glomerular endothelium is fenestrated and no basement membrane separates glomerular capillaries from the mesangium, mesangial cells represent a particularly suitable target for drug delivery by OX7-IL. After intravenous administration to rats, OX7-IL was found to specifically target mesangial cells but its kidney targeting was blocked when free OX7 $F_{(ab)2}$ fragments were coadministered. Rats injected with low-dose doxorubicin encapsulated in OX7-IL showed extensive glomerular damage whereas other parts of the kidney and other organs were spared [61].

2.3 Regulatory Aspects and Clinical Perspectives

Nanotechnology has recently revolutionized the pharmaceutical field; so, it is fair to question whether current regulatory guidelines are adequate to tackle the regulations, product approvals, policy guidance, and market surveillance in this field.

Liposomes, nanocrystals, microemulsions, and polymeric nanoparticles are few of the products already approved by FDA for targeted delivery of variety of therapeutics. The preliminary and very critical parameters considered during pharmaceutical product approval are the products' safety, efficacy, and stability.

Before acceptance and regulatory approval, a pharmaceutical nanoproduct should fulfill the following criteria:

(a) The accelerated and long-term stability data of the pharmaceutical nanoproduct
(b) The pharmacokinetic profile of the nanoproduct
(c) The scale-up design should be implemented and checked beforehand
(d) The detailed characterization of the physicochemical parameters of the nanoproduct
(e) Suitable techniques should be adopted to determine the release kinetics
(f) Robust analytical methods to ensure the reproducible quality of the product.

Table 2.1 lists the selected FDA-approved nanotechnology-based targeted drug-delivery systems.

Table 2.1 List of selected FDA-approved nanotechnology-based targeted drug-delivery systems [62].

FDA-approved nanotechnology	Description of nanoformulation	Merits of nanoformulation	Therapeutic indication	Year of approval
AmBisome® (Gilead Sciences)	Liposomal amphotericin B	Reduced nephrotoxicity, targeted delivery, stable	Fungal/protozoal infections	1997
Doxil®/Caelyx™ (Janssen)	Liposomal doxorubicin	Improved site-specific delivery; decreased systemic drug toxicity	Kaposis sarcoma; ovarian cancer; multiple myeloma	1995; 2005; 2008
Estrasorb™ (Novavax)	Micellar estradiol	Controlled delivery of therapeutic	Menopausal therapy	2003
Abraxane®/ABI-007 (Celgene)	Albumin-bound paclitaxel NPs	Improved solubility; improved delivery to tumor	Breast cancer; NSCLC; Pancreatic cancer	2005; 2012; 2013
Nanotherm® (MagForce)	Iron oxide nanocrystals	Allows cell uptake and introduces superparamag-netism	Glioblastoma	2010
Plegridy® (Biogen)	Polymer–protein conjugate (PEGylated IFN β-1a)	Improved stability of protein through PEGylation	Multiple sclerosis	2014
Zilretta	Triamcinolone acetonide with a poly lactic-*co*-glycolic acid (PLGA) matrix microspheres	Extended pain relief over 12 weeks	OA of the knee	2017

Abbreviations: NSCLS: Non-small cell lung cancer; OA: Osteoarthritis.
Source: Modified from Patra et al. [62].

Table 2.2 List of selected nanotechnology-based targeted drug-delivery systems in different stages of clinical trials [63].

Disease condition	Targeting nanocarrier	Therapeutic agent	Sponsor and location	Title of clinical study	Current status	Clinical trial identifier
Colon cancer	Polymeric NPs	Cetuximab	Ahmed A. H. Abdellatif, Al-Azhar University	Targeted polymeric nanoparticles loaded with cetuximab and decorated with somatostatin analog to colon cancer	Phase I	NCT03774680
Solid tumors	Liposomes	Doxorubicin	University Hospital, Basel, Switzerland	Anti-EGFR immunoliposomes in solid tumors	Completed	NCT01702129
Ovarian cancer, fallopian tube cancer	Albumin-stabilized NPs	Paclitaxel	GOG Foundation (Gynecologic Oncology Group)	Paclitaxel albumin-stabilized nanoparticles formulation in treating patients with recurrent or persistent ovarian epithelial cancer, fallopian tube cancer, or primary peritoneal cancer	Completed	NCT00499252
Breast cancer	Albumin-stabilized NPs	Gemcitabine hydrochloride; bevacizumab	Alliance for clinical trials in oncology	Paclitaxel albumin-stabilized nanoparticle formulation, gemcitabine, and bevacizumab in treating patients with metastatic breast cancer	Completed	NCT00662129
Breast cancer; skin cancer; skin diseases	Quantum dots	Veldoreotide	Ahmed A. H. Abdellatif, Al-Azhar University	Topical fluorescent nanoparticles conjugated somatostatin analog for suppression and bioimaging breast cancer	Recruiting	NCT04138342

Although several patents have been filed on nanotechnology-based targeted drug-delivery systems, very few have reached the shelf of pharmacy due to the challenges associated with the scale-up of nanopharmaceutical products.

Nanotechnology-based delivery systems aid the targeting ability and have better therapeutic outcomes. A number of clinical trials have been conducted in past couple of decades to study the effectiveness of nanotechnology-based targeted drug products. Table 2.2 enlists the selected nanotechnology-based targeted drug-delivery systems in different stages of clinical trials.

2.4 Conclusion and Future Outlook

The ever-emerging field of nanomedicine has major applications in the pharmaceutical, medical, biomedical, and tissue engineering sectors, where major focus would be to achieve enhanced targetability and bring safe, efficacious, and stable platforms. Although targeted drug-delivery systems offer several advantages in preclinical and clinical trials, however, it is not flawless. The immune response to antibody-mediated therapy and inadequate pharmacokinetic profiles, while translating from preclinical to clinical studies, are the major limitations of the targeted drug-delivery systems.

Still, there is lot of scope in the development of targeted drug-delivery systems for various pathophysiological conditions and needs extensive investigations for the development of innovative therapies. Successful commercialization of such targeted drug-delivery systems is the major task for the formulation scientists to bring them into the clinic, keeping in mind the cost-effectiveness of targeted drug products along with their safety, efficacy, and stability.

List of Abbreviations

9dR	nona-D-arginine
AD	Alzheimer's disease
AIDS	acquired immunodeficiency syndrome
AZT	azidothymidine
AZTP	azidothymidine palmitate
BTZ	Bortezomib
COPD	chronic obstructive pulmonary disease
CTS/PEG-GA	chitosan/poly(ethylene glycol)-glycyrrhetinic acid
DC3	dendritic cell-targeting 12-mer peptide
DOX	doxorubicin
DPI	dry powder inhalation
DS	dextran sulfate
EDC	3-(3-dimethylaminopropyl)-1-ethylcarbodiimide hydrochloride
F-PEG-HMD	folate-polyethylene glycol-hydrophobically-modified dextran
FRs	folate receptors

GLUT-1	glucose transporter-1
GN	glomerulonephritis
HAART	highly-active anti-retroviral therapy
HIV	human immunodeficiency virus
HPH	high pressure homogenization
HupA	huperzine A
ILs	immunoliposomes
Lf	lactoferrin
EPR	enhanced permeation and retention
MDR	multi drug resistance
mNNEs	mucoadhesive neuronanoemulsions
Mo/Mac	monocytes/macrophages
MPS	mononuclear phagocytic system
NHS	*N*-hydroxysuccinimide
NIR	near infrared irradiation
NNEs	neuronanoemulsions
NPs	nanoparticles
NSCLC	non-small cell lung cancer
OA	osteoarthritis
pRBC	*Plasmodium*-infected RBCs
PD	Parkinson's disease
PDA	polydopamine
PEG	polyethylene glycol
PLA	poly(D,L-lactide)
RES	reticuloendothelial system
ROPI	ropinirole hydrochloride
SLNs	solid lipid nanoparticles
SPECT	single-photon emission computed tomography
TMC	*N,N,N*-trimethyl chitosan
TMZ	temozolomide
TNF-α	tumor necrosis factor alpha
UEA-I	ulex europaeus agglutinin I
WGA	wheat germ agglutinin

References

1 Bae, Y.H. and Park, K. (2011). Targeted drug delivery to tumors: myths, reality and possibility. *J. Controlled Release* 153: 198–205.

2 Ezzell, C. (2001). Magic bullets fly again. *Sci. Am.* 285: 34–41.

3 Pattni, B.S. and Torchilin, V.P. (2015). Targeted drug delivery systems: strategies and challenges. In: *Targeted Drug Delivery: Concepts and Design*, Advances in Delivery Science and Technology (ed. P. Devarajan and S. Jain). Cham: Springer.

4 Banerjee, A., Pathak, S., Subramanium, V.D. et al. (2017). Strategies for targeted drug delivery in treatment of colon cancer: current trends and future perspectives. *Drug Discovery Today* 22: 1224–1232.
5 Aguilar, Z.P. (2013). *Nanomaterials for Medical Applications.* Academic Press.
6 Yadav, K.S., Mishra, D.K., Deshpande, A., and Pethe, A.M. (2019). Levels of drug targeting. In: *Basic Fundamentals of Drug Delivery* (ed. R.K. Tekade). Academic press.
7 Hoffman, A.S. (2008). The origins and evolution of "controlled" drug delivery systems. *J. Controlled Release* 132: 153–163.
8 Ruenraroengsak, P., Cook, J.M., and Florence, A.T. (2010). Nanosystem drug targeting: facing up to complex realities. *J. Controlled Release* 141: 265–276.
9 Mills, J.K. and Needham, D. (1999). Targeted drug delivery. *Expert Opin. Ther. Pat.* 9: 1499–1513.
10 Florence, A.T. (2007). Pharmaceutical nanotechnology: more than size. Ten topics for research. *Int. J. Pharm.* 339: 1–2.
11 Sonvico, F., Clementino, A., Buttini, F. et al. (2018). Surface-modified nanocarriers for nose to brain drug delivery: from bioadhesion to targeting. *Pharmaceutics* 10: 1–34.
12 Li, J. (2012). Surface-modified PLGA nanoparticles for targeted drug delivery to neurons. Master thesis. 47. Louisiana State University.
13 Sonawane, R.O., Bachhav, Y., Tekade, A.R., and Pardeshi, C.V. (2021). Nanoparticles for direct nose-to-brain drug delivery: implications of targeting approaches. In: *Direct Nose-to-brain Drug Delivery: Mechanism, Technological Advances, Applications, and Regulatory Updates* (ed. C.V. Pardeshi and E.B. Souto). Academic Press.
14 Pardeshi, C.V., Kulkarni, A.D., Sonawane, R.O. et al. (2019). Mucoadhesive nanoparticles: a roadmap to encounter the challenge of rapid nasal mucociliary clearance. *Indian J. Pharma. Edu. Res.* 53: 17–27.
15 Pardeshi, C.V. and Souto, E.B. (2021). Surface-modification of nanocarriers as a strategy to enhance the direct nose-to-brain drug delivery. In: *Direct Nose-to-Brain Drug Delivery: Mechanism, Technological Advances, Applications, and Regulatory Updates* (ed. C.V. Pardeshi and E.B. Souto). Academic Press.
16 Vyas, S.P. and Khar, R.K. (2008). Basis of targeted drug delivery. In: *Targeted and Controlled Drug Delivery*. CBS Publishers and Distributors.
17 Muro, S. (2012). Challenges in design and characterization of ligand-targeted drug delivery systems. *J. Controlled Release* 164: 125–137.
18 Vyas, S.P. and Sihorkar, V. (2000). Endogenous carriers and ligands in non-immunogenic site-specific drug delivery. *Adv. Drug Delivery Rev.* 43: 101–164.
19 Srinivasarao, M. and Low, P.S. (2017). Ligand-targeted drug delivery. *Chem. Rev.* 117: 12133–12164.
20 Mi, P., Cabral, H., and Kataoka, K. (2019; 1902604). Ligand-installed nanocarriers toward precision therapy. *Adv. Mater.* 1–29.
21 Christie, R.J., Matsumoto, Y., Miyata, K. et al. (2012). Targeted polymeric micelles for siRNA treatment of experimental cancer by intravenous injection. *ACS Nano* 6: 5174–5189.

22 Mura, S., Nicolas, J., and Couvreur, P. (2013). Stimuli-responsive nanocarriers for drug delivery. *Nat. Mater.* 12: 991–1003.

23 Chien, Y.H., Chou, Y.L., Wang, S.W. et al. (2013). Near-infrared light photo-controlled targeting, bioimaging, and chemotherapy with caged upconversion nanoparticles in vitro and in vivo. *ACS Nano* 7: 8516–8528.

24 Koren, E., Apte, A., Jani, A., and Torchilin, V.P. (2012). Multifunctional PEGylated 2C5-immunoliposomes containing pH-sensitive bonds and TAT peptide for enhanced tumor cell internalization and cytotoxicity. *J. Controlled Release* 160: 264–273.

25 Lee, E.S., Gao, Z.G., Kim, D. et al. (2008). Super pH-sensitive multifunctional polymeric micelle for tumor pH (e) specific TAT exposure and multidrug resistance. *J. Controlled Release* 129: 228–236.

26 Cao, J., Gao, X., Cheng, M. et al. (2019). Reversible shielding between dual ligands for enhanced tumor accumulation of ZnPc-loaded micelles. *Nano Lett.* 3: 1665–1674.

27 Li, Y., Hong, W., Zhang, H. et al. (2020). Photothermally triggered cytosolic drug delivery of glucose functionalized polydopamine nanoparticles in response to tumor microenvironment for the GLUT1-targeting chemo-phototherapy. *J. Controlled Release* 10: 232–245.

28 Ellington, A.D. and Szostak, J.W. (1990). In vitro selection of RNA molecules that bind specific ligands. *Nature* 346: 818–822.

29 Tuerk, C. and Gold, L. (1990). Systematic evolution of ligands by exponential enrichment: RNA ligands to bacteriophage t4 DNA polymerase. *Science* 249: 505–510.

30 Xing, H., Tang, L., Yang, X. et al. (2013). Selective delivery of an anticancer drug with aptamer-functionalized liposomes to breast cancer cells in vitro and in vivo. *J. Mater. Chem. B* 1: 5288–5297.

31 Wang, J., Masehi-Lano, J.J., and Chung, E.J. (2017). Peptide and antibody ligands for renal targeting: nanomedicine strategies for kidney disease. *Biomater. Sci.* 8: 1450–1459.

32 Scindia, Y., Deshmukh, U., Thimmalapura, P.R., and Bagavant, H. (2008). Anti-alpha 8 Integrin immunoliposomes: a novel system for delivery of therapeutic agents to the renal glomerulus in systemic lupus erythematosus. *Arthritis Rheumatol.* 58: 3884–3891.

33 Chen, J., Zhang, C., Liu, Q. et al. (2012). *Solanum tuberosum* lectin-conjugated PLGA nanoparticles for nose-to-brain delivery: in vivo and in vitro evaluations. *J. Drug Targeting* 20: 174–184.

34 Reynoso-Camacho, R., de Mejıa, E.G., and Loarca-Pina, G. (2003). Purification and acute toxicity of a lectin extracted from tepary bean (*Phaseolus acutifolius*). *Food Chem. Toxicol.* 41: 21–27.

35 Meng, Q., Wang, A., Hua, H. et al. (2018). Intranasal delivery of Huperzine A to the brain using lactoferrin-conjugated N-trimethylated chitosan surface-modified PLGA nanoparticles for treatment of Alzheimer's disease. *Int. J. Nanomed.* 13: 705–718.

36 Grantab, R., Sivananthan, S., and Tannock, I.F. (2006). The penetration of anticancer drugs through tumor tissue as a function of cellular adhesion and packing density of tumor cells. *Cancer Res.* 66: 1033–1039.

37 Kwon, I.K., Lee, S.C., Han, B., and Park, K. (2012). Analysis of the current status of targeted drug delivery to tumors. *J. Controlled Release* 164: 108–114.

38 Maeda, H., Bharate, G.Y., and Daruwalla, J. (2009). Polymeric drugs for efficient tumor-targeted drug delivery based on EPR-effect. *Eur. J. Pharm. Biopharm.* 71: 409–419.

39 Abouzeid, A.H., Patel, N.R., Rachman, I.M. et al. (2013). Anti-cancer activity of anti-GLUT1 antibody-targeted polymeric micelles co-loaded with curcumin and doxorubicin. *J. Drug Targeting* 21: 994–1000.

40 Zhang, L., Pornpattananangku, D., Hu, C.-M.J., and Huang, C.-M. (2010). Development of nanoparticles for antimicrobial drug delivery. *Curr. Med. Chem.* 17: 585–594.

41 Umamaheshwari, R.B. and Jain, N.K. (2003). Receptor mediated targeting of lectin conjugated gliadin nanoparticles in the treatment of *Helicobacter pylori*. *J. Drug Targeting* 11: 415–423.

42 Subramanya, S., Kim, S.S., Abraham, S. et al. (2010). Targeted delivery of small interfering RNA to human dendritic cells to suppress dengue virus infection and associated proinflammatory cytokine production. *J. Virol.* 84: 2490–2501.

43 Li, H.Z., Ma, S.H., Zhang, H.M. et al. (2017). Nano carrier mediated co-delivery of dapsone and clofazimine for improved therapeutic efficacy against tuberculosis in rats. *Biomed. Res.* 28: 1284–1289.

44 Pandey, R. and Khuller, G.K. (2005). Solid lipid particle-based inhalable sustained drug delivery system against experimental tuberculosis. *Tuberculosis* 85 (4): 227–234.

45 Marques, J., Valle-Delgado, J.J., Urban, P. et al. (2017). Adaptation of targeted nanocarriers to changing requirements in antimalarial drug delivery. *Nanomed. Nanotechnol. Biol. Med.* 13: 515–525.

46 Mosqueira, V.C., Loiseau, P.M., Bories, C. et al. (2004). Efficacy and pharmacokinetics of intravenous nanocapsule formulations of halofantrine in *Plasmodium berghei*-infected mice. *Antimicrob. Agents Chemother.* 48: 1222–1228.

47 Sierra, S., Kupfer, B., and Kaiser, R. (2005). Basics of virology of HIV-1 and its replication. *J. Clin. Virol.* 34: 233–244.

48 Gupta, U. and Jain, N.K. (2010). Non-polymeric nano-carriers in HIV/AIDS drug delivery and targeting. *Adv. Drug Delivery Rev.* 62: 478–490.

49 Weiss, R.A. (1993). How does HIV cause AIDS. *Science* 260: 1273–1279.

50 Wu, H.B., Deng, Y.H., Wang, S.N. et al. (2007). The distribution of azidothymidine palmitate galactosylated liposomes in mice. *Acta Pharm. Sinica B* 42: 538–544.

51 Garg, M. and Jain, N.K. (2006). Reduced hematopoietic toxicity, enhanced cellular uptake and altered pharmacokinetics of azidothymidine loaded galactosylated liposomes. *J. Drug Targeting* 14: 1–11.

52 Pardeshi, C.V. and Belgamwar, V.S. (2018). *N,N,N*-trimethyl chitosan modified flaxseed oil based mucoadhesive neuronanoemulsions for direct nose to brain drug delivery. *Int. J. Biol. Macromol.* 120: 2560–2571.

53 Rytting, E., Nguyen, J., Wang, X., and Kissel, T. (2008). Biodegradable polymeric nanocarriers for pulmonary drug delivery. *Expert Opin. Drug Delivery* 5: 629–639.

54 Rosiere, R., Gelbcke, M., Mathieu, V. et al. (2015). New dry powders for inhalation containing temozolomide-based nanomicelles for improved lung cancer therapy. *Int. J. Oncol.* 47: 1131–1142.

55 Brannon-Peppas, L. and Blanchette, J.O. (2004). Nanoparticle and targeted systems for cancer therapy. *Adv. Drug Delivery Rev.* 56: 1649–1659.

56 Wu, D.Q., Lu, B., Chang, C. et al. (2009). Galactosylated fluorescent labeled micelles as a liver targeting drug carrier. *Biomaterials* 30: 1363–1371.

57 Tian, Q., Zhang, C.N., Wang, X.H. et al. (2010). Glycyrrhetinic acid-modified chitosan/poly(ethylene glycol) nanoparticles for liver-targeted delivery. *Biomaterials* 31: 4748–4756.

58 Haas, M., Moolenaar, F., Meijer, D.K.F., and de Zeeuw, D. (2010). Specific drug delivery to the kidney. *Cardiovasc. Drugs Ther.* 16: 489–496.

59 Chen, Z., Peng, H., and Zhang, C. (2020). Advances in kidney-targeted drug delivery systems. *Int. J. Pharm.* 587: 119679.

60 Asgeirsdottir, S.A., Zwiers, P.J., Morselt, H.W. et al. (2008). Inhibition of proinflammatory genes in anti-GBM glomerulonephritis by targeted dexamethasone-loaded AbEsel liposomes. *Am. J. Physiol. Renal Physiol.* 294: F554–F561.

61 Tuffin, G., Waelti, E., Huwyler, J. et al. (2005). Immunoliposome targeting to mesangial cells: a promising strategy for specific drug delivery to the kidney. *J. Am. Soc. Nephrol.* 16: 3295–3305.

62 Patra, J.K., Das, G., and Fraceto, L.F. (2018). Nano based drug delivery systems: recent developments and future prospects. *J. Nanobiotechnol.* 16: 71.

63 Data obtained from www.clinicaltrials.gov. accessed October 2021.

3

Nanocarriers-Based Targeted Drug Delivery Systems: Small and Macromolecules

Preshita Desai

Department of Pharmaceutical Sciences, College of Pharmacy, Western University of Health Sciences, Pomona, CA 91766, USA

3.1 Nanocarriers (Nanomedicine) – Overview and Role in Targeted Drug Delivery

Nanomedicine has been a buzzword in pharmaceutical arena for over a decade with widespread applications in the field of prevention, diagnostics, imaging, treatment, and theranostics. From clinical perspective, nanomedicine [nanocarriers (NCs)/nanoparticles/nanoconjugates] has significantly transformed drug delivery research by showing great potential in enhancing the overall *in vivo* performance (safety and efficacy). As per the United States Food and Drug Administration (USFDA), nanomedicine-based products comprise nanoscale materials with at least one-size dimension between 1 and 100 nm [1]. However, from literature and more practical perspective, nanomedicine covers a broad range of particles between 1 and 1000 nm [1, 2]. NCs are broadly classified based on their composition and fabrication material. These include lipids (lipid NCs), polymers (polymeric NCs), inorganic materials (inorganic NCs), nanoconjugates/prodrugs, and combination of two more types of materials (hybrid NCs). These are schematically represented in Figure 3.1 (modified from [3]).

Successful translation of any drug (small/macromolecule) to market requires acceptable efficacy as well as safety and quality. In this context, it is imperative to know that drug potency at the target site (pharmacological response) is an intrinsic phenomenon. Even though many drugs are potent, they are not successful in preclinical/clinical trials, mainly due to safety and/or efficacy issues. This is because, in addition to potency, *in vivo* efficacy and safety of the drug depend upon variable factors, such as the dosage form, route of administration, ADME profiles, toxicity, non-site-specific side effects (mainly with chemotherapeutic drugs) drug stability, and drug resistance. These parameters can be tailored to achieve the desired drug product properties/performance by multiple approaches, such as development of drug conjugates/prodrugs, combination with efflux/metabolism inhibitors, combination with absorption enhancers, altering route of administration, smart

Targeted Drug Delivery, First Edition. Edited by Yogeshwar Bachhav.

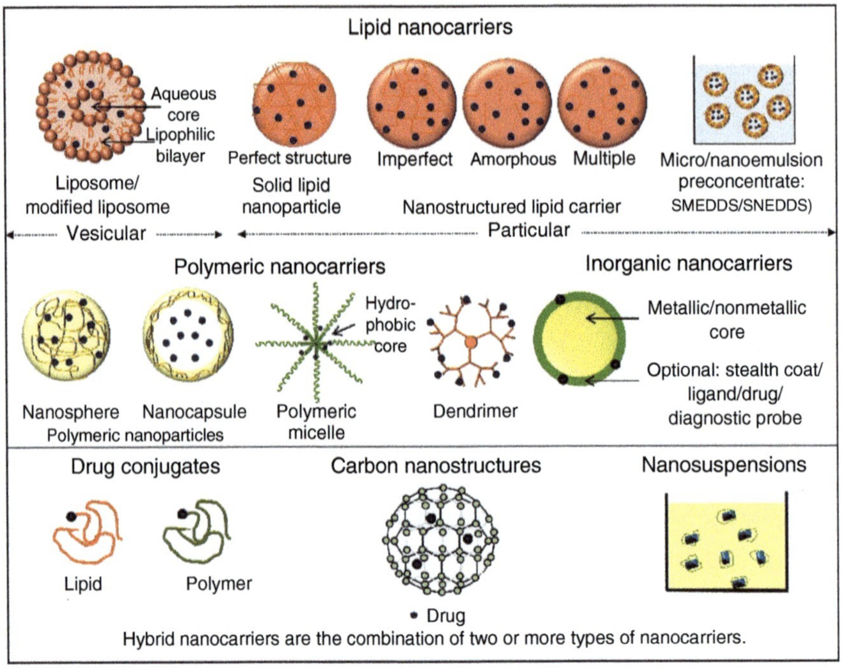

Figure 3.1 Schematic representation of various types of NCs. SMEDDS, Self microemulsifying drug delivery system; SNEDDS, Self nanoemulsifying drug delivery system. Source: Desai et al. [3]/Frontiers media/CC BY4.0.

formulation strategies to control, delay, and/or target drug delivery. In this context, among various smart formulation strategies, NCs are one widely researched and accepted strategy. This growing interest in use of NCs for drug delivery application has emerged because of following advantages: (i) small size (nano) with large surface area allowing better interaction and uptake; (ii) capability of encapsulating both hydrophilic and lipophilic drugs; (iii) offering drug stability by means of encapsulation; (iv) controlled drug delivery; (v) longer *in vivo* circulation time; (vi) enhanced absorption and efflux inhibition and most importantly; (vii) opportunity for targeted site-specific drug delivery by means of modulating targeted drug particle size, shape and surface charges, and surface functionalities; (viii) possibility of multifunctionality in terms of simultaneous treatment and diagnosis (theranostic), targeting, triggered drug release, etc. (Figure 3.2).

The need for targeted drug delivery has been extensively discussed in the Chapter 1. NCs are well suited to offer drug targeting because of above-listed advantages [schematically represented in Figure 3.2 (modified from [3])] via various mechanisms that can be broadly classified under three categories, viz. passive targeting, active targeting, and stimuli sensitive targeting, which are discussed in Sections 3.2–3.4 (schematically represented in Figure 3.3). It must be noted that all types of NCs can be developed for targeting application; however, selection and optimization of most efficient targeted NCs require meticulous rationalization w.r.t. type of NCs, targeting strategy, etc. Such selection depends upon various

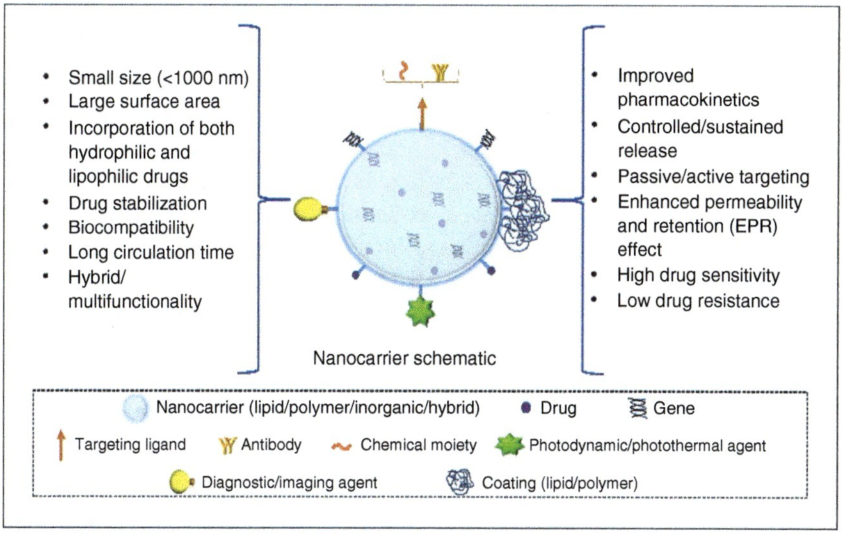

Figure 3.2 Schematic representation of multifunction NCs and their advantages. Source: Desai et al. [3]/Frontiers media/CC BY4.0.

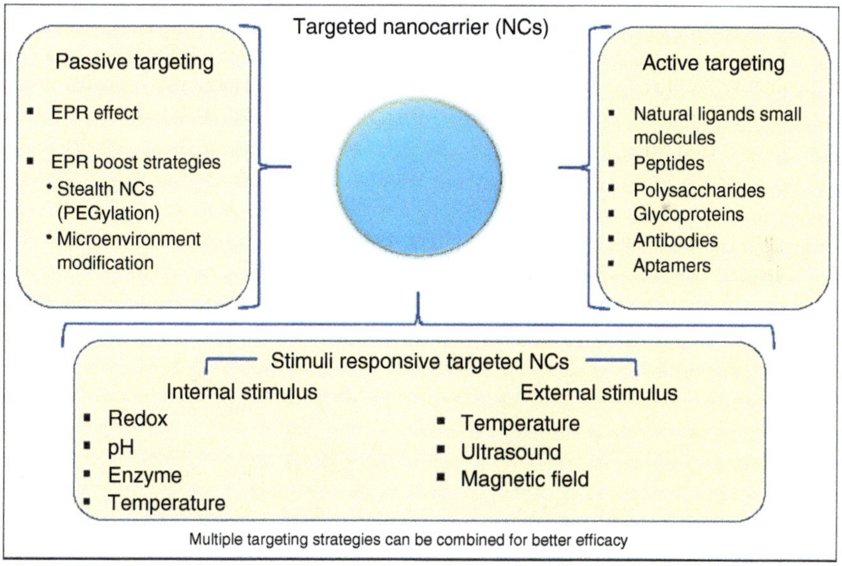

Figure 3.3 Types of targeted NCs.

parameters, such as thorough understanding of disease pathophysiology, drug properties, NCs type, and most importantly target product profile. Interestingly, NCs offer a dynamic platform wherein multiple targeting strategies, drugs, or diagnostic agents can be combined to design "multifunctional NCs" to achieve enhanced efficacy (Figure 3.2). Such multifunctional NCs are gaining wide attention in recent years due to their promising performance and future clinical translation.

Targeted NCs have been explored for diagnostics and treatment of almost all diseases and disorders, but the widest application of targeted NCs is seen in the field of cancer therapeutics for two main reasons – efficacy and safety. Firstly, to ensure the delivery of chemotherapeutic drugs at the tumor site by surpassing drug resistance/efflux, stromal barrier, etc., and secondly to avoid non-site-specific accumulation of chemotherapeutic drugs to reduce severe toxicity. Other significant applications of targeted NCs are seen in diseases associated with organs, such as brain (due to blood–brain barrier) and lungs, which are difficult to access via normal circulation [4]. Few of such targeted NCs have already been approved and have made their way from bench to bedside (Table 3.1), while many are in the clinical pipeline (Table 3.2). Various types of targeted drug delivery systems along with their applications have been discussed in Sections 3.2–3.4.

Table 3.1 Few examples of targeted nanocarriers – market scenario.

Product	Active	NCs type	Targeting strategy	Indication	References
Doxil	Doxorubicin	PEGylated Liposome	Passive: stealth, EPR effect	Ovarian cancer, Kaposi sarcoma in AIDS patients	[5]
Abraxane®	Paclitaxel	Albumin NCs	Passive	Metastatic cancer (breast)	[6]
Combidex®	Iron oxide	Iron oxide	Passive: dextran 10-coated NCs	Tumor imaging	[7]
Oncaspar®	L-Asparaginase	Polymer–protein conjugate	Passive	Leukemia	[8]
Emend®	Aprepitant	Nanocrystal	Passive	Chemotherapy side effect (vomiting/nausea)	[9]
Mylotarg®	Gemtuzumab ozogamicin	ADC	Active: against CD33 antigen	Acute myeloid leukemia	[10]
Adcetris®	Brentuximab vedotin	ADC	Active: against CD 30 antigen	Refractory Hodgkin lymphoma	[10]
Kadcyla®	Trastuzumab emtansine	ADC	Active: against HER2 receptor	HER2-positive breast cancer	[11]

NCs, nanocarrier; ADC, antibody–drug conjugate; HER, human epidermal growth factor receptor.

Table 3.2 Few examples of targeted nanocarriers – clinical investigation scenario.

Product	Active	NCs type	Targeting strategy	Indication	Clinical Phase	References
Paclitaxel Albumin-Stabilized NCs + Carboplatin	Paclitaxel + Carboplatin	Polymeric	Passive: stealth	Non-small-cell lung cancer	Phase II	[12]
AuroShell particle infusion	—	Gold nanoshells	Passive: nanoparticle-directed laser irradiation	Prostate neoplasm	Clinical study phase not listed	[13]
Cetuximab nanoparticles	Cetuximab	Polymeric NCs	Active: somatostatin receptors Stimuli: pH responsive	Colon, colorectal cancer	Phase I	[14]
PROMITI® + FOLFOX	Mitomycin-C	PEGylated liposomes	Passive: stealth, EPR effect	Gastrointestinal intraepithelial neoplasia	Phase I	[15]
ThermoDox® with radiofrequency ablation (RFA)	Doxorubicin	Thermosensitive liposomes	Passive: temperature responsive	Hepatocellular carcinoma	Phase III	[16]
NANOM-FIM	—	Silica-gold NCs, iron-bearing silica-gold NCs, CD68 targeted micro-bubbles	Passive: magnetic; Active: CD68 targeting	Atherosclerosis	Randomized study, clinical study phase not listed	[17]
(^{64}Cu)-labeled PSMA-NCS	—	Polymeric NCs	Passive: targeted imaging	Surgical treatment of prostate cancer	Phase I	[18]
NC conjugate	Somatostatin	Quantum dots coated with veldoreotide	Passive and theranostic	Breast cancer, skin cancer, skin diseases	Phase I	[19]
Micelles	Paclitaxel	PEG-polyaspartate polymeric micelle	Passive: stealth	Breast cancer	Phase III	[20]

NCs, nanocarrier; PEG, polyethylene glycol.

3.2 Passive Targeting Approaches

Certain diseases, such as inflammation and cancer, cause pathophysiological variations at disease site leading to altered vasculature, lymphatic setup, and permeability. Such alterations can be used to design targeted NCs. Such targeting strategy is based on NCs characteristics (size, shape, etc.), biological variations, and does not involve any energy-dependent processes of ligand binding/receptor-mediated transport, etc., and hence is known as "passive targeting approach." Enhanced permeation and retention (EPR) phenomenon is a unique passive targeting approach (Figure 3.4) [21].

3.2.1 Enhanced Permeability and Retention-Effect-Based Targeting

EPR effect is a hallmark of cancer and associated malignancies. During the onset of cancer disease, hyperactive, hypermetabolic, and atypical cellular activity leads to abnormal tumor tissue construct that exhibits (i) hypervascularity, (ii) leaky vasculature, and (iii) poor lymphatic system and drainage. These structural abnormalities lead to elevated permeability, uptake, and retention of biomolecules and are commonly known as EPR effect. Interestingly, this very phenomenon can be used at the benefit of targeted drug delivery to engineer macromolecules and NCs that are capable of permeating via EPR effect at the tumor site (high tumor porosity and permeability) but otherwise cannot permeate through natural healthy tissue. In the past few decades, macromolecule delivery and development of NCs for passive tumor targeting via EPR effect have gained wide attention and have

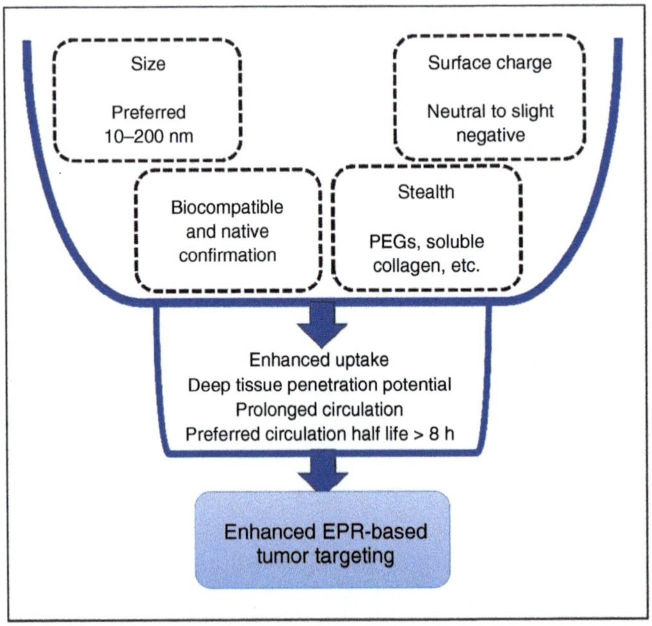

Figure 3.4 Schematic representation of factors affecting EPR passive NCs targeting.

been proven to be effective [22]. Macromolecule/NCs with high tissue penetrability and long-circulation time have been proven to show enhanced targeting via EPR effect (Figure 3.4). To achieve this, it should be biocompatible and exhibit optimum size, surface charge, and surface properties. A neutral-to-mild negative charge is recommended to ensure suitability and effectiveness via EPR effect. This is because high positive charge results in nonspecific cell binding (cells have negative charge), and high negative charge will lead to clearance via reticuloendothelial (RES) system [23]. There are varied reports on optimum NCs size for most efficient uptake across the tumor using EPR effect but on an average, a particle size of 10–200 nm is considered suitable as per majority of literature reports. Further, as per recent findings, efficiency of EPR effect lowers with increase in the particle size [23–29]. Minimum size of 40–50 kDa is reported to be useful for macromolecule targeting via EPR effect [23–27]. This becomes critical in poorly penetrable tumors, such as pancreatic cancer, and so lower particle sizes have been preferred for such targeting. For example, Cabral et al. developed polymeric micelles with varying particle sizes (30, 50, 70, and 100 nm), and their investigation in pancreatic tumors revealed that only 30 nm particles could penetrate into tumor to elicit measurable efficacy [30].

Low permeability with certain complex tumors, extensive EPR heterogeneity (cancer-to-cancer and patient-to-patient), and much-debated actual clinical benefit of passive targeting using EPR have raised some challenges on effectiveness and consistency with use of this approach. In recent years, the focus is moving toward use of this passive targeting modality in combination with complementary approaches that include stealth NCs, remodeling of tumor microenvironment, use of specific active targeting approach to ensure enhanced efficacy, and site-specific delivery (Section 3.3) or application of external stimuli (Section 3.4) [31].

As previously mentioned, long-circulation time is the key for enhanced uptake and tissue targeting via EPR effect. This is challenging because foreign materials, such as macromolecules and NCs, are prone to fast clearance via RES system. The first step in the RES clearance process is opsonization, i.e. adsorption of plasma proteins on surface of NCs, followed by their interaction and clearance by the phagocytic cells, such as monocytes, neutrophils, Kupffer cells, dendritic cells, and macrophages. Therefore, if NCs can be protected from opsonization, they will not be subjected to RES clearance resulting in increased circulation time. To achieve this, surface coating with neutral, biocompatible, and hydrophilic polymers, such as polyethylene glycol (PEG), has been identified to prevent opsonization and subsequent clearance, and the approach is known as "Stealth" NCs [30]. PEGylation is most widely used stealth technique and interestingly it is a platform technique that can be employed for any type of NCs, including liposomes, polymeric nanoparticles, micelles, and inorganic NCs. From the clinical standpoint, PEGylated NCs have already made entry into the market in the form of doxorubicin PEGylated liposomes (Doxil®, Caelyx®). Though effective, PEGylation must be designed and used with caution as it may interfere with cellular interaction, uptake, and endosomal escape [30, 32]. The PEG length has great impact on circulation time and targeting efficiency. For example, doxorubicin-conjugated PEGylated polylysine dendrimers of generation G4 and G5 with different PEG chain lengths (570 vs. 1100 Da) were developed and assessed for cathepsin-mediated triggered drug release. The studies revealed that variation in PEG length significantly impacted the performance, wherein the generation 4 dendrimers with small-chain PEG showed superior performance in terms of triggered drug release and efficacy. Hence, a meticulous study is desired to achieve balance between stealth properties and efficacy because long-chain PEG may interfere with permeability and cell internalization [33]. Other polymers, such as soluble collagen, albumin, dextran, heparin, and polyoxazolines, *N*-(2-hydroxypropyl)methacrylamide (HPMA), have also been used either to coat the NCs or used as drug-linked macromolecules with increased circulation time [30, 34]. For example, an EPR-based macromolecule Abraxane® comprising albumin-bound paclitaxel is available in market for treatment of cancer and has shown superior efficacy compared to free drug treatment.

NCs/macromolecules with long-circulation time show enhanced targeting via EPR effect. However, the challenge still persists especially in case of tumors with low permeability. Modulation of tumor microenvironment to enhance permeability has been explored to overcome this challenge. For this, nitric oxide (NO), antiangiogenic agents, matrix metalloproteinase (MMP) inhibitors, macrophage depletion, etc., strategies have been reported to enhance the permeation of long-circulating NCs [35–37]. For example, NO-releasing liposomes (size: 100 nm) were developed that showed enhanced tumor accumulation within 1 hour, which increased gradually up to 48 hours postinjection. This enhanced uptake was attributed to NO-induced vasodilation and blood flow modulation only at the tumor tissue [37]. In another study, NO-releasing, S-nitrosated human serum albumin dimer was given prior to Abraxane treatment that resulted in enhanced tumor selectivity, accumulation, and efficacy of Abraxane [36].

To summarize, EPR-based passive targeting has advantages, such as increased permeability, increased uptake, relatively easy design, formulation feasibility, and cost-efficiency. However, the targeting is limited in terms of its efficacy, specificity, and selectivity and hence often requires complementary strategies to further improve performance.

3.3 Active Targeting Approaches

More intensive approach of targeting NCs to a specific site is surface modification using cell-specific ligands. Such ligands decorated on NCs surface bind to specific cell receptors enabling receptor-mediated cell internalization. This method of selective targeting is known as "active targeting." The pathophysiological alterations during various diseases cause cell-surface variation in terms of upregulation/downregulation of certain receptors and/or antigens. If NCs are designed with a ligand specific to a receptor or antigen that is unique and upregulated exclusively for that disease, they can be effectively targeted to the disease site. The key advantages of active targeting strategy are (i) uniqueness and selectivity for the disease tissue targeting; (ii) specificity and enhanced permeability; (iii) opportunity of personalized medicine.

However, some challenges include (i) complex design and formulation, (ii) high cost, and (iii) being relatively new technology there are regulatory hurdles for market approval due to lack of specific guidelines and additional scrutiny [21, 38, 39]. Wide range of active targeting modalities and ligands are reported in the literature for this purpose (Figure 3.3) that include natural ligands – small molecules (hyaluronic acid, folic acid, Tween 80, etc.), peptides (polylysine, synthetic peptides, glutathione [GSH], lectins, etc.), polysaccharides, antibodies (monoclonal antibodies [mAb]) and aptamers. [40, 41].

Folic acid is one such natural ligand that is extensively explored to develop active targeted liposomes, exosomes, niosomes, solid NCs, polymeric NCs, gold and iron oxide, silica NCs, and carbon nanotubes (CNTs) [41, 42]. For example, folate-targeted CNTs were designed and studied for targeting lung carcinoma. The folate–chitosan conjugate was linked to CNTs and exhibited excellent biocompatibility. The targeted formulation showed enhanced cellular internalization with low IC50 values in lung cancer cell line – A549 [42]. In another investigation, folate-targeted CNTs were incorporated with methotrexate for treatment of arthritis. *In vivo* studies indicated significant accumulation at arthritic site confirming the targeting potential [43]. Interestingly, instead of using natural ligand to bind an entire receptor, a chemical variant, or a fragment capable of binding a specific site could be used to enhance targeting. In one such study, phenylboronic acid was used as a ligand to target *N*-acetylneuraminic acid, a main component of overexpressed sialic acid epitopes on cancer cells. Phenylboronic acid was grafted in micellar NCs that showed high uptake by melanoma cell line B16F10 even at slight acidic tumor microenvironment with enhanced cytotoxicity. The study further confirmed regression in lung cancer model confirming the targeting potential [44].

As small molecules, various peptides have also been employed for this purpose and offer an advantage of easy linkage to NCs surface due to reactive functional groups. For atherosclerosis treatment, arginylglycylaspartic acid (RGD) peptide-(binds to integrin receptor widely spread in atherosclerotic plaque)-linked iron oxide NCs were prepared. These peptide-targeted NCs (size: 60–80 nm) showed enhanced uptake in RAW 264.7 macrophage cells owing to upregulated integrin receptors. Further, atherosclerosis and associated blood vessels are abundant in collagen IV and, hence, collagen IV-targeting peptide was designed and used as an alternative targeting approach. The study gave valuable insight suggesting that extent of active targeting varies from one ligand to another and hence, thorough understanding of disease pathophysiology and biological variation is a must to select most optimum strategy [45]. Polysaccharides have also been explored for active targeting applications. Some examples include adipose tissue macrophage-targeted polysaccharide NCs to treat obesity [46], *Astragalus* polysaccharide-targeted NCs for breast cancer management [46], and glycoprotein transferrin (iron-binding glycoprotein) NCs for brain targeting, and tumor targeting [47, 48].

As a completely synthetic alternative, aptamers were identified as suitable approach for active targeting. Chemically, aptamers are oligonucleotide/peptide molecules that bind to a specific target molecule (receptor/antigen in this case). These are mainly synthesized by systematic evolution of ligands by exponential enrichment (SELEX) process that enables development of high-affinity aptamer for any cellular target. This opens a great avenue to target the cellular receptors that are otherwise difficult to target or have low-affinity ligands. Multitude of such aptamer-targeted NCs has been reported [49, 50]. For instance, 5TR1 aptamer-targeted chitosan-poly(lactic-*co*-glycolic acid) (PLGA) NCs were developed that not only exhibited improved targeting and uptake toward breast cancer cell line MCF-7 but also showed pH-dependent drug release owing to chitosan and hence achieving dual targeting with enhanced efficacy [51]. Such aptamers can be tailor-made based on the specific target requirements. However, biocompatibility is one of the concerns while designing along with high cost of development.

Other main classes of active targeting ligands are antibodies that have potent binding capability to cell-surface antigens. These ligands are extremely selective and, with developed technology, can be synthesized in laboratory. Such antibodies are known as specific mAb and are widespread for active targeting purpose. mAb can either be directly linked to drug-loaded NCs or the drug can be chemically linked to mAb to form targeted macromolecule known as antibody–drug conjugates (ADCs). Few of such ADCs have made entry into market and are summarized in Table 3.1. The trend is now moving toward use of antibody fragments that are capable of binding to the antigen instead of a large and complex antibody to minimize the NCs size, steric hindrance, and enhance its stability. In an interesting study, anti-folate antibody Fab fragment was used to tag iron oxide NCs which showed exhibit higher internalization with greater stability. Further, this strategy is considered to be very suitable for macromolecular delivery, such as gene and siRNA [52]. (For detailed discussion on antibody-based drug delivery refer to Chapter 5.)

3.4 Stimuli Responsive Targeted NCs

In-depth study of disease pathophysiology, underlying molecular mechanisms, and disease-associated biological changes has revealed important and unique chemical and biological variations associated with respective diseases. For example, in case of cancer, tumor tissue exhibits low pH of 6.0–7.0 (in contrast to normal physiological pH of 7.4) owing to lactic acid buildup; high temperature of about 40–42 °C (in contrast to normal physiological temperature of 37 °C); high oxidative stress; overexpression of metabolic enzymes (metalloproteinases, proteases, elastases, etc.); and varied pH of gastrointestinal (GI) tract from stomach (acidic) to rectum (neutral to slight alkaline) [53]. These alterations can be used at the benefit of designing a smart NCs system that will trigger the drug release only under the influence of altered physiological conditions and hence achieving targeted drug release only at the disease site. Such NCs are known as "internal stimuli responsive" NCs. These are further classified based on the type of stimulus used to trigger the drug delivery and these include chemical (redox, pH, and enzyme); biological (temperature); or combinations thereof (Figure 3.3).

The stimuli to trigger drug release can also be external. These include temperature, ultrasound, and magnetism which can be applied externally and in localized manner to achieve effect only at the target site. Specialized NCs that are sensitive to one or more above-mentioned triggers can be developed that would release drug in site-specific manner only in presence of respective external stimuli and the strategy is known as "external stimuli responsive" NCs. External stimuli responsive NCs are highly effective and are choicest option for targeting superficial tissues/organs. However, one must be extremely careful in selection of such triggers for deep tissue targeting (Figure 3.3).

3.4.1 Redox Stimuli Responsive Targeted NCs

High oxidative stress associated with tumors results in significant difference in redox potential of tumor tissue in contrast to normal tissue. This redox potential difference has been identified as an excellent internal stimulus to develop targeted drug-delivery systems. As a physiological response, it leads to upregulation and accumulation of body's natural antioxidant GSH at the tumor site (a natural antioxidant mechanism – physiological response) [53]. Hence, a drug NCs comprising redox-responsive material (polymer/lipid, etc.) can be designed that will undergo preferential chemical modification in presence of high reductive environment at the tumor site achieving targeted drug release. Such NCs are known as redox stimuli responsive targeted NCs.

The most widely explored targeted NCs for this application are those comprising polymers/lipids with disulfide linkage. Disulfide bond is promisingly stable under normal redox environment of healthy tissue but is susceptible to reduction (tumor environment) leading to cleavage of disulfide bond to sulfhydryl and thereby releasing the drug cargo [53]. This leads to selective and targeted delivery of drug to tumor site avoiding non-site-specific drug release that is key to reduce toxicity, especially in case of chemotherapeutic drugs.

Variety of NCs have been specially designed for redox targeted delivery [53]. For example, redox-responsive mesoporous silica nanoparticles (MSNs) were developed for simultaneous delivery of siRNA (effective gene therapy for cancer management) and doxorubicin (anticancer drug). Herein, the siRNA was conjugated with MSN to form disulfide bond and doxorubicin was loaded in the MSN. An *in vitro* study indicated higher drug release from these nanoparticles in presence of GSH than in its absence indicating cleavage of disulfide linkage under redox stimuli enhances the drug release. The studies indicated significantly high accumulation of these targeted nanoparticles compared to free doxorubicin, and the enhanced efficacy was also proven via *in vivo* studies [54]. In another investigation, redox-sensitive mesoporous organosilica nanoparticles (MONs) with thioether linkage were fabricated to deliver doxorubicin. The targeted MONs showed a particle size of 50 nm with enhanced tumor uptake and inhibition resulting from their degradation under reductive environment. Additionally, biodegradability and biocompatibility make them a lucrative strategy for such targeted treatment [55].

Polymeric NCs have also been investigated for this purpose. For instance, micelles of an amphiphile hyaluronate – tocopherol succinate (TOS) conjugate linked by disulfide bond were designed to carry anticancer drug paclitaxel. Synergistic anti-tumor activity was observed in melanoma cell lines due to dual targeting achieved by the micelles. Firstly, due to presence of hyaluronic acid, the micelles were preferentially taken up by CD-44 overexpressing cell line B16F10 (melanoma), and secondly, the enhanced cytotoxicity was observed due to redox initiated disulfide bond cleavage leading to selective drug release in the cancer cells [56].

3.4.2 pH Stimuli Responsive Targeted NCs

pH variation is observed throughout the human biological system healthy or otherwise and has been explored as an internal stimulus to develop targeted drug-delivery systems. For example, natural physiological pH of blood and tissue is 7.4, while for certain intracellular organelles, such as endosomes (pH 5–6) and lysosomes (pH 40–5), it is acidic [57, 58]. During inflammation, cancers, or oxidative stress, the pH of affected tissue might drop to 6.5 (pH 6–7) [53, 59]. This pH variation can be used as a stimulus to design targeted NCs comprising pH-responsive excipient. Such NCs either undergo conformational change or are degraded under a specific pH range, leading to targeted drug delivery.

Design and development of pH stimuli responsive NCs require meticulous selection of excipients (lipid/polymer, etc.) that are sensitive to changes in pH (ionization, cleavable bonds, solubilization, etc.) [60]. Literature has wide range of pH-responsive NCs with applications ranging from intracellular drug delivery (e.g. gene therapy) to tissue/organ targeting in inflammatory disease and cancer. Variation in GI tract has also been used at an advantage for targeted NCs delivery [61]. For example, succinylated ε-polylysine-coated MSNs were developed for colon targeting wherein the NCs were shown to be stable under acidic environment of stomach and small intestine but release the loaded drug prednisolone in colonic environment (pH 5.5–7.4) [62].

Further, MSNs with hyaluronic acid (CD44 receptor-mediated targeted cell uptake by cancer cells) and lipid bilayer coating were formulated with doxorubicin as a model anticancer drug. The formulation exhibited stable encapsulation at pH 7.4 followed by burst release under acidic pH environment. Further, the formulation was shown to undergo selective cellular uptake in HeLa cell lines with enhanced cell inhibition triggered by pH-mediated burst release of the drug arresting the tumor growth [63]. In another study, MSNs were chemically linked to dextran polymer via acylation. The chemical linkage was observed to be stable at pH 7.4 but underwent hydrolysis under acidic pH triggering drug release and was confirmed to elicit intracellular drug delivery under the influence of biological acidic environment [64]. Interestingly, pH-sensitive polymers have been widely investigated for this application in the form of various NCs, such as polymerosomes, nanoparticles, nanocapsules, and dendrimers [65–69]. For example, pH-responsive copolymer comprising branched polyethylenimine and PEG was employed to coat MSNs encapsulating tumor necrosis factor-alpha (TNF-α). This architecture not only offered targeted drug release (enhanced efficacy, reduced toxicity) but also ensured stability to TNF-α, which otherwise is unstable (short half-life) if given in free form. The copolymer can be looked upon as a gatekeeper that restricts the release of TNF-α under natural physiological pH but allows burst release under acidic pH at the target tumor site [70]. pH-responsive NCs have also been studied for targeted treatment of inflammatory diseases. For example, acute lung inflammation (ALI) is a serious condition with very limited effective treatment modalities due to lack of targeted lung delivery approaches. If untreated, it leads to rather severe and possibly lethal acute respiratory distress syndrome. For this, NCs comprising a core of acidic pH-responsive poly(β-amino esters) polymer for drug encapsulation (anti-inflammatory agent TPCA-1), allowing pH-triggered release, and a coat of PEG-biotin for bioconjugation of lung-specific targeting ligand (anti-ICAM-1 antibodies) were designed. The nanoformulation showed lung targeted delivery in ALI mouse model with significant anti-inflammatory response [71]. Hence such approach can be used in targeted treatment of inflammatory diseases, such as rheumatoid arthritis, oxidative stress, and ischemia [72, 73].

Interestingly, NCs comprising natural biomaterials, such as viral nanoparticles (VNPs), virus-like nanoparticles (VLNPs), acetalated dextran-based VLNPs, and truncated hepatitis B antigen-drug NCs, have also been developed with or without specialized polymers [74–77]. In one study, virus-like polymeric nanoparticle was designed with pH-responsive properties for cellular internalization, endosomal escape, and ability to carry macromolecule payload. For this, endosomolytic poly(lauryl methacrylate-*co*-methacrylic acid) polymer (resembling viral capsid) was grafted on acetalated dextran (resembling viral core). The NCs were synthesized using nanoprecipitation method and were observed to destabilize near late pH 5 (endosomal pH), releasing the drug cargo. Further, a tumor-targeting property was instilled by conjugation of tumor-penetrating peptide (receptor-mediated targeting). In a nutshell, the unique virus-like targeted NCs exhibited favorable properties toward tumor targeting and selective drug release [76].

To summarize, in majority of instances, pH-triggered NCs are used in conjunction with specific ligand-based active targeted strategy that ensures the enhanced uptake at disease site and pH trigger ensures the burst release.

3.4.3 Enzyme Stimuli Responsive Targeted NCs

Biological enzymes play crucial role in maintaining body homeostasis and are very selective and specific of their substrates. Under certain disease pathophysiology, some enzymes are upregulated. For example hydrolytic enzyme Cathepsin B, MMPs are overexpressed in many tumors [78–81]. If substrates for such enzymes are chemically linked to NCs, they will undergo selective and spontaneous hydrolysis at the disease site due to upregulated enzymes, leading to site-specific drug release. The key advantage of such smart NCs is that the substrates used for this are highly selective toward the enzyme ensuring targeted drug delivery, and most importantly these substrates are biocompatible and hence safe.

MSNs have been widely explored for this application as drugs can be loaded into the mesopores and enzyme-sensitive linkers can be grafted on surface as gatekeepers. At the enzyme-responsive target tissue in the body, the grafted linked would undergo hydrolysis, opening the gates to release the loaded drug cargo in controlled manner.

Polymeric NCs comprising surface-modified poly(dimethylsiloxane)-*b*-poly(methyloxazoline) was developed to carry anticancer drug paclitaxel to tumor-specific tissue. The Cathepsin B-sensitive conjugate was observed to hydrolyze in an ovarian cancer cell line, leading to targeted drug release and enhanced efficacy [80]. In another investigation, a dendrimer conjugate was designed on similar rationale. For this, dendrimer linked to Cathepsin B-cleavable peptide (Gly–Phe–Leu–Gly) was designed for site-specific delivery. The developed targeted dendrimer exhibited spherical geometry with size of less than 200 nm and was studied for delivery of anticancer drug doxorubicin using CT26 tumor xenograft model. The targeted NCs showed Cathepsin B sensitivity and significant inhibition of tumor volume compared to plain drug and nontargeted NCs [82].

Enzyme-based nanodelivery systems have also gained wide limelight in targeting specific regions of gastrointestinal tract (stomach, intestine, and colon) to treat diseases, such as cancer, uncreative colitis, inflammation, and ulcers, due to presence of selective microbial spectrum and enzyme action [61, 83, 84]. Liposomes, polymeric NCs MSNs, and their combinations have been widely studied for this application due to possibility of forming cleavable enzyme linkages [85]. For example, MSN with chemically linked chitosan via azo bond was developed to carry a model anticancer drug doxorubicin to colon sites. Azoreductase is a colon-specific enzyme produced by colonic microbes. The developed NCs were shown to undergo enzymatic cleavage in presence of colonic enzymes with almost threefold higher cell inhibitory potential [86].

3.4.4 Temperature Stimuli Responsive Targeted NCs

Thermosensitive nanocarriers are another approach to targeted delivery. Such thermal stimuli can be either internal or external. For example, tumor sites exhibit slightly high temperature (40–42 °C) compared to normal body temperature. This phenomenon can be used for temperature-triggered drug release as tumor site by using thermoresponsive NCs. However, use of internal temperature variation has limited applications due to high variability. Hence, to achieve significant tissue targeting, an external method of inducing site-specific hyperthermia is designed by means of localized photothermal, ultrasound, NIR, infrared and radiofrequency electromagnetic heating, etc. Specialized thermosensitive polymeric NCs and inorganic NCs, such as gold and iron oxide, have been widely explored for such application owing to their temperature-sensitive nature [87–90]. For example, a phase I clinical study was conducted in 10 patients with recurrent prostate cancer for thermotherapy using iron oxide NCs and alternating magnetic field. Use of alternate magnetic field resulted in prostate temperature of 55 °C and was considered suitable for treatment [91]. In another study, focused ultrasound was used to achieve targeted hyperthermia and temperature-triggered drug release from liposomes [92]. Further, a multimodality theranostic approach for tumor targeting is well appreciated and has been proven to be more effective. For example, composite NCs capable of active targeting, photothermal chemotherapy, and imaging were developed. The NCs had a core–shell geometry (size: ~230 nm) wherein the core comprised a phase change material (imaging) and shell was formed with polypyrrole and hyaluronic acid (active targeting). The NCs exhibited significant cellular internalization with high tumor inhibition in 4T1 tumor-bearing mice model [93].

3.4.5 Ultrasound Stimuli Responsive Targeted NCs

In addition to creating hyperthermia, ultrasound waves (external stimulus) have been used to achieve targeted delivery of NCs in site-specific manner. It has been shown that enhanced targeting and performance was observed with ultrasound contrast agent – antibody carrying PLGA NCs [94], iron oxide NCs [95], ultrasound augmented exosomes [96], MSNs [97], etc. For example, stent implantation is a standard procedure to treat acute myocardial infarction and blocked blood vessels. However, stent implantation may lead to future complication due to late-stage in-stent thrombosis and neointimal hyperplasia. To avoid this, thrombolytic drugs and tissue plasminogen activator (t-PA) are commonly given to patients as management protocol. As an alternate and more efficient approach, albumin NCs encapsulating t-PA plasmid gene were developed. These NCs were further crosslinked to ultrasound-sensitive albumin microbubbles. Therapeutic ultrasound aid was used to target transfection of the gene-carrying NCs at the stent-treated artery site. The expression of t-PA was observed up to eight weeks and results revealed high blood concentration of t-PA with no signs of thrombosis or hyperplasia, confirming the efficacy of targeted gene delivery by means of ultrasound [98].

Further, ultrasound has been combined with microbubbles to create transient pores in the tumor vasculature to further enhance the NCs permeation and the method is widely known as sonoporation. Various NCs, including polymeric, lipid, and inorganic, have been investigated for this application [99]. For example, *in vivo* distribution of liposomes tagged with fluorophores (for *in vivo* computed tomography and photon laser scanning) were studied in two xenograft cancer models, viz. A431 epidermoid and BxPC-3 pancreatic cancer. The studies corroborated enhanced tumor distribution in both models confirming utility of sonoporation in improving the extravasation of liposomes from the blood vessels into the tumor. Such combination of sonoporation with liposomes can be looked upon as a platform strategy to deliver any drug cargo (macromolecule, small molecule, imaging, or diagnostic agent) [100].

3.4.6 Magnetic Field Stimuli Responsive Targeted NCs

Magnetic NCs for targeted delivery are a specialized class of NCs with magnetic properties that make them highly selective. Iron-based NCs and gold NCs have been majorly explored for targeted drug delivery using the principle of externally applied magnetic field [101–103].

In an interesting study, nanobullets with iron-based (Fe_2O_3) head and mesoporous silica-based body with drug doxorubicin were prepared. These specialized NCs showed enhanced internalization and drug release in both orthotopic and xenograft liver cancer models confirming the targeting efficiency [102]. Further, magnetic macromolecule conjugates have also been reported. For instance, magnetic protein–polyelectrolyte complex for the delivery of the drug 5-fluorouracil was synthesized with folic acid conjugation to ensure targeted release [103].

Composite NCs with core–shell geometry, wherein core construct was made of gold and iron shell, were developed that offered advantages in terms of magnetism-based tissue targeting, photothermal treatment, and magnetic resonance imaging. Such multifunction NCs can further be loaded with drugs macromolecules for enhanced efficacy. Magnetism-based NCs offer high selectivity and theranostic opportunity, but can be constructed only with magnetism-responsive excipients that limit its applications. Another challenge is limited drug loading as drugs are generally tagged to the surface of these inorganic NCs.

3.5 Conclusion and Future Prospects

It must be appreciated that commendable progress has been made in design, development, and optimization of various targeted NCs. These smart NCs have shown significant improvement in performance with reduced non-site-specific toxicities due to tissue-specific targeting and few have made entry in clinics. In addition to being an excellent carrier of small-molecule cargos, these targeted NCs are exceptionally promising for delivery of macromolecules as they not only allow intracellular targeting but also offer stability to otherwise sensitive moieties, such as gene and siRNA. With advances made over the years, it is only a matter of time till many such targeted NCs will be available in the market offering great benefit and patient compliance. However, regulatory challenges and high cost are the two pressing barriers hindering this successful clinical translation roadmap and require more meticulous attention. This can very well be achieved by developing more detailed regulatory guidelines, interactive support from drug regulators, and possible fast-track approval. Further, more advances in nanoformulation scale-up techniques, robust formulation methods will definitely address the cost concerns. Most importantly, in my perspective, given the success and promise of targeted NCs to date, the future of this technology lies in development of personalized medicine wherein, a drug/macromolecule carrying targeted NCs can be custom made for patients based on their genetic makeup and molecular analysis of disease site to identify most appropriate targeting strategy.

References

1 USFDA. Considering whether an FDA-regulated product involves the application of nanotechnology. https://www.fda.gov/regulatory-information/search-fda-guidance-documents/considering-whether-fda-regulated-product-involves-application-nanotechnology (accessed 3 June 2020).

2 Desai, P.P., Date, A.A., and Patravale, V.B. (2012). Overcoming poor oral bioavailability using nanoparticle formulations – opportunities and limitations. *Drug Discovery Today: Technol.* 9: e87–e95.

3 Desai, P., Thumma, N.J., Wagh, P.R. et al. (2020). Cancer chemoprevention using nanotechnology-based approaches. *Front. Pharmacol.* 11: 323–323.

4 Rizvi, S.A.A. and Saleh, A.M. (2018). Applications of nanoparticle systems in drug delivery technology. *Saudi Pharm. J.* 26: 64–70.

5 Baxter. Doxil (doxorubicin HCl liposome injection) for hospital care. https://www.baxter.com/doxil-doxorubicin-hcl-liposome-injection-hospital-care (accessed 20 July 2021).

6 Bristol-Myers Squibb. Abraxane. https://www.abraxanepro.com (accessed 20 July 2021).

7 Fortuin, A.S., Brüggemann, R., van der Linden, J. et al. (2018). Ultra-small superparamagnetic iron oxides for metastatic lymph node detection: back on the block. *Wiley Interdiscip. Rev. Nanomed. Nanobiotechnol.* 10: e1471.

8 Servier Pharmaceuticals LLC. Oncaspar. https://www.oncaspar.com (accessed 20 July 2021).

9 Svanberg, A. and Birgegård, G. (2015). Addition of aprepitant (Emend®) to standard antiemetic regimen continued for 7 days after chemotherapy for stem cell transplantation provides significant reduction of vomiting. *Oncology* 89: 31–36.

10 Joubert, N., Beck, A., Dumontet, C., and Denevault-Sabourin, C. (2020). Antibody–drug conjugates: the last decade. *Pharmaceuticals (Basel)* 13 (9): 245.

11 Genentech, Inc. Kadcyla. https://www.kadcyla.com/hcp/early-breast-cancer.html?c=kad-16e1d6d6947&gclid=CjwKCAjwmK6IBhBqEiwAocMc8ofnMgDfJ8J-lOKiQnOCiYeSXw2Gg_HYgZw11xZMHEXF0WbK6GuSsxoC7GEQAvD_BwE&gclsrc=aw.ds (accessed 20 July 2021).

12 Otterson, G. Paclitaxel albumin-stabilized nanoparticle formulation and carboplatin in treating patients with stage IIIB, stage IV, or recurrent non-small cell lung cancer. https://clinicaltrials.gov/ct2/show/NCT00729612?term=targeted+nanoparticle&draw=5&rank=20 (accessed 20 July 2021).

13 Nanospectra Biosciences, Inc. An extension study MRI/US fusion imaging and biopsy in combination with nanoparticle directed focal therapy for ablation of prostate tissue. https://clinicaltrials.gov/ct2/show/NCT04240639?term=targeted+nanoparticle&draw=5&rank=10 (accessed 20 July 2021).

14 Abdellatif, A.A.H. Targeted polymeric nanoparticles loaded with cetuximab and decorated with somatostatin analogue to colon cancer. https://www.clinicaltrials.gov/ct2/show/NCT03774680 (accessed 20 July 2021).

15 Lipomedix Pharmaceuticals Inc. Assessment of safety and therapeutic efficacy of Promitil in combination with Folfox in patients with GI malignancies. https://clinicaltrials.gov/ct2/show/NCT04729205?term=PROMITIL&draw=2&rank=1 (accessed 20 July 2021).

16 Celsion. Phase 3 study of ThermoDox with radiofrequency ablation (RFA) in treatment of hepatocellular carcinoma (HCC). https://clinicaltrials.gov/ct2/

show/results/NCT00617981?term=ThermoDox%C2%AE&draw=2&rank=10 (accessed 20 July 2021).

17 Ural State Medical University. Plasmonic nanophotothermal therapy of atherosclerosis (NANOM-FIM). https://clinicaltrials.gov/ct2/show/NCT01270139?term=targeted+nano&draw=2&rank=7 (accessed 20 July 2021).

18 Memorial Sloan Kettering Cancer Center. The use of nanoparticles to guide the surgical treatment of prostate cancer. https://clinicaltrials.gov/ct2/show/NCT04167969?term=targeted+nanoparticle&draw=5&rank=17 (accessed 20 July 2021).

19 Al-Azhar University. Topical fluorescent nanoparticles conjugated somatostatin analog for suppression and bioimaging breast cancer. https://clinicaltrials.gov/ct2/show/NCT041383342?term=targeted+nanoparticle&draw=5&rank=23 (accessed 20 July 2021).

20 Fujiwara, Y., Mukai, H., Saeki, T. et al. (2019). A multi-national, randomised, open-label, parallel, phase III non-inferiority study comparing NK105 and paclitaxel in metastatic or recurrent breast cancer patients. *Br. J. Cancer* 120: 475–480.

21 Attia, M.F., Anton, N., Wallyn, J. et al. (2019). An overview of active and passive targeting strategies to improve the nanocarriers efficiency to tumour sites. *J. Pharm. Pharmacol.* 71: 1185–1198.

22 Jasim, A., Abdelghany, S., and Greish, K. (2017). Chapter 2 – Current update on the role of enhanced permeability and retention effect in cancer nanomedicine. In: *Nanotechnology-Based Approaches for Targeting and Delivery of Drugs and Genes* (ed. V. Mishra, P. Kesharwani, M.C.I. Mohd Amin and A. Iyer), 62–109. Academic Press.

23 Fang, J., Islam, W., and Maeda, H. (2020). Exploiting the dynamics of the EPR effect and strategies to improve the therapeutic effects of nanomedicines by using EPR effect enhancers. *Adv. Drug Delivery Rev.* 157: 142–160.

24 Baek, J.-S. and Cho, C.-W. (2017). A multifunctional lipid nanoparticle for co-delivery of paclitaxel and curcumin for targeted delivery and enhanced cytotoxicity in multidrug resistant breast cancer cells. *Oncotarget* 8: 30369–30382.

25 Albanese, A., Tang, P.S., and Chan, W.C. (2012). The effect of nanoparticle size, shape, and surface chemistry on biological systems. *Annu. Rev. Biomed. Eng.* 14: 1–16.

26 Maeda, H., Nakamura, H., and Fang, J. (2013). The EPR effect for macromolecular drug delivery to solid tumors: Improvement of tumor uptake, lowering of systemic toxicity, and distinct tumor imaging in vivo. *Adv. Drug Delivery Rev.* 65: 71–79.

27 Kang, H., Rho, S., Stiles, W.R. et al. (2020). Size-dependent EPR effect of polymeric nanoparticles on tumor targeting. *Adv. Healthcare Mater.* 9: 1901223.

28 Tong, X., Wang, Z., Sun, X. et al. (2016). Size dependent kinetics of gold nanorods in EPR mediated tumor delivery. *Theranostics* 6: 2039–2051.

29 Sykes, E.A., Chen, J., Zheng, G., and Chan, W.C. (2014). Investigating the impact of nanoparticle size on active and passive tumor targeting efficiency. *ACS Nano* 8: 5696–5706.

30 Cabral, H., Matsumoto, Y., Mizuno, K. et al. (2011). Accumulation of sub-100 nm polymeric micelles in poorly permeable tumours depends on size. *Nat. Nanotechnol.* 6: 815–823.

31 Golombek, S.K., May, J.-N., Theek, B. et al. (2018). Tumor targeting via EPR: strategies to enhance patient responses. *Adv. Drug Delivery Rev.* 130: 17–38.

32 Bennie, L.A., McCarthy, H.O., and Coulter, J.A. (2018). Enhanced nanoparticle delivery exploiting tumour-responsive formulations. *Cancer Nanotechnol.* 9: 10.

33 Mehta, D., Leong, N., McLeod, V.M. et al. (2018). Reducing dendrimer generation and PEG chain length increases drug release and promotes anticancer activity of PEGylated polylysine dendrimers conjugated with doxorubicin via a cathepsin-cleavable peptide linker. *Mol. Pharmaceutics* 15: 4568–4576.

34 Suk, J.S., Xu, Q., Kim, N. et al. (2016). PEGylation as a strategy for improving nanoparticle-based drug and gene delivery. *Adv. Drug Delivery Rev.* 99: 28–51.

35 Sakurai, Y., Hada, T., Yamamoto, S. et al. (2016). Remodeling of the extracellular matrix by endothelial cell-targeting siRNA improves the EPR-based delivery of 100 nm particles. *Mol. Ther.* 24: 2090–2099.

36 Kinoshita, R., Ishima, Y., Chuang, V.T.G. et al. (2017). Improved anticancer effects of albumin-bound paclitaxel nanoparticle via augmentation of EPR effect and albumin-protein interactions using S-nitrosated human serum albumin dimer. *Biomaterials* 140: 162–169.

37 Yoshikawa, T., Mori, Y., Feng, H. et al. (2019). Rapid and continuous accumulation of nitric oxide-releasing liposomes in tumors to augment the enhanced permeability and retention (EPR) effect. *Int. J. Pharm.* 565: 481–487.

38 Tekade, R.K., Maheshwari, R., Soni, N. et al. (2017). Chapter 1 – Nanotechnology for the development of nanomedicine. In: *Nanotechnology-Based Approaches for Targeting and Delivery of Drugs and Genes* (ed. V. Mishra, P. Kesharwani, M.C.I. Mohd Amin and A. Iyer), 3–61. Academic Press.

39 Muhamad, N., Plengsuriyakarn, T., and Na-Bangchang, K. (2018). Application of active targeting nanoparticle delivery system for chemotherapeutic drugs and traditional/herbal medicines in cancer therapy: a systematic review. *Int. J. Nanomed.* 13: 3921–3935.

40 Kapoor, M.S., D'Souza, A., Aibani, N. et al. (2018). Stable liposome in cosmetic platforms for transdermal folic acid delivery for fortification and treatment of micronutrient deficiencies. *Sci. Rep.* 8: 16122.

41 Yoo, J., Park, C., Yi, G. et al. (2019). Active targeting strategies using biological ligands for nanoparticle drug delivery systems. *Cancers (Basel)* 11 (5): 640.

42 Yang, Y., Zhao, Z., Xie, C., and Zhao, Y. (2020). Dual-targeting liposome modified by glutamic hexapeptide and folic acid for bone metastatic breast cancer. *Chem. Phys. Lipids* 228: 104882.

43 Kayat, J., Mehra, N.K., Gajbhiye, V., and Jain, N.K. (2016). Drug targeting to arthritic region via folic acid appended surface-engineered multi-walled carbon nanotubes. *J. Drug Targeting* 24: 318–327.

44 Deshayes, S., Cabral, H., Ishii, T. et al. (2013). Phenylboronic acid-installed polymeric micelles for targeting sialylated epitopes in solid tumors. *J. Am. Chem. Soc.* 135: 15501–15507.

45 Kim, M., Sahu, A., Kim, G.B. et al. (2018). Comparison of *in vivo* targeting ability between cRGD and collagen-targeting peptide conjugated nano-carriers for atherosclerosis. *J. Controlled Release* 269: 337–346.

46 Ma, L., Liu, T.W., Wallig, M.A. et al. (2016). Efficient targeting of adipose tissue macrophages in obesity with polysaccharide nanocarriers. *ACS Nano* 10: 6952–6962.

47 Clark, A.J. and Davis, M.E. (2015). Increased brain uptake of targeted nanoparticles by adding an acid-cleavable linkage between transferrin and the nanoparticle core. *Proc. Natl. Acad. Sci. U.S.A.* 112: 12486–12491.

48 Nogueira-Librelotto, D.R., Codevilla, C.F., Farooqi, A., and Rolim, C.M.B. (2017). Transferrin-conjugated nanocarriers as active-targeted drug delivery platforms for cancer therapy. *Curr. Pharm. Des.* 23: 454–466.

49 Jo, H. and Ban, C. (2016). Aptamer-nanoparticle complexes as powerful diagnostic and therapeutic tools. *Exp. Mol. Med.* 48: e230.

50 Liu, J., Wei, T., Zhao, J. et al. (2016). Multifunctional aptamer-based nanoparticles for targeted drug delivery to circumvent cancer resistance. *Biomaterials* 91: 44–56.

51 Taghavi, S., Ramezani, M., Alibolandi, M. et al. (2017). Chitosan-modified PLGA nanoparticles tagged with 5TR1 aptamer for *in vivo* tumor-targeted drug delivery. *Cancer Lett.* 400: 1–8.

52 Richards, D.A., Maruani, A., and Chudasama, V. (2017). Antibody fragments as nanoparticle targeting ligands: a step in the right direction. *Chem. Sci.* 8: 63–77.

53 Raza, A., Hayat, U., Rasheed, T. et al. (2018). Redox-responsive nano-carriers as tumor-targeted drug delivery systems. *Eur. J. Med. Chem.* 157: 705–715.

54 Zhao, S., Xu, M., Cao, C. et al. (2017). A redox-responsive strategy using mesoporous silica nanoparticles for co-delivery of siRNA and doxorubicin. *J. Mater. Chem. B* 5: 6908–6919.

55 Yu, L., Chen, Y., Lin, H. et al. (2018). Ultrasmall mesoporous organosilica nanoparticles: morphology modulations and redox-responsive biodegradability for tumor-specific drug delivery. *Biomaterials* 161: 292–305.

56 Xia, J., Du, Y., Huang, L. et al. (2018). Redox-responsive micelles from disulfide bond-bridged hyaluronic acid-tocopherol succinate for the treatment of melanoma. *Nanomed. Nanotechnol. Biol. Med.* 14: 713–723.

57 Wang, C., Zhao, T., Li, Y. et al. (2017). Investigation of endosome and lysosome biology by ultra pH-sensitive nanoprobes. *Adv. Drug Delivery Rev.* 113: 87–96.

58 Ko, M., Quiñones-Hinojosa, A., and Rao, R. (2020). Emerging links between endosomal pH and cancer. *Cancer Metastasis Rev.* 39 (2): 519–534.

59 Kato, Y., Ozawa, S., Miyamoto, C. et al. (2013). Acidic extracellular microenvironment and cancer. *Cancer Cell Int.* 13: 89.

60 Gu, M., Wang, X., Toh, T.B., and Chow, E.K.-H. (2018). Applications of stimuli-responsive nanoscale drug delivery systems in translational research. *Drug Discovery Today* 23: 1043–1052.

61 Naeem, M., Awan, U.A., Subhan, F. et al. (2020). Advances in colon-targeted nano-drug delivery systems: challenges and solutions. *Arch. Pharmacal Res.* 43: 153–169.

62 Nguyen, C.T.H., Webb, R.I., Lambert, L.K. et al. (2017). Bifunctional succinylated ε-polylysine-coated mesoporous silica nanoparticles for pH-responsive and intracellular drug delivery targeting the colon. *ACS Appl. Mater. Interfaces* 9: 9470–9483.

63 Wang, Z., Tian, Y., Zhang, H. et al. (2016). Using hyaluronic acid-functionalized pH stimuli-responsive mesoporous silica nanoparticles for targeted delivery to CD44-overexpressing cancer cells. *Int. J. Nanomed.* 11: 6485–6497.

64 Lin, Z., Li, J., He, H. et al. (2015). Acetalated-dextran as valves of mesoporous silica particles for pH responsive intracellular drug delivery. *RSC Adv.* 5: 9546–9555.

65 Indermun, S., Govender, M., Kumar, P. et al. (2018). 2 – Stimuli-responsive polymers as smart drug delivery systems: classifications based on carrier type and triggered-release mechanism. In: *Stimuli Responsive Polymeric Nanocarriers for Drug Delivery Applications*, vol. 1 (ed. A.S.H. Makhlouf and N.Y. Abu-Thabit), 43–58. Woodhead Publishing.

66 Duro-Castano, A., Talelli, M., Rodríguez-Escalona, G., and Vicent, M.J. (2019). Chapter 13 – Smart polymeric nanocarriers for drug delivery. In: *Smart Polymers and their Applications (Second Edition)* (ed. M.R. Aguilar and J. San Román), 439–479. Woodhead Publishing.

67 Pramanik, S.K., Pal, U., Choudhary, P. et al. (2019). Stimuli-responsive nanocapsules for the spatiotemporal release of melatonin: protection against gastric inflammation. *ACS Appl. Bio Mater.* 2: 5218–5226.

68 Tiwari, A., Verma, A., Panda, P.K. et al. (2019). 20 – Stimuli-responsive polysaccharides for colon-targeted drug delivery. In: *Stimuli Responsive Polymeric Nanocarriers for Drug Delivery Applications* (ed. A.S.H. Makhlouf and N.Y. Abu-Thabit), 547–566. Woodhead Publishing.

69 Thambi, T. and Lee, D.S. (2019). 15 – Stimuli-responsive polymersomes for cancer therapy. In: *Stimuli Responsive Polymeric Nanocarriers for Drug Delivery Applications* (ed. A.S.H. Makhlouf and N.Y. Abu-Thabit), 413–438. Woodhead Publishing.

70 Kienzle, A., Kurch, S., Schlöder, J. et al. (2017). Dendritic mesoporous silica nanoparticles for pH-stimuli-responsive drug delivery of TNF-alpha. *Adv. Healthcare Mater.* 6 (13).

71 Zhang, C.Y., Lin, W., Gao, J. et al. (2019). pH-responsive nanoparticles targeted to lungs for improved therapy of acute lung inflammation/injury. *ACS Appl. Mater. Interfaces* 11: 16380–16390.

72 Ahamad, N., Prabhakar, A., Mehta, S. et al. (2020). Trigger-responsive engineered-nanocarriers and image-guided theranostics for rheumatoid arthritis. *Nanoscale* 12 (24): 12673–12697.

73 Wu, W., Luo, L., Wang, Y. et al. (2018). Endogenous pH-responsive nanoparticles with programmable size changes for targeted tumor therapy and imaging applications. *Theranostics* 8: 3038–3058.

74 Biabanikhankahdani, R., Alitheen, N.B.M., Ho, K.L., and Tan, W.S. (2016). pH-responsive virus-like nanoparticles with enhanced tumour-targeting ligands for cancer drug delivery. *Sci. Rep.* 6: 37891.

75 Biabanikhankahdani, R., Ho, K.L., Alitheen, N.B., and Tan, W.S. (2018). A dual bioconjugated virus-like nanoparticle as a drug delivery system and comparison with a pH-responsive delivery system. *Nanomaterials (Basel)* 8: 236.

76 Wannasarit, S., Wang, S., Figueiredo, P. et al. (2019). A virus-mimicking pH-responsive acetalated dextran-based membrane-active polymeric nanoparticle for intracellular delivery of antitumor therapeutics. *Adv. Funct. Mater.* 29: 1905352.

77 Wannasarit, S., Wang, S., Figueiredo, P. et al. (2019). Antitumor therapeutics: a virus-mimicking pH-responsive acetalated dextran-based membrane-active polymeric nanoparticle for intracellular delivery of antitumor therapeutics (Adv. Funct. Mater. 51/2019). *Adv. Funct. Mater.* 29: 1970351.

78 Gong, F., Peng, X., Luo, C. et al. (2013). Cathepsin B as a potential prognostic and therapeutic marker for human lung squamous cell carcinoma. *Mol. Cancer* 12: 125.

79 Eiján, A.M., Sandes, E.O., Riveros, M.D. et al. (2003). High expression of cathepsin B in transitional bladder carcinoma correlates with tumor invasion. *Cancer* 98: 262–268.

80 Ehrsam, D., Porta, F., Hussner, J. et al. (2019). PDMS-PMOXA-nanoparticles featuring a Cathepsin B-triggered release mechanism. *Materials (Basel)* 12 (17): 2836.

81 Chan, Y.-C. and Hsiao, M. (2017). Protease-activated nanomaterials for targeted cancer theranostics. *Nanomedicine* 12: 2153–2159.

82 Lee, S.J., Jeong, Y.-I., Park, H.-K. et al. (2015). Enzyme-responsive doxorubicin release from dendrimer nanoparticles for anticancer drug delivery. *Int. J. Nanomed.* 10: 5489–5503.

83 Joseph, S.K., Sabitha, M., and Nair, S.C. (2020). Stimuli-responsive polymeric nanosystem for colon specific drug delivery. *Adv. Pharm. Bull.* 10: 1–12.

84 Zhang, M. and Merlin, D. (2018). Nanoparticle-based oral drug delivery systems targeting the colon for treatment of ulcerative colitis. *Inflamm. Bowel Dis.* 24: 1401–1415.

85 Gou, S., Huang, Y., Wan, Y. et al. (2019). Multi-bioresponsive silk fibroin-based nanoparticles with on-demand cytoplasmic drug release capacity for CD44-targeted alleviation of ulcerative colitis. *Biomaterials* 212: 39–54.

86 Cai, D., Han, C., Liu, C. et al. (2020). Chitosan-capped enzyme-responsive hollow mesoporous silica nanoplatforms for colon-specific drug delivery. *Nanoscale Res. Lett.* 15: 123.

87 Yoshimatsu, K., Lesel, B.K., Yonamine, Y. et al. (2012). Temperature-responsive "catch and release" of proteins by using multifunctional polymer-based nanoparticles. *Angew. Chem. Int. Ed.* 51: 2405–2408.

88 Chen, P.-M., Pan, W.-Y., Wu, C.-Y. et al. (2020). Modulation of tumor microenvironment using a TLR-7/8 agonist-loaded nanoparticle system that exerts low-temperature hyperthermia and immunotherapy for *in situ* cancer vaccination. *Biomaterials* 230: 119629.

89 Wang, J., Wang, D., Yan, H. et al. (2017). An injectable ionic hydrogel inducing high temperature hyperthermia for microwave tumor ablation. *J. Mater. Chem. B* 5: 4110–4120.

90 Tamarov, K., Xu, W., Osminkina, L. et al. (2016). Temperature responsive porous silicon nanoparticles for cancer therapy – spatiotemporal triggering through infrared and radiofrequency electromagnetic heating. *J. Controlled Release* 241: 220–228.

91 Johannsen, M., Gneveckow, U., Thiesen, B. et al. (2007). Thermotherapy of prostate cancer using magnetic nanoparticles: feasibility, imaging, and three-dimensional temperature distribution. *Eur. Urol.* 52: 1653–1662.

92 Deng, Z., Xiao, Y., Pan, M. et al. (2016). Hyperthermia-triggered drug delivery from iRGD-modified temperature-sensitive liposomes enhances the anti-tumor efficacy using high intensity focused ultrasound. *J. Controlled Release* 243: 333–341.

93 Zhao, T., Qin, S., Peng, L. et al. (2019). Novel hyaluronic acid-modified temperature-sensitive nanoparticles for synergistic chemo-photothermal therapy. *Carbohydr. Polym.* 214: 221–233.

94 Lin, J., Stevens, M., and Smith, J. (2019). Preparation of anti-HER-2 antibody PLGA polymer nano-ultrasound contrast agent in vitro targeting experiment. *bioRxiv* 619742.

95 Wang, Z., Qiao, R., Tang, N. et al. (2017). Active targeting theranostic iron oxide nanoparticles for MRI and magnetic resonance-guided focused ultrasound ablation of lung cancer. *Biomaterials* 127: 25–35.

96 Liu, Y., Bai, L., Guo, K. et al. (2019). Focused ultrasound-augmented targeting delivery of nanosonosensitizers from homogenous exosomes for enhanced sonodynamic cancer therapy. *Theranostics* 9: 5261–5281.

97 Paris, J.L., Manzano, M., Cabañas, M.V., and Vallet-Regí, M. (2018). Mesoporous silica nanoparticles engineered for ultrasound-induced uptake by cancer cells. *Nanoscale* 10: 6402–6408.

98 Ji, J., Ren, C., Tu, H. et al. (2016). Ultrasound-targeting transfection of nano-tissue-type plasminogen activator gene to prevent in-stent thrombosis and neointimal hyperplasia after stenting intervention in injured rabbit iliac artery. *Nanosci. Nanotechnol. Lett.* 8: 671–681.

99 Duan, L., Yang, L., Jin, J. et al. (2020). Micro/nano-bubble-assisted ultrasound to enhance the EPR effect and potential theranostic applications. *Theranostics* 10: 462–483.

100 Theek, B., Baues, M., Ojha, T. et al. (2016). Sonoporation enhances liposome accumulation and penetration in tumors with low EPR. *J. Controlled Release* 231: 77–85.

101 Kesavan, M.P., Kotla, N.G., Ayyanaar, S. et al. (2018). A theranostic nanocomposite system based on iron oxide-drug nanocages for targeted magnetic field responsive chemotherapy. *Nanomed. Nanotechnol. Biol. Med.* 14: 1643–1654.

102 Shao, D., Li, J., Zheng, X. et al. (2016). Janus "nano-bullets" for magnetic targeting liver cancer chemotherapy. *Biomaterials* 100: 118–133.

103 Anirudhan, T.S., Christa, J., and Binusreejayan. (2018). pH and magnetic field sensitive folic acid conjugated protein–polyelectrolyte complex for the controlled and targeted delivery of 5-fluorouracil. *J. Ind. Eng. Chem.* 57: 199–207.

4

Liposomes as Targeted Drug-Delivery Systems

Raghavendra C. Mundargi, Neetika Taneja, Jayeshkumar J. Hadia, and Ajay J. Khopade

Sun Pharmaceutical Industries Limited, R&D-Formulations, Nima Compound, Tandalja, Vadodara 390020, Gujarat, India

4.1 Introduction

The Liposomology era was originated in the 1960s at Babraham, Cambridge with the first description of swollen phospholipid systems by Alec Bangham [1–11]. He established the first evidence for model membrane systems for a cell membrane. The term "liposomes" was given by Weismann and coworkers to the spontaneously formed closed structures of phospholipid bilayers, when they were hydrated in water [12]. As new preparation methods were developed, a different class of vesicles was observed which led to the classification of liposomes based on their size and lamella; small unilamellar vesicles 20–50 nm, large unilamellar vesicles 100–1000 nm, giant unilamellar vesicles (>1000 nm), multilamellar vesicles (>500 nm), oligolamellar vesicles (100–1000 nm), and multivesicular vesicles (>1000 nm) [13].

Gregoriadis established the concept that the drugs could be encapsulated into liposomes and used as drug-delivery vehicles [14–16]. The liposomal vesicle has a distinct ability to encapsulate both hydrophilic and lipophilic compounds. The aqueous core can entrap hydrophilic molecules while the bilayer can entrap hydrophobic molecules. The large internal aqueous core of liposome and external bilayer provides distinct compartments for the packaging of chemical entities, theranostic agents, or macromolecules, such as peptides, DNA, oligonucleotides, and RNAs. The science of liposomes evolved with new methods that improved entrapment efficiency and reduced leakage, such as remote loading using counter ion [14].

The manufacturing of liposomes has advanced significantly, beginning with the thin lipid film hydration and progressing through reverse-phase evaporation, freeze-drying and scalable ethanol injection methods [17–19]. The optimization of techniques such as sonication, homogenization, and membrane extrusion through 50–200 nm pore-size polycarbonate membranes, resulted in a significant step forward in large-scale production of uniformly size-distributed liposomes suitable

Targeted Drug Delivery, First Edition. Edited by Yogeshwar Bachhav.

for commercial use [14]. The production process of liposomes has further advanced to a low-pressure process using microfluidic technology with a scalable process and improved size uniformity [19]. Another advancement has been the development of the lipids which have provided control over the rigidity and fluidity of the bilayer resulting in stability under storage and in the biological environment [20]. Natural phospholipids, for example, are unsaturated and prone to permeability and drug leakage, whereas semisynthetic-saturated lipids with long acyl chains provide a rigid impermeable bilayer structure [21]. Further advancement is related to the in vivo distribution of encapsulated drugs [20]. This was accomplished by stabilizing the drugs in liposomes, improving biodistribution of compounds to target sites through PEGylation, and overcoming barriers to cellular and tissue uptake, thereby reducing organ toxicity. With all these advancements, the liposomes have eventually risen to a status of a clinically validated delivery vehicle addressing a wide range of therapeutic needs [22].

With the current state of maturity in the liposomal field, it is critical to provide readers with a bird's eye view of liposome technology by categorising them based on the key functionality they bring on to the table. Accordingly, the liposome-based delivery systems are categorized into four main types: (i) Non-conventional liposomes, (ii) Conventional liposomes, (iii) Ligand-anchored liposomes, and (iv) Theranostic liposomes (Figure 4.1) [23–31].

The non-conventional liposomes consist of a lipid bilayer that can be composed of different lipids – neutral (phospholipids), cationic, anionic, and cholesterol [20]. These conventional liposomes enclose an aqueous volume to facilitate stabilization of therapeutic load and thereby improving the therapeutic index of the entrapped drug. The conventional liposome reduces the organ toxicity in the clinic by virtue of extending plasma half-life and better delivery to the target tissue (tumor) in comparison to nonliposomal formulations. However, these are susceptible to rapid elimination because of the phenomenon called opsonization in the plasma component of blood which eventually leads to uptake by the macrophages of the reticuloendothelial system, mainly the liver and spleen [14, 16].

To address this issue of undesired macrophage uptake, improve *in vivo* liposome stability and enhance their systemic circulation time, the concept of sterically stabilized (Conventional) liposomes was introduced as the second category of liposomes also called stealth liposomes [14]. The hydrophilic polymer, PEG-derived lipid (PEG-DSPE), is coated onto the liposomes for obtaining sterically stabilized liposomes (Figure 4.1). The *in vivo* opsonization with serum components and the rapid recognition and uptake by the RES is significantly reduced using stealth liposomes, which translate into a significant improvement in the efficacy of encapsulated drugs [14].

In the third category, ligand-anchored liposomes, various types of ligands, such as peptides/proteins, carbohydrates, and antibodies, are attached to the liposome surface for receptor-mediated distribution (Figure 4.1) [28, 29]. These ligand-targeted liposomes have the ability to deliver the encapsulated drug specifically to pathological cells or organs that over-express a specific protein at the diseased site. While the

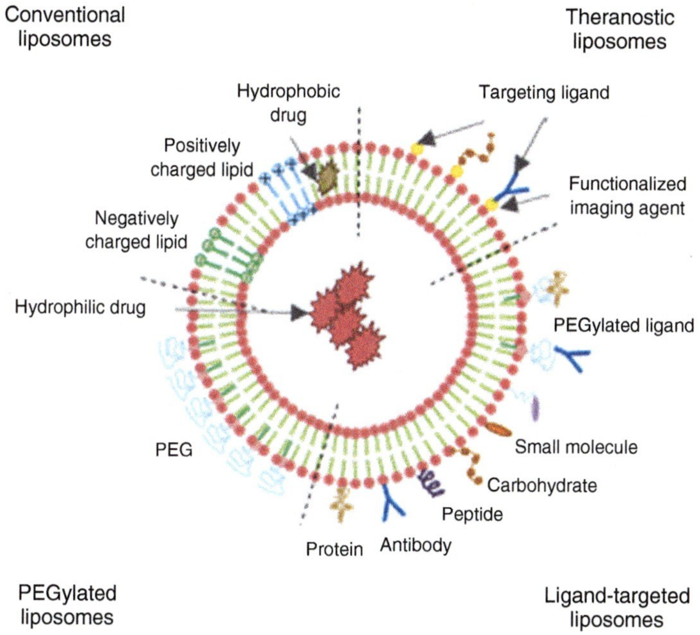

Figure 4.1 Schematic representation of different types of targeted liposomes. (i) Conventional liposomes consist of lipid bilayer with phospholipids of different charge (neutral, cationic, and anionic) and cholesterol. The lipid bilayer encloses an aqueous core with hydrophilic drug complex and bilayer with hydrophobic drug complex. (ii) Conventional liposomes comprised of PEG as hydrophilic polymer coating to the liposome surface to provide better steric stabilization. By virtue of pegylation, the pharmacokinetic properties can be modulated to get desired clinical benefits. (iii) Ligand-targeted liposomes for active-targeting. The active targeting is achieved by attaching ligands, such as peptides and antibodies, to the liposome surface by conjugation with bilayer lipids or terminal end of PEG chains. (iv) Theranostic liposomes consist of a drug, and targeting ligand with imaging agent which enables image-guided targeted drug delivery.

concept has shown pretty good success in preclinical models, it is far from proof of concept in clinic. The major challenges in translation of ligand-targeted liposomes are as follows:

- Selection of appropriate ligand to achieve ligand functionalization onto liposomes to achieve target specificity and receptor-binding efficiency [32].
- In-depth understanding of ligand-targeted liposome interactions with cellular membrane and internalization mechanism to achieve most optimum formula.
- Implementing human pathological similarity during preclinical evaluation of ligand-targeted liposomes to understand and evade the targeting barriers [33].
- Appropriate ligand conjugation chemistry to achieve optimum confirmation of ligand anchoring onto liposomes for efficient receptor binding [34].
- Comprehensive characterization of ligand-targeted liposomes to establish efficient ligand conjugation onto lipid or liposome surface, receptor binding.

- Lack of established manufacturing methods to scale up ligand-targeted liposomes and consistently produce large-scale batches.
- Clinical differentiation with desired outcomes compared to nontargeted liposomes and commercial viability.

The concept of ligand-targeted liposomes concept is currently a focus area for scientists, and industrial technologists working to bring next-generation actively targeted liposomes to market.

Targeted liposomes advanced even further, resulting in the fourth category of liposomes, which combines therapeutic and diagnostic applications (Figure 4.1). In this category, liposomes are radiolabeled by incorporating a lipid-conjugated bifunctional chelator into the liposome's bilayer for additional radiolabeling [35]. The concept is in phase 3 clinical trials [36].

4.2 Liposome Commercial Landscape

Since the first US FDA approval for liposome product-Doxil® in 1995, there have been approximately 16 liposome-based products including bilayer vesicles, drug–lipid complexes, multivesicular vesicles, and RNA–lipid nanoparticles in the market [37, 38]. The majority of these products have been approved by several regulatory organizations worldwide, including the DCGI in India, the FDA in the United States, the EMA in Europe, and Health Canada in Canada. For the convenience of the readers, Table 4.1 shows the products that passively target tissue or cells through i.v. injection and Table 4.2 shows the local delivery products that are administered to release the drug at the point of action.

The majority of the products are intended for intravenous administration due to direct availability in systemic circulation and the preservation of intact liposome structure [38]. Over the three decades, liposomes has emerged as reliable platform for the delivery of diverse molecules covering a variety of indications including cancer, infection, pain, and eye disorders [38, 53]. The increased interest in cancer therapeutics is attributed to the targeting ability of liposomes via enhanced permeation and retention (EPR) effect which allow for a relatively higher drug delivery to the tumor interstitium. Furthermore, the majority of cytotoxic molecules suffer from disadvantages, such as undesired solubility, inadequate stability, suboptimal efficacy, and organ toxicity [54, 55]. The liposomal formulations overcome these issues while also improving efficacy and safety.

The clinic has demonstrated the benefits of liposomal formulations primarily by improving the drug's pharmacokinetics and pharmacodynamics [56–60]. This was made possible by liposome technology's ability to accommodate drugs with different solubility and molecular properties, such as different chemical entity, prodrugs, peptides, antibodies and by using different processes such as complex-formation with different counter ions and lipids [49, 61–63]. Generally, hydrophobic drugs have an affinity for the phospholipid bilayer, and thus entrap in the lipid bilayer whereas hydrophilic drugs entrap in the aqueous cavity [62].

Table 4.1 Liposome products based on passive delivery by intravenous injection.

S No.	Product and approval year	Innovator	Active ingredient	Indication	Liposome technology	Lipid:lipid: drug molar ratio	Size (nm)	Clinical benefits of liposome vs nonliposome	Current IMS sale (US) ($Mn) MAT May 2020	References
1	Doxil 1995	Sequus Pharmaceuticals	Doxorubicin HCl	Kaposi's sarcoma, ovarian, breast cancer	Stealth liposomes	HSPC:Cholesterol: PEG 2000-DSPE (56 : 39.5 molar ratio)	87	Improved half-life of 46 h vs 2.5 h (non-conventional), prolonged retention, increased target distribution, equivalent efficacy, and significant reduced cardiac and GI toxicity	$131	[14]
2	Abelcet 1995	Sigma-Tau Pharmaceuticals	Amphotericin B	Invasive severe fungal infections	Drug–lipid complex	DMPC:DMPG (7 : 3 molar ratio)	3000	Improved half-life of 173.4 v 91.1 h, significantly less nephrotoxicity for Abelcet (5 mg kg^{-1} day^{-1}) than for conventional amphotericin B (0.7 mg kg^{-1} day), large volume of distribution due to macrophage uptake in the liver and spleen hence successfully used to treat macrophage Infections, such as histoplasmosis and leishmaniasis	$6.9	[39]
3	DaunoXome, 1996	NeXstar Pharmaceuticals	Daunorubicin citrate	AIDS-related Kaposi's sarcoma	Non-conventional liposomes	DSPC and cholesterol (2 : 1 molar ratio)	45	Improved half-life of 4–5.6 h (NPL) vs 0.77 h (plain drug), equivalent efficacy, and fewer side effects	$7.8	[40]
4	Amphotec, 1996	Ben Venue Laboratories Inc.	Amphotericin B	Severe fungal infections	Drug–lipid complex	Cholesteryl sulfate: amphotericin B (1 : 1 molar ratio)	~122	Large volume of distribution due to macrophage uptake in the liver and spleen hence successfully used to treat macrophage infections such as histoplasmosis and leishmaniasis	Discontinued; sales NA	[41]

Table 4.1 (Continued)

SNo.	Product and approval year	Innovator	Active ingredient	Indication	Liposome technology	Lipid:lipid: drug molar ratio	Size (nm)	Clinical benefits of liposome vs nonliposome	Current IMS sale (US) ($Mn) MAT May 2020	References
5	Ambisome, 1997	Astellas Pharma	Amphotericin B	Presumed fungal infections	Drug–lipid complex liposomes	HSPC:DSPG: cholesterol: amphotericin B (2 : 0.8 : 1 : 0.4 molar ratio)	<100	Half-life of 7–10 h, amphotericin B only released when the liposome binds to the fungus. Well tolerable, significantly safer than conventional amphotericin B (Nephrotoxicity 18.7% vs 33.7%. permitting larger doses to be administered	$120	[41]
6	Myocet, 2000	Elan Pharmaceuticals	Doxorubicin	Solid tumors	Non-conventional liposomes	EPC:Cholesterol (55 : 45 molar ratio)	190	Half-life of 2.5 h vs 0.2 h, prolonged retention, increased target distribution, equivalent efficacy, and reduced toxicity	EU: $28.4 (MAT Dec. 2019)	[42]
7	Visudyne, 2000	Valeant, Novartis	Verteporphin	PDT sensitizer	Non-conventional liposomes	Verteporphin: DMPC and EPG (1 : 8 molar ratio)	150–300	Equivalent clearance, slightly increased tissue distribution, and reduced toxicity	NA	[43]
8	Mepact, 2004	Takeda Pharmaceutical Ltd.	Mifamurtide	High-grade, resectable, non-metastatic osteosarcoma	Non-conventional liposomes	DOPS:POPC (3 : 7 molar ratio)	<100		EU: $10.6 (MAT Dec. 2019)	[44]
9	Exparel, 2011	Pacira Pharmaceuticals Inc.	Bupivacaine	Pain management	DepoFoam™	DEPC, DPPG, Cholesterol and tricaprylin	24 000 to 31 000 (injected in surgical area)	Optimal postoperative pain control with significantly reduced nausea, lower maximal pain scores at all time periods studied, and significantly decreased total opioid use in the first 72 h after injections	$385.8	[45]

10	Marqibo, 2012	Acrotech/ Talon Therapeutics, Inc.	Vincristine sulfate	Acute lymphoblastic leukemia	Non-conventional liposomes	SM:Cholesterol (60 : 40 molar ratio)	100	Half-life of 6.6 h vs 1.36 h, Increased tumor drug distribution and therapeutic index, superior efficacy, and reduced toxicity.	$5.1	[46]
11	Onivyde™, 2015	Merrimack Pharmaceuticals, Inc. Ipsen	Irinotecan	mPDAC	Stealth liposomes	DSPC:MPEG-2000-DSPE (3 : 2 : 0.015 molar ratio)	~110	Half-life of 25.8 h vs 5.8 h, Prolonged retention and increased exposure of the bioactive metabolite of irinotecan (SN-38), superior efficacy, and reduced toxicity	$120.7	[47]
12	Vyxeos, 2017	Celator	Cytarabine and daunorubicin (5 : 1)	Acute myeloid leukemia	Non-conventional liposomes	DSPC:DSPG: Cholesterol (7 : 2 : 1 molar ratio)	~100	Increased targeted exposure of daunorubicin and cytarabine in a fixed ratio, superior efficacy, and comparable toxicity	$0.7	[48]
13	Onpattro, 2018	Alnylam	Patiseran	Hereditary transthyretin-mediated amyloidosis	RNA–lipid complex -Stealth liposomes	DLin-MC3-DMA (MC3):DSPC: cholesterol:PEG-2000-DSPE 50 : 10 : 38.5 : 1.5 (molar ratio)	100	First clinically intracellular delivery of siRNA, polyneuropathy scores 75% vs 7%	$4.6	[49]

Table 4.2 Liposome products based on local delivery.

SNo.	Product and approval	Innovator	Active ingredient	Indication	Administration	Liposome technology	Lipid/lipid: drug molar ratio	Size (nm)	Clinical benefits of liposome vs nonliposome	Global Sale ($ US Mil)	References
1	Depocyt, 1999	Skypharma, Pacira	Cytarabine/ Ara-C	Lymphomatous meningitis	Intrathecal	DepoFoam™	DOPC, DPPG, cholesterol and triolein	20	Prolonged tumor exposure to cytarabine, increased response rate, The complete cytological response (CCR) rate was 41% vs 6% in the conventional cytarabine, Reduced toxicity	Discontinued; Sales NA	[50]
2	DepoDur™, 2004	Skypharma, Pacira	Morphine sulfate	Pain management	Epidural	DepoFoam™	DOPC, DPPG, cholesterol and triolein	17 000 to 23 000	Prolonged and superior analgesia, reduced peak concentration, superior efficacy, and reduced toxicity	Discontinued; Sales NA	[51]
3	Arikayce, 2018	Insmed	Amikacin	Cystic fibrosis	Oral, inhalation	Non-conventional liposomes	DPPC, cholesterol	375	Prolonged the release of localized amikacin in the lungs while limiting systemic exposure, Acceptable toxicity, only limited clinical safety and effectiveness data for Arikayce are currently available	$2.9	[52]

The commercially successful liposomal products have evolved from (i) stealth liposomes, (ii) drug–lipid complexes, (iii) non-conventional liposomes, (iv) DepoFoam™ and (v) RNA–lipid complex stealth liposomes. These liposome innovations provide optimized drug delivery by preserving the therapeutic agent's unique properties while minimizing its drawbacks, such as low plasma half-life, organ toxicity, plasma instability, inadequate target tissue concentration, higher dose, and frequency [61, 63].

Stealth technology has been extensively investigated in developing a drug-delivery system making these liposomes difficult to detect by the mononuclear phagocyte system [64]. Liposomes are known to interact with proteins in the body to trigger innate immune system responses [65]. This is beacause, once injected in the body, circulating proteins can adsorb onto the surface of the liposome, thereby creating a protein corona unique to the characteristics of the liposome. The protein corona can then induce the activation or suppression of various immune responses. Particularly, liposomes interact with complement proteins to activate the complement cascade and increase the body's response to antigens. The cationic liposomes, which are largely studied for gene delivery, are known to elicit toxicity in macrophages, macrophage-like cells, and monocyte-like cells, as well as alter the secretion of important immunomodulators [65]. Liposomes can also elicit immunogenic responses depending on their surface charge, size, and pegylation [66]. Therefore, advantages and important drawbacks must be considered during design of a liposome-delivery system for a particular disease. Several researchers demonstrated that PEGylation of the liposome surface improved liposome stability and circulation time after intravenous administration – this kind of liposome was then named "Stealth®" due to its ability to evade interception by the immune system and RES [5, 67].

PEG is commontly used in this technology and the process is known as PEGylation [68]. It is accomplished by the incubating a reactive derivative of PEG with lipid or liposomes. This covalent linkage of the liposome to a PEG chain protects the liposome from the recipient's immune system reducing immunogenicity [69]. Furthermore, it alters the physicochemical properties of liposomes, including changes in hydrodynamic size. This reduces its renal clearance even further extending its circulatory time [56]. These modifictions have no effect on efficacy and have a lower toxicity [70, 71]. In addition, because of the leaky tumor vasculature, nano-sized formulations with long circulatory times exhibit EPR and slowly accumulate in the tumor. The first successful stealth technology was the introduction of Doxil® to the US market in 1995 initially for the treatment of patients with AIDS-related Kaposi's sarcoma after the prior systemic chemotherapy failed and later extended to patients with ovarian cancer [72, 73].

Doxil® contains a drug encapsulated in the aqueous core. Developed by Barenholz, it was the first nano-sized liposomal product to obtain regulatory approval in Israel apart from the USA [14]. NeXstar Pharmaceuticals, USA developed DaunoXome® for the delivery of daunorubicin, which was approved in 1996 by the US FDA for the treatment of advanced HIV-associated Kaposi's sarcoma [74, 75]. In the year 2000, Myocet® a liposomal doxorubicin, was developed by Elan Pharmaceuticals, USA, which was approved in Europe and Canada for the management of metastatic breast cancer in combination with cyclophosphamide [61]. This liposomal doxorubicin injection out performed doxorubicin and Doxil® in terms of safety, as it not only reduced the cardiac toxicity associated with doxorubicin but also the dose-limiting toxicity related to Doxil®, such as HFS [76]. This is due to the combination of specific composition, a unique manufacturing process, and liposome size. The expected clinical outcomes depend on particle size of liposomes which in turn depends on drug physicochemical properties and target delivery site. The liposomes with diameter of 80–150 nm can easily escape from blood vessel capillaries that perfuse tissues like the lung, heart, and kidney [5, 77, 78]. Furthermore, the size range of 80–150 nm is favorable for cell uptake and reducing phagocytosis clearance. The liposome size of commercial products varies from nanometer size for cancer treatment to micron size for fungal treatment. The products used in pain control have larger liposome size so that they can persist longer at the injection site, such as Exparel® and DepoDur™ [39, 45]. Subsequently, a few more products have become available based on non-conventional liposome technology, such as Visudyne® (2000), Mepact®(2004), Marqibo® (2012), Vyxeos® (2017), and Arikayce®(2018).

Apart from stealth liposome technology, which encapsulated drugs in the internal aqueous core, other technologies that entrapped drugs into liposome bilayer via the interaction between the charged functional group of drug molecules and the negatively charged lipids, such as DSPG, were also commercialized [79]. The product was liposomal Amphotericin B, approved and commercialized as Ambisome® in 1997 for the treatment of systemic fungal infection [80]. The hydrophobic interactions with cholesterol constituents enable Amphotericin/DSPG complex to entrap in the membrane. The interactions between the sterol-binding regions provided the stabilization of drug–lipid complexes in the bilayer. Abelcet® (1995) and Amphotec® (1996) were other nonliposomal products that are commercialized based on drug–lipid complex [81]. Ambisome® provides distinct advantages over these complexes due to the reduction in nephrotoxicity.

Further development led to DepoFoam®, an extended-release liposome drug-delivery technology introduced by Pacira Pharmaceuticals, Inc., USA in 1999 [82]. DepoFoam® is the core technology behind several marketed products, such as Depocyt® (1999), DepoDur® (2004), and Exparel® (2011). This technology encapsulates drugs in its multivesicular liposomal platform [83]. The multivesicular liposomes release drug over a required period of time ranging from 1 to 30 days [84]. DepoFoam® made up of microscopic granular spheroids (3–30 μm) and single-layered lipid particles made up of a honeycomb of numerous nonconcentric internal aqueous chambers containing the bounded drug [83, 84]. Each particle contains a number of nonconcentric aqueous chambers separated by a single-bilayer lipid membrane and each chamber is separated from the others by bilayer lipid membranes made of synthetic analogues of naturally occuring lipids (DOPC, DPPG, cholesterol, and triolein). Upon administration, DepoFoam™ particles release the drug over a period of time ranging from hours to weeks due to erosion and/or reorganization of the lipid membranes. DepoFoam™ technology has significantly improved patient care by providing a remarkable solution for medications that require frequent multiple injections or have a short duration of action [84].

Liposome technology advanced even further using cationic lipids such as DOTAP and DOTMA to create siRNA–lipid complex formulations for RNA delivery [85, 86]. In 2018, the major milestone was accomplished with the approval of the first RNA-based product Onpattro®, siRNA–cationic lipid complex packaged into stealth liposomes for liver delivery [20, 86, 87]. Onpattro® approval led to a benchmark in RNA therapeutics. Naked RNA molecules have intrinsic instability due to degradation by RNAs; hence, these macromolecules need to be condensed with cationic lipids to yield stable RNA. This leads to preferential cell uptake mediated by cationic charge and efficient endosomal escape to deliver the therapeutic RNA into the cytoplasm for further processing either gene silencing or protein synthesis [69]. These cationic RNA/lipid complexes needs further packaging inside liposome for its administration into the body.

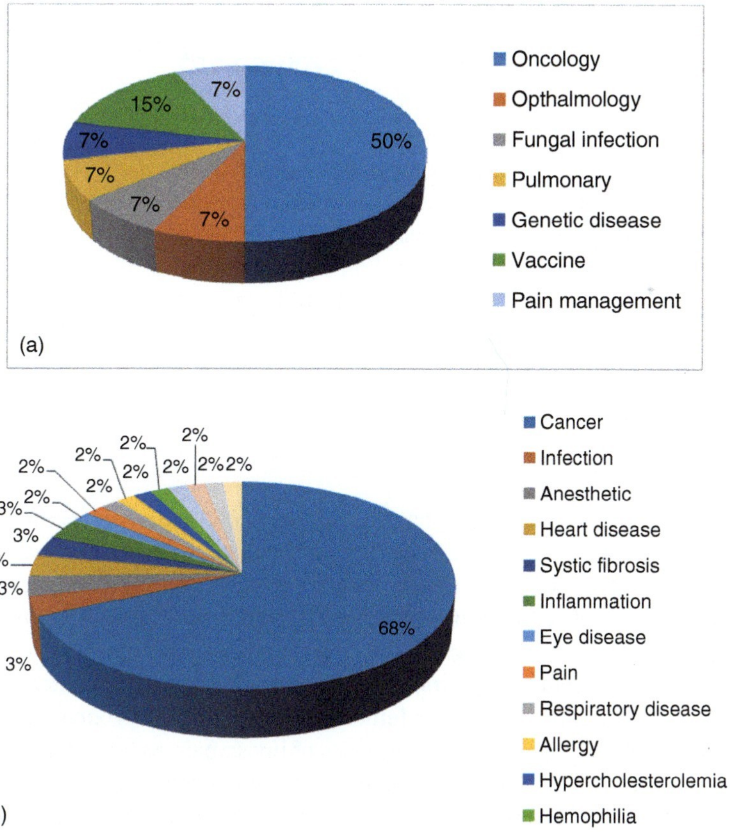

Figure 4.2 Therapeutic segment-wise distribution of liposome products: (a) approved liposomes, (b) investigational liposomes.

Looking at the current liposome product development landscape, oncology remains the major therapeutic category, accounting for half of all approved liposome products (Figure 4.2a). Other therapeutic sectors with a market share of around 7% include fungal infection, pulmonary delivery, pain management, and ophthalmology (Figure 4.2a). The improved understanding, including CMC, regulatory hurdles related to liposomes approval has led to almost 55 investigational liposome and lipid nanoparticle-based products which are under the different stages of clinical development [22, 23]. It is evident from investigational products (Figure 4.2b) that the liposomes are ahead in oncology as a major therapeutic segment with 68% share. The remaining segments add up individuals at 2–3% for infection, anesthetic, cardiology, cystic fibrosis, inflammation, eye disease, pain, respiratory disease, allergy, hypercholesterolemia, and hemophilia.

Although liposomes have demonstrated significant therapeutic advantages for a multitude of therapeutic applications; a few products still are not able to maintain business continuity due to issues related to liposome manufacturing and supply. Hence, the pharma industry recognizes liposomes as low-volume and high-value product projected with significant global sales and market capitalization.

4.3 Important Considerations in Development and Characterization of Liposomes

Liposome-based products are complex formulations; hence, small changes in the formulation may significantly affect the clinical outcomes. In this section, for the convenience of readers, important considerations are discussed in six different subsections which are as follows: (i) selection of lipids, (ii) drug:lipid ratio, (iii) pegylation, (iv) ligand anchoring, (v) drug-loading techniques, (vi) physicochemical characterizations, (vii) manufacturing processes, and (viii) product stability. Each section emphasizes how they influence critical quality attributes and therapeutic outcomes.

4.3.1 Selection of Lipids

The building blocks of liposomes comprise a combination of two or more membrane bilayer lipids. There are three key regions in a lipid that may affect its safety and stability – a hydrophilic head group, a hydrophobic tail region, and a linker that connects the hydrophilic and hydrophobic regions [88]. The commercial liposome products are composed of two or more phospholipids, except the product Marqibo® which is composed of sphingolipids [46]. Various parameters of phospholipids, including tail length, tail unsaturation, linker type (ether vs ester), and head group structure influence the overall performance characteristics of liposomal products. Liposome stability is promoted if the lipid is long chain, has a high degree of tail saturation, and has minimal ether and/or ester linkages. The size and charge of the lipid head group also influence liposome interactions and drug permeation across membranes.

In particular, during liposome manufacturing and storage, chemical degradation of lipids by hydrolysis, oxidation or reduction of double-tailed lipids (e.g. DOPC) may occur, resulting in single-tailed lysolipids. These lysolipids bind to red blood cell membranes and cause hemolysis. It also perturbs other cell membrane integrity and triggers programmed cell death [89]. In addition to phospholipid, cholesterol is almost always used to stabilize liposomes [87]. Cholesterol intercalates between the lipid chains, thereby stabilizing the liposome bilayer. It decreases the fluidity of the liposomal membrane bilayer, thereby reducing the permeability of water-soluble molecules through the liposomal membrane [18, 90]. Liposomes without cholesterol tend to interact with albumin, transferrin, macroglobulin, and high-density lipoproteins, thereby destabilizing and decreasing liposome integrity *in vivo*. Cholesterol has been used in liposome composition as high as 50 mol% and such concentration decreased the liver accumulation from 70% to 40% and extended blood circulation half-life by threefolds in rats. Also, these liposomes have shown lower macrophage uptake IgG response in mice [91].

Almost all the commercially available liposomal products, including Doxil®, Marqibo®, DaunoXome®, and AmBisome®, contain cholesterol. Other than cholesterol, the hydrophilic and flexible polymer, such as PEG-conjugated lipid is also optionally included in the liposomal bilayer. PEG molecules provide a stealth effect to liposomes by mimicking water-like structures, thereby avoiding recognition by the mononuclear phagocytic system that otherwise assists in rapid liposome clearance explained below (see Section 1.1.3) [68, 92]. Various fusogenic lipids, such as dioleylphosphatidylethanolamine (DOPE), are sometimes incorporated into liposomes to facilitate endosomal escape [93] for efficient delivery of macromolecules, such as nucleic acids, proteins, and peptides. In addition, tagging the liposomes with lipid derivative of targeting ligands, such as folate and monoclonal antibodies, onto the liposomal surface can improve the site-specific delivery [28, 62]. The selection of lipids, as well as relative ratio of lipids within a liposome, is also critical factor that determines safety, stability, and *in vivo* behavior of a liposomal drug product [53], because small changes in the liposome formulation may significantly affect pharmacokinetics and pharmacodynamics performance of the product [17].

4.3.2 Drug : Lipid Ratio

Drug-to-lipid ratio (D : L ratio) is a critical process parameter as it expresses the actual capacity of the liposome to accommodate the drug. D : L ratio is affected by the lipid composition, gradient salts (in case of active loading), and loading techniques. It is an index of the effectiveness of the drug-loading process [94, 95]. Achieving an optimum drug-loading efficiency and maintaining it during the whole-life cycle of a liposomal product is critical to achieve the optimal therapeutic activity. This is achieved by tailoring liposome composition that does not allow leakage during storage and systemic circulation. The variations of this ratio could affect the release profile of the loaded drug and consequently its effectiveness. This means that the stability of the liposomes correlates with the variability of D : L ratio in different liposomal formulations. The delivery system will not be administrable and economic if a large amount of phospholipid is needed for the encapsulation of a low dose of the drug [94]. For many drugs, the therapeutic dose is relatively high (e.g. doxorubicin dose is 50 mg m^{-2}), thus requiring a high D : L ratio in the formulation. However, in some cases, if there is a high D : L ratio, it can damage the liposomal membrane causing early release of the drug. It is possible to achieve better loading efficiency when the D : L ratio is lower (0.95). Therefore, it is crucial to identify the D : L ratio, which is required to achieve high-loading efficiency [95]. Various anticancer drugs, such as doxorubicin, idarubicin, or daunorubicin are loaded at a D : L ratio of 0.3 w/w, with nearly 100% of encapsulation efficiency and excellent *in vitro* stability. To achieve high D : L ratio, active-drug-loading technique based on different counter ions is employed. The choice of counter ion drives the formulation design in active-loading mechanism. Different counter ions, such as ammonium sulfate, ammonium citrate, sucrose octa sulfate have been used to achieve optimum encapsulation and drug release [5, 27]. The active drug loading of doxorubicin in marketed liposomal products, Doxil® and Myocet®, and Irinotecan in Onivyde® constitute as good examples with optimum D : L ratio [75]. The degree of drug loading mainly depends on the nature (i.e. physicochemical properties) of drugs and the composition of the liposomes (lipids, lipid/cholesterol ratio, drug/lipid ratio, and charge of the liposomes), as well as the drug-loading process [18, 19].

4.3.3 PEGylation

Frank Davis et al. pioneered the first chemical step of pegylation [96]. PEGylation is the chemical modification that involves both covalent and noncovalent attachment of polyethylene glycol polymer chains to drugs, therapeutic proteins, and a liposome. The conjugation of PEG allows liposomes to circulate within the body for a longer period of time, extending the circulation half-life and as a result, increasing the accumulation of liposomes within tumors. To achieve optimal clinical benefits of pegylation, it is crucial to consider all aspects of pegylation, such as drug properties, desired biological half-life, and PEG molecular weight [23, 97].

Several factors govern the development of effective stealth liposomes, such as Flory dimension (*Rf*) of PEG polymer chain represents the volume occupied by a single-polymer chain. This value is influenced by both the polymer chain length and PEG density on the liposomal surface. *D* signifies the average distance between adjacent PEG chains on the vesicles. Importantly, density of PEG on the liposomal surface also influences *D* between each individual PEG polymer [97]. Figure 4.3 depicts different confirmations of conventional liposomes; an increase in the PEG–lipid concentration in the formulation increases the PEG density and ultimately decreases *D*. When *D* is larger than *Rf*, PEG polymers will reorganize themselves and coil into a mushroom-like conformation. Whereas, when *D* is smaller than *Rf*, the lateral pressure between overcrowded PEG polymers forces each polymer chain to extend into a brush conformation. Important to note, the mushroom conformation of liposomal surface PEG was observed when the PEG concentration was below 4 mol%. In the case of brush conformation, PEG concentrations were found to be above 4 mol%. As shown in Figure 4.3, “Polymer Brush” is characterized by a dense structure that may reach up to 50 Å from liposomal surface whereas “Polymer Mushroom” is relatively more compact form with 35 Å.

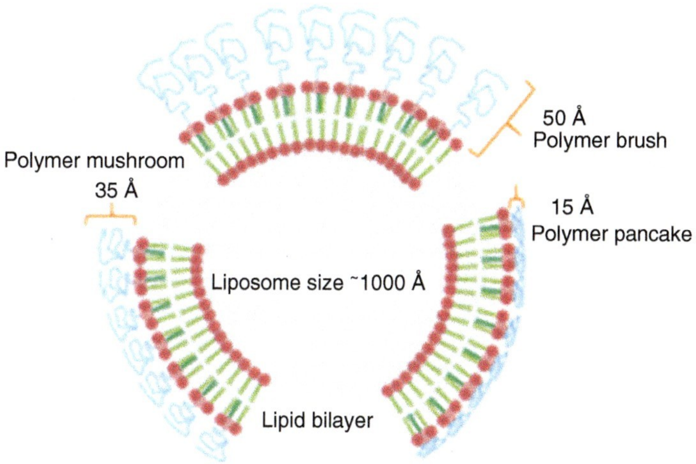

Figure 4.3 Different possible poly(ethylene glycol) conformations for the conventional liposomes.

The least space of 15 Å is attributed to "Pan Cake" with a structure firmly occupying the liposomal surface.

It is believed that prolonged systemic circulation time of liposomes is due to brush conformation of conventional liposome by providing repulsive force against proteins and other liposomes [98]. Hence, PEG chain length and concentrations are important in determining an effective liposome preparation, PEG configuration, liposome size, encapsulation of drug, membrane permeability, liposomal stability, and the effectiveness of PEG as a protective layer.

4.3.4 Ligand Anchoring

Ligand-functionalized liposomes can be manufactured in three different ways, such as preinsertion, postinsertion, and surface reaction methods (Figure 4.4). The preinsertion and postinsertion methods result in different liposomal surface topologies, such as the orientation of the ligand and its density [99, 100]. When ligand–PEG–lipids are incorporated into liposomes by the preinsertion method, only about 50% of the ligands face outwards from the liposome bilayer membrane [101, 102]. The remaining 50% of ligands face toward the inside of the liposomes. Hence, this confirmation will lead to nonfunctional ligands. Interestingly, for ligand-functionalized liposomes manufactured by the postinsertion method, all the ligands are facing outward from the liposome bilayer membrane. Though the postinsertion method is better than preinsertion in anchoring the ligands, it is important to address ligand-modified micelles which are remained in the incubation medium. This is critical to avoid ligand-modified micelles because they still have the ability to bind to target receptors and compete with functional liposomes [103]. To address the relative drawbacks with pre and postinsertion methods, surface reaction modification is developed as it can efficiently modify the

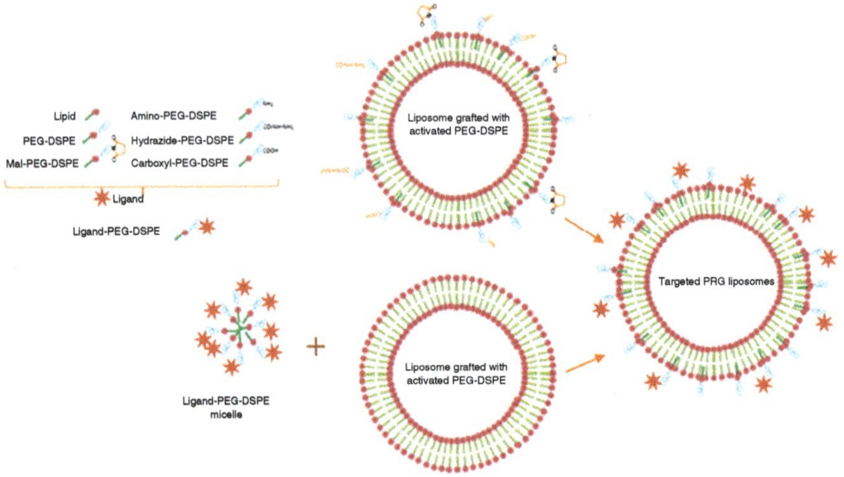

Figure 4.4 Strategies for ligand functionalization of liposomes for active-targeted delivery.

liposomal surface without micelle contamination and ensures that the ligands are oriented facing outward from the vesicle membrane [101].

4.3.5 Drug-Loading Techniques

Mainly two linearly scalable manufacturing processes are employed in producing liposomes based on passive- and active-loading techniques [17, 18] In passive-loading-based process, drug-loaded liposomes are prepared using ethanol injection method followed by extrusion, diafiltration, and aseptic filtration [53]. The passive-loading technique depends on the ability of liposomes to capture aqueous volume during vesicle formation. In the case of water-soluble drugs, the encapsulation efficiency after passive loading is proportional to the aqueous volume enclosed by the vesicles, which, in turn, depends on the phospholipid concentration and lamellarity or morphology of the liposomes. The passive loading of hydrophobic molecules occurs in the bilayer of the liposome. The amphiphilic molecules will be encapsulated in such a way that the lipid-soluble part will be embedded between the liposomal lipids while their water-soluble part will be located in the liposomal aqueous phase. Hence there will always be a limit to which a hydrophobic or an amphiphilic drug can be loaded in the bilayer without affecting the lamellar structure.

The active-drug-loading technique is largely applicable for water-soluble salts of membrane permeable molecules. It involves preparation of liposomes in an aqueous medium with desired gradient salt using ethanol injection or lipid hydration method to form MLVs. These preformed MLVs are then subjected to pressure extrusion where MLVs passed through predefined nanopore polycarbonate membranes to achieve desired size of unilamellar vesicles. The unilamellar vesicles are processed in diafiltration unit to replace external buffer [103, 104]. This establishes a gradient between internal aqueous volume and external aqueous phase. Typically, the gradient buffer is capable of interacting with the molecule and forming an insoluble gel or precipitate inside the liposome. Drug loading into empty liposomes is performed by incubating liposomes into drug solution near the glass transition of bilayer lipids. The drug permeates through the lipid bilayers into the liposomes following the concentration gradient until equilibrium between the interior of the liposome and surrounding medium is achieved. The example is a well-known illustrated mechanism of doxorubicin-loaded liposomes. The active loading of doxorubicin into liposome uses $(NH_4)_2SO_4$ concentration gradient as a driving force [16]. The doxorubicin forms an amphipathic weak base DOX-NH_2 in external phase pH, which can diffuse across the phospholipid bilayer of the liposome. Inside liposome, two molecules of DOX precipitate with SO_4^{2-} to form the $(DOX\text{-}NH_3)_2SO_4$ salt and NH_3 diffuses across liposome membrane into the external medium [13]. Another such example is the encapsulation of weakly basic, amphipathic molecule irinotecan using sucrose octasulfate triethylammonium salt as counter ions for complexation. The active-loading technique, however, is limited to a particular class of drugs that behave as weak amphipathic bases or acids and also can permeate bilayers in the unionized, but not in the ionized form [18].

Unlike small molecules, macromolecules, such as siRNA, mRNA, and DNAs, are encapsulated by *in situ* formation of lipoplex between macromolecules and cationic lipids. The ionizable cationic lipid with pKa < 7 efficiently complex with the nucleic acids at low pH and maintain a neutral or low cationic surface-charge density at pH 7.4 [78, 105]. The complex is then loaded into liposomes using a microfludic technique where spontaneous formation and encapsulation of liposome occurs. One such example is Onpattro–siRNA lipid nanoparticles based on the siRNA complexation with cationic lipid and this siRNA complex is enclosed into liposome vesicle [4].

4.3.6 Physicochemical Characterization

As liposome products involve multiple complex unit operations and processes, it is critical to monitor the manufacturing at each stage of processes against desired set of specifications, and any change in manufacturing process and composition will have a significant impact on product performance. Hence, it is important to characterize liposomes thoroughly from a physicochemical perspective for batch-to-batch reproducibility. The characterization of liposomes should meet the desired specifications against liposome product CQAs, such as lipid-phase-transition temperature, D : L ratio, free drug, lipid-degradation products, particle size distribution, zeta potential, drug-release profile [53]. The actual measurement of the PEG layer thickness is difficult to perform, as the PEG moiety does not show up in cryo-TEM, and researchers have utilized SAXS technique to establish the degree of pegylation [106]. To establish seamless liposome product development at stages of development and to set standard review process, US FDA has issued a guidance on liposome drug products that involves chemistry, manufacturing and controls, and product-specific guidance [27, 50, 107]. The initial development data, including lipid selection, optimization of D : L ratio, physicochemical characterization of liposomes, play a critical role to establish most optimized formulation. Eventually, such optimum formulation with acceptable liposome characteristics has higher probability of clinical success and provides desired clinical benefits. In addition, QbD is often recommended to implement at initial level of formulation development to arrive at most optimum formulation composition and manufacturing process as per ICH Q8, Q9, and Q10 guidance [87, 103].

Particularly, for ligand-functionalized liposomes, issues are usually encountered in evaluating both the density of the liposomal ligands, PEG, and issue of micelle contamination, controlling the orientation of the anchors are main obstacles for such evaluation [108, 109]. Hence, the surface topology of ligand-functionalized liposome is important characterization data required to produce the optimum quality of ligand-functionalized liposomes [110, 111]. For characterizing surface-ligand density, it is important to confirm that on the ligand-modified liposomal surfaces, few anchors are attached to ligands but not all. To distinguish these two scenarios during optimization, an important parameter, viz surface anchor density, is introduced to specify the amount of total anchors on the liposomal surface. Furthermore, the reaction yield is also important to identify the number of ligand-attached anchors among total anchors. Based on these two parameters, there exists incorporation ratio which is defined below

$$\text{Incorporation ratio} = \text{surface-anchor density} \times \text{reaction yield}$$

This incorporation ratio defines the surface-ligand density and it can be evaluated based on spectroscopic and spectrofluorimetric analysis.

4.3.7 Manufacturing Process

The liposome-manufacturing process that involves several complex unit operations and unit processes may significantly influence the product quality and clinical performance [18]. Irrespective of the liposome-manufacturing process adopted, it is important to understand the process and impact of critical process parameters on product quality and safety through prior knowledge and risk assessment techniques [53]. This development perceptive is necessary to have a robust liposome-manufacturing process in place that gives batch-to-batch reproducibility and maintains product quality irrespective of small deviations in the process parameters.

Some examples of deviations in the manufacturing process and their effect on liposome product performance are as follows [103]:

I. Freezing step, primary- and secondary-drying temperatures during the lyophilization process impacted morphology, residual water content, and stability of lyophilized liposomes.
II. Temperature of aqueous phase influenced liposome particle size distribution.
III. Speed, time, and temperature during homogenization influence mean liposome particle size and release profile which, in turn, could have altered product performance as the size of a liposome determines its biodistribution and mechanism of clearance.
IV. Temperatures greater than lipid-phase-transition temperature could result in drug loading or drug leakage, respectively, during the active- or passive-drug-loading process.

V. Dialfiltration- in-process product dilution, material of construction of diafilter, mol-wt cutoff size of membrane, transmembrame presurre, presuure drop etc.
VI. Extrusion: pore size of membrane, o of memberanes in a stack, no. of cycles, pressure etc

It is important to understand the effect of such variables on the manufacturing process to establish appropriate in-process controls during product development and achieve desired process performance and finished product quality. Some examples for in-process controls are as follows:

- Visual inspection for complete lipid dissolution in ethanol as the organic phase.
- Controlling MLVs size during ethanol injection step so that desired liposome size with better uniformity (low PDI) can be achieved during pressure-extrusion step.
- Checking the integrity of the extrusion membrane before and after liposome-extrusion step.
- Measuring the particle size measurement pre-extrusion and post-extrusion at every step to monitor the desired particle size and PDI.
- Measurement of conductivity post diafiltration to confirm optimum counter-ion gradient.
- Determination of membrane integrity and individual lipid composition changes after extrusion, diafiltration, and sterile filtration processes to monitor filtration/extrusion efficiency (this is important for liposomal formulations since lipids may get adsorbed to filter material causing filter blockage, formulation instability, and reduction in overall product yield).
- Determination of encapsulated and free drug to confirm encapsulation efficiency.
- Measurement of the fill weight/volume and moisture content of the lyophilized powder is monitored before packaging to evaluate the efficiency of the lyophilization process.
- Determining the particulate matter, sterility, and BET as part of injectable product compliance.

In general, failures during scale-up and commercial batches can be minimized by a thorough understanding of the manufacturing process and the use of appropriate in-process controls. A well-controlled manufacturing process may rarely experience failures [17]. The understanding and identification of appropriate controls can be achieved by applying principles of QbD in the early phases of product development. Application of QbD principles helps to define a design space that provides flexibility during the selection of formulation and manufacturing attributes for achieving desired quality attributes for a drug product. Especially, the scalability of liposomal products is one of the key challenges and it can be resolved with the help of QbD applications.

4.3.8 Product Stability

Liposome products constitute thermodynamically unstable vesicle formulations and therefore are prone to physical instabilities, such as fusion or aggregation, during storage and chemical instabilities, such as hydrolysis (lysolipids) and oxidative degradation, during storage [21]. The product stability is largely affected by (i) properties of the drug, (ii) the effect of the liposomal lipids and other formulation components, (iii) the drug-loading technique.

Stability data and post-approval stability study protocols are an essential part of liposome product development and these studies are focused to establish physical and chemical stability of liposomal products in terms of an assay, free drug (nonencapsulated part), related substances, *in vitro* drug release, particle size distribution, lysolipids as lipid degradants, phase-transition temperature, related substances, particulate matter to ensure robust product quality during storage and support product shelf life [40, 41, 87]. Irrespective of the drug to be formulated as liposome formulation, product category and its intended clinical use, it is especially important to establish product stability through the characterization of the final formulation, Chemistry, manufacturing, and controls for liposome product include acceptable specifications for lipids, drug substance, buffer, formulation composition, and manufacturing process conditions. For every liposome product before initiating GMP manufacture and clinical supply, all CMC parameters are identified and optimized. In addition, long-term storage stability has a direct impact on product quality, safety, and clinical outcome.

4.4 Targeted Delivery of Liposomes

Targeted drug delivery is a strategy that preferentially and selectively delivers the active therapeutics to the site of action, i.e. target site while concurrently minimizing the access and exposure to the nontarget site. Principally and in general, it has become an important tool to protect drugs, modulate pharmacokinetics, increase half-life, reduce dosing frequency, and reduce the side effects of drugs.

Liposome-targeted delivery systems are classified as; (i) passive-targeted delivery systems, which involves preferential accumulation at the target site, such as tumor, site of inflammation or RES rich site, and release the drug in higher quantities compared to the nonencapsulated drug. Local drug delivery has also been explored on this principle to enhance the on-site efficacy of liposomes. (ii) Active-targeted delivery systems deliver the therapeutic molecule to the site of action or target receptor with a high degree of specificity. The target specificity in active-targeted delivery is attributed to preferential attachment of ligand-anchored liposomes to the cellular target receptors and inducing endocytosis or liposome–cell membrane fusion, therefore, assisting enhanced intracellular delivery. In some of the literature, this is generally classified as the first order (organ targeting), the second order (cellular targeting), and the third order (intracellular targeting).

In the most conventional sense, the targeting was achieved through the route of administration, for example, inhalation for delivery to the lungs, topical delivery to the skin, intrathecal injection into the brain to treat different pathological conditions of the respective organs [54]. However, when the naked drug is given intravenously, it reaches each and every organ in the body wherever the vasculature reaches and beyond by diffusion. Naked drugs target preferentially to receptors wherever the receptors are highly expressed and this results in unique distribution in the body and accumulation in the different target organs. Thus, the concept of targeting becomes more relevant here when one can modulate the natural distribution of the small molecules to achieve therapeutic benefit. There are various ways to do it by taking advantage of the body's own natural mechanisms. Distribution of macromolecules is different than small molecules and the distribution of liposomes is different than either small molecules or macromolecules. There are natural mechanisms of evading the distribution of large particulates, such as RBCs prolonging their half-life [112]. There is a whole lot of large macromolecules produced by the body, such as antibodies that are highly specific and accumulate or bind only at the target site. Based on the learning from these natural mechanisms, the targeted drug-delivery strategies are classified as passive targeting and active targeting [54]. These are discussed with the specific context of the liposome structure.

4.4.1 Passive Targeting

Systemic administration of liposome into the systemic circulation triggers natural particulate-removal mechanism. Thus, it faces many obstacles before reaching the target tissue. First is protein corona formation in the blood that covers the surface of liposome in which specific protein called opsonin directs it to phagocytic cell and second is uptake by MPS organs (liver, spleen, and bone marrow) which contain abundant phagocytic cells and remove it from blood circulation hence reduce the circulation life of the liposomes. Thus, organs, such as liver, spleen, and bone marrow, are attractive targets for passive targeting of drugs using liposomes [54].

The remarkable milestone in hepatic targeting is achieved with the approval of Onpattro® the first *siRNA* product for intracellular delivery of *siRNA* prescribed for a liver disease called hereditary transthyretin-mediated amyloidosis [49]. It consists of DSPC, cholesterol, and a PEG-lipid as bilayer components in which ionizable cationic lipid dilinoleylmethyl-4-dimethylaminobutyrate (DLin-MC3-DMA) and negative *siRNA* complex is encapsulated. It contains an optimized molar percentage of modified variants of PEG lipid (PEG2000-C-DMG) that is weakly anchored to the lipid nanoparticles and is gradually lost during systemic circulation to avoid hindering membrane destabilization problems due to high surface content of PEG lipids. DLin-MC3-DMA is a novel ionizable cationic lipid with an optimized pKa value that leads to dramatically enhanced endosomal escape and *siRNA* release, leading to a third-order intracellular targeting [49].

Most of the solid tumor exhibits leaky vasculature and impaired lymphatic drainage which results in enhanced vascular permeability and retention of liposomes within tumors referred to as EPR effect. The passive-targeting mechanism is closely related to EPR effects in which liposome particles permeate through interendothelial gaps (paracellular transport) of compromised blood vessels of tumors (Figure 4.5). To achieve the paracellular transport through passive-delivery system, few prerequisites need to be met which include:

- High-concentration gradient between blood and tumor interstitium.

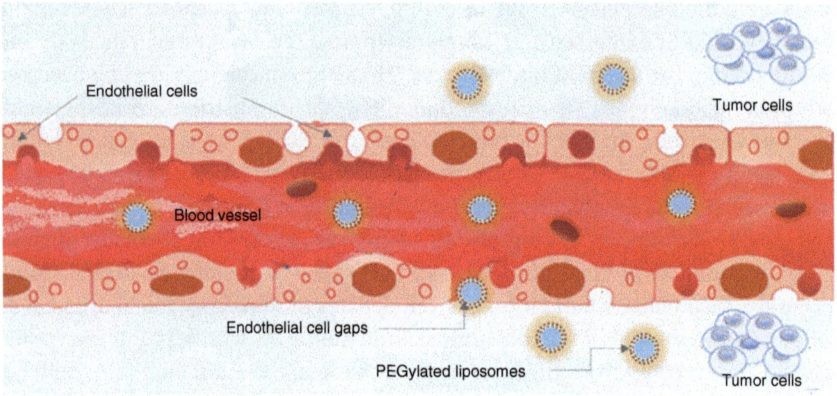

Figure 4.5 Schematic representation of passive-targeted delivery of liposome based on EPR effect.

- Long blood circulation times of liposomes.
- Smaller particle size than the cut-off size of tumor interendothelial gaps (preferable <100 nm).
- Sufficient level of drug loading.
- Stability of drug–lipid cargo to reach tumor tissues.
- Slow-release of drug into the target cells upon reaching the pathological tissue.

Passive mode of liposome drug delivery has several advantages and disadvantages, as listed in Table 4.3.

The liposome products based on passive-targeted delivery includes Ambisome®, Myocet®, DaunoXome®, Marqibo®, and Vyxeos®.

Ambisome® consists of hydrogenated soybean phosphatidylcholine (HSPC) (bilayer lipid): DSPG (drug complexation lipid in bilayer): Cholesterol (bilayer stabilizer lipid): Amphotericin B. It releases the amphotericin B only when the liposome binds to the fungus [113]. Due to modulated pharmacokinetics and fungal binding, it provides significant reduction in nephrotoxicity (18.7% vs 33.7%) compared to nonliposomal amphotericin.

DaunoXome® is a daunorubicin-loaded liposome consist of DSPC:Cholesterol (2 : 1) approved for HIV-related Kaposi sarcoma. It improved safety profile of

Table 4.3 Advantages and disadvantages of liposome based passive-targeted delivery.

Advantages	Disadvantages
• Relatively high drug distribution at therapeutic site through passive targeting compared to conventional drug product. • Reduced toxicity and increased overall safety profile. • May increase efficacy.	• Relatively gradual increase in target organ/tissue drug concentration compared to other organs, and drug accumulation in the liver and spleen. Highly variable EPR effect due to intratumoral heterogeneity, interpatient variability, and animal to human variability.

daunorubicin by reducing drug distribution in the nontarget tissues. Myocet doxorubicin NPL is composed of EPC:Cholesterol (55 : 45 molar ratio) (refer to Table 4.1). It is approved for combination therapy with cyclophosphamide in metastatic breast cancer. Based on EPR effect, it provides increased target distribution and reduced toxicities. It provides equivalent efficacy compared to doxorubicin conventional injection with low incidence of cardiac events (13% vs 29%) [114].

Marqibo® is a vincristine sulfate liposome approved as kit for acute lymphoblastic leukemia. At the time of administration, vincristine was entrapped in sphingomyelin and cholesterol empty liposome. It has improved efficacy compared to conventional injection and lower toxicity profile. Vyxeos® consists of cytarabine and daunorubicin (5 : 1) encapsulated in DSPC:DSPG:CHOL at the molar ratio of 7 : 2 : 1. It provides increased targeted exposure of daunorubicin and cytarabine in a fixed ratio, hence showing superior efficacy and comparable toxicity in high-risk acute myeloid leukemia. Vyxeos® is the first approved liposome based on dual-drug-loading technology [115].

Visudyne® was the first liposomal drug approved by US FDA in the year 2000 for the treatment of AMD. It is a non-conventional liposome encapsulating verteporfin and consists of BPD:EPG:DMPC;1.05 : 3 : 5 w/w/w. Wet AMD is an inflammatory condition of the posterior segment of the eye characterized by leaky neovascularization. Visudyne® passively accumulates in these abnormal blood vessels or neovasculature which is then destroyed by transpupillary red laser [116].

Another example is of amikacin loaded non-conventional liposomes, Arikayce® approved in 2018. Normally inhaled amikacin can be inefficient in treating mycobacterium avium complex as the bacteria produce a negatively charged substance that can trap the positively charged amikacin before it can reach bacteria [52]. The liposomal amikacin delivers amikacin directly to the site of infection in the lungs for immediate treatment and maintains the antibiotic locally in the lungs over time. In phase 3 results, 29% of patients receiving liposomal amikacin achieved culture conversion (sputum test) compared to just 8.9% of patients receiving standard care.

Nonstealth liposome products generally have low-circulation half-life due to MPS uptake. To achieve efficient passive targeting, liposomes should have stealth composition. As discussed in previous sections, the stealth liposome has an antifouling polymer layer on liposome (pegylation) which provides long-circulation half-life by avoiding the rapid clearance of liposomes from blood circulation and reducing the uptake of liposomes by macrophages [70]. Doxil® (also known as Caelyx®) is the first approved stealth liposome. The liposome bilayer of Doxil® consists of HSPC:Chol:DSPE-mPEG at a molar ratio of 56 : 39 : 5 along with optimum DOX–sulfate complexation in the internal aqueous phase. It has 46 hours long-circulation half-life because of PEGylation [14]. Doxil® significantly limits the distribution and elimination of doxorubicin, resulting in 5 to 11 times higher drug accumulation in Kaposi sarcoma lesions compared to conventional doxorubicin injection that leads to improved overall response rate up to 80%. It had a better safety profile compared to free doxorubicin in a Phase III trial of metastatic breast cancer, including reduction of cardiotoxicity (3.9% vs 18.8%), neutropenia (4% vs

10%), vomiting (19% vs 31%), and alopecia (20% vs 66%) whereas its efficacy is equivalent to free doxorubicin, PFS (6.9 vs 7.8 months), and OS (21 vs 22 months).

Doxil® is known to accumulate in the skin due to the polyethylene glycol coating because of which the principal dose-limiting side effect is the palmar–plantar erythrodysesthesia, often known as hand–foot syndrome [117]. Following administration of Doxil, small amounts of the drug can leak from capillaries in the palms of the hands and soles of the feet resulting in redness, tenderness, and peeling of the skin that can be uncomfortable and even painful. This side effect limits the Doxil dose administration as compared with doxorubicin in the same treatment regimen.

Another milestone in stealth liposome product development is the approval of Onivyde® irinotecan-loaded conventional liposomes as combination therapy for PDAC [47]. The bilayer consists of DSPC:Chol:DSPE-mPEG at molar ratio of 3 : 2 : 0.015. Onivyde® is primarily differentiated from other liposomal products in that the active (irinotecan) complexation is based on multivalent counter ions (octa valency) which provide relatively high drug to lipid ratio (1.36 : 1) that is almost 10-folds higher than other liposome products. Liposome composition protects irinotecan degradation and prolongs systemic retention, hence increasing the exposure of active metabolite (SN-38) in the tumor with reduced toxicity [47]. This liposomal formulation of irinotecan improves the pharmacokinetics of the drug by increasing drug encapsulation and loading efficiency, providing sustained release, protecting the drug in the active lactone configuration, prolonging circulation time, increasing tumor accumulation via the EPR effect, and re-routing the drug from sites of toxicity, such as the gastrointestinal tract, thereby reducing host toxicity. Based on the encouraging preclinical and clinical data available for the treatment of various solid tumors, Onivyde® was approved as a combination regimen for patients with gemcitabine-based chemotherapy-resistant metastatic pancreatic cancer by the US-FDA in October 2015. In addition, it is currently undergoing Phase II/III clinical trials for the therapy of many other solid tumors, such as small-cell lung cancer, colorectal cancer, and esophagogastric cancer. It indicates the potential antitumor activity of Onivyde (irinotecan liposomes) across a broad range of advanced solid tumors.

4.4.2 Active-Targeted Delivery

Active targeting is needed where the target is not only organ but the specific cells of the organ, such as tumor. Passive-targeted systems rely on the EPR effect and only accumulation at the tumor interstitium. The delivery inside tumor cell, therefore, depends on the liposome destruction and release of drug from liposome which then diffuses into the cell and exerts its effect. Hence, passive targeting may not be efficient. To address these limitations, ligand-targeted liposomes are under active investigation by the researchers as an area of interest and it is also termed as active targeting. The ligand that selectively targets a specific receptor is coupled to the surface of liposomes to increase tumor-selective delivery of liposome-encapsulated drugs [25]. It is suggested that the EPR effect leads to the accumulation of liposomes within tumor interstitium and the attached ligand promotes selective uptake. There are many receptors that are overexpressed on the tumor cells, such as transferrin receptors, folate receptors, epidermal growth factor receptor (EGFR), human epidermal growth receptor (HER), prostate-specific membrane antigen (PSMA), intracellular adhesion molecule-1 (ICAM-1), vascular endothelial growth receptor (VEGF), and $\alpha_v\beta_3$ integrin [20]. These receptors are being extensively explored to identify the targeting potential of ligands and the efficient delivery of drugs at site (Figure 4.6).

For successful targeting, firstly, the receptors should be overexpressed and uniformly distributed on the surface of cancer cells. Secondly, the interaction between the ligand on liposome surface and the receptor overexpressed on cancer cells must be sufficiently strong to be easily uptaken by the internalization of ligand–receptor occurs via receptor-mediated endocytosis [118]. Thirdly, the physicochemical properties of ligands, such as size, orientation, charge, and physical adsorption should be optimal. Finally, the type of ligand and its characteristics, such as specificity, density and affinity, and conjugation chemistry, are the other critical parameters that regulate the targeting efficiency of liposomes [119]. Ligands are divided into two categories – small molecules and macromolecules. Targeting moieties that have small size and low molecular weight are considered as small-molecule ligands. One of the examples of a small molecule is folic acid which has an affinity for folate receptors known to be overexpressed on tumor cells. Macromolecules for targeting, on the other hand, are large in size with a high-molecular weight, such as proteins, peptide, antibodies, and antibodies fragment [120]. There are two cellular targets for active targeting in cancer, viz. (i) cancer cells and (ii) tumor endothelium.

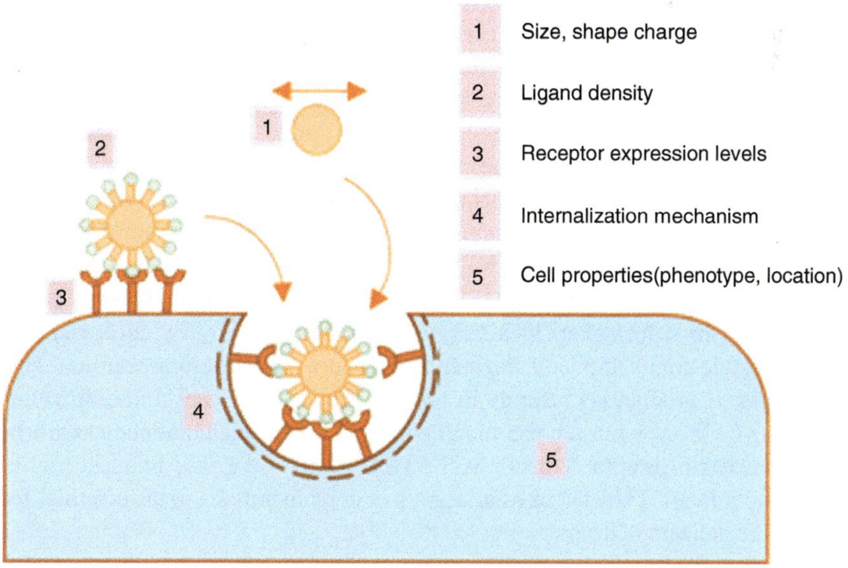

Figure 4.6 Schematic representation of active-targeting liposomes through receptor-mediated uptake.

4.4.2.1 Cancer Cell Targeting

In cancer cell targeting, there is enhanced cellular internalization of targeted liposomes rather than tumor accumulation. Cellular targeting causes direct killing of cells by releasing the encapsulated drug after internalization of targeted liposomes. The following receptors and specific cell surface moieties have been studied extensively for cell targeting, namely (i) transferrin receptor, (ii) folate receptor, (iii) EGFR, (iv) HER-2, (v) PSMA, and (vi) ICAM-1.

4.4.2.1.1 Transferrin Receptor

Transferrin (Tf) is a transmembrane glycoprotein with a molecular weight of 80 KDa responsible for transporting iron to cells throughout the body. Transferrin binds to the transferrin receptor (TfR) and internalizes into the cell via receptor-mediated endocytosis. As iron is vital for energy production, heme synthesis, and cell proliferation, Tf-R is overexpressed on cancer cells to fulfill their need to proliferate. Therefore, Tf-R is being extensively used as a target for tumor-specific drug delivery. In one study, surface modification of conventional liposomes of oxaliplatin with transferrin increased the tumor accumulation of oxaliplatin compared to untargeted liposomes, with the highest tumor inhibition growth in C26 colon carcinoma bearing mice [121]. Similarly, transferrin-conjugated doxorubicin-loaded liposomes resulted in higher doxorubicin delivery to tumors in mice study and thereby tumor growth inhibition in liver cancer compared to nontargeted doxorubicin-loaded liposomes [122]. Transferrin as a ligand has also been explored along with cell-penetrating peptides for brain delivery where crossing blood–brain barrier is a challenge for drugs. Wang et al. prepared cell-penetrating peptide and Tf-modified doxorubicin-loaded liposomes, where the survival time of U87 glioma-bearing mice was longer compared to nontargeted liposomes [123]. Similarly, paclitaxel-loaded liposomes modified with Tf and arginine-glycine-aspartic acid peptide demonstrated highest brain distribution in mice-bearing C6 orthotopic glioma [124]. Holo-Lf, a ligand of TfR, has been conjugated to the surface of liposomes and has improved the chemotherapeutic effect of drugs loaded in the liposomes [125]. Currently, there are three products based on transferrin ligand-mediated drug delivery that are under clinical trials revealing the true potential of targeting TfR. MBP-426 is a liposomal system conjugated with transferrin for the delivery of oxaliplatin is in Phase II clinical trial for treating gastric cancer [126]. SGT-94 is a systemically administered complex composed of the RB94 gene (plasmid DNA) encapsulated in a liposome. It is targeted to tumor cells by an antitransferrin receptor single-chain antibody fragment attached to the liposome surface. This investigational product is currently in Phase 1 clinical trial [127]. Liposomal formulation of wild-type human p53 modified with a single-chain antibody fragment to the transferrin receptor TfRscFv SGT-53 has completed Phase 1clinical trial in solid tumors [128]. Transferrsin as a target has demonstrated a great potential for site-specific delivery of drugs.

4.4.2.1.2 Folate Receptor

Folate receptor is a membrane protein with molecular weight of 38–40 kDa and is abundantly expressed on the surface of cancer cells. FR has two distinct isoforms (α and β) in humans. Isoform β has been found on CD34+ cells and does not show affinity for folic acid or its derivatives [129]. On the other hand, isoform α is frequently overexpressed on the surface of broad range of human tumor cells, such as brain, ovarian, breast, and lung cancer [130]. Folic acid, as a ligand, has an edge over other targeting moieties because of its various properties. Firstly, it has a strong-binding affinity (K_D = ~10^{-10} M) for FR even after conjugation. Secondly, the conjugation chemistry of folate on the surface of liposomes has been worked out [119]. Besides targeting the tumor tissues, folic acid is also evaluated to target other inflammatory diseases, (due to uptake by macrophage) such as rheumatoid arthritis, psoriasis and Crohn's disease as reported by Low et al. [131] Several studies have revealed that the folic acid can be used to deliver therapeutic and imaging agent to human cancer cells [132]. A number of studies have also confirmed that the folic acid-conjugated liposomes improve cellular uptake in various cancer cell models. Folate-targeted irinotecan-loaded liposomes showed selective targeting in MDA-MB-231 xenografts model with minimum systemic side effects [133]. In a study by Peres-Filho et al., liposome formulation co-encapsulating paclitaxel and imatinib and targeted to folate receptor had significant reduction in MCF7 and PC-3 cell viability compared to nontargeted liposomes and free drugs [134]. Folate-mediated drug targeting also has some limitations. Even though FR is overexpressed in cancer patients, this expression varies from as low as 17% in testicular cancer to 90% in ovarian cancer [132]. In the field of organ imaging, EC17 (Folate–FITC conjugate targeting folate receptor) and OTL38 Folate receptor alpha (Fra)-targeting ligand (folic acid) conjugated to NIR fluorophore S0456) are under Phase-I trial for detecting breast, lung, renal, and ovarian cancer [135]. Moreover, a vaccine multi-epitope folate receptor alpha peptide vaccine, sargramostim (GM-CSF) is in Phase II clinical trials to prevent recurrence of stage 1-3 triple-negative breast cancer [136]. Phase III clinical trial is ongoing in platinum-resistant ovarian cancer to evaluate the efficacy and safety of the combination of EC145 and conventional liposomal doxorubicin. EC145 is a drug that is specifically designed to enter cancer cells via the folate vitamin receptor [137]. Although there are no FR-targeted liposomes in the pipeline, successful preclinical activity makes it a potential candidate for clinical translation.

4.4.2.1.3 Epidermal Growth Factor Receptor

Epidermal growth receptor (EGFR) is a tyrosine kinase receptor belonging to the family of ErbB receptors. EGFR stimulates various processes, such as cell proliferation, differentiation, angiogenesis, and invasion, required for tumor growth [138]. EGFR is overexpressed in many solid tumors, including breast, colorectal, non-small-cell lung cancer, ovary, kidney, head, neck, pancreas, and prostate cancer [139]. Due to overexpression in various tumors, EGFR has become an attractive target for specific tumor drug delivery. Currently, anti-EGFR therapy includes monoclonal antibodies (cetuximab and panitumumab) and tyrosine kinase inhibitors (gefitinib and erlotinib) for the treatment of various solid tumors [140].

EGFR-targeted monoclonal antibodies conjugated to liposomal systems are being studied for active tumor targeting. Among EGFR monoclonal antibodies, cetuximab has been explored as a targeting ligand on various nanocarriers, including human serum albumin nanoparticles [141], silica nanoparticles [142], and liposomes [143]. For instance, cetuximab-modified liposomes of oxaliplatin were prepared and studied in colorectal cancer xenograft model. The result demonstrated that the EGFR-directed oxaliplatin liposomes showed an increased tumor drug delivery and efficacy over nontargeted drug-loaded liposomes and cetuximab [144]. Moreover, cetuximab as a targeting agent has also been explored for imaging in vitro. Here, cetuximab-targeted fluorescent liposomes were found to have 3.5-fold binding to A431 cells (epidermal cancer cell with EGFR overexpression) as compared to IEC-6 (normal enterocytes with physiological EGFR levels), thus likely to have good *in vivo*-targeting ability [145]. Currently, C225-ILs-dox (doxorubicin-loaded liposomes with cetuximab fragment covalently conjugated on liposome membrane) is in Phase 2 clinical trial [146]. Owing to the success of EGFR targeted monoclonal antibody therapies this seems to be a potential ligand for a successful targeting.

4.4.2.1.4 Human Epidermal Growth Factor Receptor

Human epidermal growth factor receptor (HER) family of receptors is made up of four main members: HER-1, HER-2, HER-3, and HER-4, also called ErbB1, ErbB2, ErbB3, and ErbB4, respectively. Like EGFR, HER also regulates cell proliferation, differentiation, and survival. HER2 is overexpressed in many solid tumors, including invasive breast cancers (15–30%), gastric cancer(10–30%), ovarian cancer (20–30%), endometroid carcinoma (21–47%), and esophageal cancer (0–83%) [147]. Therefore, it has attracted a large interest in targeted therapy for various cancers. For HER-2 targeting, approved monoclonal antibodies include trastuzumab and pertuzumab whereas, margetuximab is under phase III clinical trial. Trastuzumab emtansine is an antibody–drug conjugate of monoclonal antibody (trastuzumab) and the chemotherapeutic maytansinoid (emtansine) and is approved by US FDA for targeted therapy of HER-2 positive tumor [148]. Among the tyrosine kinase inhibitors, lapatinib, afatinib, neratinib, and tucatinib are approved therapies for cancer, and pyrotinib and pziotinib are in Phase III and Phase II clinical trials, respectively [149]. Various anti-HER-2 monoclonal antibodies and their fragments attached on the surface of nanocarriers are being studied for targeting ability toward tumor. For instance, Rapamycin and Paclitaxel immunoliposome functionalized with trastuzumab were prepared. These immunoliposomes were better able to control tumor growth *in vivo* compared to Paclitaxel/Rapamycin solution and non-functionalized liposomes [150]. MM-302 is a stealth liposomal formulation of doxorubicin conjugated with an anti-HER2-targeted antibody that delivers doxorubicin to tumor cells and limits exposure to healthy cells, such as cardiomyocytes. This liposomal formulation showed promising clinical activity in Phase I [151]; however, it failed in subsequent Phase II trials due to poor efficacy [152]. Following the identification of the root cause of its failure, targeting of HER2 was achieved using a lab-synthesized anti-HER2 antibody as reported by Rodallec et al. [153]

Trastuzumab (Herceptin®), anti-HER2 monoclonal antibody (approved in 1998), has considerably improved the poor clinical outcome and survival of $HER2^+$ breast cancer patients when associated with docetaxel [154]. Therefore, development of Trastuzumab-modified liposomes of taxane represents a more promising strategy for treating breast cancer, as Transzumab is well-known and fully characterized antibody to explore as ligand for liposome anchoring with strong-targeting (HER-2) evidence rather than using other investigational antibodies [155].

4.4.2.1.5 Prostate-Specific Membrane Antigen

Prostate-specific membrane antigen (PSMA) is a transmembrane protein with a molecular weight of 100 KDa and is also known as folate hydrolase 1. PSMA is weekly expressed in normal prostate; however, it is up-regulated in prostate cancer [156]. PSMA expression is also known to express in the neovasculature of many other solid tumor malignancies [157]. This unique expression of PSMA makes it an attractive target for imaging and specific delivery of therapeutic agents [158]. A range of PSMA ligands, including antibodies, aptamer, peptide, and small molecules, are being explored for tumor-specific drug delivery [159]. Prostascint® is a *radiolabeled* monoclonal antibody to PSMA approved for the imaging of prostate cancer [160]. In one study, liposomes loaded with the α-particle generator (225)Ac and functionalized with PSMA J591 antibody or with the A10 PSMA aptamer demonstrated selective killing of PSMA-expressing cells [161].

Recently, Yari et al. have worked on targeting PSMA where they synthesized lipopolymer (P3) comprising of PSMA ligand, polyethylene glycol (PEG2000), and palmitate and postinserted into the surface of liposomes followed by doxorubicin loading and tagging with 99mTc radionuclide. Cytotoxicity study of these functionalized liposomes demonstrated higher toxicity compared to nontargeted doxorubicin liposomes in LNCaP cells (PSMA expressing) [162]. Apart from liposomes, BIND-014 (polymeric nanoparticles containing docetaxel and modified with small-molecule PSMA ligand ACUPA) targets PSMA. BIND-014 showed good preclinical activity [163] and was found to be well-tolerated with predictable toxicity in Phase I clinical trial [164]. However, it failed to achieve primary endpoints in Phase-II clinical trials. Learning lessons from the failure of this novel product that are critical for achieving better therapeutic outcomes, as described by BIND therapeutics, were preselection of patients based on the extent of EPR, the presence of target receptor and tumor heterogeneity, the ability of the actively targeted NC to bind to the target receptor, and the need for companion diagnostics (measuring the level of prostate-specific antigen in blood) [164, 165]. Merrimack pharmaceutical determined the tumor accumulation of FMX iron nanoparticles as a marker to predict response to novel nanoliposomal irinotecan (nal-IRI). This study showed that the tumor with high accumulation of FMX was more responsive to nal-IRI indicating the importance of patient selection via prediagnosis [166]. Although, currently, there are no PSMA targeted liposomes under clinical trial, it is expected to have a few in the near future as evidenced by the fast pace of research in the field of PSMA targeting using liposomes.

4.4.2.1.6 Intracellular Adhesion Molecule-1

Intracellular adhesion molecule-1 (ICAM-1) is a transmembrane glycoprotein (80–115 kDa) found in endothelial and immune cells. At the site of inflammation, cytokines such as interleukin 1β (IL1-β), tumor necrosis factor-alpha (TNF-α) and lipopolysaccharide (LPS), endothelial cells up-regulate ICAM-1 on the cell surface to help the firm adhesion and extravasation of leukocytes through the inflamed endothelium into diseased tissue via transcellular and paracellular pathways. Tumors overexpress ICAM-1 due to chronic inflammation compared to the normal tissue [167] and thus it is an emerging target for specific drug delivery.

The therapeutic potential of targeting ICAM-1 was demonstrated by the synthesis of ICAM-1-targeted immunoliposomes. Guo et al. synthesized pH-sensitive liposomes of Lipocalin-2 siRNA conjugated with ICAM-1 mAb and demonstrated that these liposomes achieved 3.5-fold higher cellular uptake in the triple-negative breast cancer cell line MDA-MB-231 compared to nontargeted immunoliposomes [168]. Approach of targeting ICAM-1 using liposomes functionalized with ICAM-1 ligand holds a potential to emerge as a product in the near future.

4.4.2.2 Tumor Endothelium Targeting

Unlike targeting cancer cells via receptor on its surface, targeting of tumoral endothelium deprives cancer cell of oxygen, nutrients, and results in cell death. Inhibiting the supply of blood to tumor also reduces metastatic capabilities of tumor and thus tumor shrinks [168]. Tumor endothelium targeting has an edge as tumor vascular network is always more accessible to circulating liposomes compared to cells localized in the tumor interstitial matrix. Also, tumor endothelial cells are more stable than cancer cells, therefore, the risk of developing resistance is lower [169]. The following proteins are the main candidate for targeting tumoral endothelium by liposomes.

4.4.2.2.1 Vascular Endothelial Growth Factor

Vascular endothelial growth factor is an angiogenic factor that regulates the process of angiogenesis and is up-regulated in many tumors. VEGF and its receptors, VEGFR-1 and VEGFR-2, are overexpressed in bone marrow failure states, such as chronic and acute leukemia and a variety of solid tumors [170]. There are two approaches to target angiogenesis via the VEGF pathway. The first approach is targeting VEGFR-2, which reduces VEGF binding through endocytosis of VEGFR-2. The second approach focuses on the targeting of VEGF which inhibits its binding to VEGFR-2 [171]. Bevacizumab (a human mAB) is a VEGF inhibitor and was first approved drug for treatment of cancer. Based on the potential applications of targeting VEGF and VEGFR-2, various studies are being performed to target tumors by modifying the surface of nanocarriers with anti-VEGF ligands. For instance, Shein et al. developed liposomes that were covalently conjugated with murine monoclonal antibodies against VEGF. These anti-VEGF liposomes were extremely specific for $VEGF^+$ tumor cells *in vitro* and *in vivo*. The intravenous injection of such anti-VEGF-liposomes in rats bearing intracranial C6 glioma showed accumulation in tumor and engulfment by glioma cells, confirming brain targeting [172]. Later, Shein et al. prepared the liposomes of cisplatin analog that were functionalized with mAB against VEGF and VEGFR-2 to target glioblastoma. These targeted liposomes demonstrated high-loading capacity, prolonged circulation time, and high accumulation in glioma compared to the conventional cisplatin formulation

[173]. VEGF-targeted liposomes are being explored and are expected to enter into clinical trials in the near future due to their promising preclinical activity.

4.4.2.2.2 The $\alpha_v\beta_3$ Integrin

The integrin $\alpha_v\beta_3$, consists of a 125-kDa α_v subunit and a 105-kDa β_3 subunit. Integrin $\alpha_v\beta_3$ binds to a range of molecules present in extracellular matrices, such as fibronectin, fibrinogen, von Willebrand factor, vitronectin, and proteolyzed forms of collagen and laminin, with an Arg-Gly-Asp (RGD) triple-peptide motif [174]. In addition to interacting with a number of ECM proteins, integrin αvβ3 has been shown to be associated with fibroblast growth factor-2, metalloproteinase MMP-2, activated PDGF, insulin, and VEGF receptors. Integrin αvβ3 has diverse roles in several distinct processes, such as angiogenesis, pathological neovascularization, and tumor metastasis [175]. Therefore, critical roles of integrin αvβ3 in tumor angiogenesis have led to a promising strategy to block integrins as this would inhibit the tumor angiogenesis enhancing the efficacy of other therapeutics. The activity of αvβ3 integrin can be inhibited by monoclonal antibodies (LM609) and cyclic RGD peptide antagonists (cilengitide) [176]. Schiffelers et al. formulated polyethylene-coated doxorubicin liposomes and conjugated them with RGD peptide with the affinity for the integrin. These targeted liposomes inhibited tumor growth in a doxorubicin-insensitive murine C26 colon carcinoma model, whereas nontargeted doxorubicin liposomes failed to slow down the tumor growth [177]. Therefore, $\alpha_v\beta_3$ integrin seems to be a potential target for specific drug delivery via liposomes conjugated with $\alpha_v\beta_3$ integrin antagonists.

4.4.2.2.3 Other Important Targeted Liposomes

Phosphatidylserine is an anionic phospholipid that is present in all cells and constitutes approximately 2–10% of total cellular lipid [178]. Generally, PS is localized in the inner leaflet of plasma membrane facing cytosol mediated by a membrane protein or ATP-dependent transporter, aminophospholipid translocase [179]. However, certain conditions, such as oxidative stress and tumor microenvironment, can cause translocation of PS to the outer leaflet of plasma membrane and thus is considered as a marker of cancer cells [180]. Cancer cells have an elevated amount of PS on the outer surface and therefore, it has generated a lot of interest for PS as a target not only in drug delivery but also in imaging [181]. In the area of targeted liposomes for PS, Annexin-A5 as a ligand is being explored by many scientists. For instance, Garnier et al. functionalized conventional liposomes with Annexin-A5, and these targeted liposomes were able to specifically bind to PS-containing membrane with high efficacy [182]. In one of the other interesting studies, cationic liposomes containing phosphatidylcholine-stearylamine were synthesized that strongly bind to PS exposed on the surface of cancer cells. Anticancer drugs irinotecan and doxorubicin when encapsulated into these cationic liposomes demonstrated high efficacy compared to free drugs *in vitro* and *in vivo*. Moreover, delayed tumor growth and potential enhancement of the survival of the animals without any substantial sign of toxicity or relapse of the tumor were observed [183]. PS ligand Annexin-A5 has also been evaluated for the delivery of Avastin (bevacizumab) across corneal epithelial barrier. Here, bevacizumab-loaded phospholipid vesicles were prepared and studied for corneal epithelium permeation in association with Annexin-A5. The study demonstrated that Annexin-A5 enhances the delivery of topically administered phospholipid vesicles encapsulated Avastin to the posterior chamber of eye in rats and rabbits when compared to Avastin phospholipid vesicles in absence of Annexin-A5 [184]. These studies suggest Annexin-A5 as a potential emerging candidate for specific delivery of lipid-based vesicles.

4.4.2.2.4 siRNA and mRNA Lipid Nanoparticles

siRNA therapy has emerged as a potential mode for the treatment of diseases from the last few years. However, siRNA as such (naked) is highly unstable and prone to degradation by nucleases in biological fluids and thus, cannot cross the cell membrane and accumulate in tumor [185]. siRNA delivery is also a challenge due to its high-molecular weight, anionic charge, and hydrophilicity, which prevents passive diffusion across the plasma membrane of most cells. Various drug-delivery platforms have been studied among which, lipid nanoparticles with cationic/ionizable lipid are the leading and promising for siRNA delivery [186]. Lipids are able to self-assemble into well-ordered nanoparticle structures known as lipoplex and cationic/ionizable lipids are capable to bind electrostatically with siRNA [187]. Other hydrophobic moieties, such as cholesterol and PEG-lipids, when added to cationic/ionizable lipids significantly enhance the efficacy of the RNA delivery.

Onpattro® (Patisiran) is LNPs formulation consisting of siRNA directed against transthyretin encapsulated in lipids is the first commercially available product for the treatment of polyneuropathy of hereditary transthyretin-mediated amyloidosis in adults [78]. The translation of LNP-enabled siRNA systems for the treatment of hATTR amyloidosis in humans was established in two stages of clinical development. The first generation DLinDMA-based formulation was evaluated in a placebo-controlled phase 1 trial to determine safety and efficacy of ALNTTR01 after administration of a single dose, ranging from 0.01 to 1 $mg\,kg^{-1}$. In this phase 1 study, it was observed that NHP studies seemed to provide a reasonable prediction of human efficacy (approximately 50% mean TTR reduction observed in NHPs at 1 $mg\,kg^{-1}$). In addition, ALNTTR01 showed an encouraging safety profile with no drug-related SAEs in subset of participants. The single dose of ALNTTR01 at 1 $mg\,kg^{-1}$ led to a mean reduction in serum TTR levels of 38% compared to the placebo, with one patient achieving a substantial TTR reduction of >80%. The clinical data with Onpattro® served as validation of RNAi approach, the siRNA and the LNP platform in a real-patient clinical trial for the first time. The success of Onpattro® has shown promise in other investigational candidates ALN-VSP, TKM-PLKs, ALN-PCS [188–190].

Messenger RNA (mRNA) is a pivotal molecule of life, involved in almost all aspects of cell biology. mRNA has only recently come into the focus as a potentially powerful drug class able to deliver genetic information. Synthetic mRNA can be engineered to resemble mature and processed mRNA molecules as they occur naturally in the cytoplasm of eukaryotic cells and to transiently deliver proteins.

Recent advances addressed challenges inherent to this drug class and provided the basis for a broad spectrum of applications. A major mRNA application has emerged as a vaccine to a significant extent. Vaccines help to prepare antibodies in the body to against foreign pathogens, such as bacteria or viruses, to prevent infection. All vaccines introduce a harmless piece of a specific bacteria or virus, into the body to elicit an immune response. The harmless piece contain bacteria or viruses that have been weakened or killed. The scientists have developed a new type of vaccine that uses mRNA rather than part of an actual bacteria or virus.

mRNA vaccines function by delivering a short amount of mRNA that correlates to a viral protein, usually a protein located on the virus's outer membrane. Cells make the viral protein using this mRNA blueprint. The immune system identifies that the protein is foreign and creates specialized proteins called antibodies as part of a normal immunological response. Antibodies aid in the body's defense against infection by identifying particular viruses or other pathogens, adhering to them, and marking them for elimination. Antibodies persist in the body after the virus has been eliminated, allowing the immune system to respond promptly if it is exposed again. If a person is exposed to a virus after getting an mRNA vaccine, antibodies will immediately recognize it, attach to it, and label it for destruction before it may cause any harm.

The World Health Organization declared coronavirus disease 2019, which is caused by the coronavirus 2 that causes severe acute respiratory syndrome, a pandemic in March 2020. A vaccination was urgently needed due to the rapidly increasing number of illnesses and deaths reported around the world. Soon, the safety, tolerability, and immunogenicity data from a placebo-controlled, observer-blinded dose-escalation study among 45 healthy adults (18–55 years old) who were randomized to receive two doses of 10, 30, or 100 g of vaccine named BNT162b1 separated by 21 days were reported. This nucleoside-modified mRNA lipid nanoparticle vaccine encodes the trimerized receptor-binding domain (RBD) of the spike glycoprotein of the ***COVID-19 virus***. Local reactions and systemic events were dose dependent, generally mild to moderate, and transient. A second vaccination with 100 μg was not administered because of the increased reactogenicity and a lack of meaningfully increased immunogenicity after a single dose compared with the 30 μg dose. RBD-binding IgG concentrations and ***COVID-19 virus*** neutralizing titers in sera increased with dose level and after a second dose. Geometric mean neutralizing titers reached 1.9–4.6 fold that of a panel of COVID-19 convalescent human sera, which were obtained at least 14 days after a positive SARS-CoV-2 PCR. These results support further evaluation of this mRNA vaccine candidate and eventual vaccine approval for clinical use.

Furthermore, Phase 3 data with BioNTech and Pfizer mRNA vaccine met primary endpoints in November 2020. Primary efficacy analysis demonstrated BNT162b2 to be 95% effective against COVID-19 beginning 28 days after the first dose;170 confirmed cases of COVID-19 were evaluated, with 162 observed in the placebo group versus 8 in the vaccine group. Efficacy was consistent across age, gender, race, and ethnicity demographics; observed efficacy in adults over 65 years of age was over 94%. Safety data milestone required by US FDA for Emergency Use Authorization was achieved. Phase 3 data demonstrated that the vaccine was well-tolerated across all populations with over 43 000 participants enrolled; no serious safety concerns were observed. The only Grade 3 adverse event greater than 2% in frequency was fatigue at 3.8% and headache at 2.0%.

Another success story of mRNA vaccine based on LNPs was evidenced by **Moderna**. The mRNA-1273 vaccine is an LNP encapsulated mRNA-based vaccine that encodes the prefusion-stabilized full-length spike protein of the severe acute respiratory syndrome coronavirus 2 (COVID-19 virus). The results of a phase 3 randomized, observer-blinded, placebo-controlled trial that was conducted at 99 centers across the United States was reported by Moderna. Persons were randomly assigned in a 1 : 1 ratio to receive two intramuscular injections of mRNA-1273 (100 μg) or placebo 28 days apart who were at high risk for SARS-CoV-2 infection or its complications. The primary endpoint was to avoid COVID-19 sickness in patients who had never experienced SARS-CoV2 infection, with symptoms appearing at least 14 days following the second injection.

The trial 30 420 people were registered who were randomly assigned to receive either vaccine or placebo (in 1:1 ratio, 15 210 participants in each group). More than 96% of participants received both injections, and 2.2% had SARS-CoV-2 infection at baseline either serologically, virologically, or both. 185 participants in the placebo group (56.5 per 1000 person-years; 95% confidence interval (CI), 48.7 to 65.3) and in 11 participants in the mRNA-1273 group (3.3 per 1000 person-years; 95% CI, 1.7 to 6.0) were found to have Symptomatic COVID-19 illness. Vaccine efficacy was 94.1% (95% CI, 89.3 to 96.8%; $P < 0.001$). Key secondary analyses showing efficacy were assessed 14 days after the first dosage, included people who had SARS-CoV2 infection at baseline, and included participants 65 years of age or older. All 30 people who experienced severe COVID-19, including one fatality, were in the placebo group. The mRNA-1273 group experienced moderate, transitory reactogenicity more frequently after vaccination. Serious adverse effects were uncommon, and the incidence was comparable in the two groups

The COVID19 vaccine's effectiveness at preventing sickness, including severe disease, was 94.1%. Other than brief local and systemic reactions, there were no safety issues found. The successful clinical translation of two different mRNA vaccines based on LNPs has paved the way for exploring mRNA as innovative medicine for major global diseases, including cancer and immune disorders.

4.5 Recent Clinical Trials with Liposomes with Investigational Liposome Candidates

In the past few decades, liposomes have been extensively explored for the treatment of various diseases. There are a number of lipid-based liposomal products that have been commercialized. As evidenced by the recent clinical trials, lipid-based carriers and liposomes are very promising in a wide range of pathological conditions. Many products that have been studied in one pathological condition are also being explored for other indications. Although there are number of ongoing clinical trials based on LNPs and liposomes, focus of development has now been shifted toward targeted liposomes, siRNA and mRNA-based vaccines, and gene delivery. A few ongoing trials are listed in Table 4.4. The compiled investigational liposomes and lipid-based drug-delivery system here are classified according to the targeting mechanism, such as passive and active targeting. The products under passive targeting also include the delivery at the site of action that can be applied locally and given by nasal route. Also, lipid-based drug delivery systems have become important carrier in vaccine development and interest in liposomal vaccines have markedly increased because of the development of products Epaxal® and Inflexal® V for vaccination against hepatitis and influenza, respectively. Shingrix® (an anti-herpes zoster subunit vaccine based on a liposome, approved in 2017) and Onpattro® (siRNA-based vaccine for polyneuropathy, approved in 2018) are recent clinical successes in the field of liposomal vaccines. Many other products, such as Stimuvax® and VaxiSome®, have completed Phase 2 trials with positive outcomes paving the future of liposomal vaccines in the treatment of various diseases.

4.6 Factors Influencing the Clinical Translation of Liposomes for Targeted Delivery

The liposome products have been established as a promising nanomedicine platform in the pharma industry. It is obvious, the pharma industry's decision to perceive liposome-based preclinical assets to commercial assets is based on the business forecasts, unmet needs, the development costs, including CMC, clinical development, and marketing costs. So far, liposome-based drug products are approved only for repurposing of the molecule strategies to address issues associated with the drug molecule, such as organ toxicities or shorter half-life.

However, it is also important to use liposome-based products for investigational drugs to explore full potential of these delivery systems. US FDA has approved liposome products for diverse therapeutic classes with oncology being a major area and also it is evident from investigational products liposomes are in frontrunner when it comes to nano-based delivery systems [191]. The established clinical benefits of liposome products are based on the fact that the increased drug accumulation in pathological tissue provided by liposomes allows the effective dose of a drug to be reduced, and reduce the toxic side effects [192]. It is evident, as low as 0.01% of the injected dose of angstrom-sized agents accumulates in a target tissue when compared to 1–5% for liposomal drugs. Hence, better accumulation, as well as targeted release, will enable dose reduction and consequent side effects.

The clinical translation of liposome-based products is relatively complex, expensive, and time-consuming in comparison to conventional formulation technologies. The key points to consider for clinical translation includes proposed target indication, drug molecule of interest, expected product differentiation, cost-effectiveness, and timeline to commercialize the product [193]. Since liposomal formulation feasibility is largely driven by drug's physicochemical properties, it should fit into criteria

Table 4.4 Investigational liposome-based candidates.

Candidate	Molecule	Indication	Administration	Targeting mechanism	Sponsor name	Status
Passive-targeting encompass liposome-based candidates with conventional liposomes, non-conventional liposomes, topical, nasal						
2B3-101	Doxorubicin	Solid tumor	Intravenous	Passive PEG	2-BBB therapeutic	Phase 2
SPI-077	Cisplatin	Lung, head, and neck cancer	Intravenous	Passive, PEG	Alza Corporation	Phase 2
ATI-1123	Docetaxel	Solid tumors	Intravenous	Human serum albumin that facilitates tumor targeting	Azaya Therapeutics	Phase 2
Atragen	Tretinoin	Solid tumors	Intravenous	Passive, NPG	Antigenics	Phase 2
BP-100-1.02	Bcl-2	Lymphoma	Intravenous	NF	Bio-path Holding	Phase 1
CPX-1	Irinotecan	Solid tumors	Intravenous	Passive	Celator Pharmaceuticals	Phase 2
C-VI SA bikDD	BikDD	Pancreatic cancer	Intravenous	Cationic lipid	M.D. Anderson Cancer Center	Phase I withdrawn
EndoTAG-1	Paclitaxel	Solid tumors	Intravenous	Cationic lipid	Syncore biotechnology	Phase 2 completed
IHL-305	Irinotecan	Solid tumors	Intravenous	Passive, PEG	Yakult Honsha	Phase 1 completed
INGN-401	DOTAP: Chol-fus1	Lung cancer	Intravenous	Passive	Genprex	Phase 1
l-annamycin	Annamycin	Acute lymphocytic leukemia	Intravenous	Passive, NPEG	Aronex Pharmaceuticals	Phase 2
L-Grb-2	Grb2 antisense oligodeoxynu-cleotide	Leukemia	Intravenous	NF	Bio-Path Holdings	Phase 1
Lipoplatin	Cisplatin and its analog	Solid tumors	Intravenous	Passive PEG	Regulon and Medison Pharma	Phase 1 terminated
Lipotecan	Camptothecin	Solid tumors	Intravenous	Passive	TLC BIO	Phase 1 completed
Mepact	L-MTP-PE	Osteosarcoma	Intravenous	Local, passive	Medison Pharma; Takeda America Holdings	Phase 2, Approved
MMII	NF	Osteoarthritis	Intraarticular	Local	Sun Pharma-Moebis	Phase 2 Ongoing

Nanocort	Prednisolone	Rheumatoid arthritis	Intravenous	Passive, PEG	Enceladus Pharmaceuticals	Unknown
PNT2258	Oligonucleotide	Cancer	Intravenous	Anionic and pH tunable lipid	ProNAi Therapeutics	Phase 2 completed
SapC-DOPS	Saposin C	Solid tumors	Intravenous	SapC-DOPS targets phosphatidylserine, an anionic phospholipid preferentially exposed in the surface of cancer cells and tumor-associated	Bexion Pharmaceuticals	Phase 1 completed
S-CKD602	Camptothecin analog	Advanced malignancies	Intravenous	Passive, PEG	Chong Kun Dang	Phase 1 completed
ThermoDox	Doxorubicin, lyso-thermosensitive	Solid tumors	Intravenous	Temperature sensitive, PEG	Celsion Corporation	Phase ½ completed
Dimericine	T4N5	Precancerous condition	Topical	Local	AGI Dermatics, Inc	Phase 2 completed
NanoDOX™	Doxycycline monohydrate	Foot ulcer, diabetic	Topical	Local	Nanotherapeutics	Phase 2 completed
AeroLEF	Fentanyl	Pain relief	Aerosol	Local, rapid onset-sustained delivery	YM Bioscience	Phase 2 completed
L9NC	rubitecan	Solid tumors, Lung	Aerosol	Local	Clincosm	Phase 2 completed
pGT-1 gene liposome	pGT-1 gene	Cystic fibrosis	Nasal	Cationic liposome	University of Alabama at Birmingham	Phase 1 completed in 1999
L-CsA	Cyclosporine	Bronchiolitis obliterans	Interstitial	Inhalation use	Breath Therapeutics	Phase 3
Vaccines						
Lipovaxin-MM	Vaccine	Melanoma	Intravenous	Active, DCR ligand-bearing liposome	Lipotek	Phase 1 completed, 2011
Stimuvax	BLP25 vaccine tecemotide	Solid tumors	S.C	Peptide vaccine	Biomira	Phase 2 completed
VaxiSome	CCS/C Vaccine	Influenza	Intramuscular	Vaccine	NasVax Ltd.	Phase 2 completed
CAF01	Vaccine	Tuberculosis	Intramuscular	Vaccine	Statens Serum Institut	Phase 1 completed
Lipo-MERIT	tetravalent RNA-lipoplex cancer vaccine	Melanoma	Intravenous	Vaccine	BioNTech SE	Phase 1

Table 4.4 (Continued)

Candidate	Molecule	Indication	Administration	Targeting mechanism	Sponsor name	Status
Passive-targeting encompass liposome-based candidates with conventional liposomes, non-conventional liposomes, topical, nasal						
RTS S/AS02	FMP2.1/AS02A	Malaria	Intramuscular	Vaccine	GlaxoSmithKline	Phase 2 completed
DPX-0907	Cancer vaccine	Neoplasms	Subcutaneous	Multipeptide vaccine	ImmunoVaccine Technologies, Inc. (IMV Inc	Phase 1 completed in 2010
NF	Novel Liposomal Based Intranasal Influenza Vaccine	Influenza	Nasal	Polycationic	Hadassah Medical Organization	Phase 2
NF	MPER-656 Liposome Vaccine	HIV Infections	Intramuscular injection	Peptide vaccine	National Institute of Allergy and Infectious Diseases (NIAID)	Phase 1
Lipovaxin-MM		Melanoma	Intravenous infusion	Novel dendritic cell-targeted liposomal vaccine, the attachment of recombinant proteins to the surface of liposome	Lipotek Pty Ltd.	Phase 1 completed
Active-targeted liposome drug candidates based on ligands, LNPs with siRNA and mRNA						
IL-2 LIPO	liposomal interleukin-2	Melanoma (Skin)	Intradermally	Local	NYU Langone Health	Phase 2 completed
PDS0101	Liposomal HPV-16 E6/E7 Multipeptide Vaccine PDS0101	Locally Advanced Cervical Squamous Cell Carcinoma	SC	Cationic lipid	PDS Biotechnology Corporation	Phase 2
MCC-465	Doxorubicin	Metastatic stomach cancer	IV	Active, PEG, antibody	Mitsubishi Tanabe Pharma Corporation	Phase 1
Atu027	siRNA	Solid tumors	Intravenous	Active, PLK1 targeting	Atugen AG	Phase 2
MM-302	doxorubicin	Breast cancer	Intravenous	HER2-targeted antibody–liposomal conjugate, PEG	Merrimack Pharmaceuticals	Phase 1
EphA2 siRNA	EphA2-targeting DOPC-encapsulated siRNA	Advanced Malignant Solid Neoplasm	Intravenous	Cationic lipid	M.D. Anderson Cancer Center	Phase completed 2012

TKM-PLK1	siRNA	Solid tumors	Intravenous	Active PLK1 targeting	Tekmira Pharmaceuticals Corp	Phase 2 completed
ALN-VSP	siRNA	Solid tumors	Intravenous	Targeting the gene for vascular endothelial growth factor (VEGF)	Alnylam Pharmaceuticals	Phase 1 completed
ALN-PCS	siRNA	Hypercholesterolemia	Intravenous	Intracellular PCSK9 targeting	Alnylam Pharmaceuticals	Phase 1 completed
CFTR gene liposome	*CFTR* gene	Cystic fibrosis	Nasal	Cationic lipid	University of Alabama at Birmingham	Phase 1 completed in 2000
JVRS-100	Immunostimulatory DNA	Leukemia	Intravenous	Cationic lipid–DNA complex	Juvaris BioTherapeutics	Phase 1 completed
LE-rafAON	LErafAON-ETU	Neoplasms	Intravenous	Cationic lipid	INSYS Therapeutics, Inc.	Phase 1 completed
TKM-ApoB	siRNA	Hypercholesterolemia	Intravenous	Cationic lipid, PEG	Tekmira Pharmaceuticals Corp	Terminated (Potential for immune stimulation to interfere with further dose escalation.)
TKM-Ebola	siRNA	Ebola	Intravenous	NF	Tekmira Pharmaceuticals Corp	Terminated (Corporate decision to reformulate the investigational product.)
MRX34	miR-34a	Advanced solid tumors and hematological malignances	Intravenous	Cationic lipid	Mirna Therapeutics, Inc.	Phase 2 terminated
CALAA-01	Sirna liposomes (CALAA-01)	Solid tumors	Intravenous	transferrin receptor-targeted anti-RRM2 siRNA CALAA-01, PEG conjugate	Calando Pharmaceuticals	Phase 1 terminated
SGT-53[a)]	P53 plasmid DNA liposomes	Solid tumors	Intravenous	Cationic liposomal, tumor-targeting p53 (TP53) gene delivery	SynerGene Therapeutics, Inc.	Phase 2

a) SGT-53: Delivery system with potential anti-tumor activity. Transferrin receptor-targeted liposomal p53 cDNA contains plasmid DNA encoding the tumor suppressor protein p53 packaged in membrane-like liposome capsules that are complexed with antitransferrin receptor single-chain antibody (TfRscFv); NF information not available.

of encapsulation either by passive- or active-loading techniques. In addition to the optimum D : L ratio, target delivery needs to meet target delivery requirements [194].

Furthermore, the clinical envisaged differentiation with reference to blood and tissue pharmacokinetics, pharmacodynamics, nonclinical toxicity should meet the differentiation against the standard of care for the particular pathological condition. Irrespective of small or large molecules (e.g. approved drugs, NCEs, macromolecules, such as siRNAs and mRNAs), the target molecules should meet liposome composition and process challenges that emerge during the development and manufacturing and need to be appropriately tailored [195]. Since liposome products are thermodynamically unstable and comprise lipids and therapeutic molecules, their development evolves from the careful selection of desired lipids, optimum drug – lipid ratio based on the properties of the molecule to be encapsulated. Appropriate selection of liposome components, identification, and precise characterization of CQAs need to be established with appropriate control strategies to meet the desired product quality and in vivo performance. In addition to formulation aspects, preclinical PK study models need to be considered for establishing similarity to originator products in case of the generic product [196]. For developing a liposome-based formulation of investigational drugs in cancer indication, blood, and tumor PK are indicative of relative tumor targeting, thereby differentiating liposome products in comparison to nonliposomal formulation. Generally, liposomal products provide several-fold increase in tumor drug concentration compared to nonliposomal product at equimolar dose [197]. Hence, prolonged plasma exposure of liposomal drug is indicative of enhanced efficacy and safety compared to free drug. Furthermore, the improvement in maximum tolerable dose of liposome formulation in rodent and non-rodent models will ensure the right IND candidate for clinical translation.

The clinical success of targeted liposomes, particularly in oncology is based on how best the liposome formulation will achieve the dose escalation with acceptable adverse events [20]. This is based on the fact that, generally, the anticancer molecules are dosed at a maximum tolerable dose. The expected significant improvement in clinical tolerability of anticancer molecule with liposomal formulation will directly impact positively the clinical outcomes, such as improved safety, overall survival, progression-free survival with commercially intended liposome formulation [198]. The higher tolerability of the liposome product is expected to lead to a higher therapeutic index of molecules. Because, providing relatively higher localized delivery of drug at the target tissue can better translate to safer clinical dose and results in better clinical outcomes, such as progression-free survival, overall survival with acceptable or no toxicity.

4.7 Conclusions and Future of Prospects of Targeted Liposomal-Delivery Systems

Liposome technology has evolved to a much more mature state in the past three decades as the most successful drug-delivery platform with established chemistry, manufacturing, and control processes with better regulatory understanding. This is confirmed through the development and commercialization of pharmaceutical products, such as Abelcet®, MyoCet®, Marqibo®, Vyxeos®, Doxil®, and Onpattro®. The success is largely driven by stealth liposome's ability to escape RES and preferentially localize in the target tissue passively by EPR effect. This has translated to an overall increase in therapeutic index, especially reduced toxicity side rather than increased efficacy. Furthermore, the liposome products have proven the risk-to-benefit ratio as clinically differentiating products with improved patient's health in various therapeutic segments, including oncology, ophthalmology, dermatology, infection, and pain management. In addition to small molecules, recent success of LNP formulations of siRNA in the clinic for silencing genes holds tremendous promise for upcoming siRNA and mRNA-based therapeutics.

It is evident that the liposomes-based drug-delivery market is fast growing with an average growth rate of 13.52%. In 2017, the global revenue from liposome products is nearly 2.3 billion USD. The current (2021) global market of liposome products is valued at 3.35 billion USD with forecast to reach 7.85 billion USD by the end of 2026, growing at CAGR of 12.8% during 2021–2026. With global key players Alnylam, CSPC, Fudan-Zhangjiang, Gilead sciences, Ipsen, Johnson and Johnson, Luye Pharma, Moderna, Novartis, Pacira, Pfizer, Sigma-Tau, Sun Pharmaceutical Industries Ltd, Takeda, and Teva Pharmaceutical – the future for liposome-based products looks promising as many major milestones still need to be achieved. Despite clinical success, still liposomes are not explored to the fullest extent in addressing the clinical unmet need, in particular, reference to (i) use of liposomes for investigational drugs, (ii) ligand-functionalized liposomes for active-targeted delivery. Liposomes for investigational drugs can be translated by careful evaluation of cost-to-benefit ratio, target validation, molecule feasibility to fit into liposome formulations, and desired clinical outcomes against standard-of-care. This approach of NCE-based liposome products holds a promise, especially for undruggable moderate-to-ultra-toxic molecules. Furthermore, such clinical translation is believed as one of the viable inventive ways of increasing the therapeutic index in anticancer therapy in terms of enhanced safety and efficacy. For ligand-functionalized liposomes, the clinical translation can be achieved by identifying right target transmembrane receptors, their organ specificity, protein expression levels in a pathological condition, selection of ligand with highest binding affinity to the receptor, and finally optimization of ligand density on surface-functionalized liposomes without compromising longer systemic circulation time. Hence, future developments in targeted liposomes for the pharmaceutical segment should focus on these approaches along with additional factors, such as:

- A deep understanding of pathophysiological mechanism of the disease.
- The standardization of manufacturing, stability, and regulatory approval of ligand-functionalized liposomes.
- Identify optimum selection of criteria for ligand and its conjugation chemistry to lipid or liposome for efficient target binding and cell internalization.
- Establishing correlation of ligand-functionalized liposomes interaction with biological membranes through intensive biophysical studies.

List of Abbreviations

ACUPA	2-(3-((S)-5-amino-1-carboxypentyl)ureido) pentanedioic acid
AIDS	aquired immune deficiency syndrome
AMD	age-related macular degeneration
ATP	adenosine triphosphate
BBB	blood–brain barrier
Bcl-2	B-cell lymphoma 2
BET	bacterial endotoxin
BLP25	tecemotide
BPD	benzoporphyrin derivative monoacid
CAGR	compound annual growth rate
CCS/C	ceramide Carbamoyl Spermine/cholesterol
CH	cholesterol
CHOL	cholesterol
CMC	chemistry, manufacturing, and control
CQAs	critical quality attributes
CSA	cholesteryl sulfate
CSPC	CSPC Pharmaceutical Group Limited
DEPC	dierucoyl phosphatidyl choline
DLin-MC3-DMA	dilinoleylmethyl-4-dimethylaminobutyrate
DMPC	dimyristyl phosphatidyl choline
DMPG	dimyristyl phosphatidyl glycerol
DNA	deoxyribonucleic acid
DOPC	dioleyl phosphatidyl choline
DOPE	dioleylphosphatidyl choline
DOPS	dioleyl phosphatidyl serine
DOTAP	1,2-dioleoyl-3-trimethylammonium-propane
DOTMA	trimethyl[2,3-(dioleyloxy)propyl]ammonium Chloride
DOX	doxorubicin
DPPG	dipalmitoyl phosphatidyl glycerol
DSPC	distearoylglycero phosphocholine
DSPE-mPEG	1,2-distearoyl-sn-glycero-3-phosphoethanolamine-N-[methoxy(polyethylene glycol)-2000], sodium salt
DSPG	distearoyl phosphatidyl glycerol
EGFR	epidermal growth factor receptor
EPC	egg phosphatidylcholine
EPG	egg phosphatidylglycerol
EPR	enhanced permeation and retention
FITC	fluorescein isothiocyanate
FMX	ferumoxytol
FR	folate receptor
GM-CSF	granulocyte macrophage-colony stimulating factor
GMP	good manufacturing practices

Grb2	growth factor receptor-bound protein 2
GUV	giant unilamellar vesicles
HER-2	human epidermal growth factor
HFS	Hand–foot syndrome
HIV	human immunodeficiency virus
HSPC	hydrogenated soybean phosphatidylcholine
ICAM-1	intracellular adhesion molecule-1
ICH	The International Council for Harmonization of Technical Requirements for Pharmaceuticals for Human Use
IgG	immunoglobulin G
IL1-β	interleukin 1β
IND	investigational new drug
LErafAON	c-Raf antisense oligodeoxynucleotides
LNCaP	androgen-sensitive human prostate adenocarcinoma cells
LNPs	lipid nanoparticles
LPS	lipopolysaccharide
LUV	large unilamellar vesicles
mAb	monoclonal antibody
miR-34a	brain-enriched miRNA family
MLV	multilamellar vesicles
MMP-2	metaloproteinase-2
MPER	membrane proximal external region
MPS	mononuclear phagocyte system
MPS	mononuclear phagocytic system
mRNA	messenger ribonucleic acid
MTP-PE	muramyl tripeptide phosphatidyl ethanolamine
MVV	multivesicular vesicles
NCE	new chemical entity
NHP	non human primate
NIR	near infrared
NJ	New Jersey
NPL	non-conventional liposome
PDAC	pancreatic ductal adenocarcinoma
PDGF	platelet derived growth factor
PDI	polydispersity index
PEG 2000-DSPE	methoxyl poly(ethylene glycol) distearoyl phosphoethanolamine
PEG	poly(ethylene) glycol
PEG2000-C-DMG:	dimyristoyl methoxypolyethylene glycol
PK	pharmcokinetics
PLK1	polo-like kinase 1
POPC	palmitoyloleoylphosphatidylcholine
PS	phosphatidylserine
PSMA	prostate-specific membrane antigen
QbD	quality by design

RBC	red blood cell
RBD	receptor binding domain
RES	reticuloendothelial system
RGD	tripeptide Arg-Gly-Asp
RNA	ribonucleic acid
RNAi	small interfering ribonucleic acid
SAXS	small angle X-ray scattering
SCLC	small cell lung carcinoma
siRNA	small interfering RNA
siRNAs	small-interfering ribose nucleic acids
SM	sphingomyelin
SUV	small unilamellar vesicles
T4N5	enzyme T4-bacteriophage endonuclease V
TEM	transmission electron microscope
TfRscFv	antitransferrin receptor single-chain antibody fragment
TKM	Tekmira Pharmaceuticals Corp
TKM-PLK	company nomenclature of investigational drug candidate
TNF-alpha	tumor necrosis factor alpha
TTR	transthyretin
US FDA	United States Food and Drug Administration
USD	United States Dollar
VEGF	vascular endothelial growth factor
VEGFR1/R2	vascular endothelial growth factor receptor1/2

References

1 Zhang, J.A., Xuan, T., Parmar, M. et al. (2004). Development and characterization of a novel liposome-based formulation of SN-38. *Int. J. Pharm.* 270: 93–107.

2 Fang, Y.-P., Chuang, C.-H., Wu, Y.-J. et al. (2018). SN38-loaded< 100 nm targeted liposomes for improving poor solubility and minimizing burst release and toxicity: in vitro and in vivo study. *Int. J. Nanomed.* 13: 2789.

3 Abawi, A., Wang, X., Bompard, J. et al. (2021). Monomethyl auristatin E grafted-liposomes to target prostate tumor cell lines. *Int. J. Mol. Sci.* 22: 4103.

4 Kulkarni, J.A., Darjuan, M.M., Mercer, J.E. et al. (2018). On the formation and morphology of lipid nanoparticles containing ionizable cationic lipids and siRNA. *ACS Nano* 12: 4787–4795.

5 Tan, Y.F., Mundargi, R.C., Chen, M.H.A. et al. (2014). Layer-by-layer nanoparticles as an efficient siRNA delivery vehicle for SPARC silencing. *Small* 10: 1790–1798.

6 Elia, U., Ramishetti, S., Rosenfeld, R. et al. (2021). Design of SARS-CoV-2 hFc-conjugated receptor-binding domain mRNA vaccine delivered via lipid nanoparticles. *ACS Nano* 15: 9627–9637.

7 Thi, T.T.H., Suys, E.J., Lee, J.S. et al. (2021). Lipid-based nanoparticles in the clinic and clinical trials: from cancer nanomedicine to COVID-19 vaccines. *Vaccines* 9: 359.
8 Oberli, M.A., Reichmuth, A.M., Dorkin, J.R. et al. (2017). Lipid nanoparticle assisted mRNA delivery for potent cancer immunotherapy. *Nano Lett.* 17: 1326–1335.
9 Belfiore, L., Saunders, D.N., Ranson, M. et al. (2018). Towards clinical translation of ligand-functionalized liposomes in targeted cancer therapy: challenges and opportunities. *J. Control. Release* 277: 1–13.
10 Bangham, A. (1993). Liposomes: the Babraham connection. *Chem. Phys. Lipids* 64: 275–285.
11 Düzgüneş, N. and Gregoriadis, G. (2005). Introduction: the origins of liposomes: Alec Bangham at Babraham. In: *Methods in Enzymology*, vol. 391 (ed. N. Duzgunes), 1–3. Elsevier.
12 Sessa, G. and Weissmann, G. (1968). Phospholipid spherules (liposomes) as a model for biological membranes. *J. Lipid Res.* 9: 310–318.
13 Lasic, D.D. (1988). The mechanism of vesicle formation. *Biochem. J.* 256: 1–11.
14 Barenholz, Y.C. (2012). Doxil®—the first FDA-approved nano-drug: lessons learned. *J. Control. Release* 160: 117–134.
15 Gregoriadis, G. (1978). Liposomes in therapeutic and preventive medicine: the development of the drug-carrier concept. *Ann. N. Y. Acad. Sci.* 308: 343–370.
16 Gregoriadis, G. (2016). Liposomes in drug delivery: how it all happened. *Pharmaceutics* 8 (2): 19.
17 Mozafari, M.R. (2005). Liposomes: an overview of manufacturing techniques. *Cell. Mol. Biol. Lett.* 10: 711.
18 Maherani, B., Arab-Tehrany, E., Mozafari, M.R. et al. (2011). Liposomes: a review of manufacturing techniques and targeting strategies. *Curr. Nanosci.* 7: 436–452.
19 Kastner, E., Kaur, R., Lowry, D. et al. (2014). High-throughput manufacturing of size-tuned liposomes by a new microfluidics method using enhanced statistical tools for characterization. *Int. J. Pharm.* 477: 361–368.
20 Yingchoncharoen, P., Kalinowski, D.S., and Richardson, D.R. (2016). Lipid-based drug delivery systems in cancer therapy: what is available and what is yet to come. *Pharmacol. Rev.* 68: 701–787.
21 Grit, M. and Crommelin, D.J. (1993). Chemical stability of liposomes: implications for their physical stability. *Chem. Phys. Lipids* 64: 3–18.
22 Caracciolo, G. (2018). Clinically approved liposomal nanomedicines: lessons learned from the biomolecular corona. *Nanoscale* 10: 4167–4172.
23 Immordino, M.L., Dosio, F., and Cattel, L. (2006). Stealth liposomes: review of the basic science, rationale, and clinical applications, existing and potential. *Int. J. Nanomed.* 1: 297.
24 Jahn, F., Jordan, K., Behlendorf, T. et al. (2015). Safety and efficacy of liposomal cytarabine in the treatment of neoplastic meningitis. *Oncology* 89: 137–142.
25 Kirpotin, D.B., Drummond, D.C., Shao, Y. et al. (2006). Antibody targeting of long-circulating lipidic nanoparticles does not increase tumor localization but does increase internalization in animal models. *Cancer Res.* 66: 6732–6740.

26 Kirpotin, D., Park, J.W., Hong, K. et al. (1997). Sterically stabilized anti-HER2 immunoliposomes: design and targeting to human breast cancer cells in vitro. *Biochemistry* 36: 66–75.

27 Klimuk, S.K., Semple, S.C., Scherrer, P., and Hope, M.J. (1999). Contact hypersensitivity: a simple model for the characterization of disease-site targeting by liposomes. *Biochim. Biophys. Acta, Biomembr.* 1417: 191–201.

28 Koning, G.A. and Storm, G. (2003). Targeted drug delivery systems for the intracellular delivery of macromolecular drugs. *Drug Discov. Today* 11: 482–483.

29 Kono, K. (2001). Thermosensitive polymer-modified liposomes. *Adv. Drug Deliv. Rev.* 53: 307–319.

30 Kraft, J.C., Freeling, J.P., Wang, Z., and Ho, R.J. (2014). Emerging research and clinical development trends of liposome and lipid nanoparticle drug delivery systems. *J. Pharm. Sci.* 103: 29–52.

31 Kunstfeld, R., Wickenhauser, G., Michaelis, U. et al. (2003). Paclitaxel encapsulated in cationic liposomes diminishes tumor angiogenesis and melanoma growth in a "humanized" SCID mouse model. *J. Invest. Dermatol.* 120: 476–482.

32 Noble, G.T., Stefanick, J.F., Ashley, J.D. et al. (2014). Ligand-targeted liposome design: challenges and fundamental considerations. *Trends Biotechnol.* 32: 32–45.

33 Inglut, C.T., Sorrin, A.J., Kuruppu, T. et al. (2020). Immunological and toxicological considerations for the design of liposomes. *Nanomaterials* 10: 190.

34 Moncalvo, F., Martinez Espinoza, M.I., and Cellesi, F. (2020). Nanosized delivery systems for therapeutic proteins: clinically validated technologies and advanced development strategies. *Front. Bioeng. Biotechnol.* 8: 89.

35 Boerman, O., Laverman, P., Oyen, W. et al. (2000). Radiolabeled liposomes for scintigraphic imaging. *Prog. Lipid Res.* 39: 461–475.

36 Lee, W. and Im, H.-J. (2019). Theranostics based on liposome: looking back and forward. *Nucl. Med. Mol. Imaging* 53: 242–246.

37 Bozzuto, G. and Molinari, A. (2015). Liposomes as nanomedical devices. *Int. J. Nanomed.* 10: 975.

38 Jensen, G.M. and Hodgson, D.F. (2020). Opportunities and challenges in commercial pharmaceutical liposome applications. *Adv. Drug Deliv. Rev.* 154–155: 2–12.

39 Adedoyin, A., Swenson, C.E., Bolcsak, L.E. et al. (2000). A pharmacokinetic study of amphotericin B lipid complex injection (Abelcet) in patients with definite or probable systemic fungal infections. *Antimicrob. Agents Chemother.* 44: 2900–2902.

40 Gill, P.S., Espina, B.M., Muggia, F. et al. (1995). Phase I/II clinical and pharmacokinetic evaluation of liposomal daunorubicin. *J. Clin. Oncol.* 13: 996–1003.

41 Paterson, D.L., David, K., Mrsic, M. et al. (2008). Pre-medication practices and incidence of infusion-related reactions in patients receiving AMPHOTEC®: data from the Patient Registry of Amphotericin B Cholesteryl Sulfate Complex for Injection Clinical Tolerability (PRoACT) registry. *J. Antimicrob. Chemother.* 62: 1392–1400.

42 Chang, H.-I. and Yeh, M.-K. (2012). Clinical development of liposome-based drugs: formulation, characterization, and therapeutic efficacy. *Int. J. Nanomed.* 7: 49.

43 Group1A, Verteporfin in Photodynamic Therapy Study Group (2001). Verteporfin therapy of subfoveal choroidal neovascularization in age-related macular degeneration: two-year results of a randomized clinical trial including lesions with occult with no classic choroidal neovascularization—verteporfin in photodynamic therapy report 2. *Am. J. Ophthalmol.* 131: 541–560.

44 Kager, L., Pötschger, U., and Bielack, S. (2010). Review of mifamurtide in the treatment of patients with osteosarcoma. *Ther. Clin. Risk Manag.* 6: 279.

45 Vogel, J.D. (2013). Liposome bupivacaine (EXPAREL®) for extended pain relief in patients undergoing ileostomy reversal at a single institution with a fast-track discharge protocol: an IMPROVE Phase IV health economics trial. *J. Pain Res.* 6: 605.

46 Silverman, J.A. and Deitcher, S.R. (2013). Marqibo®(vincristine sulfate liposome injection) improves the pharmacokinetics and pharmacodynamics of vincristine. *Cancer Chemother. Pharmacol.* 71: 555–564.

47 Chang, T., Shiah, H., Yang, C. et al. (2015). Phase I study of nanoliposomal irinotecan (PEP02) in advanced solid tumor patients. *Cancer Chemother. Pharmacol.* 75: 579–586.

48 Lancet, J.E., Ritchie, E.K., Uy, G.L. et al. (2017). Efficacy and safety of CPX-351 versus 7+ 3 in older adults with secondary acute myeloid leukemia: combined subgroup analysis of phase 2 and phase 3 studies. *Blood* 130: 2657–2657.

49 Urits, I., Swanson, D., Swett, M.C. et al. (2020). A review of patisiran (ONPATTRO®) for the treatment of polyneuropathy in people with hereditary transthyretin amyloidosis. *Neurol. Ther.* 9 (2): 301–315.

50 Domínguez, A.R., Hidalgo, D.O., Garrido, R.V., and Sánchez, E.T. (2005). Liposomal cytarabine (DepoCyte®) for the treatment of neoplastic meningitis. *Clin. Transl. Oncol.* 7: 232–238.

51 Carvalho, B., Roland, L.M., Chu, L.F. et al. (2007). Single-dose, extended-release epidural morphine (DepoDur™) compared to conventional epidural morphine for post-cesarean pain. *Anesth. Analg.* 105: 176–183.

52 Zhang, J., Leifer, F., Rose, S. et al. (2018). Amikacin liposome inhalation suspension (ALIS) penetrates non-tuberculous mycobacterial biofilms and enhances amikacin uptake into macrophages. *Front. Microbiol.* 9: 915.

53 Kapoor, M., Lee, S.L., and Tyner, K.M. (2017). Liposomal drug product development and quality: current US experience and perspective. *AAPS J.* 19: 632–641.

54 Alavi, M. and Hamidi, M. (2019). Passive and active targeting in cancer therapy by liposomes and lipid nanoparticles. *Drug Metab. Personal. Ther.* 34 (1): 1–8. https://doi.org/10.1515/dmpt-2018-0032. PMID: 30707682.

55 Li, J., Elkhoury, K., Barbieux, C. et al. (2020). Effects of bioactive marine-derived liposomes on two human breast cancer cell lines. *Mar. Drugs* 18: 211.

56 Gabizon, A., Catane, R., Uziely, B. et al. (1994). Prolonged circulation time and enhanced accumulation in malignant exudates of doxorubicin encapsulated in polyethylene-glycol coated liposomes. *Cancer Res.* 54: 987–992.

57 Batist, G., Gelmon, K.A., Chi, K.N. et al. (2009). Safety, pharmacokinetics, and efficacy of CPX-1 liposome injection in patients with advanced solid tumors. *Clin. Cancer Res.* 15: 692–700.

58 Gabizon, A., Amselem, S., Goren, D. et al. (1990). Preclinical and clinical experience with a doxorubicin-liposome preparation. *J. Liposome Res.* 1: 491–502.

59 Chen, L., Chang, T., Cheng, A. et al. (2008). Phase I study of liposome encapsulated irinotecan (PEP02) in advanced solid tumor patients. *J. Clin. Oncol.* 26: 2565–2565.

60 Heinemann, V., Bosse, D., Jehn, U. et al. (1997). Pharmacokinetics of liposomal amphotericin B (Ambisome) in critically ill patients. *Antimicrob. Agents Chemother.* 41: 1275–1280.

61 Crommelin, D.J., van Hoogevest, P., and Storm, G. (2020). The role of liposomes in clinical nanomedicine development. What now? Now what? *J. Control. Release* 318: 256–263.

62 Zucker, D., Marcus, D., Barenholz, Y., and Goldblum, A. (2009). Liposome drugs' loading efficiency: a working model based on loading conditions and drug's physicochemical properties. *J. Control. Release* 139: 73–80.

63 Ostro, M.J. and Cullis, P.R. (1989). Use of liposomes as injectable-drug delivery systems. *Am. J. Health Syst. Pharm.* 46: 1576–1588.

64 Barenholz, Y. (1992). Liposome production: historic aspects. In: *Liposome Dermatics* (ed. O. Braun-Falco, H.C. Korting and H.I. Maibach), 69–81. Springer.

65 Milla, P., Dosio, F., and Cattel, L. (2012). PEGylation of proteins and liposomes: a powerful and flexible strategy to improve the drug delivery. *Curr. Drug Metab.* 13: 105–119.

66 Briuglia, M.-L., Rotella, C., McFarlane, A., and Lamprou, D.A. (2015). Influence of cholesterol on liposome stability and on in vitro drug release. *Drug Deliv. Transl. Res.* 5: 231–242.

67 Roerdink, F.H., Regts, J., Handel, T. et al. (1989). Effect of cholesterol on the uptake and intracellular degradation of liposomes by liver and spleen; a combined biochemical and γ-ray perturbed angular correlation study. *Biochim. Biophys. Acta, Biomembr.* 980: 234–240.

68 Cattel, L., Ceruti, M., and Dosio, F. (2003). From conventional to stealth liposomes a new frontier in cancer chemotherapy. *Tumori J.* 89: 237–249.

69 Lasic, D.D. and Martin, F.J. (1995). *Stealth Liposomes*, vol. 20. CRC Press.

70 Solomon, R. and Gabizon, A.A. (2008). Clinical pharmacology of liposomal anthracyclines: focus on pegylated liposomal doxorubicin. *Clin. Lymphoma Myeloma* 8: 21–32.

71 Skubitz, K.M. (2003). Phase II trial of pegylated-liposomal doxorubicin (Doxil™) in Sarcoma* ORIGINAL ARTICLE. *Cancer Invest.* 21: 167–176.

72 James, N., Coker, R., Tomlinson, D. et al. (1994). Liposomal doxorubicin (Doxil): an effective new treatment for Kaposi's sarcoma in AIDS. *Clin. Oncol.* 6: 294–296.

73 Tejada-Berges, T., Granai, C., Gordinier, M., and Gajewski, W. (2002). Caelyx/Doxil for the treatment of metastatic ovarian and breast cancer. *Expert Rev. Anticancer Ther.* 2: 143–150.

74 Rosenthal, E., Poizot-Martin, I., Saint-Marc, T. et al., DNX Study Group(2002). Phase IV study of liposomal daunorubicin (DaunoXome) in AIDS-related Kaposi sarcoma. *Am. J. Clin. Oncol.* 25: 57–59.

75 Swenson, C., Perkins, W., Roberts, P., and Janoff, A. (2001). Liposome technology and the development of Myocet™(liposomal doxorubicin citrate). *Breast* 10: 1–7.

76 Leonard, R., Williams, S., Tulpule, A. et al. (2009). Improving the therapeutic index of anthracycline chemotherapy: focus on liposomal doxorubicin (Myocet™). *Breast* 18: 218–224.

77 Beltrán-Gracia, E., López-Camacho, A., Higuera-Ciapara, I. et al. (2019). Nanomedicine review: clinical developments in liposomal applications. *Cancer Nanotechnol.* 10: 1–40.

78 Akinc, A., Maier, M.A., Manoharan, M. et al. (2019). The Onpattro story and the clinical translation of nanomedicines containing nucleic acid-based drugs. *Nat. Nanotechnol.* 14: 1084–1087.

79 Hiemenz, J.W. and Walsh, T.J. (1996). Lipid formulations of amphotericin B: recent progress and future directions. *Clin. Infect. Dis.* 22: S133–S144.

80 Torrado, J., Espada, R., Ballesteros, M., and Torrado-Santiago, S. (2008). Amphotericin B formulations and drug targeting. *J. Pharm. Sci.* 97: 2405–2425.

81 Adedoyin, A., Bernardo, J.F., Swenson, C.E. et al. (1997). Pharmacokinetic profile of ABELCET (amphotericin B lipid complex injection): combined experience from phase I and phase II studies. *Antimicrob. Agents Chemother.* 41: 2201–2208.

82 Mantripragada, S. (2002). A lipid based depot (DepoFoam® technology) for sustained release drug delivery. *Prog. Lipid Res.* 41: 392–406.

83 Mantripragada, S. (2002). DepoFoam technology. In: *Modified-Release Drug Delivery Technology* (ed. M. Rathbone and J. Hadgraft), 729–736. CRC Press.

84 Golf, M., Daniels, S.E., and Onel, E. (2011). A Phase 3, randomized, placebo-controlled trial of DepoFoam® bupivacaine (extended-release bupivacaine local analgesic) in bunionectomy. *Adv. Ther.* 28: 776.

85 Xu, L. and Anchordoquy, T. (2011). Drug delivery trends in clinical trials and translational medicine: challenges and opportunities in the delivery of nucleic acid-based therapeutics. *J. Pharm. Sci.* 100: 38–52.

86 ur Rehman, Z., Zuhorn, I.S., and Hoekstra, D. (2013). How cationic lipids transfer nucleic acids into cells and across cellular membranes: recent advances. *J. Control. Release* 166: 46–56.

87 Caillaud, M., El Madani, M., and Massaad-Massade, L. (2020). Small interfering RNA from the lab discovery to patients' recovery. *J. Control. Release* 321: 616–628.

88 Li, M., Du, C., Guo, N. et al. (2019). Composition design and medical application of liposomes. *Eur. J. Med. Chem.* 164: 640–653.

89 Ulrich, A.S. (2002). Biophysical aspects of using liposomes as delivery vehicles. *Biosci. Rep.* 22: 129–150.

90 Mochizuki, S., Kanegae, N., Nishina, K. et al. (2013). The role of the helper lipid dioleoylphosphatidylethanolamine (DOPE) for DNA transfection cooperating with a cationic lipid bearing ethylenediamine. *Biochim. Biophys. Acta, Biomembr.* 1828: 412–418.

91 Drummond, D.C., Noble, C.O., Guo, Z. et al. (2006). Development of a highly active nanoliposomal irinotecan using a novel intraliposomal stabilization strategy. *Cancer Res.* 66: 3271–3277.

92 Gasselhuber, A., Dreher, M.R., Rattay, F. et al. (2012). Comparison of conventional chemotherapy, stealth liposomes and temperature-sensitive liposomes in a mathematical model. *PLoS One* 7: e47453.

93 Tseng, Y.-C., Mozumdar, S., and Huang, L. (2009). Lipid-based systemic delivery of siRNA. *Adv. Drug Deliv. Rev.* 61: 721–731.

94 Chountoulesi, M., Naziris, N., Pippa, N., and Demetzos, C. (2018). The significance of drug-to-lipid ratio to the development of optimized liposomal formulation. *J. Liposome Res.* 28: 249–258.

95 Johnston, M.J., Edwards, K., Karlsson, G., and Cullis, P.R. (2008). Influence of drug-to-lipid ratio on drug release properties and liposome integrity in liposomal doxorubicin formulations. *J. Liposome Res.* 18: 145–157.

96 Koning, G.A. and Storm, G. (2003). Targeted drug delivery systems for the intracellular delivery of macromolecular drugs. *Drug Discov. Today* 8: 482–483.

97 Hau, P., Fabel, K., Baumgart, U. et al. (2004). Pegylated liposomal doxorubicin-efficacy in patients with recurrent high-grade glioma. *Cancer: Interdiscip. Int. J. Am. Cancer Soc.* 100: 1199–1207.

98 Wang, R., Xiao, R., Zeng, Z. et al. (2012). Application of poly (ethylene glycol)–distearoylphosphatidylethanolamine (PEG-DSPE) block copolymers and their derivatives as nanomaterials in drug delivery. *Int. J. Nanomed.* 7: 4185.

99 Yan, Z., Wang, F., Wen, Z. et al. (2012). LyP-1-conjugated PEGylated liposomes: a carrier system for targeted therapy of lymphatic metastatic tumor. *J. Control. Release* 157: 118–125.

100 Gill, K.K., Nazzal, S., and Kaddoumi, A. (2011). Paclitaxel loaded PEG5000–DSPE micelles as pulmonary delivery platform: formulation characterization, tissue distribution, plasma pharmacokinetics, and toxicological evaluation. *Eur. J. Pharm. Biopharm.* 79: 276–284.

101 Allen, T.M., Sapra, P., and Moase, E. (2002). Use of the post-insertion method for the formation of ligand-coupled liposomes. *Cell. Mol. Biol. Lett.* 7: 217–219.

102 Nogueira, E., Gomes, A.C., Preto, A., and Cavaco-Paulo, A. (2015). Design of liposomal formulations for cell targeting. *Colloids Surf., B Biointerfaces* 136: 514–526.

103 Crommelin, D.J., Metselaar, J.M., and Storm, G. (2015). Liposomes: the science and the regulatory landscape. In: *Non-Biological Complex Drugs* (ed. D.J.A. Crommelin and J.S.B. de Vlieger), 77–106. Springer.

104 Beltrán-Gracia, E., López-Camacho, A., Higuera-Ciapara, I. et al. (2019). Nanomedicine review: clinical developments in liposomal applications. *Cancer Nanotechnol.* 10: 11.

105 Adams, D., Gonzalez-Duarte, A., O'Riordan, W.D. et al. (2018). Patisiran, an RNAi therapeutic, for hereditary transthyretin amyloidosis. *N. Engl. J. Med.* 379: 11–21.

106 Varga, Z., Wacha, A., Vainio, U. et al. (2012). Characterization of the PEG layer of sterically stabilized liposomes: a SAXS study. *Chem. Phys. Lipids* 165: 387–392.

107 Gray, V., Cady, S., Curran, D. et al. (2018). In vitro release test methods for drug formulations for parenteral applications. *Dissolut. Technol.* 25: 8–13.

108 Goren, D., Horowitz, A., Zalipsky, S. et al. (1996). Targeting of stealth liposomes to erbB-2 (Her/2) receptor: in vitro and in vivo studies. *Br. J. Cancer* 74: 1749–1756.

109 Lee, S.-H., Sato, Y., Hyodo, M., and Harashima, H. (2016). Topology of surface ligands on liposomes: characterization based on the terms, incorporation ratio, surface anchor density, and reaction yield. *Biol. Pharm. Bull.* 39: 1983–1994.

110 Nicolas, J., Mura, S., Brambilla, D. et al. (2013). Design, functionalization strategies and biomedical applications of targeted biodegradable/biocompatible polymer-based nanocarriers for drug delivery. *Chem. Soc. Rev.* 42: 1147–1235.

111 Fakhari, A., Baoum, A., Siahaan, T.J. et al. (2011). Controlling ligand surface density optimizes nanoparticle binding to ICAM-1. *J. Pharm. Sci.* 100: 1045–1056.

112 Fang, R.H., Hu, C.-M.J., and Zhang, L. (2012). Nanoparticles disguised as red blood cells to evade the immune system. *Expert Opin. Biol. Ther.* 12 (4): 385–389.

113 Boswell, G., Buell, D., and Bekersky, I. (1998). AmBisome (liposomal amphotericin B): a comparative review. *J. Clin. Pharmacol.* 38: 583–592.

114 Rafiyath, S.M., Rasul, M., Lee, B. et al. (2012). Comparison of safety and toxicity of liposomal doxorubicin vs. conventional anthracyclines: a meta-analysis. *Exp. Hematol. Oncol.* 1: 10.

115 Krauss, A.C., Gao, X., Li, L. et al. (2019). FDA approval summary:(daunorubicin and cytarabine) liposome for injection for the treatment of adults with high-risk acute myeloid leukemia. *Clin. Cancer Res.* 25: 2685–2690.

116 Rogers, A.H., Duker, J.S., Nichols, N., and Baker, B.J. (2003). Photodynamic therapy of idiopathic and inflammatory choroidal neovascularization in young adults. *Ophthalmology* 110: 1315–1320.

117 Yokomichi, N., Nagasawa, T., Coler-Reilly, A. et al. (2013). Pathogenesis of hand-foot syndrome induced by PEG-modified liposomal doxorubicin. *Hum. Cell* 26: 8–18.

118 Bareford, L.M. and Swaan, P.W. (2007). Endocytic mechanisms for targeted drug delivery. *Adv. Drug Deliv. Rev.* 59: 748–758.

119 Fathi, S. and Oyelere, A.K. (2016). Liposomal drug delivery systems for targeted cancer therapy: is active targeting the best choice? *Fut. Med. Chem.* 8: 2091–2112.

120 Yu, B., Tai, H.C., Xue, W. et al. (2010). Receptor-targeted nanocarriers for therapeutic delivery to cancer. *Mol. Membr. Biol.* 27: 286–298.

121 Suzuki, R., Takizawa, T., Kuwata, Y. et al. (2008). Effective anti-tumor activity of oxaliplatin encapsulated in transferrin–PEG-liposome. *Int. J. Pharm.* 346: 143–150.

122 Li, X., Ding, L., Xu, Y. et al. (2009). Targeted delivery of doxorubicin using stealth liposomes modified with transferrin. *Int. J. Pharm.* 373: 116–123.

123 Wang, X., Zhao, Y., Dong, S. et al. (2019). Cell-penetrating peptide and transferrin co-modified liposomes for targeted therapy of glioma. *Molecules* 24: 3540.

124 Qin, L., Wang, C.Z., Fan, H.J. et al. (2014). A dual-targeting liposome conjugated with transferrin and arginine-glycine-aspartic acid peptide for glioma-targeting therapy. *Oncol. Lett.* 8: 2000–2006.

125 Song, X.-l., Liu, S., Jiang, Y. et al. (2017). Targeting vincristine plus tetrandrine liposomes modified with DSPE-PEG2000-transferrin in treatment of brain glioma. *Eur. J. Pharm. Sci.* 96: 129–140.

126 Senzer, N.N., Matsuno, K., Yamagata, N. et al. (2009). Abstract C36: MBP-426, a novel liposome-encapsulated oxaliplatin, in combination with 5-FU/leucovorin (LV): phase I results of a Phase I/II study in gastro-esophageal adenocarcinoma, with pharmacokinetics. *Mol. Cancer Ther.* 8 (Supplement 1): C36.

127 Siefker-Radtke, A. (2016). A phase I study of systemic gene therapy with SGT-94 in patients with solid tumors. *Mol. Ther.* 24 (8): 1484–1491.

128 Senzer, N., Nemunaitis, J., Nemunaitis, D. et al. (2013). Phase I study of a systemically delivered p53 nanoparticle in advanced solid tumors. *Mol. Ther.* 21: 1096–1103.

129 Reddy, J.A., Haneline, L.S., Srour, E.F. et al. (1999). Expression and functional characterization of the β-isoform of the folate receptor on CD^{34+} cells. *Blood, J. Am. Soc. Hematol.* 93: 3940–3948.

130 Cheung, A., Bax, H.J., Josephs, D.H. et al. (2016). Targeting folate receptor alpha for cancer treatment. *Oncotarget* 7: 52553.

131 Low, P.S., Henne, W.A., and Doorneweerd, D.D. (2008). Discovery and development of folic-acid-based receptor targeting for imaging and therapy of cancer and inflammatory diseases. *Acc. Chem. Res.* 41: 120–129.

132 Low, P.S. and Kularatne, S.A. (2009). Folate-targeted therapeutic and imaging agents for cancer. *Curr. Opin. Chem. Biol.* 13: 256–262.

133 Soe, Z.C., Thapa, R.K., Ou, W. et al. (2018). Folate receptor-mediated celastrol and irinotecan combination delivery using liposomes for effective chemotherapy. *Colloids Surf. B Biointerfaces* 170: 718–728.

134 Peres-Filho, M.J., Dos Santos, A.P., Nascimento, T.L. et al. (2018). Antiproliferative activity and VEGF expression reduction in MCF7 and PC-3 cancer cells by paclitaxel and imatinib co-encapsulation in folate-targeted liposomes. *AAPS PharmSciTech* 19: 201–212.

135 Barth, C.W. and Gibbs, S.L. (2020). Fluorescence image-guided surgery: a perspective on contrast agent development. In: *Progress in Biomedical Optics and Imaging – Proceedings of SPIE, 11222. Molecular-Guided Surgery: Molecules, Devices, and Applications VI*, 112220J. International Society for Optics and Photonics.

136 Kalli, K.R., Block, M.S., Kasi, P.M. et al. (2018). Folate receptor alpha peptide vaccine generates immunity in breast and ovarian cancer patients. *Clin. Cancer Res.* 24: 3014–3025.

137 Oza, A., Vergote, I., Gilbert, L. et al. (2015). A randomized double-blind phase III trial comparing vintafolide (EC145) and pegylated liposomal doxorubicin (PLD/Doxil®/Caelyx®) in combination versus PLD in participants with platinum-resistant ovarian cancer (PROCEED)(NCT01170650). *Gynecol. Oncol.* 137: 5–6.

138 Scaltriti, M. and Baselga, J. (2006). The epidermal growth factor receptor pathway: a model for targeted therapy. *Clin. Cancer Res.* 12: 5268–5272.

139 Kim, S.K. and Huang, L. (2012). Nanoparticle delivery of a peptide targeting EGFR signaling. *J. Control. Release* 157: 279–286.

140 Vokes, E.E. and Chu, E. (2006). Anti-EGFR therapies: clinical experience in colorectal, lung, and head and neck cancers. *Oncology* 20 (5 Suppl 2): 15–25.

141 Löw, K., Wacker, M., Wagner, S. et al. (2011). Targeted human serum albumin nanoparticles for specific uptake in EGFR-Expressing colon carcinoma cells. *Nanomed. Nanotechnol. Biol. Med.* 7: 454–463.

142 Zhang, X., Li, Y., Wei, M. et al. (2019). Cetuximab-modified silica nanoparticle loaded with ICG for tumor-targeted combinational therapy of breast cancer. *Drug Deliv.* 26: 129–136.

143 Pan, X., Wu, G., Yang, W. et al. (2007). Synthesis of cetuximab-immunoliposomes via a cholesterol-based membrane anchor for targeting of EGFR. *Bioconjug. Chem.* 18: 101–108.

144 Zalba, S., Contreras, A.M., Haeri, A. et al. (2015). Cetuximab-oxaliplatin-liposomes for epidermal growth factor receptor targeted chemotherapy of colorectal cancer. *J. Control. Release* 210: 26–38.

145 Portnoy, E., Lecht, S., Lazarovici, P. et al. (2011). Cetuximab-labeled liposomes containing near-infrared probe for in vivo imaging. *Nanomed. Nanotechnol. Biol. Med.* 7: 480–488.

146 Merino, M., Zalba, S., and Garrido, M.J. (2018). Immunoliposomes in clinical oncology: state of the art and future perspectives. *J. Control. Release* 275: 162–176.

147 Iqbal, N. and Iqbal, N. (2014). Human epidermal growth factor receptor 2 (HER2) in cancers: overexpression and therapeutic implications. *Mol. Biol. Int.* 2014: 852748.

148 Huszno, J. and Nowara, E. (2016). Current therapeutic strategies of anti-HER2 treatment in advanced breast cancer patients. *Contemp. Oncol.* 20: 1.

149 Pernas, S. and Tolaney, S.M. (2019). HER2-positive breast cancer: new therapeutic frontiers and overcoming resistance. *Ther. Adv. Med. Oncol.* 11: https://doi.org/10.1177/1758835919833519.

150 Eloy, J.O., Petrilli, R., Chesca, D.L. et al. (2017). Anti-HER2 immunoliposomes for co-delivery of paclitaxel and rapamycin for breast cancer therapy. *Eur. J. Pharm. Biopharm.* 115: 159–167.

151 Munster, P., Krop, I.E., LoRusso, P. et al. (2018). Safety and pharmacokinetics of MM-302, a HER2-targeted antibody–liposomal doxorubicin conjugate, in patients with advanced HER2-positive breast cancer: a phase 1 dose-escalation study. *Br. J. Cancer* 119: 1086–1093.

152 Rodallec, A., Sicard, G., Giacometti, S. et al. (2020). Tumor uptake and associated greater efficacy of anti-Her2 immunoliposome does not rely on Her2 expression status: study of a docetaxel-trastuzumab immunoliposome on Her2+ breast cancer model (SKBR3). *Anticancer Drugs* 31: 463–472.

153 Rodallec, A., Brunel, J.-M., Giacometti, S. et al. (2018). Docetaxel–trastuzumab stealth immunoliposome: development and in vitro proof of concept studies in breast cancer. *Int. J. Nanomed.* 13: 3451.

154 Marty, M., Cognetti, F., Maraninchi, D. et al. (2005). Randomized phase II trial of the efficacy and safety of trastuzumab combined with docetaxel in patients with human epidermal growth factor receptor 2–positive metastatic breast cancer administered as first-line treatment: the M77001 study group. *J. Clin. Oncol.* 23: 4265–4274.

155 Metro, G., Mottolese, M., and Fabi, A. (2008). HER-2-positive metastatic breast cancer: trastuzumab and beyond. *Expert Opin. Pharmacother.* 9: 2583–2601.

156 Silver, D.A., Pellicer, I., Fair, W.R. et al. (1997). Prostate-specific membrane antigen expression in normal and malignant human tissues. *Clin. Cancer Res.* 3: 81–85.

157 Chang, S.S., Reuter, V.E., Heston, W. et al. (1999). Five different anti-prostate-specific membrane antigen (PSMA) antibodies confirm PSMA expression in tumor-associated neovasculature. *Cancer Res.* 59: 3192–3198.

158 Ghosh, A. and Heston, W.D. (2004). Tumor target prostate specific membrane antigen (PSMA) and its regulation in prostate cancer. *J. Cell. Biochem.* 91: 528–539.

159 Yoo, J., Park, C., Yi, G. et al. (2019). Active targeting strategies using biological ligands for nanoparticle drug delivery systems. *Cancers* 11: 640.

160 Taneja, S.S. (2004). ProstaScint® scan: contemporary use in clinical practice. *Rev. Urol.* 6: S19.

161 Bandekar, A., Zhu, C., Jindal, R. et al. (2014). Anti–prostate-specific membrane antigen liposomes loaded with 225Ac for potential targeted antivascular α-particle therapy of cancer. *J. Nucl. Med.* 55: 107–114.

162 Yari, H., Nkepang, G., and Awasthi, V. (2019). Surface modification of liposomes by a lipopolymer targeting prostate specific membrane antigen for theranostic delivery in prostate cancer. *Materials* 12: 756.

163 Hrkach, J., Von Hoff, D., Ali, M.M. et al. (2012). Preclinical development and clinical translation of a PSMA-targeted docetaxel nanoparticle with a differentiated pharmacological profile. *Sci. Transl. Med.* 4: 128ra39.

164 Von Hoff, D.D., Mita, M.M., Ramanathan, R.K. et al. (2016). Phase I study of PSMA-targeted docetaxel-containing nanoparticle BIND-014 in patients with advanced solid tumors. *Clin. Cancer Res.* 22: 3157–3163.

165 Rosenblum, D., Joshi, N., Tao, W. et al. (2018). Progress and challenges towards targeted delivery of cancer therapeutics. *Nat. Commun.* 9: 1–12.

166 Ramanathan, R.K., Korn, R.L., Raghunand, N. et al. (2017). Correlation between ferumoxytol uptake in tumor lesions by MRI and response to nanoliposomal irinotecan in patients with advanced solid tumors: a pilot study. *Clin. Cancer Res.* 23: 3638–3648.

167 Hua, S. (2013). Targeting sites of inflammation: intercellular adhesion molecule-1 as a target for novel inflammatory therapies. *Front. Pharmacol.* 4: 127.

168 Guo, P., Yang, J., Di Jia, M.A.M., and Auguste, D.T. (2016). ICAM-1-targeted, Lcn2 siRNA-encapsulating liposomes are potent anti-angiogenic agents for triple negative breast cancer. *Theranostics* 6: 1.

169 Denekamp, J. (1984). Vasculature as a target for tumour therapy. In: *Angiogenesis*, vol. 4 (ed. F. Hammersen and O. Hudlicka), 28–38. Karger Publishers.

170 Bellamy, W.T., Richter, L., Sirjani, D. et al. (2001). Vascular endothelial cell growth factor is an autocrine promoter of abnormal localized immature myeloid precursors and leukemia progenitor formation in myelodysplastic syndromes. *Blood, J. Am. Soc. Hematol.* 97: 1427–1434.

171 Byrne, J.D., Betancourt, T., and Brannon-Peppas, L. (2008). Active targeting schemes for nanoparticle systems in cancer therapeutics. *Adv. Drug Deliv. Rev.* 60: 1615–1626.

172 Shein, S., Nukolova, N., Korchagina, A. et al. (2015). Site-directed delivery of VEGF-targeted liposomes into intracranial C6 glioma. *Bull. Exp. Biol. Med.* 158: 371–376.

173 Shein, S.A., Kuznetsov, I.I., Abakumova, T.O. et al. (2016). VEGF-and VEGFR2-targeted liposomes for cisplatin delivery to glioma cells. *Mol. Pharm.* 13: 3712–3723.

174 Hsu, A.R., Veeravagu, A., Cai, W. et al. (2007). Integrin αvβ3 antagonists for anti-angiogenic cancer treatment. *Recent Pat. Anticancer Drug Discov.* 2: 143–158.

175 Brooks, P.C., Clark, R.A., and Cheresh, D.A. (1994). Requirement of vascular integrin alpha v beta 3 for angiogenesis. *Science* 264: 569–571.

176 Kumar, C. (2003). Integrin αvβ3 as a therapeutic target for blocking tumor-induced angiogenesis. *Curr. Drug Targets* 4: 123–131.

177 Schiffelers, R.M., Koning, G.A., ten Hagen, T.L. et al. (2003). Anti-tumor efficacy of tumor vasculature-targeted liposomal doxorubicin. *J. Control. Release* 91: 115–122.

178 Vance, J.E. and Steenbergen, R. (2005). Metabolism and functions of phosphatidylserine. *Prog. Lipid Res.* 44: 207–234.

179 Soares, M.M., King, S.W., and Thorpe, P.E. (2008). Targeting inside-out phosphatidylserine as a therapeutic strategy for viral diseases. *Nat. Med.* 14: 1357–1362.

180 Dong, H.P., Holth, A., Kleinberg, L. et al. (2009). Evaluation of cell surface expression of phosphatidylserine in ovarian carcinoma effusions using the annexin-V/7-AAD assay: clinical relevance and comparison with other apoptosis parameters. *Am. J. Clin. Pathol.* 132: 756–762.

181 Schutters, K. and Reutelingsperger, C. (2010). Phosphatidylserine targeting for diagnosis and treatment of human diseases. *Apoptosis* 15: 1072–1082.

182 Garnier, B., Bouter, A., Gounou, C. et al. (2009). Annexin A5-functionalized liposomes for targeting phosphatidylserine-exposing membranes. *Bioconjug. Chem.* 20: 2114–2122.

183 De, M., Ghosh, S., Sen, T. et al. (2018). A novel therapeutic strategy for cancer using phosphatidylserine targeting stearylamine-bearing cationic liposomes. *Mol. Ther.-Nucl. Acids* 10: 9–27.

184 Davis, B.M., Normando, E.M., Guo, L. et al. (2014). Topical delivery of avastin to the posterior segment of the eye in vivo using annexin a5-associated liposomes. *Small* 10: 1575–1584.

185 Whitehead, K.A., Langer, R., and Anderson, D.G. (2009). Knocking down barriers: advances in siRNA delivery. *Nat. Rev. Drug Discov.* 8: 129–138.

186 De Fougerolles, A.R. (2008). Delivery vehicles for small interfering RNA in vivo. *Hum. Gene Ther.* 19: 125–132.

187 Tam, Y.Y.C., Chen, S., and Cullis, P.R. (2013). Advances in lipid nanoparticles for siRNA delivery. *Pharmaceutics* 5: 498–507.

188 Tabernero, J., Shapiro, G.I., LoRusso, P.M. et al. (2013). First-in-humans trial of an RNA interference therapeutic targeting VEGF and KSP in cancer patients with liver involvement. *Cancer Discov.* 3: 406–417.

189 Liu, X. (2015). Targeting polo-like kinases: a promising therapeutic approach for cancer treatment. *Transl. Oncol.* 8: 185–195.

190 Fitzgerald, K., Frank-Kamenetsky, M., Mant, T. et al. (2012). Phase I safety, pharmacokinetic, and pharmacodynamic results for ALN-PCS, a novel RNAi therapeutic for the treatment of hypercholesterolemia. *Arterioscler. Thromb. Vasc. Biol.* 32: A67.

191 Etheridge, M.L., Campbell, S.A., Erdman, A.G. et al. (2013). The big picture on nanomedicine: the state of investigational and approved nanomedicine products. *Nanomed. Nanotechnol. Biol. Med.* 9: 1–14.

192 Sercombe, L., Veerati, T., Moheimani, F. et al. (2015). Advances and challenges of liposome assisted drug delivery. *Front. Pharmacol.* 6: 286.

193 Hua, S., De Matos, M.B., Metselaar, J.M., and Storm, G. (2018). Current trends and challenges in the clinical translation of nanoparticulate nanomedicines: pathways for translational development and commercialization. *Front. Pharmacol.* 9: 790.

194 Akbarzadeh, A., Rezaei-Sadabady, R., Davaran, S. et al. (2013). Liposome: classification, preparation, and applications. *Nanoscale Res. Lett.* 8: 102.

195 Wahlich, J., Desai, A., Greco, F. et al. (2019). Nanomedicines for the delivery of biologics. *Pharmaceutics* 11 (5): 210.

196 Kirchhoff, C.F., Wang, X.Z.M., Conlon, H.D. et al. (2017). Biosimilars: key regulatory considerations and similarity assessment tools. *Biotechnol. Bioeng.* 114: 2696–2705.

197 Ait-Oudhia, S., Mager, D.E., and Straubinger, R.M. (2014). Application of pharmacokinetic and pharmacodynamic analysis to the development of liposomal formulations for oncology. *Pharmaceutics* 6: 137–174.

198 Bulbake, U., Doppalapudi, S., Kommineni, N., and Khan, W. (2017). Liposomal formulations in clinical use: an updated review. *Pharmaceutics* 9: 12.

5

Antibody–Drug Conjugates: Development and Applications

Rajesh Pradhan[1], Meghna Pandey[1], Siddhanth Hejmady[1], Rajeev Taliyan[1], Gautam Singhvi[1], Sunil K. Dubey[1,2], and Sachin Dubey[3]

[1] *Birla Institute of Technology and Science, Department of Pharmacy, Pilani Campus, Pilani, Rajasthan 333 031, India*
[2] *Emami Ltd., R&D Healthcare Division, 13, BT Road, Belgharia, Kolkata 700 056, India*
[3] *Ichnos Sciences SA, Chemin de la Combeta 5, La Chaux-de-Fonds 2300, Switzerland*

5.1 Introduction

The focus of drug discovery has shifted from synthetic drugs to biologics. For the past few years, the emergence of biologics has revolutionized the sector of medical care [1]. Biologics have significant advantages over synthetic drugs, especially the targeted delivery that leads to faster onset of action and reduced incidence of adverse effects. The United States Food and Drug Administration (USFDA) approved 48 drugs in 2019, out of which 10 are biologics [2]. Out of the 10 approved biologics, 3 of them are antibody–drug conjugates (ADCs). Overall, USFDA has approved seven ADCs for the treatment of cancer until now with the first approval in the 1990s. There is also a prediction that the number of ADC approvals will increase day by day considering that, more than 100 ADCs are in the clinical investigation phase [3]. Initially, the monoclonal antibodies were utilized for treating the disease, and over the years, the concept of ADCs was introduced with the help of recombinant technology. The ADC can improve the potency and the effectiveness of the treatment. It has three main components – (i) the recombinant monoclonal antibody that is covalently bound to (ii) the drug molecule or the payload with the help (iii) of synthetic linkers [4].

In the early 1900s, the German physician Paul Ehrlich coined the term "magic bullets" [5]. He came up with the theory of delivering the cytotoxic agents to the target site and provided the idea of an antibody that is conjugated with the diphtheria toxin [6]. The discovery of the hybridoma technology by Kohler and Milstein in 1975, half a decade later, accelerated the progress in this field. Murine antibodies were employed in the earlier stage, which now have been replaced by recombinant antibodies that possess low immunogenicity risk [7]. The development of ADCs has faced numerous challenges; around the 1990s, the KS1/4 antibody–methotrexate conjugate for lung cancer and the BR96 antibody–doxorubicin conjugate for breast

Targeted Drug Delivery, First Edition. Edited by Yogeshwar Bachhav.

cancer were early attempts in the field of ADC development. But these systems showcased low therapeutic benefits despite localizing to the tumor site [8]. Here the reason may be attributed to poor selection of the target antigen, and hence lower the immunogenic responses elicited by the murine antibodies, which, in turn, decreased the potency of the drugs [9]. The development of gemtuzumab ozogamicin (Mylotarg™) by Wyeth and Celltech for the treatment of acute myeloid leukemia showed improved potency followed by promising results in the clinical trials that were finally followed by USFDA approval in 2000. Though later it was withdrawn from the market due to safety issues, this is regarded as the first generation of ADCs [10]. The above problems were understood and subsequently addressed, which led to the development of the second-generation ADCs. Seattle Genetics, in collaboration with Millennium/Takeda, developed Brentuximab vedotin for the treatment of Hodgkin lymphoma and the anaplastic large-cell lymphoma. This system has shown high antimitotic activity [11]. Genentech with ImmunoGen's developed the Ado-trastuzumab emtansine for the treatment of breast cancer by targeting the human epidermal growth factor receptor 2 (HER2). These ADCs have shown more efficacy and safety as compared to the first-generation ADCs [12]. The ADCs are intended to kill the cancerous cell by binding to the specific surface antigen, and then the internalization process occurs through the endocytosis. Once, it is incorporated inside the lysosome, the ADCs undergo cleavage to release the antibody (different release mechanisms are involved; they are presented in the following Section 5.2.1), and the cytotoxic drug that leads to cell death due to the drug's cytotoxic activity. The success of ADCs is dependent on the amount of the free toxic drug that reaches inside the cell and which is driven by tumor antigen, antibody, linker, and the cytotoxic drug [13]. The chemotherapeutic agents can exert toxic effects due to their mechanism of action. They also often have limited permeability and a tendency to bind other surfaces. The monoclonal antibody mainly triggers an immunogenic reaction.

Therefore the ADCs offer various advantages as ADCs are highly tumor-specific because of the property of the monoclonal antibody that targets the specific antigen expressed on the surface of the tumor cells [14]. These can distinguish between the healthy and the cancerous tissues, so the chances of inducing adverse effects are very rare. In addition, ADCs can achieve high lethality toward the targeted cell since these systems directly deliver the drug to the tumor cells. Thus, ADCs have high efficacy and reduced systemic toxicity of the drugs that lead to an overall increase in safety and tolerability of drugs [15]. Based on the advantages discussed so far, ADCs form a successful approach in targeting the drug to the particular site. This review mainly focuses on the composition of the ADCs, mechanism of action, challenges, and the regulatory guidelines on the ADCs. It provides detailed information about different applications of the ADCs and case studies of selected ADCs that are available in the market.

5.2 Design of ADCs

The ADCs are generally prodrugs with the antibody attached to the payload with the help of a linker. The delivery of ADCs is achieved via administration through the intravenous route following which it reaches its target by recognizing

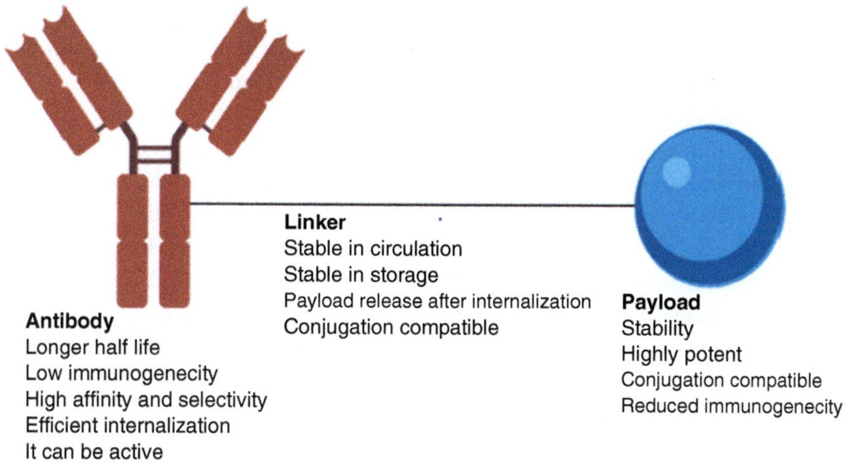

Figure 5.1 Schematic representation of an ADC and the three components and their salient features.

the target antigen [16]. ADCs are generally made up of three structurally distinct components that are the antibody, payload, and linker. The combination of all three components results in achieving greater efficacy and reduced side effects. A schematic representation of an ADC with individual components and respective properties is provided in Figure 5.1. Although an ADC is made of three components, the target antigen selection plays a vital role in the targeted delivery of the drug [17].

The target antigen is not an inherent component of the ADCs, but its selection decides the fate of the drug in the body. The selection of the target antigen is the most crucial step as it guides antibodies in recognizing target cells [18]. The antigen should be highly expressed on the target cell and have a lower or almost no expression in the healthy cells. For example, the HER2 receptor is 100 times more expressed on the surface of the tumor cell as compared to the normal cell. Ideally, the antigen should be prominently present at the surface of the target cell, which ensures the recognition by the monoclonal antibody easily and conveniently. To facilitate the internalization process, the antigen should have internalization properties that lead to increased efficacy of the payload. The bystander effect is referred to as the toxicity that is caused by the non-internalized ADCs. The bystander effect refers to the process when the drug is taken up and kills the surrounding bystander cells, which themselves may not express the antigen [17, 19]. Therefore, it is highly important to perform the careful selection of the proper antigen against which the antibody will be selected/generated. Furthermore, the recent research trends also suggest that sooner the classification of tumors could be based on the antigens present in the tumor cells.

5.2.1 Antibody

The antibody should have high specificity for the cell surface of the target molecule (target antigen) that undergoes expression in the tumor cells. The antibody mainly used in ADCs is the immunoglobulin G (IgG). The antibody has one constant fragment (Fc) and two antigen-binding fragments (Fabs) [20]. The Fc is utilized in the connection of the IgG to the immune cells, and the Fabs are focused on antigen specificity. The antibodies are molecules with higher molecular weight and have longer half-lives. They are utilized in long-lasting immunological functions. Mostly, the IgG1 isotype is used as the monoclonal antibody, and sometimes the IgG2 and IgG4 may also be utilized [21]. On the other hand, IgG3 did not find much commercial exploration due to its allotypic polymorphism and shorter half-life. However, it has a more powerful cell lysis property as compared to other antibodies. On the other hand, IgG4 has the property of the formation of hybrids in the circulation by exchanging one-half of it with another IgG4 antibody. In addition, IgG2 has the propensity of forming covalent dimers. This, in turn, helps to increase the avidity as well as the internalization of the antibody [22]. The most suitable choice is often IgG1 or IgG4 isotype antibodies.

The monoclonal antibodies should have certain ideal properties that include [23] the intracellular uptake of monoclonal antibodies; uptake is accomplished with the help of receptor-mediated endocytosis or by other processes that include the lysosomal and endosome system. So, the monoclonal antibody is designed in such a way that it only binds with the surface antigen. It is modified concerning its affinity, binding, internalization kinetics, and specificity. The shedding of the receptor cell surface should be minimized to reduce the binding during circulation. The conjugation with the payload should not affect its stability, binding, and overall pharmacokinetics (PKs). It should have a long half-life for its accumulation in the specific cell and the immunogenic response should be as low as possible. The range of 0.1–1 nm is the ideal binding affinity range of the antibodies. The first-generation ADCs are composed of murine antibodies and have displayed severe immune reactions. They have additional challenges related to the inability to efficiently penetrate inside the cells and the requirement of a more potent payload. To overcome this drawback, the second-generation antibodies came into the market and these are made via bioengineering processes, such as the humanized antibody [24]. For that reason, nowadays, most of the marketed antibodies are made up of the humanized antibody and the development of the potent payload. The therapeutic monoclonal antibodies show linear pharmacokinetics at high doses and nonlinear pharmacokinetics at low doses [25]. The target-assisted clearance is the primary reason for the nonlinear elimination [26]. The antibody has acted as a carrier system that carries the payload toward the site by responding toward the target antigen and the antibody itself can be active. Despite all the advantages provided by the monoclonal antibody, it has a profound limitation in the route of administration of the ADC [27, 28]. Poor tissue penetration results in degradation and poor bioavailability if these are given through the oral route, so the most favorable route of administration is by the parenteral route [29]. Research in this area is required to develop orthogonal delivery, formulation systems, and route of administration [30–37].

5.2.2 Linker

The chemistry of the linker plays a vital role in the efficacy as well as the safety of the ADCs [38–40]. Generally, it should be stable during the circulation in the bloodstream and it should release the payload at the target site. It should also be nontoxic and inactive when it is conjugated with the antibody [41]. The nature of the linker molecule should be hydrophobic for better targeting of the ADCs [42].

The linkers are classified into two major categories, cleavable and non-cleavable. The cleavable linkers are mainly responsive to physiological stimuli, which are further divided into three major groups that are (i) acid-labile linkers, (ii) protease-cleavable linkers, and (iii) disulfide linkers [43].

I. The acid-labile linker is hydrolyzed in the lysosomes at pH 4.4–5 and in the endosome at pH 5–6.5. They are stable in alkaline conditions. The acid-labile linkers show the nonspecific release of the drug after reaching inside the lysosome [44, 45]. The IMMU-110 is an example of the acid-labile linker ADC design in which the antibody is the humanized anti-CD74 monoclonal antibody, and the payload is doxorubicin conjugated with hydrazone linker [46].
II. The protease-cleavable linkers are the most common linker. These linkers are stable in different pH conditions and the presence of various serum protease inhibitors. They possess the ability to release the drug only at the target cell in a specific manner after the action of the lysosomal proteases, such as cathepsin B. The valine citrulline is an example of a peptide linker [47, 48]. Also, the β-glucuronide linker is a type of protease linker. The tumor cells, as well as lysosomes, are abundant in β glucuronidase [49]. So, it allows the cleavage of the β-glucuronide linker in the presence of β glucuronidase and thereby releasing the payload [50]. It has high hydrophobicity and shows high solubility and an increase in the efficacy of the payload [51].
III. The glutathione disulfide linker is another widely used linker [52]. Glutathione is released in a more considerable amount during stress conditions, such as in tumor cells [53]. The tumor cell has an enzyme that reduces the disulfide bond in the cells. The reduced glutathione is then responsible for the release of the payload in the disulfide linkers [54].

The non-cleavable linkers are dependent on the degradation of the lysosome. The non-cleavable linkers, such as the thioether-based bonds, have better stability in the blood, and these improve the therapeutic index of the ADCs [15]. The mechanism is based on the internalization of the ADC and, after this, the degradation takes place in the lysosome, releasing the payload. The non-cleavable linker can modify the chemical properties and, in turn, improve the efficacy of the payload [55]. Figure 5.2 provides a visual illustration of the classification of linkers in ADCs.

The conjugation chemistry is the vital factor in determining the therapeutic window of ADCs [56]. Many conjugation strategies have been suggested, such as the conjugation via lysine amino acids, by the cysteine residues, or by the site-specific conjugation that increases the homogeneity of the ADCs. Other approaches are enzyme-mediated conjugation or the incorporation of the unnatural amino acid [23]. The chemical conjugation process has been utilized for the past few years. The amino acids, such as the amino-terminal of the lysine or the sulfhydryl group of the cysteine conjugate with the antibody part, are used in the chemical conjugation. The enzymatic conjugation provides site-specific conjugation and is

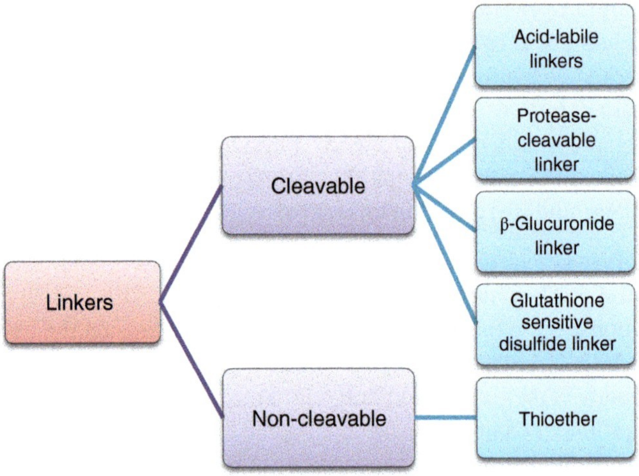

Figure 5.2 Classification of the commonly used linkers in ADCs. Figure demonstrates the key examples of different types of linkers. Source: Modified from Tsuchikama and An [55].

often considered a reproducible technology [57]. These approaches allow enhanced specificity, homogenous nature, and the narrow distribution of the drug–antibody ratio (DAR) [29]. However, the most commonly used approach is the chemical conjugation process.

5.2.3 Payload

The payload is the vital component that plays an essential role in the success of the ADCs. The payload is a drug molecule that is small and highly potent, but it lacks specificity or has a poor absorption/permeation profile. The ideal payload has sufficient solubility in the antibody and has an IC50 value in the sub-nanomolar range. The immunogenicity of the payload should be minimum, and it should have a longer half-life. It should allow the conjugation with the linker and the antibody.

The payloads are divided into two major categories for the conjugation with the monoclonal antibody [23]. The first one is the radionuclide antibody conjugates, which are composed of radionuclides that emit radiation. These are capable of penetrating the target cells with the help of antibodies and produce a response [58, 59]. The second category is the highly potent natural or synthetic drug molecule antibody conjugates. The mechanisms of these ADCs depend on the type of drug that is conjugated, often an anticancer molecule [60]. Some of the payloads used in the development of the ADCs are Auristatins, Maytansinoids, Gemtuzumab, and Inotuzumab. In addition, many other anti-rheumatoid, anti-inflammatory drugs having a smaller molecular size (typically less than 1000 Da) have been employed in ADCs. Furthermore, the ADC payload shows higher potency when it has an optimal DAR (the ratio of the number of drug molecules per molecule of antibody) of the potent molecule [61]. The higher DAR leads to higher clearance of the molecule, resulting in a decrease of efficacy and, in turn, leading to heterogeneity in the distribution [62]. Thus, the DAR should be around four for the optimal activity [63].

5.3 Mechanism of Action

The ADCs are generally administered by the intravenous route to bypass the degradation caused by the gastric acid in the stomach [64]. Then the ADCs bind to the specific target antigen that is highly expressed in the target cell, and not on the normal cell [65]. The antibody recognizes the antigen and binds to it. After the binding, the ADC–antigen complex is internalized by the receptor-mediated endocytosis process. The early endosome containing the ADC–antigen complex is coated with clathrin. Then the influx of the proton ions occurs due to the acidic environment in the endosome. This leads to the interaction of the human neonatal receptors (FcRn) and the monoclonal component of the ADCs [66]. The small fraction of ADC binds to the FcRn and is recycled back outside the cells, and this mechanism prevents the death of the healthy cells. Now, the remaining ADCs in the endosome enter the late endosome stage [67]. The late lysosome is fused with the lysosome that contains the protease enzyme that lowers the pH. The cleaving of the ADC takes place along with the subsequent release of the payload. This payload interferes with the cellular pathway, potentiates apoptosis, and consequently leads to cell death. The pathway of death of the cell depends on the type of payload used [15]. A scheme of the mechanism of action of ADCs is depicted in Figure 5.3.

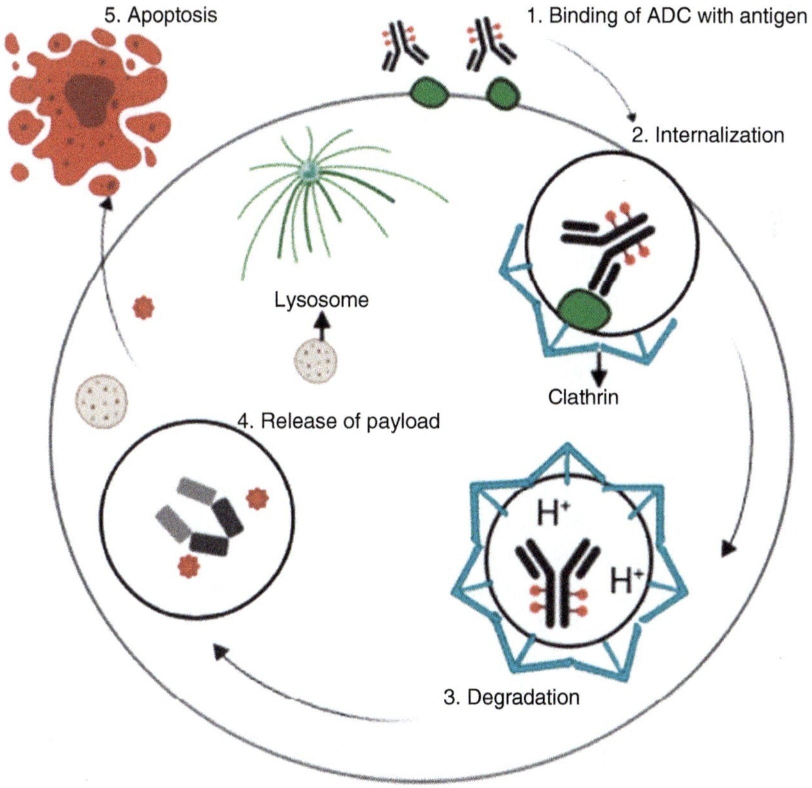

Figure 5.3 Schematic diagram of the mechanism of action of ADCs.

5.4 Pharmacokinetic Considerations for ADCs

The binding of ADCs to the target antigen present on the surface of the cells occurs followed by internalization via the receptor-mediated endocytosis. The next processes include the lysosomal action and the release of the cytotoxic drug leading to cell death, which is the primary mechanism of the action for ADCs. However, there is a possibility that it may be taken up by other nonspecific cells via pinocytosis that will lead to the undesired release of the drug causing toxicity to the normal cells [68].

In an ADC, the major portion is contributed by the antibody component, which is around 98% of the total molecular weight of the ADC (typically in the range of 150 kDa). Therefore, the antibody component has a strong influence on the pharmacokinetics (PK) parameters of the ADC. The binding of the molecule to the target, Fc receptor-dependent recycling, and the Fc effector functions are majorly influenced by the antibody component. The slow clearance, long half-life, proteolysis-mediated catabolism, and low volume of distribution of the antibody play a major role in the absorption, distribution, metabolism, and excretion (ADME) properties of the ADCs. Although the antibody has major advantages in the PK parameters of the ADCs (by virtue of long half-life), it also contributes to some of the disadvantages, such as poor oral bioavailability and incomplete absorption on subcutaneous or intramuscular administration (mainly by virtue of their large size). Also, it may induce immunogenicity, and sometimes it shows the nonlinear distribution and elimination profile [69].

The unconjugated (non-linked drug and antibody) and the conjugated (properly linked drug and antibody) show a close similarity, but these differ in some of the attributes that are the heterogeneous or nonuniform mixtures of the drug species. This is particularly in the case of the conjugation of the multiple molecules at different positions. Moreover, the ADCs contain two pharmacologically active molecules that are the antibody and the payload or the cytotoxic small drug molecule, this will decide the fate and the *in vivo* behavior of the components [70].

5.4.1 Heterogeneity of ADCs

The heterogeneity in the ADC molecule may arise due to the manufacturing process acting as the first source. The conjugation of the antibody to the drug is the main step of manufacturing, and it is controlled by chemical reactions that involve amino acid residues, such as lysine and cysteine on the antibody [71]. The cysteine and lysine residues are commonly chosen for the conjugation process as these have an immense ability for specific chemical modification. The number of sites of the linkage of the drugs or payload as well as the number of drugs or payload attached to the antibody, i.e. DAR, will differ in the different antibodies based on the conjugation [62, 71]. Generally, the ADCs that are conjugated with the help of cysteines have the DAR ranging from 0 to 8, which indicates that the drug has conjugated to a few or all the cysteines present in the unconjugated antibody and formed a disulfide linkage [71]. ADCs conjugated with the help of lysine residue have the chances of larger variability for the number of conjugated drugs and as well as the positions. The heterogeneity arising due to the chemical or biological processes after the *in vivo* administration is the second major source of heterogeneity. The processes will lead to the loss of the drug from the ADC during the systemic circulation, and this is known as deconjugation [72]. The deconjugation mechanism will depend on the type of linker used as well as the site at which the conjugation occurs by either of the processes that include an enzymatic or chemical process. This deconjugation leads to the loss of the drugs from the ADC and also lowers the average DAR which, in turn, shows an impact on the efficacy as well as the behavior of the ADC [73].

The heterogeneity of the ADC poses challenges for the characterization, quantitation, and optimization of the ADC. Various approaches have been used to reduce heterogeneity concerning the manufacturing process. Junutula et al. reported a new class of THIOMAB–drug conjugates that have enhanced site specificity by the use of engineered cysteines that show low heterogeneity, defined DAR, and enhanced therapeutic index [74, 75]. Also, McDonagh et al. have specifically engineered cysteine and their interchain disulfides that result in the antibody with fewer cysteine conjugation and having low heterogeneity [76]. The heterogeneity due to the *in vivo* process still exists [63]. Complete knowledge of the site of drug conjugation or the DAR effect on the ADC PK is needed for further studies. Furthermore, it can be concluded that DAR plays a major role in the design as well as the optimization of the ADCs [68].

5.4.2 Bioanalytical Considerations for ADCs

In an ADC, both the antibody and the drug molecule are essential components for the activity and hence require to be assayed. The components provide information on the *in vivo* behavior of a drug either in the combination or alone. The assays include the estimation of total antibody (Tab) (unconjugated and conjugated antibody), conjugated drug, conjugated antibody, unconjugated antibody, and free drug [68, 77, 78].

The Tab concentration is usually measured by the enzyme-linked immunosorbent assay (ELISA), and it contains the combination of both the unconjugated and the conjugated forms of ADCs [79] The Tab gives the idea about the PK profile related to the antibody and the *in vivo* stability. The Tab PK profile of the ADC has a vital role in the optimization, and impact of the conjugation and the drug load selection in the ADC. The ELISA assay can also be used to determine the conjugated antibody amount in the systemic circulation. The conjugated antibody amount is used as an approximate measure of active ADC concentration [80]. Furthermore, the elucidation of this amount is difficult and tiresome. The assay may have different sensitivity toward the change in drug load. The ADCs will have different potencies, and the concentration measured will not provide accurate results, and that will be reflected in the pharmacologic activity [77, 81]. Moreover, the means for the measurement of the conjugated drug is attained by cleaving the drug from the antibody. Then it is quantified, which provides a total quantity of the drug that is bound to the antibody [82]. The drug concentrations are measured by the liquid chromatography–mass spectrometry (LC–MS) or the ELISA method. These assays do not provide any insights into the concentration of the antibody that is bound with the drug. Therefore there will be no differentiation between the high concentrations and the low concentration of the DAR species [79]. Also, the unconjugated drug and the antibody may be measured with the help of the LC–MS or the ELISA method. LC–MS is extremely sensitive and specific as compared to the ELISA method. The unconjugated cytotoxic drug gives an idea about systemic exposure. Until recently, there is no single assay that can determine all the aspects of the *in vivo* characterization of the ADC. But, with the advancements in technology, more detailed and specific analysis for the estimation of the ADCs is possible. This would provide a piece of greater knowledge about the PK behavior of ADCs [68, 77–79].

5.4.3 Pharmacokinetic Parameters of ADCs

The ADCs are composed of both antibody and drug molecules and therefore, they will have the characteristics of both components. Lin and Tibbitts explained the pharmacokinetic disposition of ADCs wherein, conjugation alters the pharmacokinetic parameters exhibiting a rapid decrease in concentration-time profile, and the Tab shows a typical multi-exponential profile as of antibody. The conjugated drug starts at a higher concentration reflecting the DAR and then rapidly decreases than a Tab. The concentration of the free drug is found to be less as well [68].

5.4.3.1 Absorption

The ADCs are mainly administered by the intravenous route, and the absorption is the same as that of unconjugated antibodies [68, 83]. After the intravenous administration, the ADC enters the blood circulation and then reaches the target site depending mainly on the target antigen expression and other factors.

5.4.3.2 Distribution

The major portion of an ADC is the antibody component and therefore, the distribution is somewhat influenced by the antibody component. The primary mechanism responsible for transporting the ADC from plasma to interstitial fluid is the diffusion of the ADC through the vascular endothelial cells. The expression of the target antigen and the internalization of the bound ADC into the cells also affect the distribution of ADC [83–85]. The distribution of the ADC to the nonspecific tissues via antigen-specific or nonspecific processes has a significant effect on the pharmacologic or the toxic effect. There is a possibility of the antigen shedding by the target cell in the systemic circulation [86]. In that case, the ADC can bind to the soluble shredded antigen, and it will result in the clearance of ADC through the liver and this has the chance to cause liver toxicity [87]. The conjugation of the antibody with the drug affects the distribution to the tissues. Therefore, the distribution of the ADC affects the pharmacologic effect as well as the therapeutic index of the ADC.

5.4.3.3 Metabolism and Elimination

The ADC's can be deconjugated into the cytotoxic drug and the antibody by the enzymatic or chemical process [26, 88]. Another pharmacokinetic process followed by an ADC is the metabolism that includes the formation of the antibody and the drug-containing metabolites (drug amino conjugates or free drug). The process of formation of the metabolic factors depends on linker stability, site of conjugation, and total drug load. The ADCs made up of a linker that is susceptible to chemical or enzyme predominately displays the deconjugation process. In the metabolism step, endocytosis or the pinocytosis processes occur, which is helped by the lysosome that is followed by enzymatic degradation [70]. The size and structure of metabolites of ADCs possess similar properties to a small therapeutic molecule. The nature and behavior of the final metabolites drive the potential toxicities of the ADCs. The toxic nature of ADCs is due to the payload metabolites that signify a nonspecific nature to ADCs. Their metabolism and elimination mainly occur by the cytochrome P450 enzymes and transporters. The metabolites have the chance to engage in drug–drug interactions and may lead to toxicity [68, 85]. Future research in the field of chemical nature of final metabolites and the next generation of ADCs would help attain better-targeted antibodies, stable linkers, and lesser payload-specific toxicities.

5.5 Applications of ADCs

Currently, there are 7 approved ADCs, and more than 100 are in the clinical trials [64]. The approved ADCs have mainly found application in cancer therapy. New ADCs are now being developed for the treatment of rheumatoid arthritis, bacterial infections, ocular diseases, and for the treatment of tuberculosis. The research is focused on this field because of the ability to deliver the drug to the target along with fewer side effects. Section 5.5.1 discusses approved ADCs available in the market and a detailed compilation of key information has been presented in Table 5.1.

5.5.1 Approved ADCs in the Market

5.5.1.1 Gemtuzumab Ozogamicin

Gemtuzumab ozogamicin is the first approved ADC that is marketed by Wyeth under the brand name Mylotarg. It was approved by the US-FDA, however, withdrawn after a few years, this was mainly due to the high mortality rate when compared to conventional treatment, it had fewer benefits [96]. It is composed of cytotoxic agent *N*-acetyl gamma calicheamicin that is covalently bound with anti-CD33 humanized IgG4 monoclonal antibody by the hydrazone linker. It was used in the treatment of CD33-positive acute myeloid leukemia in pediatric patients (above the age of 2) and adults. It acts by binding to the CD33 antigen followed by internalization and delivering the *N*-acetyl gamma calicheamicin in the cell and leading to cell death [89].

It was again brought back to the market in the year 2017 by Pfizer after an open-labeled phase III trial in 280 older patients after the meta-analysis study of the previous trial [97]. In April 2018, it was approved for the treatment of acute myeloid leukemia. The US-FDA has reported hepatotoxicity as the major side effect of these medications. The therapy is comprised of one induction cycle and two consolation therapies [98]. The ADC might be administered along with the daunorubicin and the cytarabine as the combination therapy. In the market, it is available as an injection dosage with a dose of 4.5 mg as a lyophilized powder. This formulation needed to be reconstituted before use [97].

Table 5.1 Summary table of antibody–drug conjugates.

Sr. no.	Generic name	Manufacturer	Brand name	Linker	Payload	Target antigen	Use	References
1	Gemtuzumab ozogamicin	Wyeth/Pfizer	Mylotarg	Hydrazone	*N*-acetyl gamma calicheamicin	CD33	Acute myeloid leukemia	[89]
2	Brentuximab vedotin	Seattle Genetics	Adcretris®	Protease cleavable	Monomethyl auristatin E	CD30	Relapsed Hodgkin's lymphoma and relapsed anaplastic large cell lymphoma	[90]
3	Ado-trastuzumab emtansine	Roche, Genentech	Kadcyla	Non-cleavable linker	Maytansinoid DM1	HER2	Metastatic breast cancer	[91]
4	Inotuzumab ozogamicin	Pfizer/Wyeth	Besponsa	Acid labile linker	*N*-acetyl γ-calicheamicin 1,2-dimethyl hydrazine dichloride	CD22	Relapsed or refractory B cell malignancies	[92]
5	Polatuzumab vedotin	Roche, Genentech	Polivy	Protease cleavable linker	Dolastatin 10 analog monomethyl auristatin	CD79b	Relapsed or refractory (R/R) diffuse large B cell lymphoma	[93]
6	Enfortumab vedotin	Astellas Pharma/Seattle Genetics	Padcev	Protease linker	Monomethyl auristatin E	Nectin-4	Metastatic urothelial cancer	[94]
7	Trastuzumab deruxtecan	Daiichi Sankyo/AstraZeneca	Enhertu™	Cleavable linker	Deruxtecan	HER2	Metastatic HER2 positive breast cancer	[95]

5.5.1.2 Brentuximab Vedotin

Brentuximab vedotin received the marketing approval in August 2011, and it was marketed by Seattle Genetics under the brand name Adcetris® [99]. It is the second approved ADC by the USFDA and is composed of an anti-CD30 antibody, which is conjugated with monomethyl auristatin E with the help of a protease-cleavable linker. It is used in the treatment of anaplastic large-cell lymphoma and Hodgkin's lymphoma. It received accelerated approval based on the result of the phase II trial in the relapsed Hodgkin's lymphoma and relapsed anaplastic large-cell lymphoma [90, 100]. It acts by entrapping monomethyl auristatin E in the CD30-positive cancer cell and hindering cell division. This leads to cell death by creating an environment in which the cell will not replicate further. It is well tolerated in the patients but shows some adverse effects, such as peripheral sensory neuropathy, neutropenia, fatigue, nausea, anemia, cough, thrombocytopenia, diarrhea, pyrexia, rash, and upper respiratory infection. It is administered as an intravenous infusion and is available in the market as an injection with a dose of 50 mg as a lyophilized powder that needs to be reconstituted before use. Doxorubicin, vinblastine, and dacarbazine can be given in combination with the Brentuximab vedotin [101].

5.5.1.3 Ado-Trastuzumab Emtansine (T-DM1)

Ado-trastuzumab emtansine (T-DM1) gained approval in February 2013, and it is marketed by Roche, Genentech, under the brand name of Kadcyla® [102, 103]. It is the first ADC approved in non-hematological malignancies. It is composed of a trastuzumab (humanized anti-HER2) monoclonal antibody that is conjugated with several molecules of maytansinoid DM1 via a non-cleavable linker [succinimidyl *trans*-4-(maleimidylmethyl) cyclohexane-1-carboxylate]. It is used in the treatment of late-stage (metastatic) HER2-positive breast cancer [91].

It has received fast-track approval by the US-FDA for breast cancer in patients who had received taxane and trastuzumab either in combination or in monotherapy. It was approved based on the results of a randomized trial study, i.e. phase III EMILIA. It acts by disrupting microtubule synthesis after targeting the HER2 antigen, causing cell-cycle arrest, and ultimately leading to apoptosis. It shows adverse events, such as transaminitis, fatigue, nausea, and thrombocytopenia. It is marketed as an injection dosage form at a dose of 100 and 160 mg lyophilized powder meant to be reconstituted before use. It is used as a monotherapy for metastatic breast cancer [104].

5.5.1.4 Inotuzumab Ozogamicin

It was approved in August 2017, and it is marketed by Pfizer/Wyeth under the brand name Besponsa™, which targets the CD22 antigen [105]. It is composed of cytotoxic drug *N*-acetyl γ-calicheamicin 1,2-dimethyl hydrazine dichloride that is conjugated with the humanized monoclonal IgG4 antibody with the help of acid-labile linker [consisting of the condensation product of 3-methyl-3-mercaptobutane hydrazide and 4-(4′-acetylphenoxy)-butanoic acid (AcBut)]. It is used in the treatment of relapsed or refractory B cell malignancies [92]. It acts by disrupting double-strand DNA that causes cell cycle and results in apoptosis. It shows hepatotoxicity as the major side effect of the treatment [106].

It can be used either in combination with other chemotherapeutic agents or as monotherapy. It is available as an injection dosage form with a dose of 0.9 mg as a lyophilized powder that is needed to be reconstituted and diluted before use.

5.5.1.5 Polatuzumab Vedotin-piiq

Polatuzumab vedotin-piiq has received marketing approval in the year of June 2019, and it is marketed by Roche-Genentech under the brand name Polivy™ [107]. It is first in the class of ADCs that targets the CD79b and is given in combination with rituximab and bendamustine for the treatment of relapsed or refractory (R/R) diffuse large B cell lymphoma, which is a type of non-Hodgkin lymphoma in the adults. It is composed of synthetic dolastatin 10 analog monomethyl auristatin, a microtubule inhibitor that is conjugated with a humanized monoclonal antibody that targets the B cell antigens, and it is conjugated with protease-cleavable linker (maleimidocaproylvaline-citrulline-*p*-aminobenzoyloxycarbonyl) [93].

It has been granted accelerated approval based on the results of the Phase Ib/II GO29365 randomized study [108]. It shows its action by disrupting the microtubule and causing cell-cycle arrest and leading to cell death. It shows adverse effects, such as low white blood cell count, low platelet levels, numbness, diarrhea, and decreased appetite. Every single dose of the vial contains 140 mg as a lyophilized powder that needs to the reconstituted before use [109].

5.5.1.6 Enfortumab Vedotin

Enfortumab vedotin was approved in December 2019, and it is marketed by Astellas Pharma/Seattle Genetics under the brand name Padcev™ [110]. It is composed of IgG1 kappa monoclonal antibody (targeted toward the antigen Nectin-4) that is conjugated with monomethyl auristatin E with the help of protease linker (maleimidocaproyl-valyl-citrullinyl-*p*-aminobenzyloxycarbony). It is the first in the class of ADCs that is directed toward the Nectin-4. It is used in the therapy of metastatic urothelial cancer that has received the programmed death-ligand-1 or programmed death receptor-1 inhibitor and platinum-containing chemotherapy as an adjuvant [94].

It binds to the Nectin-4 antigen, and after internalization, it binds to tubulin and inhibits its polymerization leading to cell death. It is available as 30 or 20 mg as a lyophilized powder in the vial that needs to be reconstituted before use. It has been granted accelerated approval based on the rate of tumor response. Some of the adverse effects of ADC include sepsis, urinary tract infection, dyspnoea, and rash [111].

5.5.1.7 Trastuzumab Deruxtecan

Trastuzumab deruxtecan was granted US-FDA approval in December 2019, and it is marketed by Daiichi Sankyo/AstraZeneca under the brand name Enhertu® [112]. It has been used in the treatment of metastatic HER2-positive breast cancer that has previously received an anti-HER2-based regimen. It has received accelerated approval based on the duration of response and the tumor response rate. It is composed of the monoclonal antibody trastuzumab that is directed toward HER2 antigen, and it is conjugated with deruxtecan, which is a topoisomerase I inhibitor via a cleavable linker [95].

The antibody on binding to the HER2 antigen undergoes internalization and the cleaving of the bond releases the drug. This, in turn, causes DNA damage and, ultimately, cell death. The adverse effect that is seen with this therapy is nausea, fatigue, alopecia, vomiting, constipation, anemia, and cough. It is given as intravenous infusion at a dose of 5.4 mg kg^{-1} once every three weeks [113].

5.5.2 Use of ADCs in Rheumatoid Arthritis

Rheumatoid arthritis is a chronic autoimmune disorder that results in the inflammation of the synovial membrane and the destruction of bone. In these conditions, there are overexpressed pro-inflammatory cytokines like IL-1, IL-6, and tumor necrosis factor (TNF) in the serum and the synovial fluid [114]. There are many other options available for the symptomatic relief of rheumatoid arthritis, but ADCs are being developed because of their good targeting and fewer side effects [115]. The Tocilizumab alendronate ADC complex has been developed for the treatment of this disorder. The ADC is composed of the humanized monoclonal antibody Tocilizumab that is conjugated with the alendronate drug via the cleavable disulfide linker [3-(2-pyridyldithio) propionyl hydrazide–poly(ethylene glycol)-*N*-hydroxysuccinimide] [116]. The ADC blocks the IL-6 and decreases the symptoms of rheumatoid arthritis and alendronate, which acts as an anti-inflammatory agent and decreases the inflammation by inhibiting the macrophages. Thus the conjugate was found to be an effective option in the treatment of rheumatoid arthritis [116, 117].

5.5.3 Use of ADCs in Bacterial Infections

The concept of ADC can be applied to treat bacterial infection by delivering the antibiotic to the bacteria with the help of the monoclonal antibody [118]. It can be used in the therapy of *Staphylococcus aureus* bacteria and other bacterial infections, such as tuberculosis infection. One such ADC is composed of a β-*N*-acetylglucosamine cell wall teichoic acid antibody that is conjugated with rifampicin via a linker (MC-ValCit-PABQ) [119]. The ADC binds to the antigen present on the surface of the bacteria, and then it is internalized within the cell. This causes the release of the drug inside the bacteria that cause the death of the cell. This approach can be utilized to overcome antibiotic resistance and offers a great opportunity for the treatment of infectious diseases. It also has the potential to deliver the drug with poor antibiotic efficacy and reduce disease relapse [118, 120].

5.5.4 Use of ADCs in Ophthalmology

The development of ADC has a wide scope in the field of ophthalmology for achieving efficacious drug concentration in the eye. The ADC is found to be a promising approach for the treatment of choroidal neovascularization [121]. In choroidal neovascularization, there is abnormal growth and formation of new blood vessels that are leaky underneath the retinal pigment epithelium and neural retina [122]. An ADC can be developed to target the antigen that is unique in this disease. The anti-vascular endothelial growth can be used as an antigen of this disease, but it is soluble, so there are limitations for the delivery of the drug. Research is ongoing to find a target that is present in the abnormal cell and responsible for the choroidal neovascularization [123].

In the treatment of the posterior capsule opacification, the ADC has found its application. The opacity gradually develops due to the formation of multiple layers of epithelial cells [121]. Many ADCs have been in the clinical trial phases for the treatment of this disease. For example, an ADC was developed containing the murine antibody that is conjugated to polypeptide ricin toxin via a linker [124]. The ADC targets the glycoprotein epitope that presents on the surface of the epithelial cell, and it is internalized. After that, the ADC kills the cell by inhibiting protein synthesis. However, to date, laser capsulotomy remains the best choice for the treatment [125]. The advancement in the field of ADC is going on to find a viable option for diseases related to ophthalmology.

5.6 Resistance of ADC

The resistance to monoclonal-based therapeutics could become a concern in the treatment of cancer and other diseases [38]. The resistance arises due to the host or tumor-related factors [126]. The efflux of the drug due to the P-glycoprotein is the main mechanism of drug resistance [127, 128]. The cell-cycle mechanism may also be responsible for the resistance of ADC with cells in the G0 phase being less sensitive to the drug [129]. The mechanism of resistance against ADCs has been represented in Figure 5.4.

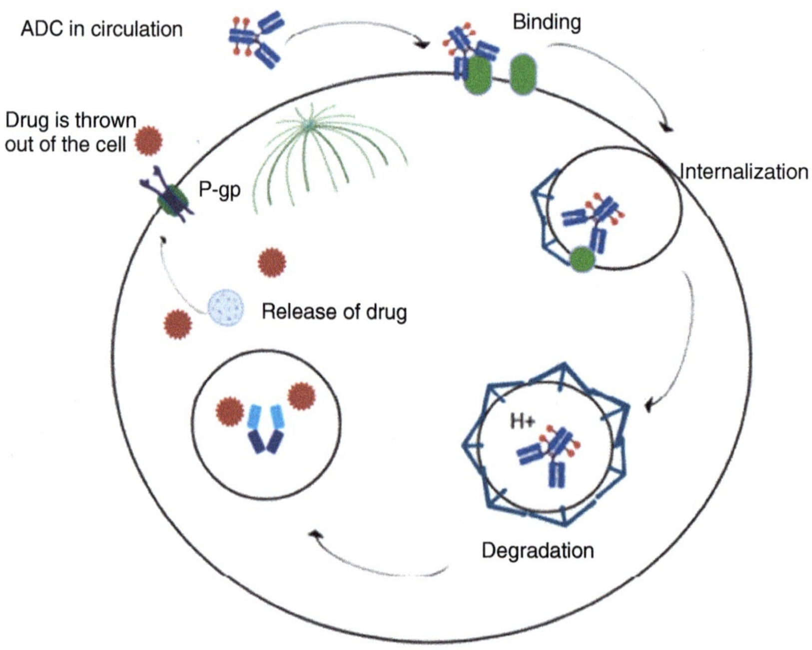

Figure 5.4 Mechanism of resistance against ADCs.

The resistance can prevail over by the prodrug strategy. The other approach is the use of the other same class molecules that are poor substrates to the P-glycoprotein or the use of large molecules ADC [38, 130]. The use of a linker that is non-cleavable will be a potential approach for overcoming the resistance [131].

5.7 Regulatory Aspects for ADCs

ADCs are evaluated by the FDA's Center for Drug Evaluation and Research (CDER), whereas monoclonal antibodies are evaluated by the Center for Biological Evaluation and Research (CBER). However, from 2003, the monoclonal antibodies, as well as the ADCs, are evaluated in the CDER. The monoclonal antibody, the linker–drug, and the ADC, as whole, are needed to be characterized for the chemistry, manufacturing, and control (CMC) section in the regulatory submission for the quality. The FDA involves two offices for the review process of the ADC; one is the Office of Biological Products (OBP), and the other is the Office of New Drug Quality Assessment (ONDQA). The monoclonal antibodies are reviewed in the OBP, and the linker and drug are evaluated in the ONDQA [132]. An overview of the regulatory aspects of ADCs is provided in Figure 5.5.

5.7.1 Role of ONDQA

The drug–linker is evaluated based on the optical chirality, impurities (related to the product, process, free drug, residual solvents, and heavy metals), polymorphism, biological activity (target specificity and binding affinity), and also the potency. In addition, the starting material source, that is the fermentation source for the microbial strains, modification of the structure of the peptides, natural products, and the synthetic compound, is to be evaluated. The linker–drug intermediates are manufactured by several conjugation reactions. Therefore, the quantification of the impurities is important. The linker–drugs are characterized by employing a series of orthogonal techniques, e.g. UV spectroscopy, IR spectroscopy, mass spectrometry, NMR spectroscopy, and elemental analysis. The impurity levels higher than 1% are typically structurally characterized [133].

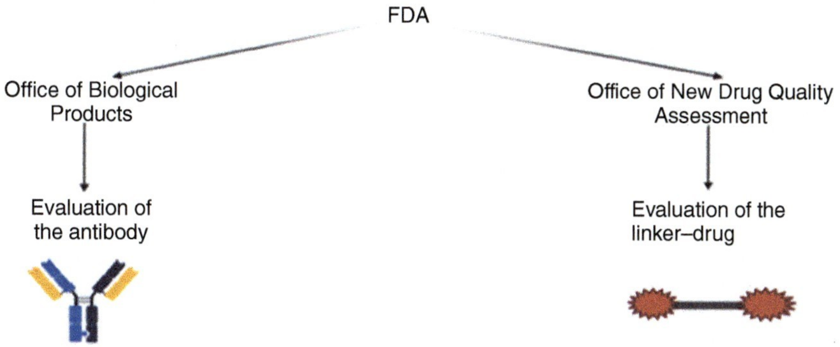

Figure 5.5 Regulatory aspect for ADCs.

5.7.2 Role of OBP

It focuses on the assessment of the monoclonal antibody. These include the methods that measure the primary structure (peptide mapping and N-terminal/C-terminal sequence analysis), conformational structure (charge, size, and molecular weight), an assay of the presence of fragments, specificity toward target antigen, and post-translation modification (sialic acid determination, the content of monosaccharide, and oligosaccharide profile analysis). The peptide mapping is used for the primary sequence of the monoclonal antibody. It also provides information about the batch-to-batch consistency of the ADCs. The techniques for the characterization should also analyze the effector function. Assay methods should characterize impurities originating from a monoclonal antibody perspective, including process (host cell DNA and microbial contaminants) and product (dimers, aggregates, and the degradation products). The impurities should be assessed for their impact on biological activity. The validation should be done after removing the process-related impurities [134]. Then finally, the OBP and ONDQA are jointly reviewed under one reviewer, and then the approval may or may not be granted depending on the report [132].

5.8 Conclusion and Future Direction

As the interest of the research in medicine is shifting toward biologics, the concept of ADCs has flourished. The approval of ADCs is increasing day by day because of their good targeting ability. This review has provided knowledge and insights about the basic aspects of the ADCs. The history of the ADCs focuses on how the concept of ADC came into the picture and some of the rewards of the therapy. The importance of every component necessary for the design of the ADC and the mechanism of action after internalization have been discussed in detail. The ADC is a hybrid system of the antibody and the payload or drug, so it has the amalgamation of both the properties on the pharmacokinetic parameters. The heterogeneity and the *in vivo* disposition are the most important parameters that are to be characterized. As this concept is emerging, the need for a better analytical system will play a crucial role in determining the pharmacokinetic parameters. Until now, there are seven USFDA approved ADCs, out of which three drugs are approved recently in 2019, proving it a revolutionary year in the field of ADCs. Most of the ADCs have been approved for utilization in the field of cancer. But the studies are going on to expand their application in other indications as well. Promising studies have already been done in the field of inflammation, bacterial infections, and ophthalmology. The chances of resistance are less in the case of the ADCs, but some events have shown the development of resistance in the cell, and hence additional strategies have also been built to overcome such challenges. The FDA has also supported the approval of the ADCs by issuing appropriate regulatory guidelines. The evaluation of the ADC is done by the two offices OBP and ONDQA. The field of ADCs is an emerging area with ongoing to resolve issues related to resistance and design improvement. The research has been done to widen the scope of their application beyond cancer therapy so that, approval in the treatment of other diseases in the future can be a reality.

References

1 McCamish, M., Yoon, W., and McKay, J. (2016). Biosimilars: biologics that meet patients' needs and healthcare economics. *Am. J. Manag. Care* 22 (13 Suppl.): S439–S442. http://www.ncbi.nlm.nih.gov/pubmed/28719221.

2 U.S. FDA. Drugs@FDA: FDA-Approved Drugs. https://www.accessdata.fda.gov/scripts/cder/daf/index.Cfm?event=overview.process&ApplNo=021409 (accessed Apr. 21, 2020).

3 de la Torre, B.G. and Albericio, F. (2019). The pharmaceutical industry in 2018. An analysis of FDA drug approvals from the perspective of molecules. *Molecules* 24 (4): 809. https://doi.org/10.3390/molecules24040809.

4 Weiner, G.J. (2015). Building better monoclonal antibody-based therapeutics. *Nat. Rev. Cancer* 15 (6): 361–370. https://doi.org/10.1038/nrc3930.

5 Schwartz, R.S. (2004). Paul Ehrlich's magic bullets. *N. Engl. J. Med.* 350 (11): 1079–1080. https://doi.org/10.1056/NEJMp048021.

6 Himmelweit, F. (ed.) (1960). *The Collected Papers of Paul Ehrlich*, 1e. Pergamon. https://www.elsevier.com/books/the-collected-papers-of-paul-ehrlich/himmelweit/978-0-08-009056-6 (accessed 21 April 2020).

7 Köhler, G. and Milstein, C. (1975). Continuous cultures of fused cells secreting antibody of predefined specificity. *Nature* 256 (5517): 495–497. https://doi.org/10.1038/256495a0.

8 Saleh, M.N., Sugarman, S., Murray, J. et al. (2000). Phase I trial of the anti–Lewis Y drug immunoconjugate BR96-doxorubicin in patients with Lewis Y–expressing epithelial tumors. *J. Clin. Oncol.* 18 (11): 2282–2292. https://doi.org/10.1200/JCO.2000.18.11.2282.

9 Tolcher, A.W., Sugarman, S., Gelmon, K.A. et al. (1999). Randomized phase II study of BR96-doxorubicin conjugate in patients with metastatic breast cancer. *J. Clin. Oncol.* 17 (2): 478–478. https://doi.org/10.1200/JCO.1999.17.2.478.

10 Petersdorf, S., Kopecky, K., Stuart, R.K. et al. (2009). Preliminary results of Southwest Oncology Group Study S0106: an international intergroup phase 3 randomized trial comparing the addition of gemtuzumab ozogamicin to standard induction therapy versus standard induction therapy followed by a second randomization to post-consolidation gemtuzumab ozogamicin versus no additional therapy for previously untreated acute myeloid leukemia. *Blood* 114 (22): 790–790. https://doi.org/10.1182/blood.V114.22.790.790.

11 Pd, S. and El, S. (2012). The discovery and development of brentuximab vedotin for use in relapsed Hodgkin lymphoma and systemic anaplastic large cell lymphoma. *Nat. Biotechnol.* 30 (7): 631–637. https://doi.org/10.1038/nbt.2289.

12 Barok, M., Joensuu, H., and Isola, J. (2014). Trastuzumab emtansine: mechanisms of action and drug resistance. *Breast Cancer Res.* 16 (2): 209. https://doi.org/10.1186/bcr3621.

13 Burris, H.A. (2013). Developments in the use of antibody–drug conjugates. *Am. Soc. Clin. Oncol. Educ. Book* https://doi.org/10.1200/EdBook_AM.2013.33.e99.

14 Lambert, J.M. and Morris, C.Q. (2017). Antibody–drug conjugates (ADCs) for personalized treatment of solid tumors: a review. *Adv. Ther.* 34 (5): 1015–1035. https://doi.org/10.1007/s12325-017-0519-6.

15 Nagayama, A., Ellisen, L.W., Chabner, B., and Bardia, A. (2017). Antibody–drug conjugates for the treatment of solid tumors: clinical experience and latest developments. *Targeted Oncol.* 12 (6): 719–739. https://doi.org/10.1007/s11523-017-0535-0.

16 Diamantis, N. and Banerji, U. (2016). Antibody–drug conjugates – an emerging class of cancer treatment. *Br. J. Cancer* 114 (4): 362–367. https://doi.org/10.1038/bjc.2015.435.

17 Donaghy, H. (2016). Effects of antibody, drug and linker on the preclinical and clinical toxicities of antibody–drug conjugates. *mAbs* 8 (4): 659–671. https://doi.org/10.1080/19420862.2016.1156829.

18 Damelin, M., Zhong, W., Myers, J., and Sapra, P. (2015). Evolving strategies for target selection for antibody–drug conjugates. *Pharm. Res.* 32 (11): 3494–3507. https://doi.org/10.1007/s11095-015-1624-3.

19 Strohl, W.R. (2018). Current progress in innovative engineered antibodies. *Protein Cell* 9 (1): 86–120. https://doi.org/10.1007/s13238-017-0457-8.

20 Hoffmann, R.M., Coumbe, B.G.T., Josephs, D.H. et al. (2018). Antibody structure and engineering considerations for the design and function of antibody drug conjugates (ADCs). *OncoImmunology* 7 (3): e1395127. https://doi.org/10.1080/2162402X.2017.1395127.

21 Yoo, E.M., Wims, L.A., Chan, L.A., and Morrison, S.L. (2003). Human IgG2 can form covalent dimers. *J. Immunol.* 170 (6): 3134–3138. https://doi.org/10.4049/jimmunol.170.6.3134.

22 van der Neut Kolfschoten, M., Schuurman, J., Losen, M. et al. (2007). Anti-inflammatory activity of human IgG4 antibodies by dynamic Fab arm exchange. *Science* 317 (5844): 1554–1557. https://doi.org/10.1126/science.1144603.

23 Bakhtiar, R. (2016). Antibody drug conjugates. *Biotechnol. Lett.* 38 (10): 1655–1664. https://doi.org/10.1007/s10529-016-2160-x.

24 Scott, A.M., Wolchok, J.D., and Old, L.J. (2012). Antibody therapy of cancer. *Nat. Rev. Cancer* 12 (4): 278–287. https://doi.org/10.1038/nrc3236.

25 Han, T.H. and Zhao, B. (2014). Absorption, distribution, metabolism, and excretion considerations for the development of antibody–drug conjugates. *Drug Metab. Dispos.* 42 (11): 1914–1920. https://doi.org/10.1124/dmd.114.058586.

26 Vugmeyster, Y., Xu, X., Theil, F.-P. et al. (2012). Pharmacokinetics and toxicology of therapeutic proteins: advances and challenges. *World J. Biol. Chem.* 3 (4): 73–92. https://doi.org/10.4331/wjbc.v3.i4.73.

27 Dubey, S. and Giovannini, R. (2021). Stability of biologics and the quest for polysorbate alternatives. *Trends Biotechnol.* 39 (6): 546–549.

28 Nongkhlaw, R., Patra, P., Chavrasiya, A. et al. (2020). Chapter 6 – Biologics: delivery options and formulation strategies. In: *Drug Delivery Aspects*, vol. 4, 115–155.

29 Chari, R.V.J., Miller, M.L., and Widdison, W.C. (2014). Antibody–drug conjugates: an emerging concept in cancer therapy. *Angew. Chem. Int. Ed.* 53 (15): 3796–3827. https://doi.org/10.1002/anie.201307628.

30 Dubey, S., Perozzo, R., Scapozza, L., and Kalia, Y.N. et al. (2021). Stability of insulin like growth factor 1 (IGF-1) in the presence of dermis and epidermis. *Eur. J. Pharm. Biopharm.* 158: 379–381.

31 Dubey, S., Perozzo, R., Scapozza, L. et al. (2020). Specific protein–protein interactions limit the cutaneous iontophoretic transport of interferon β-1B and a poly-ARG interferon β-1B analogue. *Int. J. Pharm.: X* 2: 100051. https://doi.org/10.1016/j.ijpx.2020.100051.

32 Yu, J., Dubey, S., and Kalia, Y.N. (2018). Needle-free cutaneous delivery of living human cells by Er:YAG fractional laser ablation. *Expert Opin. Drug Delivery* 15 (6): 559–566.

33 Dubey, S. and Kalia, Y.N. (2014). Understanding the poor iontophoretic transport of lysozyme across the skin: when high charge and high electrophoretic mobility are not enough. *J. Controlled Release* 183: 35–42.

34 Dubey, S., Perozzo, R., Scapozza, L. et al. (2011). Noninvasive transdermal iontophoretic delivery of
biologically active human basic fibroblast growth factor. *Mol. Pharmaceutics* 8 (4): 1322–1331.

35 Ibeanu, N., Egbu, R., Onyekuru, L. et al. (2020). Injectables and depots to prolong drug action of proteins and peptides. *Pharmaceutics* 12 (10): https://doi.org/10.3390/pharmaceutics12100999.

36 Dubey, S. and Kalia, Y.N. (2011). Electrically assisted delivery of an anionic protein across intact skin: cathodal iontophoresis of biologically active ribonuclease T1. *J. Controlled Release* 152 (3): 356–362.

37 Dubey, S. and Kalia, Y.N. (2010). Non-invasive iontophoretic delivery of enzymatically active ribonuclease A (13.6 kDa) across intact porcine and human skins. *J. Controlled Release* 145 (3): 203–209.

38 Shefet-Carasso, L. and Benhar, I. (2015). Antibody-targeted drugs and drug resistance – challenges and solutions. *Drug Resist. Updates* 18: 36–46. https://doi.org/10.1016/j.drup.2014.11.001.

39 Hughes, B. (2010). Antibody–drug conjugates for cancer: poised to deliver? *Nat. Rev. Drug Discovery* 9 (9): 665–667. https://doi.org/10.1038/nrd3270.

40 Flygare, J.A., Pillow, T.H., and Aristoff, P. (2013). Antibody–drug conjugates for the treatment of cancer. *Chem. Biol. Drug Des.* 81 (1): 113–121. https://doi.org/10.1111/cbdd.12085.

41 Teicher, B.A. and Chari, R.V.J. (2011). Antibody conjugate therapeutics: challenges and potential. *Clin. Cancer Res.* 17 (20): 6389–6397. https://doi.org/10.1158/1078-0432.CCR-11-1417.

42 Sassoon, I. and Blanc, V. (2013). Antibody–Drug Conjugate (ADC) Clinical Pipeline: a review. *Methods Mol. Biol.* 1045: 1–27. Humana Press Inc.. SpringerLink. https://link.springer.com/protocol/10.1007%2F978-1-62703-541-5_1 (accessed 21 April 2020).

43 McCombs, J.R. and Owen, S.C. (2015). Antibody drug conjugates: design and selection of linker, payload and conjugation chemistry. *AAPS J.* 17 (2): 339–351. https://doi.org/10.1208/s12248-014-9710-8.

44 Senter, P.D. (2009). Potent antibody drug conjugates for cancer therapy. *Curr. Opin. Chem. Biol.* 13 (3): 235–244. https://doi.org/10.1016/j.cbpa.2009.03.023.

45 Dan, N., Setua, S., Kashyap, V.K. et al. (2018). Antibody–drug conjugates for cancer therapy: chemistry to clinical implications. *Pharmaceuticals* 11 (2): 32. https://doi.org/10.3390/ph11020032.

46 Lu, J., Jiang, F., Lu, A., and Zhang, G. (2016). Linkers having a crucial role in antibody–drug conjugates. *Int. J. Mol. Sci.* 17 (4): 561. https://doi.org/10.3390/ijms17040561.

47 Bryden, F., Martin, C., Letast, S. et al. (2018). Impact of cathepsin B-sensitive triggers and hydrophilic linkers on *in vitro* efficacy of novel site-specific antibody–drug conjugates. *Org. Biomol. Chem.* 16 (11): 1882–1889. https://doi.org/10.1039/C7OB02780J.

48 Wang, Y.-J., Li, Y.-Y., Liu, X.-Y. et al. (2017). Marine antibody–drug conjugates: design strategies and research progress. *Mar. Drugs* 15 (1): 18. https://doi.org/10.3390/md15010018.

49 Jaracz, S., Chen, J., Kuznetsova, L.V., and Ojima, I. (2005). Recent advances in tumor-targeting anticancer drug conjugates. *Bioorg. Med. Chem.* 13 (17): 5043–5054. https://doi.org/10.1016/j.bmc.2005.04.084.

50 de Graaf, M., Boven, E., Scheeren, H.W. et al. (2002). Beta-glucuronidase-mediated drug release. *Curr. Pharm. Des.* http://www.eurekaselect.com/64491/article (accessed 21 April 2020).

51 Bargh, J.D., Isidro-Llobet, A., Parker, J.S., and Spring, D.R. (2019). Cleavable linkers in antibody–drug conjugates. *Chem. Soc. Rev.* 48 (16): 4361–4374. https://doi.org/10.1039/C8CS00676H.

52 Jeffrey, S.C., Andreyka, J.B., Bernhardt, S.X. et al. (2006). Development and properties of β-glucuronide linkers for monoclonal antibody–drug conjugates. *Bioconjugate Chem.* 17 (3): 831–840. https://doi.org/10.1021/bc0600214.

53 Wu, G., Fang, Y.-Z., Yang, S. et al. (2004). Glutathione metabolism and its implications for health. *J. Nutr.* 134 (3): 489–492. https://doi.org/10.1093/jn/134.3.489.

54 Sapra, P., Hooper, A.T., O'Donnell, C.J., and Gerber, H.-P. (2011). Investigational antibody drug conjugates for solid tumors. *Expert Opin. Invest. Drugs* 20 (8): 1131–1149. https://doi.org/10.1517/13543784.2011.582866.

55 Tsuchikama, K. and An, Z. (2018). Antibody–drug conjugates: recent advances in conjugation and linker chemistries. *Protein Cell* 9 (1): 33–46. https://doi.org/10.1007/s13238-016-0323-0.

56 Lyon, R.P., Meyer, D.L., Setter, J.R., and Senter, P.D. (2012). Conjugation of anticancer drugs through endogenous monoclonal antibody cysteine residues. *Methods Enzymol.* 502: 123–138. https://doi.org/10.1016/B978-0-12-416039-2.00006-9.

57 Junutula, J.R., Raab, H., Clark, S. et al. (2008). Site-specific conjugation of a cytotoxic drug to an antibody improves the therapeutic index. *Nat. Biotechnol.* 26 (8): 925–932. https://doi.org/10.1038/nbt.1480.

58 Kitson, S.L., Cuccurullo, V., Moody, T.S., and Mansi, L. (2013). Radionuclide antibody-conjugates, a targeted therapy towards cancer. *Curr. Radiopharm.* 6 (2): 57–71. https://doi.org/10.2174/1874471011306020001.

59 Singh, S.K., Luisi, D.L., and Pak, R.H. (2015). Antibody–drug conjugates: design, formulation and physicochemical stability. *Pharm. Res.* 32 (11): 3541–3571. https://doi.org/10.1007/s11095-015-1704-4.

60 Lambert, J.M. (2013). Drug-conjugated antibodies for the treatment of cancer. *Br. J. Clin. Pharmacol.* 76 (2): 248–262. https://doi.org/10.1111/bcp.12044.

61 Beckley, N.S., Lazzareschi, K.P., Chih, H.-W. et al. (2013). Investigation into temperature-induced aggregation of an antibody drug conjugate. *Bioconjugate Chem.* 24 (10): 1674–1683. https://doi.org/10.1021/bc400182x.

62 Hamblett, K.J., Senter, P.D., Chace, D.F. et al. (2004). Effects of drug loading on the antitumor activity of a monoclonal antibody drug conjugate. *Clin. Cancer Res.* 10 (20): 7063–7070. https://doi.org/10.1158/1078-0432.CCR-04-0789.

63 McDonagh, C.F., Turcott, E., Westendorf, L. et al. (2006). Engineered antibody–drug conjugates with defined sites and stoichiometries of drug attachment. *Protein Eng. Des. Sel.* 19 (7): 299–307. https://doi.org/10.1093/protein/gzl013.

64 Peters, C. and Brown, S. (2015). Antibody–drug conjugates as novel anti-cancer chemotherapeutics. *Biosci. Rep.* 35 (4): https://doi.org/10.1042/BSR20150089.

65 Li, F., Emmerton, K.K., Jonas, M. et al. (2016). Intracellular released payload influences potency and bystander-killing effects of antibody–drug conjugates in preclinical models. *Cancer Res.* 76 (9): 2710–2719. https://doi.org/10.1158/0008-5472.CAN-15-1795.

66 Roopenian, D.C. and Akilesh, S. (2007). FcRn: the neonatal Fc receptor comes of age. *Nat. Rev. Immunol.* 7 (9): 715–725. https://doi.org/10.1038/nri2155.

67 Gorovits, B. and Krinos-Fiorotti, C. (2013). Proposed mechanism of off-target toxicity for antibody–drug conjugates driven by mannose receptor uptake. *Cancer Immunol. Immunother.* 62 (2): 217–223. https://doi.org/10.1007/s00262-012-1369-3.

68 Lin, K. and Tibbitts, J. (2012). Pharmacokinetic considerations for antibody drug conjugates. *Pharm. Res.* 29 (9): 2354–2366. https://doi.org/10.1007/s11095-012-0800-y.

69 Carter, P. (2001). Improving the efficacy of antibody-based cancer therapies. *Nat. Rev. Cancer* 1 (2): 118–129. https://doi.org/10.1038/35101072.

70 Kamath, A.V. and Iyer, S. (2015). Preclinical pharmacokinetic considerations for the development of antibody drug conjugates. *Pharm. Res.* 32 (11): 3470–3479. https://doi.org/10.1007/s11095-014-1584-z.

71 Singh, R. and Erickson, H.K. (2009). Antibody–cytotoxic agent conjugates: preparation and characterization. In: *Therapeutic Antibodies*, Methods in Molecular Biology™ (Methods and Protocols), vol. 525 (ed. A. Dimitrov), 445–467. Totowa, NJ: Humana Press.

72 Wang, L., Amphlett, G., Blättler, W.A. et al. (2005). Structural characterization of the maytansinoid-monoclonal antibody immunoconjugate, huN901-DM1, by mass spectrometry. *Protein Sci.* 14 (9): 2436–2446. https://doi.org/10.1110/ps.051478705.

73 Sun, X., Widdison, W., Mayo, M. et al. (2011). Design of antibody–maytansinoid conjugates allows for efficient detoxification via liver metabolism. *Bioconjugate Chem.* 22 (4): 728–735. https://doi.org/10.1021/bc100498q.

74 Junutula, J.R., Bhakta, S., Raab, H. et al. (2008). Rapid identification of reactive cysteine residues for site-specific labeling of antibody–Fabs. *J. Immunol. Methods* 332 (1–2): 41–52. https://doi.org/10.1016/j.jim.2007.12.011.

75 Dornan, D., Bennett, F., Chen, Y. et al. (2009). Therapeutic potential of an anti-CD79b antibody–drug conjugate, anti-CD79b-vc-MMAE, for the treatment of non-Hodgkin lymphoma. *Blood* 114 (13): 2721–2729. https://doi.org/10.1182/blood-2009-02-205500.

76 McDonagh, C.F., Kim, K.M., Turcott, E. et al. (2008). Engineered anti-CD70 antibody–drug conjugate with increased therapeutic index. *Mol. Cancer Ther.* 7 (9): 2913–2923. https://doi.org/10.1158/1535-7163.MCT-08-0295.

77 Alley, S.C. and Anderson, K.E. (2013). Analytical and bioanalytical technologies for characterizing antibody–drug conjugates. *Curr. Opin. Chem. Biol.* 17 (3): 406–411. https://doi.org/10.1016/j.cbpa.2013.03.022.

78 Gorovits, B., Alley, S.C., Bilic, S. et al. (2013). Bioanalysis of antibody–drug conjugates: American Association of Pharmaceutical Scientists Antibody–Drug Conjugate Working Group position paper. *Bioanalysis* 5 (9): 997–1006. https://doi.org/10.4155/bio.13.38.

79 Kaur, S., Xu, K., Saad, O.M. et al. (2013). Bioanalytical assay strategies for the development of antibody–drug conjugate biotherapeutics. *Bioanalysis* 5 (2): 201–226. https://doi.org/10.4155/bio.12.299.

80 Xu, K., Liu, L., Dere, R. et al. (2013). In vivo drug-linker stability characterization of the drug-to-antibody ratio distribution for antibody–drug conjugates in plasma/serum. *Bioanalysis* 5 (9): 1057–1071. https://doi.org/10.4155/bio.13.66.

81 Sanderson, R.J., Hering, M.A., James, S.F. et al. (2005). *In vivo* drug-linker stability of an anti-CD30 dipeptide-linked auristatin immunoconjugate. *Clin. Cancer Res.* 11 (2 Pt 1): 843–852.

82 Hengel, S.M., Sanderson, R., Valliere-Douglass, J. et al. (2014). Measurement of *in vivo* drug load distribution of cysteine-linked antibody–drug conjugates using microscale liquid chromatography mass spectrometry. *Anal. Chem.* 86 (7): 3420–3425. https://doi.org/10.1021/ac403860c.

83 Boswell, C.A., Mundo, E.E., Zhang, C. et al. (2011). Impact of drug conjugation on pharmacokinetics and tissue distribution of anti-STEAP1 antibody–drug conjugates in rats. *Bioconjugate Chem.* 22 (10): 1994–2004. https://doi.org/10.1021/bc200212a.

84 Mould, D.R. and Green, B. (2010). Pharmacokinetics and pharmacodynamics of monoclonal antibodies: concepts and lessons for drug development. *BioDrugs* 24 (1): 23–39. https://doi.org/10.2165/11530560-000000000-00000.

85 Tabrizi, M.A., Tseng, C.-M.L., and Roskos, L.K. (2006). Elimination mechanisms of therapeutic monoclonal antibodies. *Drug Discovery Today* 11 (1–2): 81–88. https://doi.org/10.1016/S1359-6446(05)03638-X.

86 Pastuskovas, C.V., Mallet, W., Clark, S. et al. (2010). Effect of immune complex formation on the distribution of a novel antibody to the ovarian tumor antigen CA125. *Drug Metab. Dispos.* 38 (12): 2309–2319. https://doi.org/10.1124/dmd.110.034330.

87 Alley, S.C., Zhang, X., Okeley, N.M. et al. (2009). The pharmacologic basis for antibody–auristatin conjugate activity. *J. Pharmacol. Exp. Ther.* 330 (3): 932–938. https://doi.org/10.1124/jpet.109.155549.

88 Lobo, E.D., Hansen, R.J., and Balthasar, J.P. (2004). Antibody pharmacokinetics and pharmacodynamics. *J. Pharm. Sci.* 93 (11): 2645–2668. https://doi.org/10.1002/jps.20178.

89 Baron, J. and Wang, E.S. (2018). Gemtuzumab ozogamicin for the treatment of acute myeloid leukemia. *Expert Rev. Clin. Pharmacol.* 11 (6): 549–559. https://doi.org/10.1080/17512433.2018.1478725.

90 van de Donk, N.W.C.J. and Dhimolea, E. (2012). Brentuximab vedotin. *mAbs* 4 (4): 458–465. https://doi.org/10.4161/mabs.20230.

91 Lambert, J.M. and Chari, R.V.J. (2014). Ado-trastuzumab emtansine (T-DM1): an antibody–drug conjugate (ADC) for HER2-positive breast cancer. *J. Med. Chem.* 57 (16): 6949–6964. https://doi.org/10.1021/jm500766w.

92 Kantarjian, H.M., DeAngelo, D.J., Stelljes, M. et al. (2019). Inotuzumab ozogamicin versus standard of care in relapsed or refractory acute lymphoblastic leukemia: Final report and long-term survival follow-up from the randomized, phase 3 INO-VATE study. *Cancer* 125 (14): 2474–2487. https://doi.org/10.1002/cncr.32116.

93 Deeks, E.D. (2019). Polatuzumab vedotin: first global approval. *Drugs* 79 (13): 1467–1475. https://doi.org/10.1007/s40265-019-01175-0.

94 Enfortumab vedotin drug description. ADC Review. https://www.adcreview.com/enfortumab-vedotin-drug-description/ (accessed 22 April 2020).

95 Modi, S., Jacot, W., Yamashita, T. et al. (2020). Trastuzumab deruxtecan in previously treated HER2-positive breast cancer. *N. Engl. J. Med.* 382 (7): 610–621. https://doi.org/10.1056/NEJMoa1914510.

96 ADC Review (2017). Gemtuzumab ozogamicin (Mylotarg®; Pfizer/Wyeth). *ADC Review*. https://www.adcreview.com/gemtuzumab-ozogamicin-mylotarg/ (accessed 22 April 2022).

97 U.S. FDA (2017). MYLOTARGTM (gemtuzumab ozogamicin) for injection. https://www.accessdata.fda.gov/drugsatfda_docs/label/2017/761060lbl.pdf (accessed 22 April 2022).

98 Mylotarg™ | Pfizer for Professionals. https://mylotarg.pfizerpro.com/ (accessed 22 April 2022).

99 U.S. FDA (2014). ADCETRIS® (brentuximab vedotin) for injection. https://www.accessdata.fda.gov/drugsatfda_docs/label/2014/125388_s056s078lbl.pdf (accessed 22 April 2020).

100 Brentuximab Vedotin (SGN35) Drug Description. ADC Review. https://www.adcreview.com/brentuximab-vedotin-sgn35/ (accessed 22 April 2020).

101 ADCETRIS. ADC therapy – CD30-directed ADCETRIS® (brentuximab vedotin). https://www.adcetris.com/ (accessed 22 April 2020).

102 ADC Review, Journal of Antibody-drug Conjugates. Trastuzumab Emtansine | T-DM1 | Kadcyla®. ADC Review. https://www.adcreview.com/drugmap/trastuzumab-emtansine-t-dm1-kadcyla (accessed 22 April 2020).

103 KADCYLA® (ado-trastuzumab emtansine) in HER2+ breast cancer. https://www.kadcyla.com (accessed 22 April 2020).

104 Peddi, P.F. and Hurvitz, S.A. (2014). Ado-trastuzumab emtansine (T-DM1) in human epidermal growth factor receptor 2 (HER2)-positive metastatic breast cancer: latest evidence and clinical potential. *Ther. Adv. Med. Oncol.* 6 (5): 202–209. https://doi.org/10.1177/1758834014539183.

105 BESPONSA™ (inotuzumab ozogamicin) | R/R B-Cell ALL | Safety Info. https://www.pfizerpro.com/product/besponsa/hcp (accessed 22 April 2020).

106 ADC Review, Journal of Antibody-drug Conjugates. Inotuzumab ozogamicin (CMC-544) drug description. ADC Review. https://www.adcreview.com/inotuzumab-ozogamicin-cmc-544-drug-description/ (accessed 22 April 2020).

107 U.S. FDA (2019). FDA approves polatuzumab vedotin-piiq for diffuse large B-cell lymphoma. https://www.fda.gov/drugs/resources-information-approved-drugs/fda-approves-polatuzumab-vedotin-piiq-diffuse-large-b-cell-lymphoma (accessed 22 April 2020).

108 ADC Review, Journal of Antibody-drug Conjugates. Polatuzumab vedotin (drug description). ADC Review. https://www.adcreview.com/polatuzumab-vedotin-drug-description (accessed 22 April 2020).

109 POLIVY™ (polatuzumab vedotin-piiq) for R/R DLBCL. https://www.polivy.com (accessed 22 April 2020).

110 U.S. FDA (2019). FDA grants accelerated approval to enfortumab vedotin-ejfv for metastatic urothelial cancer. https://www.fda.gov/drugs/resources-information-approved-drugs/fda-grants-accelerated-approval-enfortumab-vedotin-ejfv-metastatic-urothelial-cancer (accessed 22 April 2020).

111 Seagen®. Enfortumab vedotin (ASG-22ME). Seattle Genetics. https://www.seattlegenetics.com/pipeline/enfortumab-vedotin (accessed 22 April 2020).

112 U.S. FDA (2019). FDA approves fam-trastuzumab deruxtecan-nxki for unresectable or metastatic HER2-positive breast cancer. https://www.fda.gov/drugs/resources-information-approved-drugs/fda-approves-fam-trastuzumab-deruxtecan-nxki-unresectable-or-metastatic-her2-positive-breast-cancer (accessed 22 April 2020).

113 AstraZeneca. Enhertu (trastuzumab deruxtecan) approved in the US for HER2-positive unresectable or metastatic breast cancer following two or more prior anti-HER2 based regimens. https://www.astrazeneca.com/media-centre/press-releases/2019/enhertu-trastuzumab-deruxtecan-approved-in-the-us-for-her2-positive-unresectable-or-metastatic-breast-cancer-following-2-or-more-prior-anti-her2-based-regimens.html (accessed 22 April 2020).

114 Schett, G. and Gravallese, E. (2012). Bone erosion in rheumatoid arthritis: mechanisms, diagnosis and treatment. *Nat. Rev. Rheumatol.* 8 (11): 656–664. https://doi.org/10.1038/nrrheum.2012.153.

115 Smolen, J.S., Aletaha, D., Koeller, M. et al. (2007). New therapies for treatment of rheumatoid arthritis. *Lancet* 370 (9602): 1861–1874. https://doi.org/10.1016/S0140-6736(07)60784-3.

116 Lee, H., Bhang, S.H., Lee, J.H. et al. (2017). Tocilizumab–alendronate conjugate for treatment of rheumatoid arthritis. *Bioconjugate Chem.* 28 (4): 1084–1092. https://doi.org/10.1021/acs.bioconjchem.7b00008.

117 Davignon, J.-L., Hayder, M., Baron, M. et al. (2013). Targeting monocytes/macrophages in the treatment of rheumatoid arthritis. *Rheumatology* 52 (4): 590–598. https://doi.org/10.1093/rheumatology/kes304.

118 Mariathasan, S. and Tan, M.-W. (2017). Antibody–antibiotic conjugates: a novel therapeutic platform against bacterial infections. *Trends Mol. Med.* 23 (2): 135–149. https://doi.org/10.1016/j.molmed.2016.12.008.

119 Lehar, S.M., Pillow, T., Xu, M. et al. (2015). Novel antibody–antibiotic conjugate eliminates intracellular *S. aureus*. *Nature* 527 (7578): 323–328. https://doi.org/10.1038/nature16057.

120 Zhou, C., Lehar, S., Gutierrez, J. et al. (2016). Pharmacokinetics and pharmacodynamics of DSTA4637A: A novel THIOMAB™ antibody antibiotic conjugate against *Staphylococcus aureus* in mice. *mAbs* 8 (8): 1612–1619. https://doi.org/10.1080/19420862.2016.1229722.

121 Shen, J. and Attar, M. (2015). Antibody–drug conjugate (ADC) research in ophthalmology – a review. *Pharm. Res.* 32 (11): 3572–3576. https://doi.org/10.1007/s11095-015-1728-9.

122 Shen, W.Y., Yu, M.J., Barry, C.J. et al. (1998). Expression of cell adhesion molecules and vascular endothelial growth factor in experimental choroidal neovascularisation in the rat. *Br. J. Ophthalmol.* 82 (9): 1063–1071. https://doi.org/10.1136/bjo.82.9.1063.

123 Kamizuru, H., Kimura, H., Yasukawa, T. et al. (2001). Monoclonal antibody-mediated drug targeting to choroidal neovascularization in the rat. *Invest. Ophthalmol. Visual Sci.* 42 (11): 2664–2672.

124 Tarsio, J.F., Kelleher, P.J., Tarsio, M. et al. Inhibition of cell proliferation on lens capsules by 4197X-ricin A immunoconjugate. *J. Cataract Refract. Surg.* 23 (2): 260–266.

125 Crosson, C.E., Kelleher, P.J., and Lam, D.M. (1992). Ocular pharmacokinetics of lens epithelial cell-specific immunotoxin 4197X-RA. *Exp. Eye Res.* 55 (1): 87–91. https://doi.org/10.1016/0014-4835(92)90096-b.

126 Kruser, T.J. and Wheeler, D.L. (2010). Mechanisms of resistance to HER family targeting antibodies. *Exp. Cell. Res.* 316 (7): 1083–1100. https://doi.org/10.1016/j.yexcr.2010.01.009.

127 Reslan, L., Dalle, S., and Dumontet, C. (2009). Understanding and circumventing resistance to anticancer monoclonal antibodies. *mAbs* 1 (3): 222–229.

128 Kovtun, Y.V. and Goldmacher, V.S. (2007). Cell killing by antibody–drug conjugates. *Cancer Lett.* 255 (2): 232–240. https://doi.org/10.1016/j.canlet.2007.04.010.

129 Jedema, I., Barge, R.M.Y., van der Velden, V.H.J. et al. (2004). Internalization and cell cycle-dependent killing of leukemic cells by gemtuzumab ozogamicin: rationale for efficacy in CD33-negative malignancies with endocytic capacity. *Leukemia* 18 (2): 316–325. https://doi.org/10.1038/sj.leu.2403205.

130 Zhao, R.Y., Erickson, H.K., Leece, B.A. et al. (2012). Synthesis and biological evaluation of antibody conjugates of phosphate prodrugs of cytotoxic DNA alkylators for the targeted treatment of cancer. *J. Med. Chem.* 55 (2): 766–782. https://doi.org/10.1021/jm201284m.

131 Kovtun, Y.V., Audette, C.A., Mayo, M.F. et al. (2010). Antibody–maytansinoid conjugates designed to bypass multidrug resistance. *Cancer Res.* 70 (6): 2528–2537. https://doi.org/10.1158/0008-5472.CAN-09-3546.

132 Kommineni, N., Pandi, P., Chella, N. et al. Antibody drug conjugates: development, characterization, and regulatory considerations. *Polym. Adv. Technol.* 31 (6): 1177–1193. https://doi.org/10.1002/pat.4789.

133 Shapiro, M.A. and Chen, X.-H. (2012). Regulatory Considerations When Developing Assays for the Characterization and Quality Control of Antibody-Drug Conjugates. *Am. Lab.* http://www.americanlaboratory.com/913-Technical-Articles/119843-Regulatory-Considerations-When-Developing-Assays-for-the-Characterization-and-Quality-Control-of-Antibody-Drug-Conjugates/ (accessed 22 April 2020).

134 Miksinski, S.P. and Shapiro, M. (2012). Regulatory Considerations for Antibody Drug Conjugates. https://fda.report/media/85350/Regulatory-Considerations-for-Antibody-Drug-Conjugates--Sarah-Pope-Miksinski--Ph.D.-%28ONDQA%29--Marjorie-Shapiro--Ph.D.-%28OBP%29--October-18--2012--AAPS-Annual-Meeting.pdf (accessed 22 April 2020).

6

Gene-Directed Enzyme–Prodrug Therapy (GDEPT) as a Suicide Gene Therapy Modality for Cancer Treatment

Prashant S. Kharkar and Atul L. Jadhav

Institute of Chemical Technology, Department of Pharmaceutical Sciences, Nathalal Parekh Marg, Matunga, Mumbai 400 019, India

6.1 Introduction

Over the last two decades, significant breakthroughs have been achieved in the virology and molecular biology domains, alongside research on the genetic abnormalities found in various types of cancers, including solid tumors and hematologic malignancies. The information gathered from various sources has been put to use in gene therapy, wherein a piece of genetic material (typically DNA) is transferred into the receiver's cells for a particular therapeutic purpose. On similar lines, a related term – suicide gene therapy, points to a technique by which a viral or bacterial gene is inserted into tumor cells. The transformed tumor cells thus gain the capability to convert a nontoxic chemical into "killer" active molecules. The inserted gene, following transfection/transduction, is popularly called transgene [1]. The transgene performs its intended function via mRNA and then the protein product (Figure 6.1). The transgenic material is a complementary DNA (cDNA)-abbreviated version of the native gene devoid of the sections that the cell eventually eliminates through splicing or posttranscriptional modifications, finally generating the mRNA template to be used for translation.

The "non-self DNA" fragment can then choose between two options – either to integrate with the host genomic DNA (insertional mutagenesis) or remain outside the cells (episomal), once at the site of action. In the first case, the reading frame of the downstream DNA is modified, in most cases, as a result of the genomic insertion. This is actually a worrisome event in exploring various potential applications of gene therapy for the reason that such a mutation, if targeted in the germline and not in the somatic cells, could be inherited from the patient's children [2]. The gene therapy for cancer has been extensively researched for quite a lot of time, mostly since 1990s. Various reviews, published during last three decades, outlined the applications of gene therapy in cancer [3]. In general, the main strategies for successful gene therapy in cancer include – (i) gene-directed enzyme–prodrug (suicide gene) therapy, i.e. gene-directed enzyme–prodrug therapy (GDEPT) (suicide gene or RNA

Targeted Drug Delivery, First Edition. Edited by Yogeshwar Bachhav.

strand enters in the patient's cells); (ii) gene therapy for potentiating the activity of the immune cells against cancer cells (gene for an antigen or cytokine enters in the patient's tumor cells); (iii) oncolytic virus therapy (virus enters in the patient's tumor cells); (iv) transfer of potential therapeutic genes, such as tumor suppressor genes, into the aberrant cancer cells; and (v) antisense therapy (antisense oligonucleotides enter in the patient's tumor cells) [4].

The GDEPT uses – (i) a prodrug which must be activated by enzymatic reaction(s), (ii) the enzyme required for prodrug activation and most importantly, and (iii) the gene delivery system [5, 6]. The transgene is introduced into the tumor cells using suitable gene delivery vector. The prodrug is then administered concomitantly. The suicide gene product(s) transforms the prodrug into cytotoxic moiety, which then kills the tumor cells (Figure 6.1). Various chemotypes and chemistries have been explored for the prodrug design alongside several viral and nonviral vectors for gene delivery [7].

Two types of enzymes have been proposed for GDEPT, which include (i) nonmammalian enzymes with or without human equivalents and, (ii) enzymes of human origin that are either not at all expressed or expressed at low levels in the tumor cells. Herpes simplex virus thymidine kinase (HSV-TK), bacterial cytosine deaminase (CD), bacterial carboxypeptidase G2 (CPG2), purine nucleoside phosphorylase (PNP), thymidine phosphorylase (TP), nitroreductase (NTR), D-amino acid oxidase

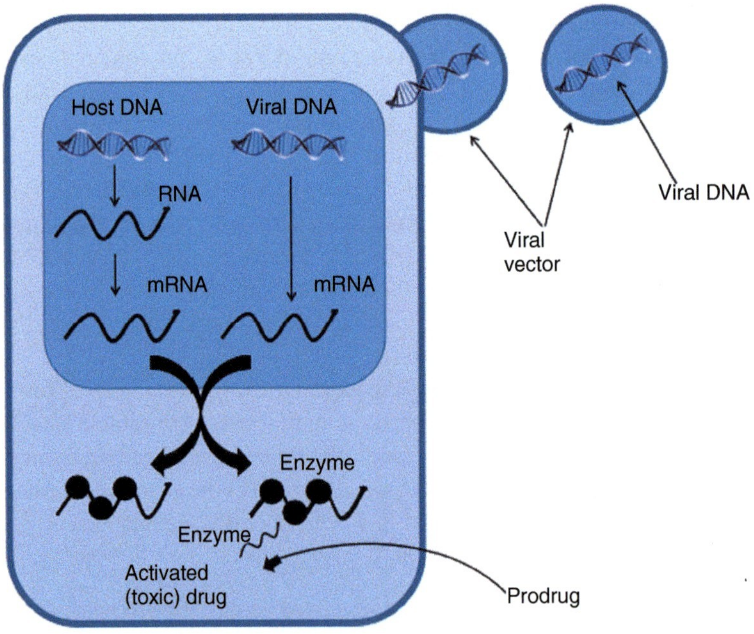

Figure 6.1 Schematic representation of gene-directed enzyme–prodrug (suicide gene) therapy, i.e. GDEPT. The DNA fragment (suicide gene) is inserted in the target cell using a suitable vector, which then produces the intended protein (enzyme), which then metabolically transforms a prodrug (inactive moiety) into its active (toxic) form. The subsequent events culminate into target cell-killing effect.

(DAAO), xanthine–guanine phosphoribosyl transferase (XGPRT), penicillin-G amidase (PGA), β-lactamase (β-L), multiple drug activation enzyme (MDAE), β-galactosidase (β-Gal), horseradish peroxidase (HRP), and deoxyribonucleotide kinase (DRNK) are examples representing the first category. Due to their nonhuman origin, these enzymes present issues related to immunogenicity. Examples of the second category include deoxycytidine kinase (dCK), carboxypeptidase A (CPA), β-glucuronidase (β-Glu), and cytochrome P450 (CYP450). These are devoid of immunogenicity-related problems due to human origin.

In GDEPT, two major types of prodrugs were used – directly-linked and self-immolative prodrugs. The first category includes pharmacologically inactive moiety that requires selective chemical transformation to transform into an active drug, while the second one comprises molecules that produce an unstable intermediate that, following activation, extrudes the active drug in subsequent stages. HSV-TK/Ganciclovir (**1**, Chart 6.1), CD/5-fluorocytosine (**2**), NTR/CB1954 (**3**), analogs, CPG2/CMDA (**4**), ZD-2767P (**5**), CYP450/cyclophosphamide (**6**), ifosfamide (**7**), and PNP/6-methylpurine-2′-deoxyriboside (**8**) are extensively studied pairs of enzyme/prodrugs [8]. Additional examples of prodrugs include penciclovir (**9**), acyclovir (**10**), fludarabine (**11**), irinotecan (**12**), and 6-thioxanthine (**13**). In a recent review article, Sheikh et al. have described the latest trends in prodrugs and prodrug-activated systems in gene therapy [9]. The authors elaborated on the biodistribution of the prodrug, efficacy, and adverse effects of the GDEPT approaches, including systemic toxicity of the active moiety. The modifications in the existing systems, such as HSV-TK/**1** and others, to avoid issues during clinical trials, such as development of resistance to **1** in genetically modified cells due to mutated HSV-TK variants, were highlighted

The vectors in GDEPT approach are viral or nonviral vehicles that are used to deliver genes to the target host cells. An optimal vector for gene delivery should be able to – (i) distribute the load in the specified cells (specificity/targetability); (ii) demonstrate stability and resistance to metabolic breakdown during its systemic residence; (iii) demonstrate safety and minimal adverse effects; and (iv) express in a controlled manner (expressibility). Retroviruses, adenoviruses, adeno-associated viruses (parvoviruses), herpes viruses, poxviruses, lentiviruses, and others have been successfully used as gene delivery vehicles. Physical methods, such as electroporation, sonoporation, laser irradiation, microinjection, and gene gun technologies, as well as chemical methods, such as cationic liposomes, cationic polymers, and lipid polymers, have shown promise as nonviral delivery systems [7–9].

The challenges to this approach, i.e. GDEPT, are manifold. The prodrug should be able to reach the site of action, i.e. the tumor tissue, in sufficient concentration following the administration and the ensuing metabolic reactions. To get recognized and activated by the activating enzyme, the prodrug must possess a trigger entity, which, following the enzymatic process, is converted into its effector counterpart destined to be released. Both the entities may be coupled with a linker if demanded by the chemistry. The generated effector must then be able to kill the tumor cells as well as exhibit bystander effect, i.e. the ability to kill the nearby tumor cells. Due to absorption of the active drug through cellular gap junctions (intercellular

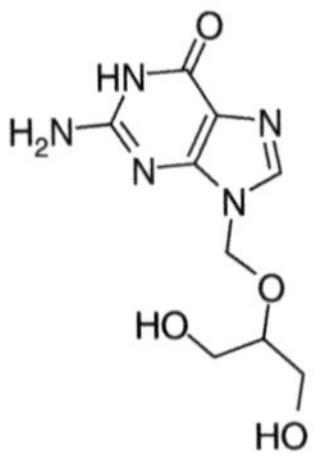

(1)

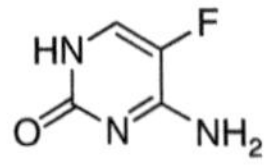

(2)

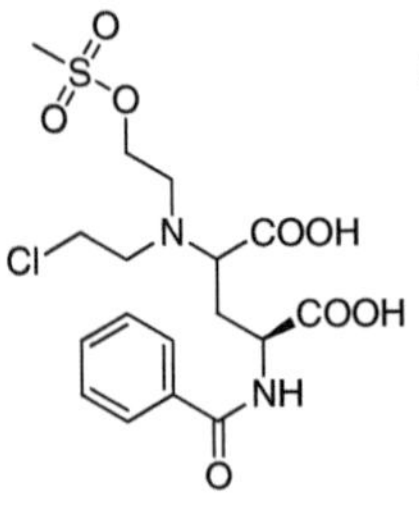

(3)

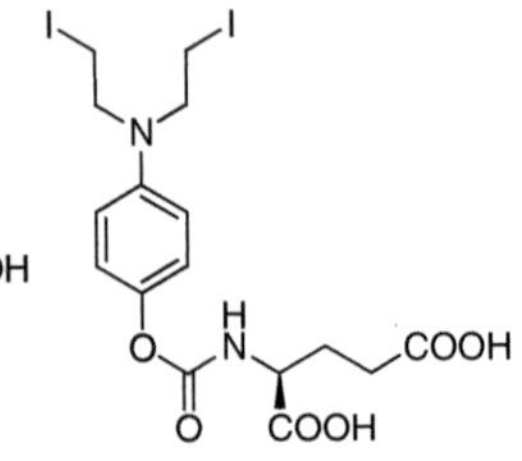

(4)

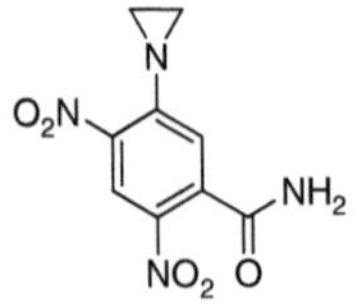

(5)

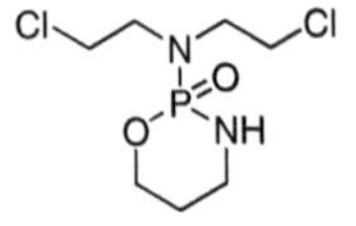

(6)

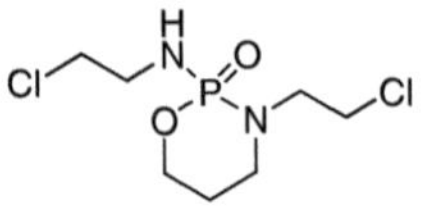

(7)

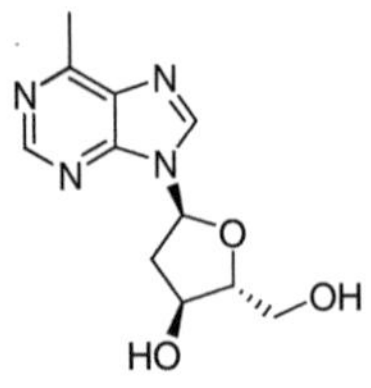

(8)

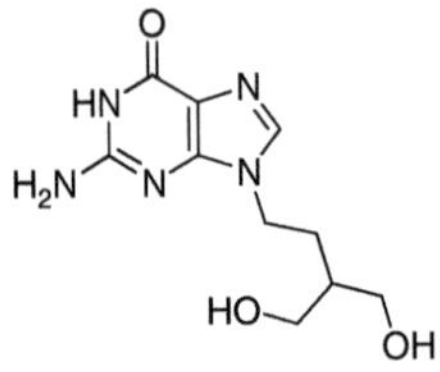

(9)

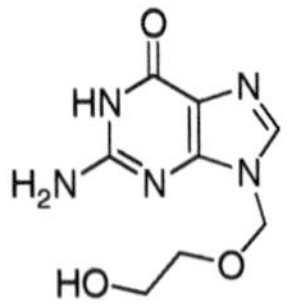

(10)

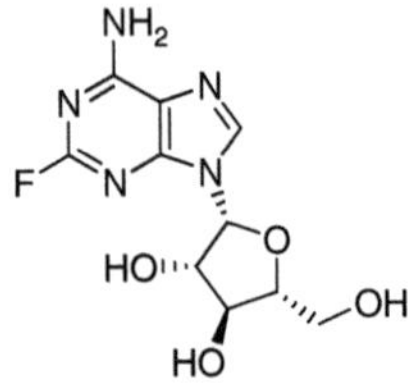

(11)

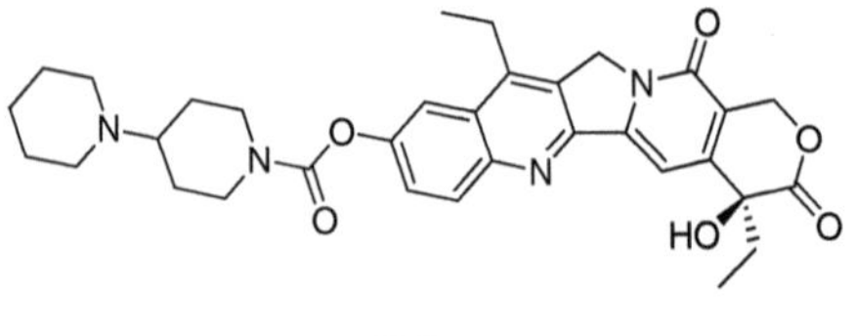

(12)

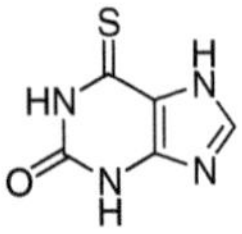

(13)

Chart 6.1 Molecular structures of representative prodrugs used in GDEPT.

transfer) and/or endocytosis, adjacent tumor cells (bystanders) are also destroyed once the transgene is introduced into a few tumor cells. Because not all tumor cells are effectively transfected with the transgene, the bystander effect is critical for the therapeutic outcome. Despite the fact that the gene transfer procedure is, most of the time, inefficient, affecting just a tiny fraction of the bulk tumor cells, the effect allows for a substantial therapeutic result [10]. Curious readers may dig the literature extensively for more such cases. In addition, there are other issues, such as slower prodrug–drug conversion kinetics, transfection/transduction efficiency of the delivery vectors, and the nonspecific toxicity issues related to the participating members, i.e. the prodrug, delivery vector, plasmid DNA, and the enzymes [7]. In short, the choice of prodrug chemistry, activating enzyme reaction kinetics, fine balance among several physicochemical and molecular properties of the prodrugs, delivery of the transgene, and related properties influence the therapeutic outcome significantly. Despite the preclinical and early-phase clinical trial success, none of the GDEPTs has reached the clinic yet.

The present chapter features the latest technological developments in the GDEPT field with particular emphasis on cancers and viral infections. Attempts have been made to cover the literature in the last five years mainly. Redundant information, e.g. the aspects already covered by the elaborate review articles on the topic, is avoided for the obvious reasons and the reader's benefit.

6.2 GDEPT for Difficult-to-Treat Cancers

6.2.1 High-Grade Gliomas (HGGs)

The high-grade gliomas (HGGs) have poor prognosis and are quite difficult to treat surgically due to the deeper spread of the tumor in the complex brain tissue. Considerable efforts have been devoted to address the lack of promising therapeutic options, such as small-molecule drugs [11], gene therapy [12], radiosensitizers for potentiating radiotherapy [13], and others. Giotta et al. [12] extensively reviewed the topic for relevant articles and the clinical trials and concluded that the gene therapy approaches, including GDEPT, were promising options for the HGGs, in view of the translational challenges related mostly to the delivery and accessibility of the transgene and the prodrug entities, alongside newer and effective target gene selection. Extensive search regarding the reports on GDEPT revealed the use of herpes simplex virus thymidine kinase (HSV-TK), cytosine deaminase (CD) and purine nucleoside phosphorylase (PNP) genes for this approach. A representative illustration of the mechanism of action of the HSV-TK/Ganciclovir (**1**) system is provided in Figure 6.2.

The ClinicalTrials.gov search (conducted on 4 November 2021, using keyword "suicide gene therapy") listed several active trials involving HGGs (Table 6.1). In addition, the findings of previously conducted GDEPT-based HGG trials have now been reported. Suicide gene therapy for brain tumor patients has been proven to be feasible and safe, but its efficacy needs to be yet proven. The lack of transgene delivery to a sufficient number of tumor cells, particularly those that are deeply invasive, and the inability of **1** (which is water-soluble) to penetrate the blood–brain barrier (BBB) in regions of the brain where the BBB is intact are two potential drawbacks of this type of therapy. Numerous research papers tried to address these issues. In a recently published review article, Tamura et al. have summarized the recent progress in the GDEPT for malignant glioma [14]. In addition to the relatively explored approaches, the authors discussed the stem cell-based strategies, particularly the neural, mesenchymal, and other stem cells as delivery vehicles owing to their migratory ability. In addition, a comprehensive list of previously conducted clinical trials involving GDEPT and related "suicide" approaches for malignant

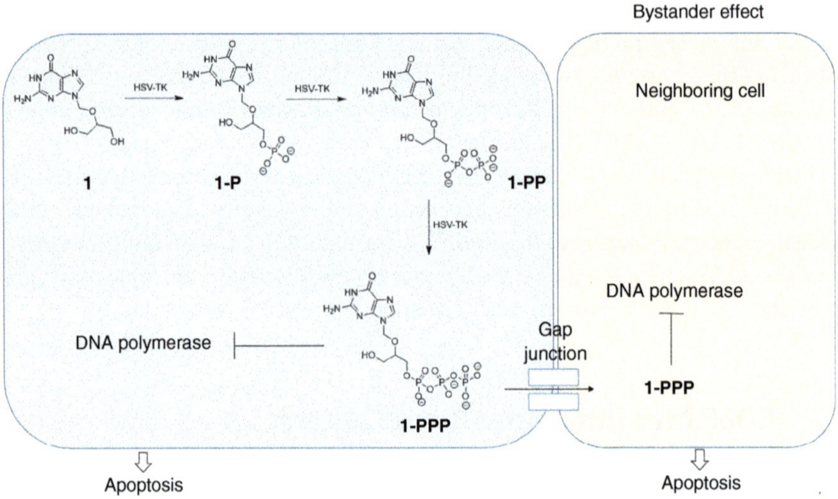

Figure 6.2 Mechanism of action of HSV-TK/Ganciclovir (**1**) system. The conversion of **1** to its triphosphorylated form **1-PPP** via **1-P** and **1-PP** leads to inhibition of DNA polymerase subsequently arresting DNA synthesis (S phase of cell cycle) and ensuing apoptosis. The escape of 1-PPP to the neighboring cells via cellular gap junctions leads to similar effect (bystander effect). Source: Adapted from Tamura et al. [14].

Table 6.1 GDEPT-based active clinical trials for HGG treatment.

NCT No.	Phase	Indication	Treatment	References
NCT04657315	1/2	Recurrent HGG	bCD/MSC11FCD	[15a]
NCT03596086	1/2	Recurrent HGG	HSV1-TL/Valacyclovir	[15b]
NCT03603405	1/2	Newly diagnosed HGG	HSV1-TK/Valacyclovir	[15c]
NCT03576612	1	Newly diagnosed HGG	HSV1-TK/Valacyclovir	[15d]
NCT02015819	1	Recurrent HGG	bCD/5FC+ Leucovorin	[15e]
NCT01811992	1	Newly diagnosed HGG	HSV1-TK/Valacyclovir	[15f]
NCT02192359	1	Recurrent HGG	hCE1m6/irinotecan	[15g]

glioma was given for reiterating the interest of the drug-discovery community in this approach. On similar lines, Hossain et al., in their review article, revisited the mixed-bag outcomes so far from clinical trials focused on evaluating GDEPT for cancer treatment and provided an overview of these modalities for glioma treatment [16]. Despite the initial setbacks in the clinic, the strategy is gaining momentum for HGG treatment and related research with main emphasis on better prodrugs, efficient delivery vectors, newer suicide genes, and newer combinatorial strategies.

The weak antitumor immune response shown by AdV-HSV-TK (adenoviral HSV-TK)/**1** system needs additional immunoboosting to demonstrate appreciable efficacy. Addition of an immune checkpoint inhibitor to the antitumor repertoire usually leads to robust antitumor immune response by the AdV-HSV-TK/**1** system [16]. Similar strategies may be explored for overcoming the previously described issues with GDEPT for HGG. This thought process is crucial for the desired therapeutic outcome in light of the fact that the HGG microenvironment is highly immunosuppressive. The complex interplay among various pathways leading to necrotic and immunogenic cell death (ICD) in some cancers and apoptotic and non-ICD in others following the GDEPT and related therapeutic strategies further complicates the understanding of the biological mechanisms, which may help us in rationalizing optimal combinations for better therapeutic outcomes. Nonetheless, novel *in vitro* and clinical findings with better strategies will pave the way for uncovering the true potential GDEPT holds for treating difficult cancers, such as HGG.

6.2.2 Triple-Negative Breast Cancer (TNBC)

The treatment of triple-negative breast cancer (TNBC) is a clinical challenge due to poor prognosis, higher recurrence rate, lack of standard chemotherapy, and involvement of cancer stem cells (CSCs) in its progression and maintenance. Discovery of new drugs for TNBC treatment is an active research area. Kuo et al. investigated the utility of STAT3/NF-κB pathway for developing an HSV-TK/**1**-based GDEPT for TNBC [17]. The novelty was to exploit the STAT3/NF-κB-responsive element-driven HSV-TK/**1** system for enhancing tumor selectivity of the system owing to the overexpression of STAT3/NF-κB-signaling pathway in bulk TNBC cells as well as CSCs. The authors used lentiviral vectors for transgene delivery. This study unequivocally supported the initial hypothesis that the tumor selectivity would be achieved by administering the STAT3/NF-κB – regulated HCV-TK/**1** therapy. There was a significant reduction in the expression of CSC markers and the invasiveness. Interestingly, the treatment sensitized the tumor cells to cisplatin, both *in vitro* and *in vivo*. The combination with cisplatin was quite effective in demonstrating effective tumor regression *in vivo*. However, the study had limitations, such as intratumoral injection of the lentiviral system which was shown to be capable of proto-oncogene activation and the local administration rather than the systemic one, precluding its usefulness in the metastatic cancer setting. Overall, the use of tumor microenvironment-specific promoters led to higher tumor selectivity (under restricted experimental settings), thereby establishing the potential of the novel GDEPT approach.

On similar lines, Abbaspour et al. used vascular endothelial growth factor (VEGF) promoter coupled with PNP/fludarabine (**11**, Chart 6.1) system and a nonviral polyplex-based pEGFP-N1 vector for achieving tumor selectivity [18]. The VEGF promoter was derived from 4T1 (murine mammary cancer cell line sharing substantial molecular features with human TNBC) and the PNP gene from the *Escherichia coli* genomic DNA, which was cloned into the vector. The transfection method used a cationic polymer. The gene delivery system could achieve transfection efficiency of ~30% with both VEGF-PNP-pEGFP-N1 and original CMV-PNP-pEGFP-N1 plasmids. Interestingly, the authors observed that the bystander effect significantly compensated for the low efficiency of the nonviral polyplex-based pEGFP-N1 gene delivery system. The study reinforced the thought process involving tumor-specific

Table 6.2 GDEPT approaches for invasive breast cancers.

Transgene	Breast cancer cell line used	Delivery system	Treatment	References
iC9	MCF-7	Pyridine-functionalized multi-walled carbon nanotubes	Suicide gene iC9 alone and in combination with actinomycin D and doxorubicin	[20a]
Yeast CD::uracil phosphoribosyl transferase suicide fusion gene (yCD::UPRT-MSCs)	MDA-MB-231	Mesenchymal stem cells (MSCs) and their exosomes	5-Fluorocytosine (**2**)	[20b]
HSV-TK driven by human FAS promoter (Ad-FAS-TK)	SK-Br3, MCF-7, MDA-MB-231	Recombinant adenovirus	Ganciclovir (**1**)	[20c]
HSV-TK	MCF-7, MDA-MB-231	Graphene oxide-hydroxyapatite nanocomposites	Ganciclovir (**1**)	[20d]
E gene	MCF-7	FuGENE 6 transfection reagent (nonliposomal, multicomponent reagent	E Suicide gene alone and in combination with Paclitaxel, Docetaxel, and Doxorubicin	[20e]

gene expression system for achieving tumor selectivity for a GDEPT. Raza and Ghosh redesigned CD activity for GDEPT in human breast cancer cell line MCF-7 [19]. The brilliant idea of using the limitations of earlier approach based on *E. coli* CD/5-fluorocytosine (**2**, Chart 6.1) wherein the combination of CD mutant F186W with gap junction forming protein connexin-43 (Cx43) was tested and compared with CD-Cx43 and CD F186W mutant alone. The prodrug **2** exhibits poor binding affinity for CD, while its mutant F186W has improved binding affinity for **2**. In addition, the expression of Cx43 has enhanced efficacy of the CD/**2** system. The authors then used a robust approach wherein the combination of the CD F185W mutant and Cx43 was tested for reduced cell viability. As expected, the combination was far superior over the individual systems. The brilliant piece of work opened up newer avenues for such multipronged approaches to fine-tune the modified GDEPT versions. Table 6.2 lists additional reports featuring GDEPT versions for invasive breast cancers.

6.2.3 Other Cancers

Faneca et al. used GDEPT based on HSV-TK/**1** system for *in vitro* and *in vivo* evaluations featuring oral squamous-cell carcinoma [21]. The authors used SCC-VII (murine oral squamous-cell carcinoma) and HSC-3 (human oral squamous-cell carcinoma) cells lines were employed for *in vitro* antitumor activity, while female C3H/HeJ mice were used for *in vivo* efficacy study. The cationic liposomes/DNA lipoplexes were used for transfection. The effect of gap junction modulation for potential bystander effect was also investigated. The studies clearly confirmed the superior *in vitro* and *in vivo* activities of the GDEPT against oral cancer cell lines. Further investigations involving different types of oral cancers and other systems are warranted for optimal therapeutic outcomes in the clinic. Working on similar lines, Gwak and Lee evaluated the utility of amphiphilic copolymer, poly(lactide-*co*-glycolide)-graft-polyethylenimine (PgP) as a nanocarrier for both the drug and the gene in cell culture and *in vivo* animal efficacy for spinal cord tumors in a rat [22]. Rat glioma cell line, C6, was used for *in vitro* assays and a rat T5 spinal cord tumor model was used for efficacy studies. The improved activity of the treatment, both *in vitro* and *in vivo*, was attributed to the ability of the nanocarrier to efficiently deliver the transgene. Further applications of the nanocarrier in delivering other transgenes in other solid tumors are warranted.

Zhang et al. explored the application of a nanobubble-assisted PNP/fludarabine (**11**, Chart 6.1) system for the potential treatment of hepatocellular carcinoma (HCC) [23]. The nanobubbles were prepared and subjected to ultrasonic irradiation for gene transfer into HCC cells – HepG2 and SMMC7721. The nontoxic, stable gene delivery system exhibited significant cell growth inhibitory activities, induced cellular apoptosis supplemented by appreciable bystander effect at much lower concentration of the prodrug. Such novel modalities for gene delivery are crucial for extending the applicability of the GDEPT. Hsiao et al. investigated the osteonectin (ON) promoter-assisted GDEPT based on HSV-TK/1 system in prostate cancers [24]. The ON, a glycoprotein, also known as secreted protein acidic and rich in cysteine (SPARC) or basement-membrane protein 40 (BM-40), binds to calcium in bone and also has affinity for collagen. The authors used the novel GDEPT for improved tumor selectivity, particularly in the prostate tumor epithelium and supporting bone stromal cells to control the local and metastatic prostate tumors. Human prostate cancer cell lines – LNCaP, PC3, and DU145 were used for the biological assays.

In a recent article, Tokay et al. have described novel prodrugs for nitroreductase as anticancer agents [25]. The dinitroaniline chemotype was designed and several members were synthesized and characterized. Human hepatoma Hep3B cells, prostate cancer PC-3 cells, and human umbilical vein endothelial cells (HUVEC), as normal cells, were used for the *in vitro* cytotoxicity evaluations (MTT assay) of the designed prodrugs. The next set of studies evaluated the prodrug activation by a novel nitroreductase previously identified from *Staphylococcus saprophyticus*. The LC-MS analyses of a representative compound identified the metabolites resulting from the NR reaction. The generated active metabolites were further evaluated for the cytotoxic effects against PC-3 cells. Overall, newer chemotypes were discovered with potential applications in the corresponding GDEPT based on NR. Metastatic castration-resistant prostate cancer (mCRPC) poses a clinical situation and is incurable. Shen et al. attempted GDEPT strategy based on prostate-specific antigen (PSA) promoter-assisted HSV-TK in androgen receptor-positive cells. The authors used JC polyomavirus-like particles as the gene delivery vehicle [26]. The transgene could effectively inhibit the growth of bone-metastasized prostate cancer cells in the used animal model, affirming the utility of the modality for treating mCRPC with promising therapeutic outcomes.

The literature cites countless articles, particularly in the last five years, on GDEPT and related suicide gene therapy approaches. Unlike GDEPT, wherein genes encoding the prodrug-activating enzymes are delivered into the tumor cells, one could think of delivering the prodrug-activating enzymes directly in the tumor cells with the help of specific carriers, such as (i) tumor-targeting monoclonal antibodies (antibody-directed enzyme–prodrug therapy, ADEPT) [27] or (ii) solid matrices (immobilized-directed enzyme–prodrug therapy, IDEPT) [28]. The latter two modalities present major challenges to cell membrane penetration due to the larger size of the antibody-conjugated enzyme as well as the immobilized enzyme derivatives.

Despite the availability of model enzyme/prodrug pairs, the research on novel suicide genes, activating enzymes, and prodrug chemistry continue. The "old" concept is being revisited with newer, modified, and more efficacious methods, including the combinations with gap junction proteins for enhanced bystander effects. Newer vectors, either viral or nonviral, are being investigated for increasing the transduction and transfection efficiency. The complex nature of the prodrug activation in the clinic adds to the chaos. Nonetheless, there has been consistent rise in the number of articles, patents on GDEPT approaches. The moderate success in recently concluded early clinical trials is crucial for continued interest in the approach. Once appreciable number of clinical candidates are approved, the scientific community will take the field to the next level.

6.3 Novel Enzymes for GDEPT

The purine and pyrimidine salvage enzymes – CD, TP, and PNP, and the associated prodrugs have been explored extensively for various cancer indications. Recently, Acosta et al. have described nucleoside 2′-deoxyribosyltransferases (NDTs) or *N*-deoxyribosyltransferases or trans-*N*-deoxyribosylases from *Lactobacillus delbrueckii* (*Ld*NDT) for GDEPT applications targeted at cancer [29]. These enzymes take part in the transglycosylation reaction of the 2′-deoxyribose entity of the nucleobases. Alternatively, the glycosylated enzyme undergoes hydrolytic reaction in the absence of the second nucleobase to be transglycosylated. Human cervical cancer (HeLa) cell line was used for the transfection along with 2′-deoxy-2-fluoroadenosine (dFAdo) as the prodrug. The GDEPT system remarkably reduced the cell viability at therapeutically achievable concentrations. The prodrug was relatively nontoxic at the concentrations tested in the wild-type cells (without the expression of the activating enzyme). Despite the lower transfection efficiency, the bystander effect contributed significantly to the overall cytotoxicity.

In this connection, US FDA approved Gedeptin®, a replication-deficient adenoviral vector expressing *E. coli* PNP gene (orphan drug status), which converts nucleoside prodrugs, such as fludarabine (**11**, Chart 6.1) to generate active metabolites responsible for killing the tumor cells [30]. The vector is injected directly into the tumor thrice over a period of two days, followed by intravenous administration of **11** for next three days. The enzymatic reaction yields 2-fluoroadenine (Fade), a highly toxic compound, in higher local concentrations.

Geldanamycin is a highly potent anticancer drug acting via inhibition of heat shock protein 90 (Hsp90) affecting the degradation of mutated or overexpressed proteins in many cancer cells. Its clinical utility is limited due to extreme hepatotoxicity. Carruthers et al. designed a novel prodrug of geldanamycin which could be used for GDEPT based on β-galactosidase prodrug-activating enzyme [31]. A recombinant adeno-associated virus (rAAV) was used as the delivery vector for β-galactosidase (LacZ) gene. The *in vivo* efficacy studies clearly demonstrated the successful outcome with no hepatoxicity, reiterating the fact that the effects of the activated drug were local. Farquhar et al. used similar system based on *E. coli* β-galactosidase for various galactoside prodrugs of anthracyclines, *N*-[4″-(β-D-galactopyranosyl)-3″- nitrobenzyloxycarbonyl]daunomycin (Daun02) and *N*-[(4″R, S)-4″-ethoxy-4″-(1′-O-β-D-galactopyranosyl) butyl]daunorubicin (gal-DNC4) [32]. Daun02 was superior over gal-DNC4.

6.4 Conclusions

The suicide gene therapy based on GDEPT modality has demonstrated astonishing outcomes for breakthroughs in cancer treatment approaches. The major advantage is the localized, i.e. tumor-specific cell-killing effects, with minimal toxicity to nontarget organs and tissues. Since its debut in 1984, it has piqued the curiosity of scientists, and substantial progress has been achieved in this sector. HSV-TK/GCV, CD/5-FU, and a variety of other suicide gene therapy methods have all been studied. The absence of an appropriate delivery method, short-term and low expression of transgenes, poor prodrug conversion rate, and restricted bystander impact must all be addressed for this strategy to be translated from bench-to-bedside. As a result, the need of the hour is to redirect research forward into enzyme engineering and prodrug advancement to accelerate the development of stable/high-affinity enzymes as well as safe prodrugs with effective bystander effect. Rigorous research is required to address these issues and make suicide gene therapy truly advantageous to patients.

References

1 Düzgüneş, N. (2019). Origins of suicide gene therapy. *Methods Mol. Biol.* 1895: 1–9. https://doi.org/10.1007/978-1-4939-8922-5_1.

2 Tiberghien, P. (1994). Use of suicide genes in gene therapy. *J. Leukocyte Biol.* 56 (2): 203–209. https://doi.org/10.1002/jlb.56.2.203.

3 (a) Cross, D. and Burmester, J.K. (2006). Gene therapy for cancer treatment: past, present and future. *Clin. Med. Res.* 4 (3): 218–227. https://doi.org/10.3121/cmr.4.3.218; (b) Collins, M. and Thrasher, A. (2015). Gene therapy: progress and predictions. *Proc. Biol. Sci.* 282 (1821): 20143003. https://doi.org/10.1098/rspb.2014.3003; (c) Shim, G., Kim, D., Le, Q.V. et al. (2018). Nonviral delivery systems

for cancer gene therapy: strategies and challenges. *Curr. Gene Ther.* 18 (1): 3–20. https://doi.org/10.2174/1566523218-666180119121949. (d) Redd Bowman, K.E., Lu, P., Vander Mause, E.R., and Lim, C.S. (2020). Advances in delivery vectors for gene therapy in liver cancer. *Ther Deliv.* 11 (1): 833–850. https://doi.org/10.4155/tde-2019-0076.

4 Sun, W., Shi, Q., Zhang, H. et al. (2019). Advances in the techniques and methodologies of cancer gene therapy. *Discov. Med.* 27 (146): 45–55.

5 Denny, W.A. (2003). Prodrugs for gene-directed enzyme-prodrug therapy (suicide gene therapy). *J. Biomed. Biotechnol.* 2003 (1): 48–70. https://doi.org/10.1155/-S1110724303209098.

6 Dachs, G.U., Tupper, J., and Tozer, G.M. (2005). From bench to bedside for gene-directed enzyme prodrug therapy of cancer. *Anticancer Drugs* 16 (4): 349–359. https://doi.org/10.1097/00001813-200504000-00001.PMID: 15746571.

7 Malekshah, O.M., Chen, X., Nomani, A. et al. (2016). Enzyme/prodrug systems for cancer gene therapy. *Curr. Pharmacol. Rep.* 2 (6): 299–308. https://doi.org/10.1007/s40495-016-0073-y.

8 Springer, C.J. and Niculescu-Duvaz, I. (2000). Prodrug-activating systems in suicide gene therapy. *J. Clin. Invest.* 105 (9): 1161–1167. https://doi.org/10.1172/JCI10001.

9 Sheikh, S., Ernst, D., and Keating, A. (2021). Prodrugs and prodrug-activated systems in gene therapy. *Mol. Ther.* 29 (5): 1716–1728. https://doi.org/10.1016/j.ymthe.2021.04.006.

10 Kramm, C.M., Sena-Esteves, M., Barnett, F.H. et al. (1995). Gene therapy for brain tumors. *Brain Pathol.* 5 (4): 345–381. https://doi.org/10.1111/j.1750-3639.1995.-tb00615.x.

11 Nayak, L. and Reardon, D.A. (2017). High-grade gliomas. *Continuum (Minneap Minn)* 23 (6): 1548–1563. https://doi.org/10.1212/CON.0000000000000554.

12 Giotta Lucifero, A., Luzzi, S., Brambilla, I. et al. (2020). Gene therapies for high-grade gliomas: from the bench to the bedside. *Acta Biomed.* 91 (7-S): 32–50. https://doi.org/10.23750/abm.v91i7-S.9953.

13 Shah, H.M., Jain, A.S., Joshi, S.V., and Kharkar, P.S. (2021). Crocetin and related oxygen diffusion-enhancing compounds: review of chemical synthesis, pharmacology, clinical development, and novel therapeutic applications. *Drug Dev. Res.* 82 (7): 883–895. https://doi.org/10.1002/ddr.21814.

14 Tamura, R., Miyoshi, H., Yoshida, K. et al. (2021). Recent progress in the research of suicide gene therapy for malignant glioma. *Neurosurg. Rev.* 44 (1): 29–49. https://doi.org/10.1007/s10143-019-01203-3.

15 (a) NCT04657315: Evaluation of maaximum tolerated dose, safety and efficiency of MSC11FCD therapy to recurrent glioblastoma patients (MSC11FCD-GBM). https://clinicaltrials.gov/ct2/show/NCT04657315 (accessed 5 November 2021); (b) NCT03596086: HSV-tk + valacyclovir + SBRT + chemotherapy for recurrent GBM. https://clinicaltrials.gov/ct2/show/-NCT03596086 (accessed 05 November 2021); (c) NCT03603405: HSV-tk and XRT and chemotherapy for newly diagnosed GBM. https://clinicaltrials.gov/ct2/show/NCT03603405 (accessed 05 November 2021); (d) NCT03576612: GMCI, Nivolumab, and radiation therapy in treating patients with newly diagnosed high-grade gliomas (GMCI). https://clinicaltrials.gov/-ct2/show/NCT03576612 (accessed 05 November 2021); (e) NCT02015819: Genetically modified neural stem cells, flucytosine, and leucovorin for treating patients with recurrent high-grade gliomas. https://clinicaltrials.gov/ct2/show/NCT02015819 (accessed 05 November 2021); (f) NCT01811992: Combined cytotoxic and immune-stimulatory therapy for glioma. https://clinicaltrials.gov/ct2/show/NCT01811992 (accessed 05 November 2021); (g) NCT02192359: Carboxylesterase-expressing allogeneic neural stem cells and irinotecan hydrochloride in treating patients with recurrent high-grade gliomas. https://clinicaltrials.gov/ct2/show/NCT02192359 (accessed 05 November 2021).

16 Hossain, J.A., Marchini, A., Fehse, B. et al. (2020). Suicide gene therapy for the treatment of high-grade glioma: past lessons, present trends, and future prospects. *Neurooncol. Adv.* 2 (1): vdaa 013. https://doi.org/10.1093/noajnl/vdaa013.

17 Kuo, W.Y., Hwu, L., Wu, C.Y. et al. (2017). STAT3/NF-κB-regulated lentiviral TK/GCV suicide gene therapy for cisplatin-resistant triple-negative breast cancer. *Theranostics* 7 (3): 647–663. https://doi.org/10.7150/thno.16827.

18 Abbaspour, A., Esmaeilzadeh, A., and Sharafi, A. (2021). Suicide gene therapy-mediated purine nucleoside phosphorylase/fludarabine system for in vitro breast cancer model with emphasis on evaluation of vascular endothelial growth factor promoter efficacy. *3 Biotech* 11 (3): 140. https://doi.org/10.1007/s13205-021-02692-0.

19 Raza, A. and Ghosh, S.S. (2019). Connexin-43 enhances the redesigned cytosine deaminase activity for suicide gene therapy in human breast cancer cells. *Biochem Insights* 12: 1178626418818182. https://doi.org/10.1177/11786264188-18182.

20 (a) Mohseni-Dargah, M., Akbari-Birgani, S., Madadi, Z. et al. (2019). Carbon nanotube-delivered iC9 suicide gene therapy for killing breast cancer cells in vitro. *Nanomedicine (London)* 14 (8): 1033–1047. https://doi.org/10.2217/nnm-2018-0342; (b) Altanerova, U., Jakubechova, J., Benejova, K. et al. (2019). Prodrug suicide gene therapy for cancer targeted intracellular by mesenchymal stem cell exosomes. *Int. J. Cancer* 144 (4): 897–908. https://doi.org/10.1002/ijc.31792; (c) Yan, C., Wen-Chao, L., Hong-Yan, Q. et al. (2007). A new targeting approach for breast cancer gene therapy using the human fatty acid synthase promoter. *Acta Oncol.* 46 (6): 773–781. https://doi.org/10.1080/02841860601016070; (d) Cheang, T.Y., Lei, Y.Y., Zhang, Z.Q. et al. (2018). Graphene oxide-hydroxyapatite nanocomposites effectively deliver HSV-TK suicide gene to inhibit human breast cancer growth. *J. Biomater. Appl.* 33 (2): 216–226. https://doi.org/10.1177/0885328218788242. (e) Rama, A.R., Prados, J., Melguizo, C. et al. (2011). Synergistic antitumoral effect of combination E gene therapy and Doxorubicin in MCF-7 breast cancer cells. *Biomed. Pharmacother.* 65 (4): 260–270. https://doi.org/10.1016/j.biopha.2011.01.002.

21 Faneca, H., Düzgüneş, N., and Pedroso de Lima, M.C. (2019). Suicide gene therapy for oral squamous cell carcinoma. *Methods Mol. Biol.* 1895: 43–55. https://doi.org/10.1007/978-1-4939-8922-5_4.

22 Gwak, S.J. and Lee, J.S. (2019). Suicide gene therapy by amphiphilic copolymer nanocarrier for spinal cord tumor. *Nanomaterials (Basel)* 9 (4): 573. https://doi.org/10.3390/nano9040573.

23 Zhang, B., Chen, M., Zhang, Y. et al. (2018). An ultrasonic nanobubble-mediated PNP/fludarabine suicide gene system: a new approach for the treatment of hepatocellular carcinoma. *PLoS One* 13 (5): e0196686. https://doi.org/10.1371/journal.pone.0196686.

24 Hsiao, W.C., Sung, S.Y., Chung, L.W.K., and Hsieh, C.L. (2019). Osteonectin promoter-mediated suicide gene therapy of prostate cancer. *Methods Mol. Biol.* 1895: 27–42. https://doi.org/10.1007/978-1-4939-8922-5_3.

25 Tokay, E., Güngör, T., Hacıoğlu, N. et al. (2020). Prodrugs for nitroreductase-based cancer therapy-3: antitumor activity of the novel dinitroaniline prodrugs/Ssap-Ntr B enzyme suicide gene system: synthesis, in vitro and in silico evaluation in prostate cancer. *Eur. J. Med. Chem.* 187: 111937. https://doi.org/10.1016/j.ejmech.2019.111937.

26 Shen, C.H., Lin, M.C., Fang, C.Y. et al. (2021). Suppression of bone metastatic castration-resistant prostate cancer cell growth by a suicide gene delivered by JC polyomavirus-like particles. *Gene Ther.* https://doi.org/10.1038/s41434-021-00280-8.

27 Sharma, S.K. and Bagshawe, K.D. (2017). Antibody directed enzyme prodrug therapy (ADEPT): trials and tribulations. *Adv. Drug Delivery Rev.* 118: 2–7. https://doi.org/10.1016/j.addr.2017.09.009.

28 Nemani, K.V., Ennis, R.C., Griswold, K.E., and Gimi, B. (2015). Magnetic nanoparticle hyperthermia induced cytosine deaminase expression in microencapsulated *E. coli* for enzyme-prodrug therapy. *J. Biotechnol.* 203: 32–40. https://doi.org/10.1016/j.jbiotec.2015.03.008.

29 Acosta, J., Pérez, E., Sánchez-Murcia, P.A. et al. (2021). Molecular basis of NDT-mediated activation of nucleoside-based prodrugs and application in suicide gene therapy. *Biomolecules* 11 (1): 120. https://doi.org/10.3390/biom11010120.

30 Gedeptin Technology overview. https://www.geovax.com/our-technology/gedeptin-technology-overview (accessed 07 November 2021).

31 Carruthers, K.H., Metzger, G., During, M.J. et al. (2014). Gene-directed enzyme prodrug therapy for localized chemotherapeutics in allograft and xenograft tumor models. *Cancer Gene Ther.* 21 (10): 434–440. https://doi.org/10.1038/cgt.2014.47.

32 Farquhar, D., Pan, B.F., Sakurai, M. et al. (2002). Suicide gene therapy using *E. coli* β-galactosidase. *Cancer Chemother. Pharmacol.* 50 (1): 65–70. https://doi.org/10.1007/s00280-002-0438-2.

7

Targeted Prodrugs in Oral Drug Delivery

Milica Markovic, Shimon Ben-Shabat, and Arik Dahan

Ben-Gurion University of the Negev, Faculty of Health Sciences, School of Pharmacy, Department of Clinical Pharmacology, Ben-Gurion Blvd. 1, Beer-Sheva 8410501, Israel

7.1 Introduction

Numerous drug molecules have unwanted features that lead to pharmaceutical, pharmacological, or pharmacokinetic obstacles in clinical application. Derivatization of the drug molecule is a useful way to overcome these obstacles while maintaining therapeutic effect. This approach is shown to enhance the efficacy of the drug molecule as well.

Prodrugs are inactive drug derivatives that are converted into the active drug moiety in the body through enzymatic and/or chemical transformations (Figure 7.1) [2]. The prodrug approach can improve drugs' chemical stability, absorption, distribution, metabolism, elimination (ADME) characteristics, site-specificity, safety, and patient compliance by improving taste/odor [3]. In the past, prodrugs were discovered by chance, and this approach was used as a last course of action; nowadays, the prodrug approach is purposefully designed early in the drug development process. From 2008 to 2017, US Food and Drug Administration (FDA) approved 30 prodrugs (nearly 10% of all approved drugs), and 10% of all drugs on the market are indeed prodrugs [2]. Oftentimes, the prodrug design is faster and more effective than some formulation approaches and is more advantageous than pursuing an entirely new compound.

Molecular revolution in the past two decades advanced the drug development process in a way that it no longer depends on empirical fitting grounded on plasma levels, rather, the novel ADME research takes into account molecular and cellular parameters (e.g. membrane influx and/or efflux transporters, expression and distribution of cellular proteins, and other molecular level factors) [4]. This molecular revolution improved the process of drug design and delivery, as well as increased the need for the use of prodrug approach, which is emphasized in this chapter.

Targeted Drug Delivery, First Edition. Edited by Yogeshwar Bachhav.

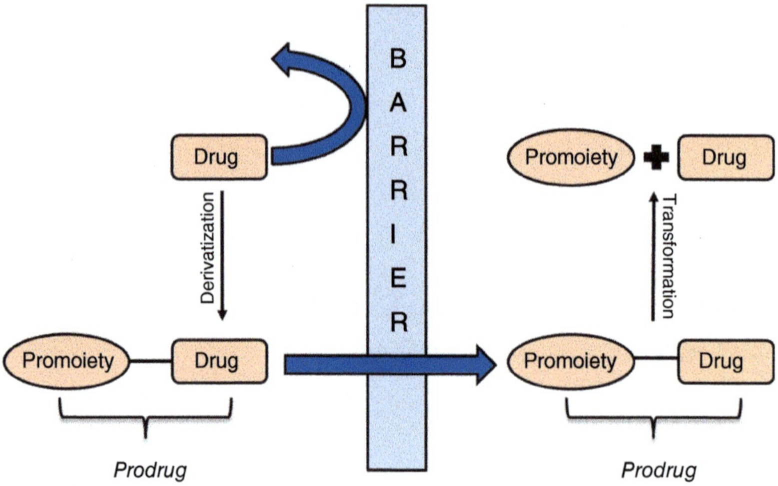

Figure 7.1 Prodrug activation *in vivo*. Source: Markovic et al. [1] / MDPI / CC BY-4.0.

7.1.1 Classic vs. Modern Prodrug Approach

Prodrugs are oftentimes used to improve drug absorption following oral administration. The traditional prodrug approach is frequently used to improve drug solubility by covalently binding the drug to the hydrophilic functional groups (e.g. sulfates and phosphates) [5], or, to increase passive permeability by binding the drug to lipophilic groups (e.g. esters) [6, 7]. The traditional prodrug approach has low site-specificity, and the prodrug itself will eventually be converted into the parent drug by enzymatic hydrolysis. The use of prodrug approach for various purposes can be achieved by different routes of administration, which influences the way a prodrug is converted into an active parent drug [8]. The main goal for using this approach is improving physicochemical drug properties and biopharmaceutical obstacles that drug faces following oral administration. This approach enables design of less reactive and less cytotoxic drugs [9]. The specific goals are improving oral absorption, blood–brain barrier (BBB) permeation, skin penetration, overcoming deficiencies in metabolic pathways, and reducing toxicity.

More recently, a modern, targeted prodrug approach is becoming increasingly employed, which exploits enzymes and transporters to improve absorption of drugs. Good knowledge of molecular and functional properties of transporters is necessary to utilize this approach. In addition, targeted proteins could have specific preference for drugs with certain pharmacokinetic profiles, as was recently studied [10]. Prodrug activation is a necessary step for the free drug to appear and exhibit therapeutic effect. This activation can be nonspecific, nevertheless, insight into the enzymes that are accountable for this activation may aid the rational prodrug design; enzyme–substrate specificity considerations can accomplish improved site-specificity, efficacy and decrease drug toxicity.

Whereas the goal of the classical prodrug approach is to alter physicochemical drug features (e.g. lipophilicity, charged functional groups), the modern prodrug approach takes into account molecular/cellular factors that influence prodrug fate in the body [11]. Section 7.2 will provide further input on targeted prodrug approach.

7.2 Modern, Targeted Prodrug Approach

The prodrug approach can be used to overcome physiological barriers or improve various drug properties by targeting particular enzymes or transporters through enzyme–substrate specificity or transporter–substrate specificity. Molecular features of enzymes/transporters are elucidated through numerous techniques, including controlled gene expression and/or cloning, which provides better understanding of requirements for rational prodrug design, and therefore improved drug targeting. This section provides an overview of transporter- and enzyme-targeted prodrug approaches.

7.2.1 Prodrug Approach-Targeting Enzymes

Enzymes are responsible for *in vivo* presystemic conversion from the prodrug to the active drug. If the aim is to reduce presystemic metabolism, enzyme targeting is more successfully accomplished by irreversible drug conversion (mainly through chemical modification), and not the prodrug approach. The aim of prodrugs, however, is to enhance oral drug absorption, as well as targeted drug delivery. Here we highlight enzymes as targeting sites for conversion of prodrugs to the parent drugs.

Hydrolysis of the prodrug to the active parent drug is the key step for accomplishing therapeutic effect. Prodrugs, with an ester functional group, are often activated by cholinesterase, acetylcholinesterase, carboxylesterase, and paraoxonase [12]. Usually, the prodrug hydrolysis is not specific. The aim of the traditional prodrug approach is the prodrug activation itself, whereas the mechanism of enzyme activation is usually disregarded. However, having a good understanding of the enzyme responsible for prodrug activation can greatly influence the rational prodrug design, and even targeting of specific enzymes involved in the prodrug metabolism. Identification of specific enzymes accountable for hydrolysis of different prodrug species is crucial for determining novel targets for the design of improved drug moieties.

When it comes to drug absorption following oral administration, the enzymes that appear in the gastrointestinal tract (GIT) are the targeting sites considered in the prodrug design; using the nutrients (lipids, peptides, and sugars) as the functional groups in the specific prodrug design increases targeting ability of such prodrugs toward gastrointestinal enzymes, and in doing so oral absorption can be improved [13, 14]. Additional advantage of such prodrugs is that once they are hydrolyzed to the active drug moiety *in vivo* such prodrugs yield safe nutrients as by-products (e.g. lipids, peptides, and sugars) [15]. However, this has been described in detail before [16, 17], therefore the focus of this chapter is targeted prodrug approach for site-specific drug delivery. To achieve efficient site-specific drug delivery, three main parameters need to be optimized – (i) prodrug transport and uptake to the desired site of action need to be quick, (ii) selective activation of the prodrug to the active parent drug at the specific site, and (iii) maintaining the drug at the targeted site [18, 19]. Targeting specific sites is made possible by use of enzymes that are either overexpressed or occur only in those specific tissues (e.g. glycosidase in colonic microflora allows colon-targeted drug delivery [20, 21]). Colonic microflora contains glycosidases, which cleave the glycoside prodrugs (hydrophilic, and difficult to be absorbed from the small intestine [SI]) to produce the parent drug; the prodrug can be designed to undergo absorption as an intact conjugate as well. An example of this strategy is orally administered dexamethasone 21-β-D-glucoside prodrug which allowed 60% of the initial dose to reach the cecum [22]. Besides prodrugs, a successful was to accomplish colonic targeting is through a codrug. Unlike prodrugs which consist of a parent drug and a carrier, the codrug is an entity that contains two synergistic drugs chemically merged together into one molecule, designed to improve the drug-delivery feature of one/both drugs. An example includes a codrug of 5-aminosalicylic acid (5-ASA) and prednisolone, linked through 1,4-self-immolative spacer by azo and ester bond, activated through colonic azoreductases and lactamization of the 1,4 elimination spacer [23].

7.2.1.1 Valacyclovirase-Mediated Prodrug Activation

A 5′valyl ester prodrug of acyclovir, valacyclovir is a potent antiviral agent, with the ability to increase oral bioavailability of acyclovir by three–five fold, due to binding to the hPEPT1 transporter [24]. However, one of the most important steps in the valacyclovir activation and consequently effect is its conversion to acyclovir in the body; it has been shown that this conversion depends on the human enzyme valacyclovirase, serine hydrolase, and one of the main enzymes responsible for specific activation of amino acid ester prodrugs, whose high affinity for amino acid esters comes from residue D123 that creates electrostatic interactions with α-amino functional group of the prodrug [25]. The enzyme valacyclovirase has a bulky pocket for the leaving groups, which accommodates many leaving groups, along with nucleoside analogs [26–28]. Even though amino acid ester approach in designing nucleoside analogs has been implemented many times, and the appropriate targeting of PEPT1 greatly improved oral absorption, activation of such analogs was considered nonspecific until the discovery of valacyclovirase. This example clearly shows the importance of identifying enzymes, responsible for prodrug transformation to the free drug, as one of the main steps in the prodrug activation, and ultimate therapeutic effect (Figure 7.2).

Guanidino group has a significant physiological role due to its positive charge and strong electrostatic interaction with carboxylic group (negatively charged). Many receptors exhibit affinity toward the L-arginine, containing the guanidino group and carboxylate from the receptor active site; therefore, prodrugs or inhibitors can be designed in a way to mimic guanidine functional group and exhibit drug targeting to the desired site [29]. Drug moieties containing guanidino group often show/exhibit low bioavailability following oral administration, due to decreased passive diffusion attributable to their positively charged groups and polarity. Masking the guanidino groups in the traditional prodrug design was somewhat successful [30, 31]. A model prodrug of amino acid esters, [3-(hydroxymethyl)phenyl]guanidine (3-HPG), was used in a double-prodrug approach, using the guanidino group to activate

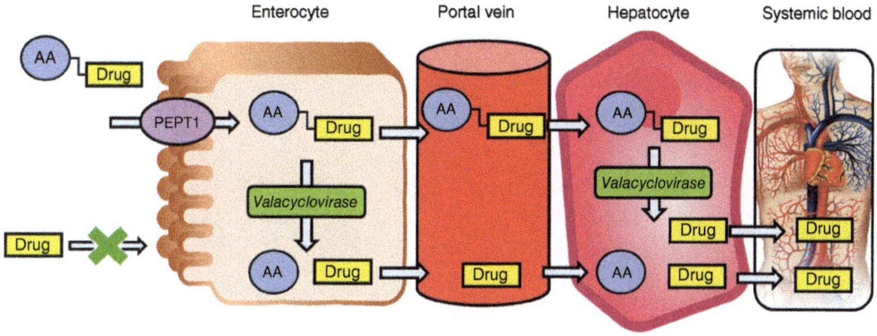

Figure 7.2 Double-targeted prodrug strategy illustration – activation through both transporter PEPT1 and enzyme valacyclovirase. PEPT1, peptide transporter 1; AA, amino acid. Source: Adapted from Milica Markovic et al. (2020)/MDPI/CC BY 4.0.

valacyclovirase and target PEPT1 intestinal absorption, thereby increasing absorption. This modern double-prodrug approach was used to evaluate both transport and activation of series of 3-HPG prodrugs [32]. Isoleucine and valine esters of the prodrugs showed considerable increase in permeability across the Caco-2 cells and rat intestinal permeability, compared to the parent drug, contributable to PEPT1 transport. Valacyclovirase successfully activated prodrugs of 3-HPG, as per K_m values that are within the range of valacyclovir, used as a positive control.

Such innovative approach, where both activation and transport steps are taken into account in the initial stages of development, is the future of prodrug approach. By doing so, the empirical steps can be reduced, and resulting prodrugs may achieve better bioperformance.

7.2.1.2 Phospholipase A_2-Mediated Prodrug Activation

Enzyme phospholipase A_2 (PLA_2) hydrolyzes the *sn*-2-positioned fatty acid (FA) of the phospholipid (PL), thus liberating absorbable products – free FA and *sn*-1 lysophospholipid (LPL). PLA_2 is overexpressed in the tissues of many inflammatory and malignant diseases [33–36]. The authors have designed and studied a type of lipidic prodrugs [37], phospholipid-based prodrugs, a strategy where the *sn*-2-positioned FA of the PL is substituted by the active drug through a carbonic linker [38–41]. This specific design is aimed at targeting PLA_2 as the activating enzyme for these prodrugs, which would cleave the *sn*-2-positioned drug (Figure 7.3) [42, 43].

Even though PLA_2 requires an *sn*-2-positioned FA [44], it was demonstrated that depending upon a linker of a certain length connecting the *sn*-2 position of the PL and the drug, the enzyme PLA_2 will be able to recognize and liberate the drug from the PL prodrug [45, 46]. This was first shown on two PL prodrugs of anti-inflammatory drugs diclofenac and indomethacin. *In vitro* activation of the PL-indomethacin prodrugs demonstrated significant decrease when linker length was shortened from 5 to 2 CH_2 units (*in vitro* activation up to 63% for 5 and up to 7.7% for 2-CH_2 units) [38]. Similarly, for PL-diclofenac prodrugs, the conjugates with lower linker lengths (up to 6-carbons) showed a linear correlation between

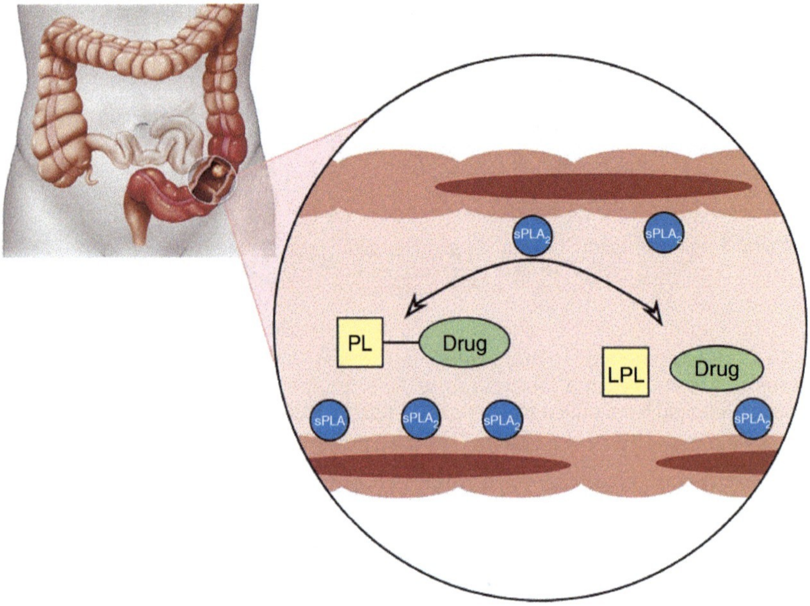

Figure 7.3 Illustration of targeted delivery with PL-drug conjugates, where the prodrug is activated by the enzyme PLA_2. PL-phospholipid; PLA_2, phospholipase A_2. Source: Markovic et al. [41] / MDPI / CC BY-4.0.

the linker length and the degree and rate of PLA_2-mediated hydrolysis. After 90 minutes post-incubation with bee venom PLA_2, for 6-carbon linker prodrug showed 95% decrease, as opposed to 2-carbon conjugate that was 20%, whereas longer linker lengths (8-carbon atoms) failed to further improve the prodrug activation, and was actually worse than the 6-carbons linker [40]. Rat *in vivo* studies on PL-indomethacin prodrugs demonstrated that the intact prodrug was not absorbed following oral administration, yet free indomethacin was liberated from the prodrug by PLA_2 in the gut lumen, giving a controlled-release profile of the free parent drug in the blood [38].

Another example of using PLA_2 as a prodrug-activating enzyme is the antiangiogenesis agent fumagillin. To overcome poor physicochemical and metabolic properties of fumagillin, the prodrug approach was utilized, and the prodrug was fused into an $\alpha_v\beta_3$-targeted perfluorocarbon (PFC) nanocarrier system which permitted the endocytosis and drug delivery followed by passive transfer to the targeted cell. Once inside the cell, the intracellular PLA_2 was responsible for prodrug activation and release of the free drug from the *sn*-2 position. *In vivo* studies demonstrated decreased angiogenesis, when compared to control nanoparticles without the drug [47, 48].

This section highlights the importance of the linker/spacer between the drug moiety and the PL to accomplish successful activation by PLA_2, as well as the fact that combining the prodrug approach, and its careful design, alongside an optimal delivery system, may offer effective drug targeting and adequate amount of free drug at the site of action. This approach also allows better control of the liberation of the free drug from the intact prodrug.

7.2.1.3 Antibody, Gene, and Virus-Directed Enzyme–Prodrug Therapy

Site-specific prodrug delivery is very important for treating cancerous tissues; however, it was shown that there are only a few human cancer types that contain correspondingly high levels of activating enzymes, and even if this is the case, these enzymes are not exclusive to cancerous tissues alone [49]. Site-specific prodrug activation relies on either using endogenous enzymes or physiological environment (hypoxia, pH of targeted tissues) (i), using and delivering genes that encode for exogenous enzymes responsible for prodrug activation to the specific sites, so-called gene-directed enzyme–prodrug therapy (GDEPT), also called the suicide gene therapy, or virus-directed enzyme–prodrug therapy (VDEPT) (ii), or using monoclonal antibodies directed at a tumor-associated antigen to deliver exogenous prodrug-activating enzymes to the site of action, and once the enzyme has been eliminated from the body, to administer a prodrug that is a substrate for the enzyme, so-called antibody-directed enzyme–prodrug therapy (ADEPT). ADEPT and GDEPT can overcome the above-mentioned limitations of prodrug targeting to the cancer sites, by focusing the prodrug-activating enzymes to that site, prior to prodrug administration. The idea behind this approach is to direct the action of cytotoxic drugs to cancerous tissues alone.

As far as ADEPT is concerned, the antitumor antibody is bound to an exogenous enzyme, not present in the physiological conditions, which places the complexes in the cancerous surroundings through intravenous infusion. Following the clearance of these conjugates from the circulation, a prodrug is introduced, which would usually be inactive, but is now activated by enzymes delivered in the cancerous tissues. This approach has been shown to be effective in cases where human tumor xenografts are resistant to chemotherapy [50–52], with the potential to become an efficient treatment option for solid tumors [53], as shown in clinical trials [54, 55].

Cancerous tissues were also targeted by genes responsible for coding enzymes that activate prodrugs, through viral vectors (adenoviral/retroviral) which transport the enzyme gene to both cancerous and noncancerous tissues [56]. Tumor-targeted expression of the exogenous enzymes can be obtained by binding the gene downstream of the tumor-specific transcription units [57, 58]. This approach is the basis for VDEPT or GDEPT.

Choosing the suitable enzyme for ADEPT and GDEPT approach is the key step for successful prodrug activation; the preferable characteristics are low-molecular weight, monomeric enzymes, that lack the need for glycosylation, which would facilitate protein transformations and overall behavior [49, 59]. The obvious choice is nonhuman enzymes, predominantly from microbiological sources, since they could bind to substrates not present in humans, and their immunogenic properties can be controlled, allowing the main focus to be on their specificity toward the substrate (i.e. prodrug) [51]. The choices of enzymes for ADEPT and GDEPT are different, since, in the ADEPT approach, the activation of the prodrug via enzyme occurs outside of the cell, whereas in GDEPT the activation occurs within the cell.

Usually, enzymes used in ADEPT approach contain a certain charge and bifunctional alkylating moieties, which makes it difficult for enzymes to go inside the cell and react with DNA, making them safer for use [60, 61]. Following prodrug activation in the cancerous tissues, the active drug is free to enter the cell and accomplish its pharmacological effect. Mammalian enzymes used for ADEPT are human β-glucuronidase, either a fusion protein with humanized antibody fragment [62] or human single-chain Fv [63] in combination with different prodrugs [64, 65]. Nonmammalian enzymes containing mammalian homolog (e.g. nitroreductase) [66] and without mammalian homolog, including (e.g. carboxypeptidase G2 and β-lactamase) [67], are in widespread use.

Enzymes used in GDEPT are usually monomeric and of viral or bacterial origin, with high specificity toward numerous substrates [68]. Bacterial enzyme nitroreductase can transform a monofunctional alkylating agent to a 10 000-fold more potent bifunctional agent [61]. Since nitroreductases involve NADH/NADPH as a required reductant, the prodrug activation can occur only within the cell, thus making this enzyme superior for GDEPT, rather than ADEPT. Chemotherapeutic prodrugs, cyclophosphamide and ifosfamide, are metabolized in the liver by CYP450, which is needed for activity. Chemosensitivity of these agents can be increased through the use of P450 gene transfer (P450 GDEPT). This approach can largely improve the therapeutic effect of P450-activated chemotherapeutic prodrugs without increasing toxicity caused by systemic distribution of active drug metabolites formed in the liver [69]. Clinical trial phase I/II of the P450-based gene therapy trial for pancreatic cancer patients highlights the importance of this approach [70].

The prodrug activation step, in which the free drug is liberated from the prodrug complex, is fundamentally the sole most important process for a prodrug to exhibit its therapeutic effect. Smart activating-enzyme-targeted design presents a significant advancement for the modern prodrug approach. This design can be planned and optimized through computational simulations, which is described later on in Section 7.3.

7.2.2 Prodrug Approach Targeting Transporters

Traditional prodrug approach is based on binding the drug to the lipophilic moiety to improve passive permeability, and consequently absorption, whereas the modern prodrug approach uses membrane transporters and/or enzymes to improve the transport of highly polar and/or charged moieties (e.g. amino acids and peptides). This way, intestinal epithelial transporters can enable drug absorption and enhance bioavailability of drugs with poor absorption. The prodrug design can mimic the structure of intestinal nutrients, and by doing so, exploit the endogenous pathways and allow successful absorption through carrier proteins, as well as releasing by-products that are safe [71].

The molecular revolution gave insight into the expression and functioning of many transporters so far. Intestinal enterocytes contain several transporters [72], that are often selectively targeted via prodrug approach, such as peptide transporter 1 (PEPT1) [16, 73], family of organic anion transporters (OATs) [74], L-type amino acid transporter 1 (LAT1) [75], proton-coupled amino acid transporter 1 (PAT1) [76, 77], amino acid transporter ATB^{0+} [78–80], proton-coupled folate transporter (PCFT) [81, 82], family of monocarboxylate transporters (MCTs) [83, 84], family of sodium-dependent glucose transporters (SGLTs) [85–87], family of apical sodium-dependent bile acid transporters (ASBTs) [88], and family of organic cation transporters (OCTs) [89, 90].

Exploiting the knowledge regarding substrate specificity of those transporters, the prodrug design can be adjusted to improve absorption of drugs following oral administration, through transporters in the intestinal tissue. The main focus of this section is going to be on intestinal transporters.

7.2.2.1 Peptide Transporter 1

PEPT1 is broadly distributed along the small intestine and is responsible for absorption of di- and tripeptides following digestion of proteins [91]. This transporter has wide substrate specificity, low affinity, and big capacity for substrate binding, which together with its high expression in the small intestine, makes PEPT1 a desirable target for modern prodrug approach. PEPT1 has become a prominent prodrug-designing target since some poorly absorbable drugs can be derivatized into peptidomimetic prodrugs that would target intestinal PEPT1 to enhance membrane permeability, and consequently, oral absorption of the active drug (e.g. angiotensin-converting enzyme [ACE] inhibitors, renin inhibitors, and β-lactam antibiotics), is demonstrated in Table 7.1.

A good example of modern prodrug approach targeting transporters is neuraminidase inhibitors, used in the treatment of influenza infections, oseltamivir, and zanamivir. Oseltamivir is the ethyl ester prodrug of the oseltamivir carboxylate; in humans, the oral bioavailability of oseltamivir carboxylate is less than 5%, whereas the bioavailability of oseltamivir is roughly 80%, which makes it a very good example of the classic prodrug approach [97]. The issue with oseltamivir is the resistance, which was not shown for zanamivir [98], however as opposed to oseltamivir, zanamivir exhibits very low oral bioavailability (2%) [99]. In an attempt to improve zanamivirs' oral absorption, acyloxy ester zanamivir conjugates linked to amino acids were designed, synthesized, and characterized to determine their stability, ability of binding to PEPT1 transporters, and enzyme metabolism (Table 7.1) [92]. L-Valyl zanamivir conjugate showed three times higher binding to PEPT1 in the cell culture overexpressing PEPT1; amino acid conjugates demonstrated higher permeability of zanamivir through intestinal epithelial cells after prodrug administration, as opposed to the free zanamivir, according to the Caco-2 permeability and rat perfusion studies in the jejunum. This targeted prodrug approach, designed to improve oral absorption via intestinal membrane transporters, such as PEPT1, followed by prodrug activation to the active pharmaceutical ingredient, and increasing the membrane permeability, is likely to deliver the increase in oral bioavailability required for oral zanamivir therapy [92].

Prodrugs containing amino acid, dipeptides, and tripeptides are found to be good substrates for PEPT1, as shown in the studies including antiretroviral agent didanosine [100], antiviral nucleoside analogs zidovudine, oseltamivir, ganciclovir and acyclovir [24, 101, 102], anticancer drug floxuridine [103–105], gemcitabine [106, 107]; these studies elucidated the importance of PEPT1 for improved drug oral absorption and consequent bioavailability. Whereas the capacity of peptide-like drugs to bind to intestinal PEPT1 was demonstrated before, a recent study included various drugs targeting the transport toward PEPT1 using a hydrolysis-resistant carrier – modified dipeptides containing a thioamide bond conjugated to several approved drug moieties and demonstrated significant binding to PEPT1 [108].

Table 7.1 Substrates (both natural and prodrug) of peptide transporters.

<table>
<tr><th colspan="4">Peptide transporter 1 (PEPT1)</th></tr>
<tr><td colspan="4">Natural substrates</td></tr>
<tr><td colspan="2">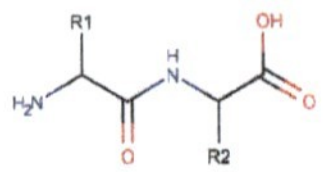
Dipeptides</td><td colspan="2">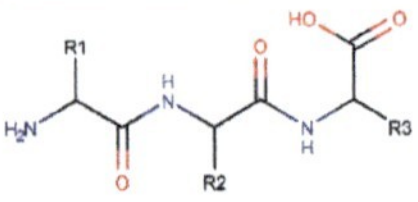
Tripeptides</td></tr>
<tr><td colspan="4">Prodrug substrates</td></tr>
<tr><td>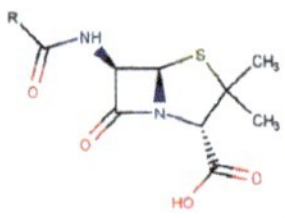
Penicillins eptides</td><td>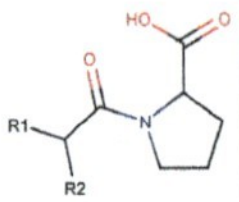
ACE inhibitor</td><td>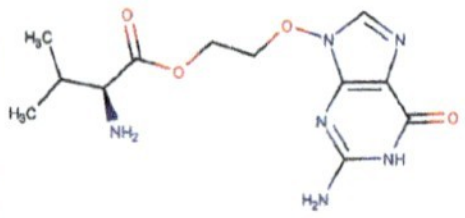
Valacyclovir</td><td>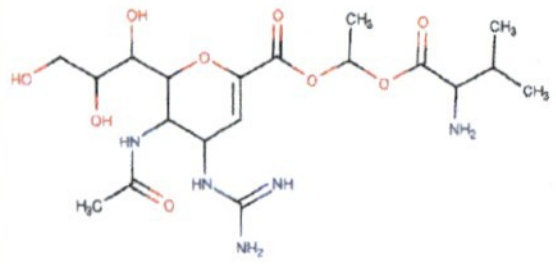
L-valyl zanamivir</td></tr>
<tr><th colspan="4">Monocarboxylate transporter 1 (MCT1)</th></tr>
<tr><td colspan="4">Natural substrates</td></tr>
<tr><td colspan="2">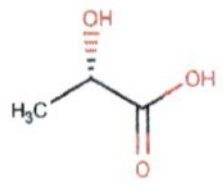
Lactate</td><td colspan="2">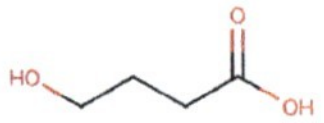
γ-hydroxybutyrate</td></tr>
<tr><td colspan="4">Prodrug substrates</td></tr>
<tr><td colspan="2">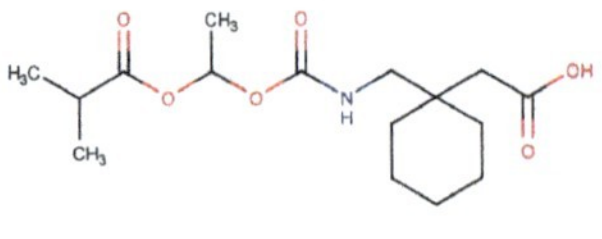
XP13512</td><td colspan="2">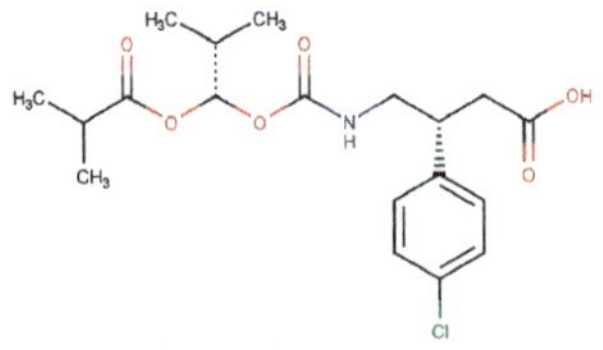
Arbaclofen placarbil</td></tr>
<tr><th colspan="4">Bile acid transporter (ASBT)</th></tr>
<tr><td colspan="4">Natural substrates</td></tr>
<tr><td colspan="2">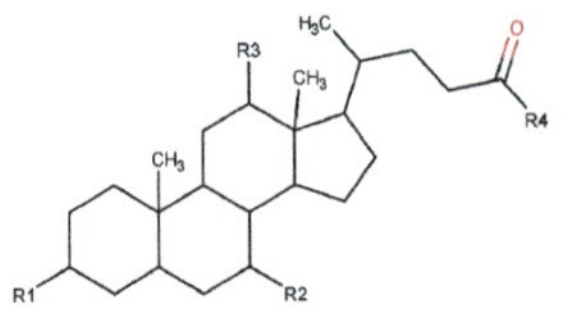
Bile salts</td><td colspan="2">R1: –H, –OH
R2: –H, –OH
R3: –H, –OH
R4: –H, –OH, $-NHCH_2COO^-$, $-NHCH_2CH_2SO_3^-$</td></tr>
</table>

(Continued)

Table 7.1 (Continued)

Prodrug substrates	
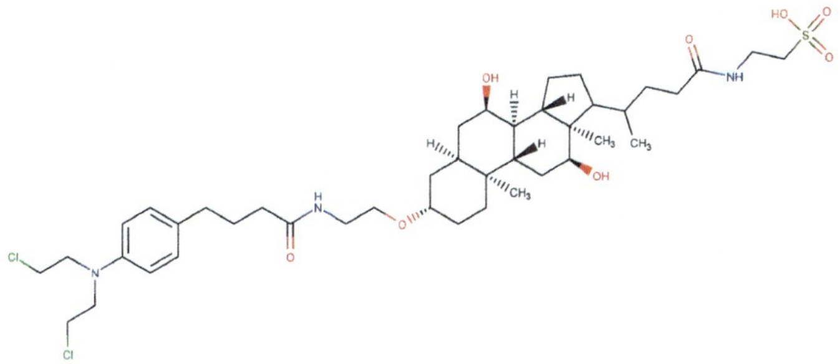 Chlorambucil-taurocholate prodrug	
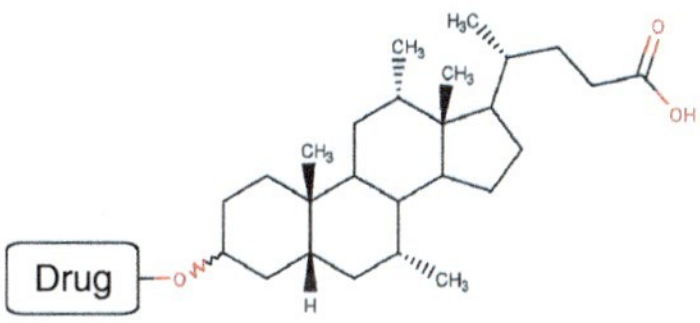 HMG-CoA reductase inhibitor prodrug	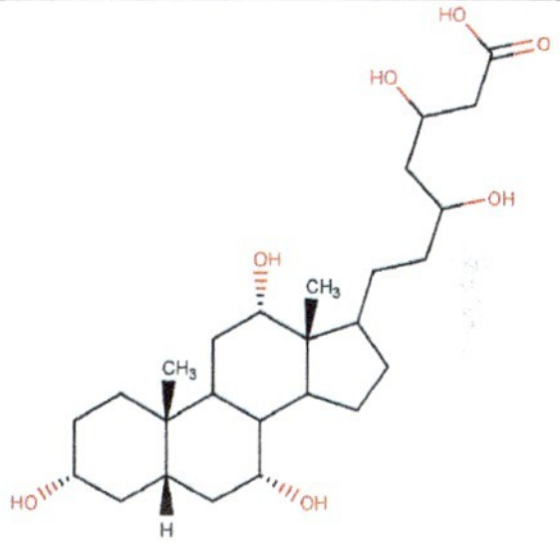 HMG-CoA reductase inhibitor hybrid

Source: Based on Refs. [92–96].

7.2.2.2 Monocarboxylate Transporter Type 1

Commonly, another transporter targeted in the modern prodrug approach is monocarboxylate transporter 1 (MCT1, SLC16A1). MCT1 has high capacity and low affinity for substrates, it is highly distributed along with the GIT (it is localized apically, with predominant expression in the colon [109]), and it mainly transports unbranched aliphatic monocarboxylic substrates [110, 111]. The carbamate prodrug of gabapentin (XP13512) was designed to overcome variable absorption between patients, dose-dependent absorption, and rapid drug clearance (Table 7.1) [93]. Following XP13512 oral administration, the bioavailability of gabapentin in the circulation was significantly increased which was also demonstrated in the clinical studies [112]; the prodrug also led to an increase in dose proportionality and improved colonic absorption. The major role in this advancement is due to exploiting MCT1-mediated transport and sodium-dependent multivitamin transporter (SMVT) [93, 112].

Arbaclofen placarbil is a prodrug of R-baclofen, designed to exploit MCT1, by competing with natural MCT1 substrate, lactate (Table 7.1) [113]. The prodrug demonstrated high uptake by cells expressing MCT1, and 5-times increased exposure to the active drug in rats and 12-times in monkeys, when compared to free baclofen; additional studies in dogs following administration of sustained-release prodrug formulation demonstrated continuous exposure to baclofen with bioavailability ~68%; blood was drawn at intervals, during 48 hours after dosing [113].

7.2.2.3 Bile Acid Transporters

Natural substrates for bile acid transporters (ASBT) are bile salts (cholic acid, chenodeoxycholic acid, deoxycholic acid, lithocholic acid, taurocholic acid) (Table 7.1). Following digestion, bile is secreted in the intestine and gets absorbed through active and passive transport, followed by recirculation to liver. ASBT are found on the apical enterocyte membrane of the terminal ileum and show wide substrate specificity, capacity, and high affinity for successful absorption, making them good drug targets [94, 114].

A bile acid prodrug of cytostatic agent chlorambucil was synthesized to overcome insufficient accumulation of chlorambucil in the cancerous tissues [115]. Chlorambucil–taurocholate (S2676) conjugate was absorbed from the ileal part of the intestine, and later on secreted from the liver while maintaining the alkylating properties of chlorambucil (Table 7.1). It was demonstrated that bile acid transporters mediate the high-affinity uptake of S2676 to the cancerous tissues (human hepatocellular carcinoma), making the bile acid conjugate of the cytostatic agent a successful strategy for targeting this type of cancer [115].

Liver-specific targeting of bile acid prodrugs was demonstrated through statins, group of drugs which inhibit the enzyme HMG-CoA reductase, responsible for cholesterol biosynthesis. Enterohepatic circulation of bile salts allows bile acid–statin conjugates to incorporate into endogenous bile salt processing pathways, thereby making them a perfect candidate for drug targeting of the cholesterol biosynthesis in the liver [94–96]. This could possibly decrease the side effects caused by cholesterol biosynthesis outside the liver. The studies present hybrid molecules, containing similar structural properties, such as statins and bile salts, and the prodrug molecule consisting of conjugated statins and bile salts (Table 7.1) [94–96]. Intact prodrug was not able to inhibit HMG-CoA reductase, nevertheless, upon hydrolysis of the bile salt, the prodrug released the active drug in the form of metabolite, with considerably greater liver concentrations when compared to extrahepatic tissues, demonstrating good targeting potential of the prodrug; on the other hand, the hybrid molecule was able to inhibit the enzyme. Following oral administration of the prodrug, the cholesterol biosynthesis inhibition was evident, and the hybrid was not active [94, 95].

Other transporters were used as potential targets in the modern prodrug approach as well, and demonstrated significantly improved intestinal permeability [116, 117]. The modern prodrug approach of targeting transporters allows enhanced absorption of parent drugs with unfavorable drug-like properties. It allows mechanistic, smart approach to the prodrug design that could improve the absorption of poorly permeated drugs. Higher permeability and thereby higher absorption, therefore, lead to easier development of future oral drug-delivery options.

7.3 Computational Approaches in Targeted Prodrug Design

Computer-aided drug design (CADD) is an *in silico* simulation of the interactions that arise between the drug and the protein (receptor or enzyme), to optimize the drug molecule [1, 118]. The key obstacle for using CADD is the inaccuracy of the simulation – the computationally attained affinity of binding between the drug molecule and the protein does not always relate to the experimental data. For example, the molecular docking method is a rapid routine way for optimizing drug design and/or drug screening, nevertheless, the accuracy of this method is low (~20%), mostly due to suboptimal size of molecule databases, the incorrect choice of docking pose, the inappropriate protein binding site, should all be taken into consideration prior to the studies [119]. Physics-based methods of *in silico* simulations are complicated and time-costly, however, they lead to more precise calculations and estimations (~80%), than the empirical methods, such as docking. Physics-based methods include quantum mechanics/molecular mechanics (QM/MM) or free-energy perturbation (FEP) methods [118].

A commonly used FEP method is MD simulation, which can elucidate the structure-to-function relationship of macromolecules. As an example, these simulations can reveal the structural requirements of the phospholipid-drug conjugates that are designed to target phospholipase A_2 [40, 41, 120]; in many cases, the conjugate design requires linkers or spacers between the PL and the drug to attain activation by PLA_2 [48, 121, 122], described in detail elsewhere [43]. Molecular modeling simulations are an essential tool for acquiring the structural information of the conjugate that would exhibit the highest degree of prodrug activation by the PLA_2 [123]. Different molecular modeling simulations were used for evaluating the activity of PLA_2 enzymes toward PL-prodrugs (both molecular docking and MD simulations) [120, 124, 125]. Modern *in silico* simulations based on thermodynamic integration and WHAM/umbrella-sampling method were established to follow the changes in the PLA_2 transition-state binding free energy for the PL-drug conjugates in terms of reducing/extending the linker length between the PL and indomethacin/diclofenac molecules [120, 124]. The results of the simulations showed an excellent correlation between computational predictions and the experimental results [38, 40, 120, 124], and it was evidently shown that the linker length has a key role in the PLA_2-mediated rate of activation, and consequently the amount of the free parent drug liberated at the site of action. Based on computational and experimental studies, an optimal linker length of 5–6 carbons was determined for PL-indomethacin and PL-diclofenac prodrugs, respectively [38, 40, 120, 124]. The simulations also concluded that both isoforms of PLA_2, either from bee venom or from a human source result in comparable activation rates, signifying that the PLA_2 from the bee venom can be used as an alternative to the human isoform of the enzyme in experimental conditions [120, 124]. In conclusion, the linker length can greatly impact the capacity of PLA_2 to activate PL-prodrugs, nevertheless, the optimal length of the linker has to be evaluated on a case-by-case basis, since the size and/or the volume of the drug moiety significantly impacts the steric interactions between the protein and the prodrug. The combined *in silico*/*in vitro* approach has the potential to guide the design of the optimal PL-drug conjugate, and consequently to decrease the number of experimentations required. Another simulation method, density functional theory (DFT), is also frequently used for evaluation of various prodrugs that contain linkers; some examples include acyclovir [126], atovaquone [127], and dopamine [128].

7.4 Discussion

The prodrug approach is no longer viewed as a last resort option, rather it is considered an effective strategy already in the very early stages of prodrug development. Deeper knowledge of molecular and functional characteristics of the enzymes and membrane influx and efflux transporters leads to a molecular revolution that changed the way the prodrugs are used. Prodrug efficacy was shown to be largely advanced by allowing activation via specific enzymes or improving absorption and thus bioavailability by membrane transporters. This approach could allow easier translation from preclinical to clinical design, as well as predicting the possibility of drug–drug interactions.

ADEPT and GDEPT, as a successful targeted prodrug approach, include gene delivery precise enzyme expression, and carrier proteins are a promising way to improve site-specific drug delivery (especially in cancer) and enhance the therapeutic efficacy. Intestinal transporters offer an important target in prodrug design for improving absorption of poorly-absorbable drugs, as presented in Section 7.2.2. Sometimes, the membrane transporters can also aid in organ/tissue targeting and selective drug uptake, as well as use the transporter gene polymorphism for developing drugs for specific population.

Another promising prodrug strategy is the double-targeted prodrug approach, where both activation enzymes and membrane transporters for improving drug permeability are used, which was shown to be an efficacious path for exploiting the molecular revolution in oral drug delivery, as demonstrated for valacyclovir.

Computational simulations using molecular-orbital methods (DFT) could enable the pathway for designing linkers used in prodrug approach, which could improve drugs bioavailability and control the drugs-release profile. FEP simulations can predict the affinity changes for substrates very fast and accurate. This field can greatly impact prodrug design, and modern computational techniques are being more and more exploited and recognized as useful.

Concurrent use of all three approaches (prodrug targeting of enzymes, transporters, and *in silico* prediction methods), in a way to plan the prodrug design by computational studies, to target-specific activating enzyme or transporter, that would increase drug absorption is an increasingly employed direction for developing new prodrug candidates.

This field is expected to grow, as greater knowledge of intestinal transporters and activating enzymes becomes known; therefore, the prodrug approach is no longer only chemical drug derivatization, but a new way of exploiting physiological targets to accomplish optimal effect.

7.5 Future Prospects and Clinical Applications

In terms of prodrug design for targeted delivery, future directions include the use of different linkers/spacers that can significantly improve drug delivery, targeting, or control the rate of prodrug conversion to the active parent drug [40, 122, 125]. In addition, emerging computational methods play an important role in prodrug optimization and can significantly cut the costs of prodrug development process. Increasing knowledge of the enzymes and membrane transporters in terms of structure, mechanism of action and role, can lead to the future development of targeted prodrug delivery [129]. For instance, in 2011, FDA approved gabapentin enacarbil, a prodrug of gabapentin and a substrate for two high-capacity nutrient transporters, MCT-1 and SMVT with wide distribution throughout the human GIT. The prodrug design increased the capacity of transport and allowed the delivery of higher gabapentin doses in appropriate dose proportion, circumventing uptake saturation at clinical doses [130].

Currently, clinical trials are conducted for evaluating the efficacy of anticancer prodrug NUC-1031 in patients with advanced biliary tract cancer [131]. NUC1031 is produced as a pre-activated monophosphate form of gemcitabine; owing to the protective phosphoramidate transformation, NUC1031 has the ability to overcome resistance mechanisms of gemcitabine, including resistance to human equilibrative nucleoside transporters based (hENT1), nucleoside activation resistance (deoxycytidine kinase, dCK deficiency) and pyrimidine deamination via deoxycytidine deaminase (CDA) [132]. Clinical trials are also underway for the prodrug AKR1C3-activated prodrug OBI-3424 for patients with relapsed/refractory T-cell acute lymphoblastic leukemia [133]; OBI-3424 also entered trials for hepatocellular carcinoma and castrate-resistant prostate cancer [134]. OBI-3424 selectively releases a potent DNA alkylating agent in the presence of the aldo-keto reductase family 1 member C3 (AKR1C3), overexpressed in cancerous tissues; it is distinguished from traditional alkylating agents (i.e. cyclophosphamide, ifosfamide) by its selective mode of activation [135]. Enalapril, a model of angiotensin-converting enzyme inhibitors (ACEIs), and a prodrug activated by carboxylesterase 1 are currently under clinical investigation to determine the impact of genetic variability of the enzyme on the enalapril activation, pharmacokinetics, and pharmacodynamics of in a multiple-dose healthy volunteer setting. The conclusions from this study could be a major stepping stone for a more personalized use of ACEI prodrugs [136].

7.6 Conclusion

Drug derivatization was used to improve undesirable drug features and improve drug therapy, however nowadays, molecular revolution provide insight into molecular and functional features of enzymes and transporters, hence facilitating prodrug development and targeted drug-delivery systems. Targeted prodrug approach, where transporters are responsible for intestinal permeability or enzymes for activation, together with computational simulations leading the design itself, is a modern direction for designing prodrugs and an important future direction for developing prodrugs for oral drug delivery.

References

1 Markovic, M., Ben-Shabat, S., and Dahan, A. (2020). Computational simulations to guide enzyme-mediated prodrug activation. *Int. J. Mol. Sci.* 21.

2 Rautio, J., Meanwell, N.A., Di, L., and Hageman, M.J. (2018). The expanding role of prodrugs in contemporary drug design and development. *Nat. Rev. Drug Discovery* 17: 559–587.

3 Stella, V.J. (2004). Prodrugs as therapeutics. *Expert Opin. Ther. Pat.* 14: 277–280.

4 Dahan, A., Zimmermann, E.M., and Ben-Shabat, S. (2014). Modern prodrug design for targeted oral drug delivery. *Molecules* 19: 16489–16505.

5 Stella, V.J. and Nti-Addae, K.W. (2007). Prodrug strategies to overcome poor water solubility. *Adv. Drug Delivery Rev.* 59: 677–694.

6 Beaumont, K., Webster, R., Gardner, I., and Dack, K. (2003). Design of ester prodrugs to enhance oral absorption of poorly permeable compounds: challenges to the discovery scientist. *Curr. Drug Metab.* 4: 461–485.

7 Testa, B. (2009). Prodrugs: bridging pharmacodynamic/pharmacokinetic gaps. *Curr. Opin. Chem. Biol.* 13: 338–344.

8 Ettmayer, P., Amidon, G.L., Clement, B., and Testa, B. (2004). Lessons learned from marketed and investigational prodrugs. *J. Med. Chem.* 47: 2393–2404.

9 Serafin, A. and Stanczak, A. (2009). Different concepts of drug delivery in disease entities. *Mini-Rev. Med. Chem.* 9: 481–497.

10 Bocci, G., Benet, L.Z., and Oprea, T.I. (2019). Can BDDCS illuminate targets in drug design? *Drug Discovery Today*.

11 Dahan, A., Khamis, M., Agbaria, R., and Karaman, R. (2012). Targeted prodrugs in oral drug delivery: the modern molecular biopharmaceutical approach. *Expert Opin. Drug Delivery* 9: 1001–1013.

12 Liederer, B.M. and Borchardt, R.T. (2006). Enzymes involved in the bioconversion of ester-based prodrugs. *J. Pharm. Sci.* 95: 1177–1195.

13 Amidon, G.L., Leesman, G.D., and Elliott, R.L. (1980). Improving intestinal absorption of water-insoluble compounds: a membrane metabolism strategy. *J. Pharm. Sci.* 69: 1363–1368.

14 Markovic, M., Ben-Shabat, S., Keinan, S. et al. (2019). Lipidic prodrug approach for improved oral drug delivery and therapy. *Med. Res. Rev.* 39: 579–607.

15 Markovic, M., Ben-Shabat, S., Aponick, A. et al. (2020). Lipids and lipid-processing pathways in drug delivery and therapeutics. *Int. J. Mol. Sci.* 21.

16 Bai, J.P. and Amidon, G.L. (1992). Structural specificity of mucosal-cell transport and metabolism of peptide drugs: implication for oral peptide drug delivery. *Pharm. Res.* 9: 969–978.

17 Fleisher, D., Stewart, B.H., and Amidon, G.L. (1985). Design of prodrugs for improved gastrointestinal absorption by intestinal enzyme targeting. *Methods Enzymol.* 112: 360–381.

18 Stella, V.J. and Himmelstein, K.J. (1980). Prodrugs and site-specific drug delivery. *J. Med. Chem.* 23: 1275–1282.

19 Stella, V.J. and Himmelstein, K.J. (1985). Prodrugs: a chemical approach to targeted drug delivery. In: *Directed Drug Delivery: A Multidisciplinary Problem* (ed. R.T. Borchardt, A.J. Repta and V.J. Stella), 247–267. Totowa, NJ: Humana Press.

20 Friend, D.R. and Chang, G.W. (1984). A colon-specific drug-delivery system based on drug glycosides and the glycosidases of colonic bacteria. *J. Med. Chem.* 27: 261–266.

21 Gorbach, S.L. (1971). Intestinal microflora. *Gastroenterology* 60: 1110–1129.

22 Laine, K. and Huttunen, K.M. (2010). Enzyme-activated prodrug strategies for site-selective drug delivery. In: *Prodrugs and Targeted Delivery: Towards Better ADME Properties*, vol. 47 (ed. J. Rautio), 231–252.

23 Ruiz, J.F.M., Radics, G., Windle, H. et al. (2009). Design, synthesis, and pharmacological effects of a cyclization-activated steroid prodrug for colon targeting in inflammatory bowel disease. *J. Med. Chem.* 52: 3205–3211.

24 Han, H., de Vrueh, R.L., Rhie, J.K. et al. (1998). 5'-Amino acid esters of antiviral nucleosides, acyclovir, and AZT are absorbed by the intestinal PEPT1 peptide transporter. *Pharm. Res.* 15: 1154–1159.

25 Kim, I., Chu, X.Y., Kim, S. et al. (2003). Identification of a human valacyclovirase: biphenyl hydrolase-like protein as valacyclovir hydrolase. *J. Biol. Chem.* 278: 25348–25356.

26 Kim, I., Crippen, G.M., and Amidon, G.L. (2004). Structure and specificity of a human valacyclovir activating enzyme: a homology model of BPHL. *Mol. Pharmaceutics* 1: 434–446.

27 Kim, I., Song, X., Vig, B.S. et al. (2004). A novel nucleoside prodrug-activating enzyme: substrate specificity of biphenyl hydrolase-like protein. *Mol. Pharmaceutics* 1: 117–127.

28 Sun, J., Dahan, A., Walls, Z.F. et al. (2010). Specificity of a prodrug-activating enzyme hVACVase: the leaving group effect. *Mol. Pharmaceutics* 7: 2362–2368.

29 Sun, J., Miller, J.M., Beig, A. et al. (2011). Mechanistic enhancement of the intestinal absorption of drugs containing the polar guanidino functionality. *Expert Opin. Drug Metab. Toxicol.* 7: 313–323.

30 Huttunen, K.M., Mannila, A., Laine, K. et al. (2009). The first bioreversible prodrug of metformin with improved lipophilicity and enhanced intestinal absorption. *J. Med. Chem.* 52: 4142–4148.

31 Saulnier, M.G., Frennesson, D.B., Deshpande, M.S. et al. (1994). An efficient method for the synthesis of guanidino prodrugs. *Bioorg. Med. Chem. Lett.* 4: 1985–1990.

32 Sun, J., Dahan, A., and Amidon, G.L. (2010). Enhancing the intestinal absorption of molecules containing the polar guanidino functionality: a double-targeted prodrug approach. *J. Med. Chem.* 53: 624–632.

33 Haapamaki, M.M., Gronroos, J.M., Nurmi, H. et al. (1999). Phospholipase A2 in serum and colonic mucosa in ulcerative colitis. *Scand. J. Clin. Lab. Invest.* 59: 279–287.

34 Kennedy, B.P., Soravia, C., Moffat, J. et al. (1998). Overexpression of the non-pancreatic secretory group II PLA2 messenger RNA and protein in colorectal adenomas from familial adenomatous polyposis patients. *Cancer Res.* 58: 500–503.

35 Laye, J.P. and Gill, J.H. (2003). Phospholipase A2 expression in tumours: a target for therapeutic intervention? *Drug Discovery Today* 8: 710–716.

36 Minami, T., Tojo, H., Shinomura, Y. et al. (1993). Elevation of phospholipase A2 protein in sera of patients with Crohn's disease and ulcerative colitis. *Am. J. Gastroenterol.* 88: 1076–1080.

37 Milica Markovic, S.B.-S. and Dahan, A. (2020). Lipidic prodrugs for drug delivery: opportunities and challenges. In: *Recent Advancement in Prodrugs* (ed. D.N.C. Kamal Shah, N.S. Chauhan and P. Mishra), 113–132. CRC Press.

38 Dahan, A., Duvdevani, R., Dvir, E. et al. (2007). A novel mechanism for oral controlled release of drugs by continuous degradation of a phospholipid prodrug along the intestine: in vivo and in vitro evaluation of an indomethacin–lecithin conjugate. *J. Controlled Release* 119: 86–93.

39 Dahan, A., Duvdevani, R., Shapiro, I. et al. (2008). The oral absorption of phospholipid prodrugs: in vivo and in vitro mechanistic investigation of trafficking of a lecithin-valproic acid conjugate following oral administration. *J. Controlled Release* 126: 1–9.

40 Dahan, A., Markovic, M., Epstein, S. et al. (2017). Phospholipid-drug conjugates as a novel oral drug targeting approach for the treatment of inflammatory bowel disease. *Eur. J. Pharm. Sci.* 108: 78–85.

41 Markovic, M., Dahan, A., Keinan, S. et al. (2019). Phospholipid-based prodrugs for colon-targeted drug delivery: experimental study and in silico simulations. *Pharmaceutics* 11 (4): 186.

42 Dahan, A., Markovic, M., Aponick, A. et al. (2019). The prospects of lipidic prodrugs: an old approach with an emerging future. *Future Med. Chem.* 11: 2563–2571.

43 Markovic, M., Ben-Shabat, S., Keinan, S. et al. (2018). Prospects and challenges of phospholipid-based prodrugs. *Pharmaceutics* 10 (4): 210.

44 Kurz, M. and Scriba, G.K. (2000). Drug-phospholipid conjugates as potential prodrugs: synthesis, characterization, and degradation by pancreatic phospholipase A_2. *Chem. Phys. Lipids* 107: 143–157. https://doi.org/10.3390/pharmaceutics14030675, https://doi.org/10.3390/ijms23052673

45 Markovic, M., Ben-Shabat, S., Nagendra Manda, J. et al. (2022). PLA2-triggered activation of cyclosporine-phospholipid prodrug as a drug targeting approach in inflammatory bowel disease therapy. *Pharmaceutics.* 14 (3): 675. https://doi.org/10.3390/pharmaceutics14030675.

46 Markovic, M., Abramov-Harpaz, K., Regev, C. et al. (2022). Prodrug-based targeting approach for inflammatory bowel diseases therapy: mechanistic study of phospholipid-linker-cyclosporine PLA2-mediated activation. *Int. J. Mol. Sci.* 23 (5): 2673. https://doi.org/10.3390/ijms23052673.

47 Pan, D., Pham, C.T., Weilbaecher, K.N. et al. (2016). Contact-facilitated drug delivery with Sn2 lipase labile prodrugs optimize targeted lipid nanoparticle drug delivery. *Wiley Interdiscip. Rev. Nanomed. Nanobiotechnol.* 8: 85–106.

48 Pan, D., Sanyal, N., Schmieder, A.H. et al. (2012). Antiangiogenic nanotherapy with lipase-labile *Sn*-2 fumagillin prodrug. *Nanomedicine (London)* 7: 1507–1519.

49 Han, H.K. and Amidon, G.L. (2000). Targeted prodrug design to optimize drug delivery. *AAPS Pharm. Sci.* 2: E6.

50 Bagshawe, K.D. (1987). Antibody directed enzymes revive anti-cancer prodrugs concept. *Br. J. Cancer* 56: 531–532.

51 Sharma, S.K. and Bagshawe, K.D. (2017). Antibody directed enzyme prodrug therapy (ADEPT): trials and tribulations. *Adv. Drug Delivery Rev.* 118: 2–7.

52 Tietze, L.F., Feuerstein, T., Fecher, A. et al. (2002). Proof of principle in the selective treatment of cancer by antibody-directed enzyme prodrug therapy: the development of a highly potent prodrug. *Angew. Chem. Int. Ed.* 41: 759–761.

53 Maji, D., Lu, J., Sarder, P. et al. (2018). Cellular trafficking of *Sn*-2 phosphatidylcholine prodrugs studied with fluorescence lifetime imaging and super-resolution microscopy. *Precis. Nanomed.* 1: 128–145.

54 Francis, R.J., Sharma, S.K., Springer, C. et al. (2002). A phase I trial of antibody directed enzyme prodrug therapy (ADEPT) in patients with advanced colorectal carcinoma or other CEA producing tumours. *Br. J. Cancer* 87: 600–607.

55 Springer, C.J., Poon, G.K., Sharma, S.K., and Bagshawe, K.D. (1993). Identification of prodrug, active drug, and metabolites in an ADEPT clinical study. *Cell Biophys.* 22: 9–26.

56 Sandmair, A.M., Loimas, S., Puranen, P. et al. (2000). Thymidine kinase gene therapy for human malignant glioma, using replication-deficient retroviruses or adenoviruses. *Hum. Gene Ther.* 11: 2197–2205.

57 Huber, B.E., Richards, C.A., and Austin, E.A. (1994). Virus-directed enzyme/prodrug therapy (VDEPT). Selectively engineering drug sensitivity into tumors. *Ann. N.Y. Acad. Sci.* 716: 104–114; discussion 140–103.

58 Niculescu-Duvaz, D., Negoita-Giras, G., Niculescu-Duvaz, I. et al. (2010). Directed enzyme prodrug therapies. In: *Prodrugs and Targeted Delivery: Towards Better ADME Properties* (ed. J. Rautio), 271–344. Wiley.

59 Knox, R.J., Friedlos, F., Jarman, M. et al. (1995). Virtual cofactors for an *Escherichia coli* nitroreductase enzyme: relevance to reductively activated prodrugs in antibody directed enzyme prodrug therapy (ADEPT). *Biochem. Pharmacol.* 49: 1641–1647.

60 de Graaf, M., Boven, E., Scheeren, H.W. et al. (2002). Beta-glucuronidase-mediated drug release. *Curr. Pharm. Des.* 8: 1391–1403.

61 Haisma, H.J., Boven, E., van Muijen, M. et al. (1992). A monoclonal antibody-β-glucuronidase conjugate as activator of the prodrug epirubicin-glucuronide for specific treatment of cancer. *Br. J. Cancer* 66: 474–478.

62 Bosslet, K., Czech, J., and Hoffmann, D. (1994). Tumor-selective prodrug activation by fusion protein-mediated catalysis. *Cancer Res.* 54: 2151–2159.
63 de Graaf, M., Boven, E., Oosterhoff, D. et al. (2002). A fully human anti-Ep-CAM scFv-β-glucuronidase fusion protein for selective chemotherapy with a glucuronide prodrug. *Br. J. Cancer* 86: 811–818.
64 Prijovich, Z.M., Burnouf, P.A., Chou, H.C. et al. (2016). Synthesis and antitumor properties of BQC-glucuronide, a camptothecin prodrug for selective tumor activation. *Mol. Pharmaceutics* 13: 1242–1250.
65 Tranoy-Opalinski, I., Legigan, T., Barat, R. et al. (2014). β-Glucuronidase-responsive prodrugs for selective cancer chemotherapy: an update. *Eur. J. Med. Chem.* 74: 302–313.
66 Gwenin, V.V., Poornima, P., Halliwell, J. et al. (2015). Identification of novel nitroreductases from *Bacillus cereus* and their interaction with the CB1954 prodrug. *Biochem. Pharmacol.* 98: 392–402.
67 Wang, H., Zhou, X.-L., Long, W. et al. (2015). A fusion protein of RGD4C and β-lactamase has a favorable targeting effect in its use in antibody directed enzyme prodrug therapy. *Int. J. Mol. Sci.* 16: 9625–9634.
68 Zhang, J., Kale, V., and Chen, M. (2015). Gene-directed enzyme prodrug therapy. *AAPS J.* 17: 102–110.
69 Chen, L. and Waxman, D.J. (2002). Cytochrome P450 gene-directed enzyme prodrug therapy (GDEPT) for cancer. *Curr. Pharm. Des.* 8: 1405–1416.
70 Lohr, M., Hoffmeyer, A., Kroger, J. et al. (2001). Microencapsulated cell-mediated treatment of inoperable pancreatic carcinoma. *Lancet* 357: 1591–1592.
71 Warren, M.S. and Rautio, J. (2010). Prodrugs designed to target transporters for oral drug delivery. In: *Prodrugs and Targeted Delivery: Towards Better ADME Properties*, vol. 47 (ed. J. Rautio), 133–151.
72 Varma, M.V., Ambler, C.M., Ullah, M. et al. (2010). Targeting intestinal transporters for optimizing oral drug absorption. *Curr. Drug Metab.* 11: 730–742.
73 Lee, V.H., Chu, C., Mahlin, E.D. et al. (1999). Biopharmaceutics of transmucosal peptide and protein drug administration: role of transport mechanisms with a focus on the involvement of PepT1. *J. Controlled Release* 62: 129–140.
74 Tamai, I. (2012). Oral drug delivery utilizing intestinal OATP transporters. *Adv. Drug Delivery Rev.* 64: 508–514.
75 Puris, E., Gynther, M., Auriola, S., and Huttunen, K.M. (2020). L-Type amino acid transporter 1 as a target for drug delivery. *Pharm. Res.* 37: 88.
76 Anderson, C.M., Grenade, D.S., Boll, M. et al. (2004). H^+/amino acid transporter 1 (PAT1) is the imino acid carrier: an intestinal nutrient/drug transporter in human and rat. *Gastroenterology* 127: 1410–1422.
77 Thwaites, D.T. and Anderson, C.M. (2011). The SLC36 family of proton-coupled amino acid transporters and their potential role in drug transport. *Br. J. Pharmacol.* 164: 1802–1816.
78 Ganapathy, M.E. and Ganapathy, V. (2005). Amino acid transporter ATB0,+ as a delivery system for drugs and prodrugs. *Curr. Drug Targets Immune Endocr. Metab. Disord.* 5: 357–364.

79 Hatanaka, T., Haramura, M., Fei, Y.J. et al. (2004). Transport of amino acid-based prodrugs by the Na^+- and Cl^- -coupled amino acid transporter $ATB^{0,+}$ and expression of the transporter in tissues amenable for drug delivery. *J. Pharmacol. Exp. Ther.* 308: 1138–1147.

80 Kovalchuk, V., Samluk, Ł., Juraszek, B. et al. (1866). Trafficking of the amino acid transporter $B^{0,+}$ (SLC6A14) to the plasma membrane involves an exclusive interaction with SEC24C for its exit from the endoplasmic reticulum. *Biochim. Biophys. Acta, Mol. Cell Res.* 2019: 252–263.

81 Qiu, A., Jansen, M., Sakaris, A. et al. (2006). Identification of an intestinal folate transporter and the molecular basis for hereditary folate malabsorption. *Cell* 127: 917–928.

82 Visentin, M., Diop-Bove, N., Zhao, R., and Goldman, I.D. (2014). The intestinal absorption of folates. *Annu. Rev. Physiol.* 76: 251–274.

83 Pierre, K. and Pellerin, L. (2009). Monocarboxylate transporters. In: *Encyclopedia of Neuroscience* (ed. L.R. Squire), 961–965. Oxford: Academic Press.

84 Tsuji, A. (1999). Tissue selective drug delivery utilizing carrier-mediated transport systems. *J. Controlled Release* 62: 239–244.

85 Cao, X., Gibbs, S.T., Fang, L. et al. (2006). Why is it challenging to predict intestinal drug absorption and oral bioavailability in human using rat model. *Pharm. Res.* 23: 1675–1686.

86 Harada, N. and Inagaki, N. (2012). Role of sodium-glucose transporters in glucose uptake of the intestine and kidney. *J. Diabetes Invest.* 3: 352–353.

87 Wright, E.M., Loo, D.D.F., and Hirayama, B.A. (2011). Biology of human sodium glucose transporters. *Physiol. Rev.* 91: 733–794.

88 Balakrishnan, A. and Polli, J.E. (2006). Apical sodium dependent bile acid transporter (ASBT, SLC10A2): a potential prodrug target. *Mol. Pharmaceutics* 3: 223–230.

89 Jonker, J.W. and Schinkel, A.H. (2004). Pharmacological and physiological functions of the polyspecific organic cation transporters: OCT1, 2, and 3 (SLC22A1-3). *J. Pharmacol. Exp. Ther.* 308: 2–9.

90 Terada, T. and Inui, K. (2008). Physiological and pharmacokinetic roles of H^+/organic cation antiporters (MATE/SLC47A). *Biochem. Pharmacol.* 75: 1689–1696.

91 Zhang, Y., Sun, J., Sun, Y. et al. (2013). Prodrug design targeting intestinal PepT1 for improved oral absorption: design and performance. *Curr. Drug Metab.* 14: 675–687.

92 Gupta, S.V., Gupta, D., Sun, J. et al. (2011). Enhancing the intestinal membrane permeability of zanamivir: a carrier mediated prodrug approach. *Mol. Pharmaceutics* 8: 2358–2367.

93 Cundy, K.C., Annamalai, T., Bu, L. et al. (2004). XP13512 [(±)-1-([(α-isobutanoyloxyethoxy)carbonyl] aminomethyl)-1-cyclohexane acetic acid], a novel gabapentin prodrug: II. Improved oral bioavailability, dose proportionality, and colonic absorption compared with gabapentin in rats and monkeys. *J. Pharmacol. Exp. Ther.* 311: 324–333.

94 Kramer, W., Wess, G., Neckermann, G. et al. (1994). Intestinal absorption of peptides by coupling to bile acids. *J. Biol. Chem.* 269: 10621–10627.

95 Kramer, W., Wess, G., Enhsen, A. et al. (1994). Bile acid derived HMG-CoA reductase inhibitors. *Biochim. Biophys. Acta, Mol. Basis Dis.* 1227: 137–154.

96 Wess, G., Kramer, W., Han, X.B. et al. (1994). Synthesis and biological activity of bile acid-derived HMG-CoA reductase inhibitors. The role of 21-methyl in recognition of HMG-CoA reductase and the ileal bile acid transport system. *J. Med. Chem.* 37: 3240–3246.

97 Doucette, K.E. and Aoki, F.Y. (2001). Oseltamivir: a clinical and pharmacological perspective. *Expert Opin. Pharmacother.* 2: 1671–1683.

98 Gupta, R.K. and Nguyen-Van-Tam, J.S. (2006). Oseltamivir resistance in influenza A (H5N1) infection. *N. Engl. J. Med.* 354: 1423–1424; author reply 1423–1424.

99 Cass, L.M., Efthymiopoulos, C., and Bye, A. (1999). Pharmacokinetics of zanamivir after intravenous, oral, inhaled or intranasal administration to healthy volunteers. *Clin. Pharmacokinet.* 36 (Suppl. 1): 1–11.

100 Yan, Z., Sun, J., Chang, Y. et al. (2011). Bifunctional peptidomimetic prodrugs of didanosine for improved intestinal permeability and enhanced acidic stability: synthesis, transepithelial transport, chemical stability and pharmacokinetics. *Mol. Pharmaceutics* 8: 319–329.

101 Incecayir, T., Sun, J., Tsume, Y. et al. (2016). Carrier-mediated prodrug uptake to improve the oral bioavailability of polar drugs: an application to an oseltamivir analogue. *J. Pharm. Sci.* 105: 925–934.

102 Li, F., Maag, H., and Alfredson, T. (2008). Prodrugs of nucleoside analogues for improved oral absorption and tissue targeting. *J. Pharm. Sci.* 97: 1109–1134.

103 Landowski, C.P., Vig, B.S., Song, X., and Amidon, G.L. (2005). Targeted delivery to PEPT1-overexpressing cells: acidic, basic, and secondary floxuridine amino acid ester prodrugs. *Mol. Cancer Ther.* 4: 659–667.

104 Tsume, Y., Hilfinger, J.M., and Amidon, G.L. (2008). Enhanced cancer cell growth inhibition by dipeptide prodrugs of floxuridine: increased transporter affinity and metabolic stability. *Mol. Pharmaceutics* 5: 717–727.

105 Tsume, Y., Hilfinger, J.M., and Amidon, G.L. (2011). Potential of amino acid/dipeptide monoester prodrugs of floxuridine in facilitating enhanced delivery of active drug to interior sites of tumors: a two-tier monolayer in vitro study. *Pharm. Res.* 28: 2575–2588.

106 Song, X., Lorenzi, P.L., Landowski, C.P. et al. (2005). Amino acid ester prodrugs of the anticancer agent gemcitabine: synthesis, bioconversion, metabolic bioevasion, and hPEPT1-mediated transport. *Mol. Pharmaceutics* 2: 157–167.

107 Tsume, Y., Drelich, A.J., Smith, D.E., and Amidon, G.L. (2017). Potential development of tumor-targeted oral anti-cancer prodrugs: amino acid and dipeptide monoester prodrugs of gemcitabine. *Molecules* 22 (8): 1322.

108 Foley, D.W., Pathak, R.B., Phillips, T.R. et al. (2018). Thiodipeptides targeting the intestinal oligopeptide transporter as a general approach to improving oral drug delivery. *Eur. J. Med. Chem.* 156: 180–189.

109 Gill, R.K., Saksena, S., Alrefai, W.A. et al. (2005). Expression and membrane localization of MCT isoforms along the length of the human intestine. *Am. J. Physiol. Cell Physiol.* 289: C846–C852.

110 Halestrap, A.P. and Meredith, D. (2004). The SLC16 gene family-from monocarboxylate transporters (MCTs) to aromatic amino acid transporters and beyond. *Pflugers Arch.* 447: 619–628.

111 Jones, R.S. and Morris, M.E. (2016). Monocarboxylate transporters: therapeutic targets and prognostic factors in disease. *Clin. Pharmacol. Ther.* 100: 454–463.

112 Cundy, K.C., Sastry, S., Luo, W. et al. (2008). Clinical pharmacokinetics of XP13512, a novel transported prodrug of gabapentin. *J. Clin. Pharmacol.* 48: 1378–1388.

113 Lal, R., Sukbuntherng, J., Tai, E.H. et al. (2009). Arbaclofen placarbil, a novel R-baclofen prodrug: improved absorption, distribution, metabolism, and elimination properties compared with R-baclofen. *J. Pharmacol. Exp. Ther.* 330: 911–921.

114 Kramer, W., Burckhardt, G., Wilson, F.A., and Kurz, G. (1983). Bile salt-binding polypeptides in brush-border membrane vesicles from rat small intestine revealed by photoaffinity labeling. *J. Biol. Chem.* 258: 3623–3627.

115 Kullak-Ublick, G.A., Glasa, J., Boker, C. et al. (1997). Chlorambucil-taurocholate is transported by bile acid carriers expressed in human hepatocellular carcinomas. *Gastroenterology* 113: 1295–1305.

116 Rais, R., Fletcher, S., and Polli, J.E. (2011). Synthesis and in vitro evaluation of gabapentin prodrugs that target the human apical sodium-dependent bile acid transporter (hASBT). *J. Pharm. Sci.* 100: 1184–1195.

117 Tolle-Sander, S., Lentz, K.A., Maeda, D.Y. et al. (2004). Increased acyclovir oral bioavailability via a bile acid conjugate. *Mol. Pharmaceutics* 1: 40–48.

118 Keinan, S., Frush, E.H., and Shipman, W.J. (2018). Leveraging cloud computing for in silico drug design using the Quantum Molecular Design (QMD) framework. *Comput. Sci. Eng.* 20: 66–73.

119 Chen, Y.C. (2015). Beware of docking! *Trends Pharmacol. Sci.* 36: 78–95.

120 Dahan, A., Ben-Shabat, S., Cohen, N. et al. (2016). Phospholipid-based prodrugs for drug targeting in inflammatory bowel disease: computational optimization and in vitro correlation. *Curr. Top. Med. Chem.* 16: 2543–2548.

121 Pedersen, P.J., Adolph, S.K., Subramanian, A.K. et al. (2010). Liposomal formulation of retinoids designed for enzyme triggered release. *J. Med. Chem.* 53: 3782–3792.

122 Rosseto, R. and Hajdu, J. (2014). Peptidophospholipids: synthesis, phospholipase A2 catalyzed hydrolysis, and application to development of phospholipid prodrugs. *Chem. Phys. Lipids* 183: 110–116.

123 Markovic, M., Ben-Shabat, S., Keinan, S. et al. (2019). Molecular modeling-guided design of phospholipid-based prodrugs. *Int. J. Mol. Sci.* 20 (9): 2210.

124 Dahan, A., Markovic, M., Keinan, S. et al. (2017). Computational modeling and in vitro/in silico correlation of phospholipid-based prodrugs for targeted drug delivery in inflammatory bowel disease. *J. Comput.-Aided Mol. Des.* 31: 1021–1028.

125 Linderoth, L., Fristrup, P., Hansen, M. et al. (2009). Mechanistic study of the sPLA2-mediated hydrolysis of a thio-ester pro anticancer ether lipid. *J. Am. Chem. Soc.* 131: 12193–12200.

126 Karaman, R., Dajani, K.K., Qtait, A., and Khamis, M. (2012). Prodrugs of acyclovir – a computational approach. *Chem. Biol. Drug Des.* 79: 819–834.

127 Karaman, R. and Hallak, H. (2010). Computer-assisted design of pro-drugs for antimalarial atovaquone. *Chem. Biol. Drug Des.* 76: 350–360.

128 Karaman, R. (2011). Computational-aided design for dopamine prodrugs based on novel chemical approach. *Chem. Biol. Drug Des.* 78: 853–863.

129 Giacomini, K.M., Galetin, A., and Huang, S.M. (2018). The international transporter consortium: summarizing advances in the role of transporters in drug development. *Clin. Pharmacol. Ther.* 104: 766–771.

130 Agarwal, P., Griffith, A., Costantino, H.R., and Vaish, N. (2010). Gabapentin enacarbil – clinical efficacy in restless legs syndrome. *Neuropsychiatr. Dis. Treat.* 6: 151–158.

131 Comparing NUC-1031 lus Cisplatin to Gemcitabine Plus Cisplatin in Patients With Advanced Biliary Tract Cancer (NuTide: 121); ClinicalTrials.gov Identifier: NCT04163900. https://clinicaltrials.gov/ct2/show/NCT04163900?term=NUC-1031&draw=2&rank=3 (accessed 14 June 2020).

132 Arora, M., Bogenberger, J.M., Abdelrahman, A. et al. (2020). Evaluation of NUC-1031: a first-in-class ProTide in biliary tract cancer. *Cancer Chemother. Pharmacol.*

133 Study to Test AKR1C3-Activated Prodrug OBI-3424 (OBI-3424) in Patients With Relapsed/Refractory T-Cell Acute Lymphoblastic Leukemia (T-ALL); ClinicalTrials.gov Identifier: NCT04315324. https://clinicaltrials.gov/ct2/show/NCT04315324?term=OBI-3424&draw=2&rank=1 (accessed 13 June 2020).

134 This Study is to Evaluate OBI-3424 Safe and Effective Treatment Dose in Subjects With Hepatocellular Carcinoma or Castrate Resistant Prostate Cancer. ClinicalTrials.gov Identifier: NCT03592264. https://clinicaltrials.gov/ct2/show/NCT03592264?term=OBI-3424&draw=2&rank=2 (accessed 11 June 2020).

135 Evans, K., Duan, J., Pritchard, T. et al. (2019). OBI-3424, a Novel AKR1C3-activated prodrug, exhibits potent efficacy against preclinical models of T-ALL. *Clin. Cancer Res.* 25: 4493–4503.

136 Genetic Determinants of ACEI Prodrug Activation; ClinicalTrials.gov Identifier: NCT03051282. https://clinicaltrials.gov/ct2/show/NCT03051282?term=NCT03051282&draw=2&rank=1 (accessed 16 June 2020).

8

Exosomes for Drug Delivery Applications in Cancer and Cardiac Indications

Anjali Pandya, Sreeranjini Pulakkat, and Vandana Patravale

Institute of Chemical Technology, Department of Pharmaceutical Sciences and Technology, N. P. Marg, Matunga (E), Mumbai, Maharashtra 400 019, India

8.1 Extracellular Vesicles: An Overview

The chemical envoys for intracellular communication comprise predominantly the extracellular vesicles (EV). EVs are the spherical entities enclosed by phospholipid bilayer with a wide size range of 50 nm up to 10 μm. The term EV is a representative of vesicles originating from all cells irrespective of the donor cells, the mechanism of biogenesis, size, composition, cargo, etc. [1]. The year 1981 marked the first observation of EVs in the form of membrane-bound enzyme containing particles shed from tumor cell lines [2]. The observation led to the indication of the cells being able to take up the released EVs and the possibility of carrying cargo thereof. Numerous EVs released by the cells are being studied for their applications as a tool for targeted drug delivery. The source and size of the EVs differentiate them into three groups viz. (i) exosomes (diameter: 30–150 nm), (ii) microvesicles/ectosomes (diameter: 50 nm 1 μm), and (iii) apoptotic bodies (diameter: 50 nm 5 μm) [3]. Additionally, the terminology "oncosomes" has been described by Jaiswal and Sedger as 100–400 nm vesicles, which carry abnormal transforming macromolecules known as large oncosomes, produced from both malignant and benign tumor tissues [1]. The origin-based classification is most widely accepted but an official nomenclature has not yet been reported. The intracellular distinction of EVs is the factor, which governs their applicability. The morphological specifications of exosomes make them conducive to organ and site-specific drug delivery. The release of exosomes is a result of the fusion of multivesicular endosomes (MVEs) and plasma membrane. Microvesicles or ectosomes are larger in size and directly bud from the plasma membrane whereas the process of cell apoptosis leads to the formation of apoptotic bodies [4]. The characteristics and the abundance of EVs in the human biological fluids, especially in conditions like cancer, are influenced by external or internal stimuli, type of cell, presence or absence of disease condition, the pathway of biogenesis, etc. [1].

Anjali Pandya and Sreeranjini Pulakkat share equal authorship.

Targeted Drug Delivery, First Edition. Edited by Yogeshwar Bachhav.

8.1.1 Evolution of Exosomes

The origin of exosomes goes around 30 years back in the year 1983 which marked the evolution of exosomes via biogenesis involving exocytosis of MVEs. Rose Johnstone, who coined the term "exosome" for the EVs, described the discovery with the phrase "Alice in Blunderland." The breakthrough goes back to the observations made with the lysis of red blood cells yielding vesicles as by-products. Johnstone and coworkers stumbled on the exosomes while working on the identification of an amino acid transport protein [5]. Exosomes are released as a result of exocytosis of multivesicular bodies (MVBs) and the cell types include hematopoietic cells like reticulocytes, lymphocytes, dendritic cells, mast cells, platelets, macrophages, alveoli, etc. [6]. Cells from even non-hematopoietic origin have now been reported to release exosomal vesicles [7].

The biogenesis of exosomes (Figure 8.1) is contributed via several pathways and follows the following steps: (i) early endosome formation through endocytosis, (ii) intraluminal vesicle (ILV) formation, (iii) late endosome or MVB formation, and (iv) exosome release from the cell [9]. The biogenetic pathway starts with the formation of MVBs involving the contribution of the membrane compartments leading to the internalization of some cellular components or extracellular ligands. The membrane vesicles transform into early endosomes, which mature to become late endosomes, which are often referred to as MVEs. The MVEs further fuse with the lysosomes leading to degradation and the ones, which fuse, with the cell surface, are secreted in the form of exosomes. Thus, the subpopulations of MVEs are typically responsible for the generation of exosomes. Several vesicles showing characteristics similar to those of exosomes have been isolated from numerous body fluids including blood,

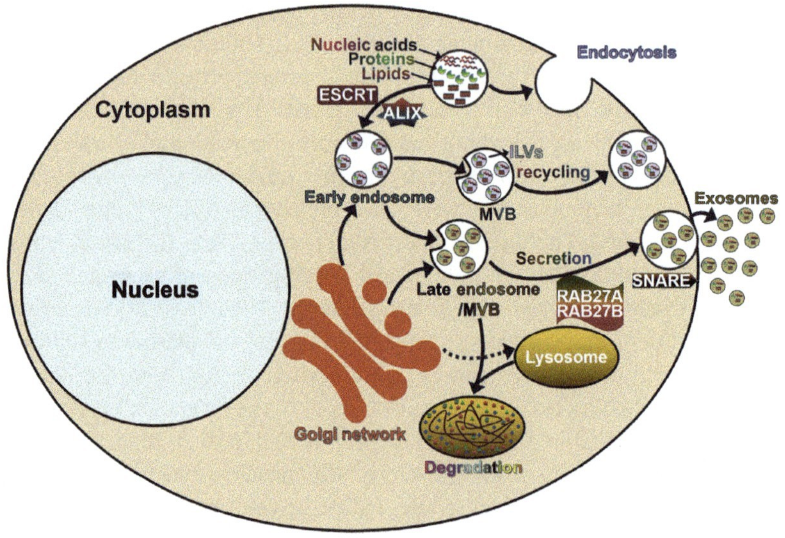

Figure 8.1 Biogenesis of Exosomes. Source: Reprinted with permission from Wang et al. [8].

urine, saliva, semen, milk, amniotic fluid, cerebrospinal fluid, ascites fluid, and bile [7, 10]. Several molecules are involved in the completion of these steps, which makes it difficult to differentiate them and their involvement in specific steps.

The fusion of MVEs with the plasma membrane is a factor dependent on the type of cells being utilized and hence the differentiation of molecules involved in each step is important. For example, MVEs with high cholesterol content or the ones secreted from the apical and basolateral sides of polarized cells give rise to different MVE populations [11]. The composition of MVEs in turn is reflected in the exosomes for which proteins associated with MVEs like endosomal sorting complex required for transport (ESCRT) or CD63 are used as indicators to recognize the exosomal molecular composition [12]. ESCRT machinery consists of ESCRT – 0, I, II, III protein complexes along with AAA ATPase VPS4 complex, and exosome biogenesis is majorly been reported using the HeLa, HEK293, MCF-7 cells [11]. The biogenesis can be dependent or independent of the ESCRT but may have similar pathways [13]. Different sub-populations of exosomes are a result of the individual machinery responsible for the execution of the pathways. The machinery is in turn a result of the cell type and its allied homeostasis used for the exosome production [11].

Exosomes have a dimension in the range of 20–150 nm and derive their origin from the endosomal pathway, which led to their production. They possess a special ability to carry and transmit information intracellularly. The ability is a result of the presence of RNA, proteins, lipids, etc. within the exosomes, which help in their communication with the cells. Exosomes play a role in normal physiological processes including immunity surveillance, tissue repair, maintenance of stem cells, and blood coagulation pathways [14]. Exosomes, along with the normal physiological processes, also form a part of the pathological conditions in multiple diseases. The exosomal components including lipids, RNA, and proteins are vital in conducting the intercellular communication with the recipient cells. The lipids present in the exosomes, including cholesterol, sphingomyelins, and phosphatidylserine, are responsible for their uptake. Proteins, on the other hand, help in the adhesion of exosomes, regulate transcription, modulate signaling pathways, and thereby assist in recipient cell interaction [15]. The exosome-mediated processes tend to get altered by the presence of viruses and aid in their spread; whereas in the case of tumors, these altered processes contribute to cancer growth and metastasis. The exosomes are also derived from tumors, which become a part of the tumor microenvironment and support its growth and survival. Numerous effects of exosomes with respect to different physiological systems and pathological conditions described by Lee et al. have been listed in Table 8.1 [14].

8.1.2 Exosomes as Delivery Vehicles for Therapeutics

Exosomes are constantly released into the extracellular space and systemic circulation transporting cellular materials including proteins, micro RNA (miRNA), mRNA, and DNA. Their innate therapeutic ability, low immunogenicity, and cell-to-cell communication triggered the concept of exploiting them as carriers for delivering bioactives. The natural phospholipid bilayer in exosomes assures the

Table 8.1 Effects of exosomes on the different physiological systems and pathological conditions.

Physiological system	Exosomal effect	References
Stem cell plasticity	• Modulation of phenotype	[16]
Immune system	• Transfer of inflammatory molecules during immune response • Activation and maturation of immune cells like T cells, dendritic cells, etc. for information transfer. • Modulation of immune response via breast milk in newborn	[17, 18]
Nervous system	• Axonal support • Transfer of neuroprotective factors • Information transmission through proteins and mRNA in neural circuit	[19, 20]
Pathological condition	**Exosomal effect**	
Cancer	• Involvement in growth, invasion and metastasis of tumor • Transfer of miRNA and oncogenic proteins	[8]
Cardiovascular diseases	• Improving heart failure, cardiac dysfunction, and angiogenesis • Attenuation of hypertrophy and fibrosis • Progression of atherosclerotic plaque, induction of cell death, and cardiac dysfunction	[21]
Viral infection	• Transfer of viral receptors • Evasion of immune system	[22]
Neurodegenerative diseases	• Transfer of neurodegenerative factors to different brain areas • Amyloid deposition with transfer of toxic proteins by exosomes in Alzheimer's disease • Assist to transfer toxic components in the propagation of Parkinson's disease • Initiation of prion protein transmission in the neuronal cells	[23, 24]

Source: Adapted from Lee et al. [14].

retention of biological molecules or therapeutics and protects them from enzymatic degradation during the transit from donor to recipient cells. The exact mechanisms of cargo transfer followed by these vesicles are not completely understood but give a hint about their application in designing specific molecule delivery at the desired site for catering to numerous physiological conditions. Exosomes may help in overcoming some of the limitations of the non-biological drug delivery systems like poor cellular uptake, immunogenicity, and low circulation time. Further, the surface of the exosomes can be engineered to target cells or tissues resulting in their enhanced uptake and efficient drug delivery. Few recent studies have explored the surface modification of exosomes with specific peptides to improve their efficient

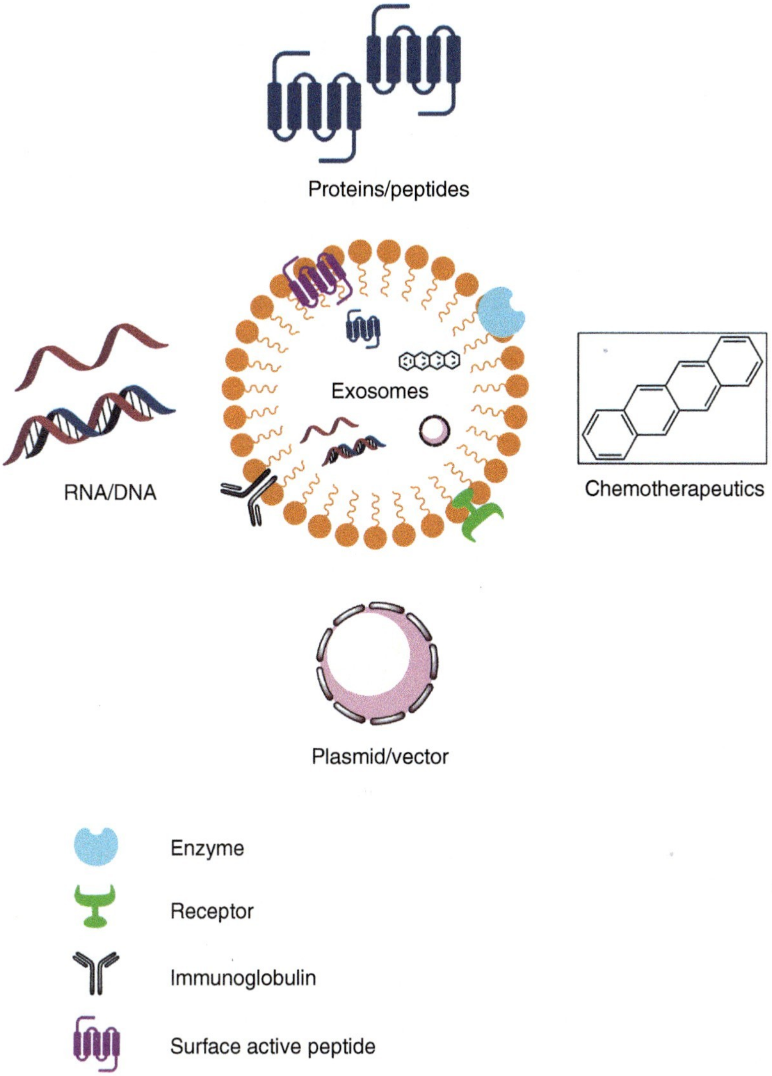

Figure 8.2 Exosomes for drug delivery applications. The versatility of exosomes as a platform for delivery of various types of biological and nonbiological cargos has been schematically represented here. Exosomes can be harnessed for the delivery of RNAs, proteins/peptides as well as small molecules. They are intrinsically characterized by the presence of multiple membrane proteins or can be engineered to express surface molecules aiding in efficient targeting and cellular uptake.

delivery to the desired target site. In general, therapeutics can be loaded within exosomes using either exogenous or endogenous methods. As every loading strategy has its own pros and cons, the preferred approach can be selected depending on the drug candidate, target site, and the feasibility of scaleup. Figure 8.2 provides a schematic representation of the various deliverable cargos via exosomes. Exosomes can be employed for delivering genetic material (siRNA, mRNA, miRNA, and

DNA), recombinant proteins/peptides, antigens, and therapeutic drugs, which are prone to RNase degradation in systemic circulation or unable to move across biological membranes to reach their intended site of action.

8.1.2.1 Endogenous Loading Methods

Endogenous loading refers to systems in which the donor cells are modified by genetic engineering or incubated with drugs so that the desired cargo is deposited directly when the exosomes are generated. This method is relatively convenient and requires minimal steps. It is mostly used for the generation of exosomes loaded with RNA or proteins from donor cells, which have been transfected earlier to overexpress a gene product using a vector. High miRNA expressions were observed by transfecting the donor cells with pre-miRNAs [25]. The same strategy has been applied to load proteins as well as overexpress specific proteins on the surface of the generated exosomes [26, 27]. The donor cells can also be extruded through filters of different pore sizes and reconstituted to generate exosome mimetics with similar biological and physical characteristics. This method facilitates the production of higher quantities of exosomal carriers [28–30]. However, the unstable nature of RNA and degradation of the peptides by endosomal proteases are some of the limitations associated with the endogenous loading of exosomes. Further, cell engineering required for loading large amounts of cargo as well as surface modification with targeting peptides can prove to be laborious and time-consuming.

8.1.2.2 Exogenous Loading Methods

Exogenous methods involve the loading of therapeutics/biomolecules after isolation of exosomes and can be further classified as passive or active loading. Passive loading is the simplest loading method where the therapeutic cargo/drug is incubated with the isolated exosomes and the concentration gradient drives the loading process. Other factors influencing the exosomal loading are the hydrophilicity of the active, zeta potential or electrokinetic mobility of the exosomes, surfactant concentration, and the size of the biomolecule being loaded. However, limited loading efficiency is one of the drawbacks of co-incubation. Only small molecules were found to pass efficiently across the exosome membrane. Further, the low transfection efficiency of these exosomes greatly affects the cargo delivery potential of passively loaded exosomes [31–33].

In the case of active loading, disruption of the exosomal membrane by electroporation, extrusion, sonication, saponin permeabilization, or freeze–thaw cycles aids in the loading of the desired cargo. Using the electroporation method, small pores are created in the phospholipid bilayer by applying an electrical field to aid in the passage of cargo into the exosomes. The integrity of the membrane is then recovered using a suitable media or buffer to obtain cargo-loaded exosomes. This method is employed mostly for the loading of large molecules like siRNA, which cannot spontaneously diffuse (passive loading) owing to their hydrophilicity. Small hydrophilic drug molecules were also successively loaded into exosomes via electroporation. Although the optimal parameters for exosome electroporation depend on the donor cell, they can be easily controlled and hence, it poses as the most feasible loading method for large-scale production [34, 35]. However, loss of exosome integrity and siRNA aggregation are some of the issues encountered in this method [36].

Simultaneous sonication of exosomes and drug mixture using a probe sonicator has also been employed for exosome loading. Sonication-induced deformation and subsequent reformation of exosomal membranes aid in the passage of drugs across lipid bilayers. However, this sonication method is not preferred for loading RNAs owing to aggregation or degradation. In the extrusion method, the exosomes-drug mixture is loaded into a syringe-based lipid extruder and extruded through extruder membranes of 100–400 nm pore size, during this passage the exosomal membrane gets disrupted and mixed with the drug and permits loading. In saponin-assisted loading, upon incubation with the exosomes, pores are generated in their membrane leading to enhanced permeability and cargo encapsulation. High encapsulation efficiencies have been observed using this method, however, saponins cause *in vivo* hemolysis and hence should be used in low amounts. Co-incubation of exosomes with saponin concentration of only 0.01% (w/v) was reported to significantly increase the loading efficiency of hydrophobic porphyrin into exosomes, indicating the potency of saponin as a surfactant [37]. In the freeze–thaw cycle method, exosomes and drugs are incubated at room temperature followed by freezing at −80 °C or in liquid nitrogen, and re-thawing at room temperature. However, it is not preferred due to exosome aggregation and low drug loading efficiency in comparison with the other active loading methods [29, 38]. The various endogenous and exogenous methods of loading exosomes have been schematically represented in Figure 8.3.

In this chapter, we discuss the applications of exosome-based drug delivery vehicles for cancer and cardiac indications. A lot of research has been carried out to deliver a host of cargo (drug candidates) including proteins, peptides, nucleic acids, and small molecules for cancer therapy using exosomal carriers using both exogeneous and endogenous loading methods. However, the use of exosomal vehicles for cardiac indications is focused on delivering nucleic acid therapeutics following the endogenous method of incorporation.

8.2 Exosomes as Cancer Therapeutics

The ability of exosomes to transfer cellular information in the normal physiological system as well as in various pathological conditions is evident from the previous sections. This ability led to the investigation of exosomes for their therapeutic potential by the virtue of their numerous characteristic features. The fact that exosomes are a part of the pathological processes establishes their potential as therapeutics. The role of exosomes in cancer is evident from the release of exosomes from the cancer cells. These secreted exosomes play a role in key events such as tumor growth, metastasis/cancer progression, angiogenesis, remodeling of the extracellular matrix, suppression of the immune system, chemotherapy/drug resistance, and alteration of the tumor microenvironment, etc. [39]. The exosomes, therefore, house extensive information about the tumor present and in turn about

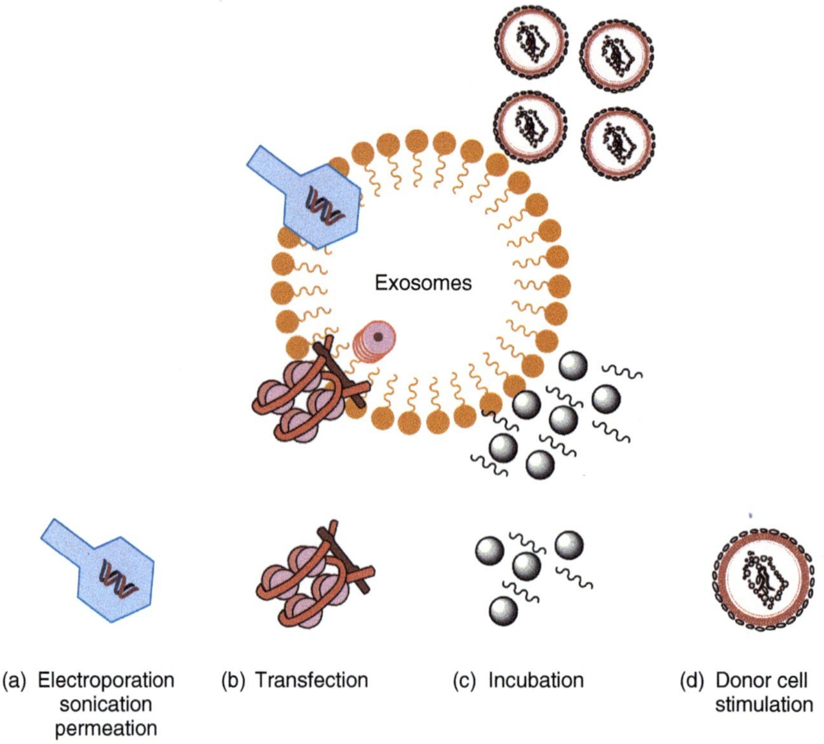

Figure 8.3 Schematic representation of various loading strategies employed in the development of exosomal delivery vehicles. (a) Electroporation, sonication, and permeation are the most common methods of exosomal loading due to their enhanced loading efficiency and scalability; (b) transfection technique is used for the delivery of RNA, DNA, nucleic acids, and plasmid taking advantage of the inherent ability of exosomes to internalize with host cells; (c) incubation is one of the mechanical techniques used for cargo loading (usually small molecules) in exosomes and it is preferred because of its simplicity and possibility of *ex vivo* loading; (d) donor cells stimulation is a viable strategy for exosome loading due to the natural ability of donor cells to impart target specificity and avoid host cell rejection.

allied cancer. This information is very helpful to study the disease pathogenesis at molecular levels and identifying biomarkers for cancer detection as well as for the discovery of unknown mechanisms in cancer biology [39]. Progression and metastasis of tumor is regulated by the intracellular communication of the cancerous and non-cancerous cells. The invasion of cancerous cell secretions including exosomes into the healthy cells is responsible for the multiplication of cancer cells. The tumor growth is dependent on the tumor microenvironment which majorly comprises blood vessels, immune cells, fibroblasts, endothelial cells, and extracellular matrix [40]. The cancer-borne exosomes are therefore strong moderators for the promotion of metastasis by the conditioning of the tumor microenvironment (Figure 8.4).

The development of exosomes as therapeutic vehicles is an attractive alternative because of their inherent biocompatibility. The major advantages of exosomes for

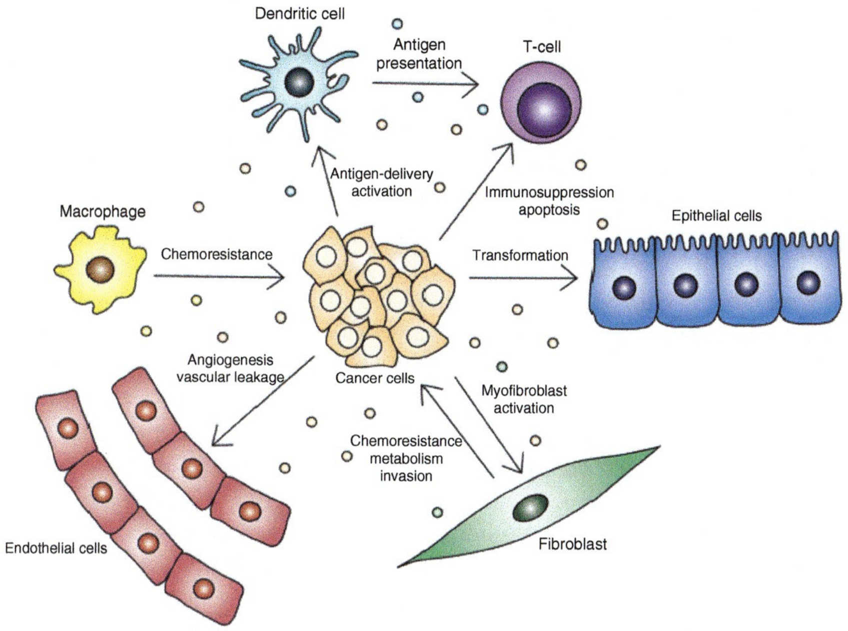

Figure 8.4 Schematic representation of the various effects of tumor-cell derived exosomes on noncancerous cells and vice versa within the tumor microenvironment. Tumor-derived exosomes inhibit T-cell activation and induce immunosuppression, deliver antigens to dendritic cells for cross-presentation, and induce vascular leakage. Further, they initiate fibroblast activation, and the fibroblast-derived exosomes can in turn contribute to chemoresistance. Source: McAndrews and Kalluri [41] / Springer Nature / CC BY-4.0.

drug delivery are their biocompatibility, non-toxic nature, and resistance to immune attack. Compared to the other EVs, exosomes have a nano-size range making their intracellular delivery more efficient. The cell internalization potential of exosomal imparts tumor invasion ability along with enhanced specificity. The tumor specificity can be improved by complementing the native exosome cargo by targeting ligands in the form of peptides, proteins, or antibodies [42]. Exosomes have been compared with the conventional liposomes, wherein, the rate at which liposomes are cleared from the systemic circulation is said to be much higher than that of exosomes which are highly stable in biological fluids [43]. The internalization and adherence of exosomes within the tumor cells are 10-fold greater as compared to liposomes of the same size indicating their superiority to target the tumors [44]. In addition, allogenic exosomes can be derived from the bone marrow, monocytes or even macrophages of the patient to attain an immune privilege to reduce the rate of drug clearance, even lower than that of the PEGylated nanoformulations [45]. These properties of exosomes enable them to stay protected in the systemic circulation and avail a longer circulation time therein. The prolonged residence time also facilitates higher accumulation in the abnormal vasculature of the tumor and thereby provides a higher drug load at the site of action.

8.2.1 Influence of Donor Cells

Cancer therapy is determined by the type of tumor and the mechanism of allied cancer biology. The nature and type of donor cell is a detrimental factor while designing an exosomal-based therapy for the treatment of cancer. The donor cells are the source of exosome generation and therefore must produce exosomes, which are non-immunogenic in nature. Different cell lines have been used as models for the generation of exosomes, of which the most common ones include the human origin HeLa, HEK-293, MCF-7, mesenchymal stem cells (MSCs), and the murine cell lines like B16-F10, B16-F1, and B16-BL6 [46]. The surface protein components of the donor cells majorly govern the immunogenicity of the exosomes and immature dendritic cells fulfill this condition [47]. MSCs have been extensively studied in this regard because of their ability to generate considerably greater quantity of exosomes. The efficiency of MSCs to produce exosomes makes them a suitable candidate for large-scale production of exosomes. A study utilized MSC-derived exosomes as delivery vehicles to load miRNA for catering to malignant glioma in a murine model. The miRNA-loaded exosomes were injected at the tumor site and resulted in a reduction of the gliomal xenograft growth in the rat brain [25]. The donor cells used for the production of exosomes have a tendency to stimulate tumorigenesis because of their inherent properties hence a validation of the safety and efficacy of the exosomes is an essential prerequisite. This phenomenon of tumorigenesis is owed to the fact that exosomes derived from tumor cells are actively associated with tumor cells and their allied microenvironment. Citing the speculations, it has been observed that exosomes being molecular transporters are influenced by both donor and recipient cells and hence have the possibility to induce beneficial or obstructive tumor progression [48]. Research in this direction is constantly evolving and only future evidence is expected to clear the picture.

8.2.2 Different Therapeutic Cargo Explored in Cancer Therapy

The capability of exosomes to carry and deliver therapeutic moieties defines their eligibility as drug delivery vehicles. The nature of therapeutic cargos differs depending on their physiological applications in different cancer types. Structurally, the presence of a lipid bilayer membrane helps protect the therapeutic moiety and enables its delivery to the target site at efficacious concentrations. Efficient drug loading into the exosomes is very important to achieve the aforementioned goals.

Diverse approaches have been utilized for an efficient drug loading such as (i) *ex vitro* loading of lipophilic small molecules via co-incubation, (ii) pre-loading of donor cells with the drug followed by the direct release of drug-loaded exosomes, and transfection of donor cells with nucleic acid encoded therapeutics to be then released in the resultant exosomes [45]. The physicochemical properties of the exosomes help them cross the cellular barriers and deliver the enclosed therapeutic cargo in their active form. The exosome-based drug delivery systems have been majorly explored for the delivery of proteins, peptides, nucleic acids, and some anticancer molecules.

8.2.2.1 Delivery of Proteins and Peptides

Delivery of proteins and peptides using exosomes has been in research for the last decade. Tumor-specific antigens, antibodies, anti-apoptosis proteins, peptides specific to the target tissue and tumor-targeting ligands like transferrins and lactoferrins, are some of the therapeutic proteins and peptides delivered by the means of exosomes [49]. Dendritic cells are used to generate exosomes for peptide loading. Wherein, the derived exosomes, containing peptides (for example major histocompatibility complex [MHC]), are complexed with a membrane protein for antigen presentation. These complexes are used for designing T cell-mediated immunotherapy where the peptide is pre-loaded in the dendritic cells followed by exosome production. In this case, the dendritic cells derived exosomes contained MHC–peptide complex along with certain stimulatory molecules on the exosome membrane to prolong the antigen presenting action and enhance the effect of immunotherapy [50]. Apart from the cells of human origin, pathogenic cells are also a source of exosomes. Certain parasites like Leishmania species are reported to produce exosomes, which can interact with the host cells. Using a similar mechanism, intestinal epithelial cells can respond to the Cryptosporidium infection through the increased release of an antimicrobial peptide containing exosomes [51]. Role of anti-apoptotic proteins is considered important when apoptosis suppression occurs in cancer cells. Survivin is an example of an anti-apoptotic protein that, in its inactive mutated form i.e. surviving-T34A, activates caspase enzyme followed by apoptosis induction. Survivin-T34A loaded in the form of therapeutic cargo into the exosomes has been found to stimulate apoptosis in pancreatic adenocarçinoma cell lines and which further enhances the action of gemcitabine [52].

Peptides are widely utilized for the targeting potential through fabrication of target-specific exosomal delivery vehicles. In this case, the targeting peptide is present on the surface of the exosomes, which directs the therapeutic cargo to the site of action. Occasionally, modified exosomes which themselves express a target-specific peptide on their surface, are produced. Overexpression of proteins can be achieved in the donor cells which is followed by further production of exosomes bearing proteins on their surface. In one of the studies, the class II transactivator gene (CIITA gene) was transfected on murine melanoma cells to induce overexpression of MHC-II protein, the cells in turn led to the production of MHC-II overexpressed exosomes. These vesicles rich in MHC-II on their surface helped to carry the cargo toward T-cells to elicit the desired action on cancer cells. In this case, the protein served both as a targeting ligand and a therapeutic agent [53]. Use of targeting peptides is important to deliver therapeutic agents which are known to induce adverse effects, the specific examples are discussed in the subsection for anticancer agent delivery (Section 8.2.2.2). Because of the specific physicochemical properties, exosomes contain functional proteins which play a significant role in targeting the cancer cells. The presence of proteasomes in stem cell-derived exosomes is an example of this phenomenon. The 20S proteasome and immunoproteasome were found to be naturally present in exosomes derived from MSCs. They play an important role in tumor progression and suggest the potential to target cancer cells using exosome-delivered proteasomes [54].

8.2.2.2 Delivery of Chemotherapeutic Cargo

The conventional and highly explored cancer therapies employ the use of chemotherapeutic agents. Along similar lines, exosomes loaded with anticancer drugs are perceived to be the best alternative to cater the tumorigenic conditions. Comparing the antitumor effect by delivery of drug alone versus their delivery as a therapeutic cargo in exosomes has led to the inclination of researchers toward the latter. A study was performed to determine the effect of liposomal formulation doxorubicin compared to that of exosomes loaded formulation on the tumor. The results depicted that the exosomal delivery was superior in reducing the tumor size in the mouse model of colon adenocarcinoma compared to the liposomal formulation [28]. The cardiotoxicity of doxorubicin, a serious side effect, has been found to be measurably reduced when the administration was in the form of the exosomal vehicle due to avoidance of contact with myocardial cells [55]. These results have been promising to deliver doxorubicin with a low incidence of cardiac adverse effects for the treatment of breast and ovary cancers. Paclitaxel is another anticancer drug, which has been extensively studied for exosomal delivery. In an extensive study by Kim et al., the feasibility of paclitaxel-loaded exosomes has been studied to treat multiple drug resistant cancer. More precisely, the potential of paclitaxel-loaded exosomes was examined in Lewis Lung Carcinoma metastasis murine model. Autologous macrophages of murine origin were used as the donor cells for the production of paclitaxel loaded exosomes which exhibited high drug loading, sustained release profile, elevated accumulation in the cancer cells and thereby higher cytotoxicity, compared to drug alone [56]. Paclitaxel-loaded exosomes derived from prostate cancer cell lines, PC-3 and LNCaP, have been evaluated in vitro to study their viability and efficacy. A major finding revealed that the use of donor cells of cancerous origin helped to improve the target specificity and reduced cytotoxicity of the delivery system. The endocytosis mechanism was used by the exosomes to deliver the therapeutic cargo inside the cancerous cells; however, the increased cellular viability could not be answered. A PoC was established which exhibited the delivery of therapeutic cargo leading to an improved cytotoxicity profile, in an exosomal system derived from the patient's own cells [57].

Researchers have also studied the delivery of chemopreventive agents by means of exosomes. Delivery of chemopreventives such as steroidal lactone called withaferin A, curcumin, and anthocyanidins using exosomes derived from bovine milk, was tried by R. Munagala et al. Folic acid was used in this case to enhance the targeting profile of the exosomal carriers. The antitumor effect showed by the withaferin A loaded exosomes was considerably higher compared to the drug in its native form [58]. The magnitude of research being carried out for designing delivery of chemotherapeutics clearly indicates the potential of this exosomal delivery cargo in the future.

8.2.2.3 Delivery of RNA

Nucleic acids form the basic components of the naturally derived exosomes because of their biological nature, RNA being the highest amongst them. Small interfering RNAs (siRNAs) and miRNA are the major therapeutic cargos in the nucleic acid category which have been explored considerably. Their role in gene therapy makes them a potential therapeutic candidate for exosomal cargo. Due to the very labile nature of RNAs in their native form, it is highly important to protect them using an appropriate carrier system.

Exosomal delivery can be very useful to avoid the degradation of these biomolecules upon administration.

The functional involvement of RNAs in restoring normal cellular processes has led to their numerous applications in exosomal delivery. siRNA delivery to the mice brain using rabies virus glycoprotein (RVG) as the targeting peptide has been demonstrated [34]. The dendritic cells for exosome preparation were derived from the mice to avoid immunogenicity; post-production, electroporation was used to load siRNA followed by IV administration of the formulation. The delivery potential was evident from the knockdown of 60% mRNA and 62% protein of the enzyme BACE1, a typical target protein in Alzheimer's therapy. Generally, siRNA delivery is associated with adverse effects in other organs due to the non-specific knockdown effect but this particular study indicated absence of such adverse effects indicating sound improvement [34]. This study of exosomal delivery of RVG, by Alvarez-Erviti et al. is one of the most successful pilot studies of exosome-based RNA delivery across the complex blood brain barrier (BBB). Citing this pilot study, Yang et al. could also deliver the exosomal cargo carrying siRNA across the BBB in a zebra fish glioblastoma model. bEND.3 brain endothelial cells were grown *in vitro* to source the exosomes and siRNA of vascular endothelial growth factor (VEGF) origin was loaded using Lipofectamine® as the transfection reagent; while rhodamine 123 was used as a fluorescent marker [59]. This is the first study to establish successful delivery of exosomes (from the origin of brain cells), loaded with siRNA, in a zebrafish brain cancer model. The exosomes could deliver the fluorescent marker and siRNA across BBB and also inhibited aggregation of the xenotransplanted cancer cells in the *in vivo* model [59]. miRNA delivery to breast cancer cells expressing epidermal growth factor (EGFR) receptors has been explored by Ohno et al. The donor cells (derived *in vitro*) were engineered so as to express transmembrane domain of EGFR fused to GE11 peptide and the miRNA loaded exosome cargo was delivered intravenously; the cargo was found to target the breast cancer cells expressing EGFR [60]. Intratracheal delivery of exosomes in the lung macrophages has been explored for siRNA, miRNA mimics, and inhibitors. Lipopolysaccharide (LPS) lung inflammation was induced in murine model and exosome cargo containing siRNA and miRNA was found to be beneficial toward improvement of the condition. Again, in this case, exosomes were derived from the mice serum to avoid triggering of the immune system. The treatment of pulmonary disorders using exosome mediated gene therapy via delivery to lung macrophages has been postulated as a future directive by the authors [61].

In summary, exosomes pose as promising drug carriers, which can overcome the limitations of current therapies for the successful treatment of deadly malignancies such as glioblastoma, pancreatic, prostate, and lung cancers. The biological origin of the exosomes renders them with higher biocompatibility and minimized clearance. Further enhancements in target specificity can be achieved by means of specific surface proteins and surface modifications. However, the heterogeneity of the encapsulated and surface molecules and the potential risk of metastasis promotion are some of the challenges to be addressed before the use of exosomes as a drug carrier for human solid tumors.

8.3 Exosome Based Drug Delivery for Cardiovascular Diseases

Another major focus area for exosome-based drug delivery is in the field of cardiovascular disease (CVD). CVD is a widespread, chronic disorder, the most prevalent categories of which include atherosclerosis, myocardial infarction (MI), and stroke. Despite the advances in therapy and management of CVD, it has the highest global mortality rate, poor prognosis, and poses a grave socio-economic burden [62, 63]. Owing to the limited regenerative capacity of cardiac cells and the absence of early diagnostic markers, there are significant gaps that need to be addressed to improve the management and prognosis of CVD. Recently, cardiac stem cells were found to be capable of differentiating into cardiomyocytes and vascular structures. However, their number is too low to restore severely impaired cardiac function. Several adult stem cells, including MSCs, were explored for regenerative therapy of MI and were successful in reducing infarct size and restore heart functions in the *in vitro* and rodent animal models. However, they failed in human cardiac regenerative studies and were associated with risks such as immunoreactivity and oncogenicity [64].

Recently, a lot of attention has been paid to the factors released from the remnant cardiac stem cells and may mediate paracrine effects. EVs from cardiac and non-cardiac stem and progenitor cells, somatic cells, and body fluids have been proposed as potential paracrine factors that play pathophysiological roles in the progression of various CVD. It has been established that they transport molecules like inflammatory cytokines and miRNA, from diseased cardiomyocytes to immune cells, fibroblasts, and endothelial cells etc. to regulate inflammation, angiogenesis, and resolution of the injured tissues [65, 66]. The fact that exosomes are involved in a multitude of cardiovascular processes makes them an apt alternative to whole-cell therapies (Figure 8.5). In addition to their advantage such as biocompatibility and non-immunogenicity, exosomes exhibit better stability and resistance to cryo-conservation. However, the methods for collection, isolation, and purification of exosomes are not yet standardized and validated and hence it leads to discrepancies with respect to batch to batch consistency.

Further, the best-suited mode of administration of exosomes for cardiovascular therapy is also to be decided. The ideal modality would be an IV injection; however, in a study to visualize and track IV administered exosomes done by Takakura et al., predominant accumulation in liver cells was observed [68]. Another study compared intramyocardial and intracoronary injections of exosomes derived from cardiosphere-derived cells in pig models of acute MI. It was found that intramyocardial injections led to higher myocardial retention causing a significant decrease in scar size and in microvascular obstruction [69].

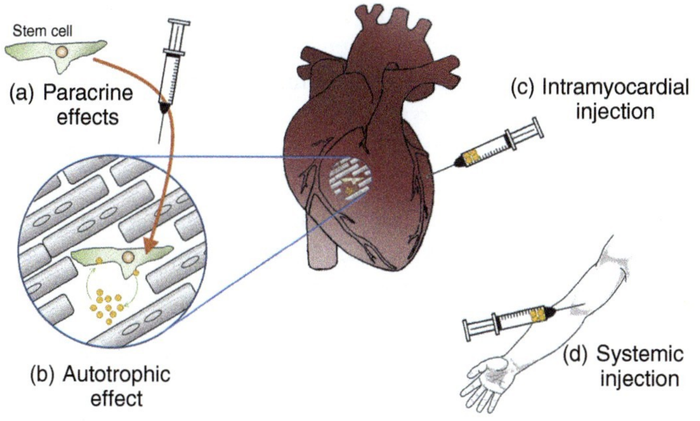

Figure 8.5 Effects of exosomes on the heart. (a) Paracrine effects of exosomes produced from intramyocardially injected stem cells, (b) autotrophic stimulation of resident cardiac stem cells or other cardiac cells, (c) intramyocardial injections of exosomes may affect different cardiac cells, and (d) interaction of systemically administered exosomes with cells of the cardiovascular system. Source: Davidson and Yellon [67] / Elsevier / CC BY-4.0.

A host of studies have explored the use of exosomes as a therapeutic agent or investigated the pathophysiological roles of EVs secreted by cardiac and extracardiac cells after MI or a cardiac dysfunction. However, the studies exploring engineered exosomes as drug delivery systems for various CVD are limited. The following sections describe the various attempts in utilizing exosomes as delivery vehicles for cardiovascular applications. Majority of the research on exosomal delivery for CVD pertains to the delivery of miRNA with specific cardioprotective or therapeutic roles. Few other studies have explored targeting exosomes to disease sites by surface modifications using specific cardiac targeting peptides.

8.3.1 Delivery of Cardioprotective RNAs

Exosomal delivery for the transport of cardioprotective miRNAs has been mainly explored to combat CVDs such as atherosclerosis and MI. Atherosclerosis refers to endothelial dysfunction at the predisposed sites along blood vessels leading to a variety of CVDs, including MI and ischemic stroke. Due to the inflammatory nature of endothelial cells and increased vascular permeability of atherosclerotic plaques, drug carriers can be designed to specifically target atherosclerotic plaques for diagnosis or treatment purposes. Atheroprotective miRNAs, especially miR-143 and miR-145, which can prevent smooth muscle cell de-differentiation were identified and enriched in exosomes derived from shear-stress stimulated endothelial cells. These miR-143 and miR-145-rich exosomes were administered IV in ApoE$^{-/-}$ mice fed with a high-fat diet. It was observed that these exosomes reduced aortic fatty lesion by approximately twofolds, thereby averting the formation of atherosclerotic lesions [70]. Another study explored miRNA-126 enriched exosomes derived from endothelial cells undergoing apoptosis. Upon IV administration of these exosomes in a mouse model of carotid artery injury, significant re-endothelization was achieved [71].

Exosomes loaded with cardioprotective miRNAs have also been explored to minimize cardiac damage and induce regeneration of functional cardiomyocytes after MI. The blockage of coronary arteries causes massive oxidative stress and inflammatory responses leading to irreversible cardiac cell death. Exosomes can target the inflamed myocardium via the enhanced permeability and retention (EPR) effect due to their nanosize. For instance, a cardioprotective miRNA, miR-93-5p, was loaded into exosomes by genetic modification of adipose-derived stromal cells. *In vitro* experiments showed inhibition of inflammatory cytokine response and hypoxia-induced autophagy. Further, IV administration of miR-93-5p-loaded exosomes in a rat model of acute MI showed reduced myocardial damage and infarct size [72]. Two other studies utilized miR-126 or miR-146a modified exosomes derived from adipose-derived stem cells, which were found to attenuate acute MI damage in a rat model. The treatment with miR-126-enriched exosomes decreased cardiac fibrosis and inflammatory cytokine expression but increased angiogenesis [73, 74].

Exosomes carrying miRNA have also been used to treat stroke. In the event of atherosclerotic plaque rupture, narrowing or blockage of cerebral vessels occur leading to ischemic stroke, which is therefore classified as a type of CVD. In such cases, the systemic drug delivery is limited by the presence of BBB. However, it has been established *in vitro* and *in vivo* that the exosomes are capable of passing through the BBB, although the exact mechanisms by which they are transported across the BBB are yet to be elucidated [75–77]. One of the early studies employing engineered exosomes to treat stroke used miR-17-92 cluster enriched exosomes to enhance the neurological recovery. MSC-derived exosomes transfected with miR-17-92 cluster were administered IV to rats subjected to middle cerebral artery occlusion. It was observed that the engineered exosomes improved neural differentiation and plasticity, and neurological recovery post stroke compared with the control group with unengineered exosomes and liposomes [78]. In addition to deliver miRNA, exosomes were also employed to deliver siRNA into primary endothelial cells isolated from mouse aorta. This *in vitro* study further demonstrated the potential of endothelial-derived exosomes enriched with therapeutic nucleic acids as an efficacious approach to treat the inflamed endothelium in atherosclerosis [79].

8.3.2 Exosomes Modified with Cardiac Targeting Peptides

Most of the studies discussed so far employed passive targeting of miRNA-loaded exosomes to the site of tissue inflammation relying completely on the enhanced permeability and retention effect. The inflammation and vasoactive mediators in the infarct site alter vascular permeability, which facilitates accumulation in the inflamed tissues. The *in vivo* selective targeting can be enhanced by identifying and modifying the exosomes with disease-specific targeting peptides that can recognize infarcted cardiac tissue. Zahid et al. incubated a cardiomyoblast cell line, H9C2 with a M13 phage peptide display library and the internalized phage was recovered, amplified and subjected to three rounds of *in vivo* biopanning to identify a cardiac targeting peptide (APWHLSSQYSRT). The mice injected with the identified peptide showed significantly higher transduction in heart as compared to other organs including liver, lung, kidney, skeletal muscle, spleen, and brain. In the next stage, this peptide was fused with an exosome surface protein Lamp2b to enhance the targeting efficacy of exosomes. However, the peptide could not differentiate between healthy and damaged cardiac cells thereby posing the potential risk of affecting the healthy cardiac tissue [80]. A similar approach was then used to identify the peptide sequence CSTSMLKAC, which exhibited preferential binding to ischemic heart tissue compared to normal heart. The peptide was then fused with two other proteins, Sumo and mCherry and the bio-distribution of Sumo-mCherry-CSTSMLKAC was studied using quantitative enzyme-linked immunosorbent assay (ELISA) after an IV injection in a mouse model of myocardial ischemia–reperfusion injury. The targeting peptide enabled a significant increase in homing to ischemic left ventricle as compared to non-ischemic left ventricle and the other organs [81]. In another study, exosomes were generated from cardiosphere-derived stem cells (CDCs), which were then modified with cardiac targeting peptides to enhance their accumulation at the injured site of the myocardium in a rat myocardial ischemia–reperfusion injury model. Compared with unmodified exosomes, the peptide conjugated exosomes showed better therapeutic efficacy against MI by reducing fibrosis and scar size while increasing cellular proliferation and angiogenesis [82]. The peptide CSTSMLKAC was also fused with the surface protein Lamp2b to modify murine MSC-derived exosomes via molecular cloning and lentivirus packaging to target the ischemic heart and treat MI-induced cardiac dysfunction [83]. A similar approach was used to generate an exosomal delivery system derived from CDCs, which were engineered to express Lamp2b fused to a 20 amino acid cardiomyocyte specific peptide (WLSEAGPVVTVRALRGTGSW). The exosomes isolated from these engineered CDCs, with the targeting peptide expressed on their surface, exhibited enhanced uptake by cardiomyocytes and higher cardiac retention when administered via intramyocardial injection. However, additional surface modifications may be required to facilitate targeted cytosolic delivery of these exosomes with minimal clearance from the circulation [84].

Improvements in brain targeting efficacy have also been attempted by surface functionalization of exosomes. For instance, Tian et al. used a cyclo(Arg-Gly-Asp-D-Tyr-Lys) peptide [c(RGDyK)] expressing high affinity to integrin $\alpha v\beta 3$ in reactive cerebral vascular endothelial cells after ischemia for functionalizing MSC-derived exosomes by click chemistry. Specifically click chemistry or azide alkyne cycloaddition involves the reaction between an alkyne moiety and an azide group to form a stable triazole linkage. This method can be utilized to directly attach molecules to the exosome surfaces via covalent bonds and since the azide functional group is absent in natural biomacromolecules, the chances of off-target binding are negligible. Click chemistry has emerged as a popular method of bioconjugation due to its simplicity, relatively mild experimental conditions, and high yield [85]. Thus, the peptide modified exosomes fabricated using this method efficiently crossed the BBB and targeted the ischemic cells in the brain [86]. The same group also developed c(RGDyK) functionalized exosomes to deliver curcumin and miR-210 to the ischemic brain and were able to demonstrate significant suppression in the inflammatory response and apoptosis in the lesion region in comparison to curcumin or exosomes treatment alone [87]. Another study reported that exosomes from miR-30d-5p overexpressing adipose-derived stem cells were able to suppress ischemia-induced neuronal damage by inhibiting the autophagy-mediated inflammatory responses in murine models of acute ischemic stroke [88]. However, majority of these studies involving exosome-based drug delivery are in pre-clinical stage and require further validation in large animal models and clinical trials.

8.4 Clinical Evaluations and Future Aspects

Design and development of gene therapy has been a challenge since its inception and with invent of exosome-based gene delivery, the task seems a tad more difficult. The inherent ability of exosomes to traverse one of the most complex biological barriers like brain gives a boost toward research in this direction. Exosomes have also shown promising evidences for their role as novel diagnostic tools. Be it discovery of unknown biomarkers or using exosomes as new targets for therapy, they have generated many avenues for research and development of exosome-based diagnostics and therapeutics. The ongoing investigations are postulated to drive the field of exosome-based therapeutics to a new level, enabling their efficient entry into preclinical and clinical trials, ultimately followed by commercial launch. The ExoDx Prostate (IntelliScore) test is one of the first diagnostic tools available in the market based on detection of exosomes specific to high-grade prostate cancer in urine. This test is now marketed as a non-invasive risk assessment test that indicates if a prostate biopsy is recommended or not [89]. Earlier clinical trials investigated the use of dendritic cell derived exosomes pulsed with human melanoma associated antigens (MAGEs) for immunotherapy of patients with advanced non-small cell lung cancer [90] and stage III/IV melanoma patients [91]. These Phase I clinical studies established the safety of exosomal therapy, their ability to activate immune effectors and the feasibility of large-scale production of exosomes. In the Phase II study, they developed second generation exosomes derived from interferon (IFN)-γ-maturated dendritic cells and investigated their capability to improve the rate of progression-free survival at four months after chemotherapy. Although the treatment was well tolerated and resulted in enhancement of natural killer cell functions, the primary endpoint of 50% progression-free survival was not achieved. This could be due to the heterogeneity of the patient cohorts rendering the loaded tumor antigens clinically non-relevant and the lack of adaptive immune responses [92]. Another Phase I study demonstrated the feasibility and safety of ascites-derived exosomes along with granulocyte-macrophage colony-stimulating factor as the adjuvant for the immunotherapy of colorectal cancer [93]. One of the first exploratory clinical trials evaluated circulating tumor cells and exosomes to diagnose resectable pancreatic cancers and possibly indicate the prognostic value and eligibility of patients for neoadjuvant treatment before surgery [94]. Several other clinical trials involving exosomes have been registered in the field of cancer research and are mostly in recruiting stage. Most of them are focused on identifying exosomes specific to the type of cancer, understanding the underlying mechanisms, and validating the exosomal content as a biomarker or indicator for diagnosis, disease recurrence, metastatic potential and prognosis of various types of cancers including sarcomas, ovarian cancer, lung cancer, pancreatic cancer, renal cell carcinoma, gallbladder carcinoma, prostate cancer, gastric cancer, melanoma, colorectal cancer, breast cancer, and thyroid cancer [95]. Clinical trials investigating exosomes as delivery vehicles are limited. In a Phase I clinical trial, the ability of plant exosomes to deliver curcumin and the effect on immune modulation, cellular metabolism, and phospholipid profile of normal and malignant colon cells in newly diagnosed colon cancer patients undergoing surgery are being studied [96]. Another recent Phase I study explores the maximum tolerated dose and the dose-limiting side effects of MSCs-derived exosomes loaded with KrasG12D siRNA (iExosomes) in treating patients having metastatic pancreatic ductal adenocarcinoma with KrasG12D mutation [97, 98].

In the field of CVDs, few clinical trials have been carried out in recognizing exosomal proteins as biomarkers for early detection, classification, and determination of treatment efficacy. For instance, observational studies involving identification of urinary exosomal proteins to classify and monitor difficult-to-treat arterial hypertension [99] and autoimmune thyroid heart disease [100] have been initiated. Clinical studies exploring the expression and role of peripheral blood exosomal miRNA in patients with MI [101] and that of epicardial fat derived exosomes in atrial fibrillation have been registered [102]. Exosomal markers are also studied among many other factors to determine treatment efficacies like effect of exercise and diet in obese children and adolescents [103], efficacy of time-restricted feeding [104], and remote ischemic conditioning [105] in patients with acute ischemic stroke. Recently, a randomized, single-blind, placebo-controlled, Phase I, II trial involving the use of MSC derived exosomes for neurovascular remodeling and functional recovery in patients with acute ischemic stroke was completed and the results are awaited [106]. Another study exploring the use of MSC-derived exosomes to treat multiple organ dysfunction syndrome in patients who have undergone surgical repair of acute type A aortic dissection has also been registered but has not started recruiting [107]. As of now, there are no clinical trials exploring exosomes as delivery vehicles for cardiovascular interventions.

8.5 Conclusion

With the rising concerns about the scenario for drug resistance and delivery system incompatibility, the advent of highly biocompatible systems like exosomes becomes a boon. The ability of exosomes to adapt to surface modification, incorporate different therapeutic cargos and retain the efficacy while surpassing complex physiological conditions, is a driving factor for exploration of these systems as a suitable drug delivery candidate for both anticancer as well as cardiovascular interventions. Further, combination of diagnostic agents and therapeutic molecules may broaden the application of exosomes as a theranostic platform allowing clinicians for in situ evaluation of the efficacy of the treatment. Citing the growing demand for personalized therapy, especially in anticancer therapy, exosomes can become a versatile platform for facilitating such endeavors.

The complexities of exosome-based therapeutics need urgent consideration and thorough validation. The properties of exosomes depend upon their cell of origin and the conditions of formation may result in heterogeneity in harvested exosomes and variability between sources. Therefore, establishing methods for mass production, characterization, and long-term storage of exosomes is crucial for obtaining high yields necessary for clinical trials. In addition, detailed investigations regarding the exact mechanisms of action of the biological activities of exosomes would provide further insights as to how they can be engineered as precision medicine. However, the constant advancements in drug delivery science would give an impetus to the evolution of these systems as the next generation therapeutics.

Acknowledgments

We would like to acknowledge University Grants Commission, Government of India for providing D.S. Kothari Postdoctoral Fellowship to Sreeranjini Pulakkat and Department of Science and Technology (DST) INSPIRE, Government of India, for providing fellowship to Anjali Pandya.

References

1 Jaiswal, R. and Sedger, L.M. (2019). Intercellular vesicular transfer by exosomes, microparticles and oncosomes – implications for cancer biology and treatments. *Front. Oncol.* 9: https://doi.org/10.3389/fonc.2019.00125.

2 Trams, E.G., Lauter, C.J. Jr., Salem, N., and Heine, U. (1981). Exfoliation of membrane ecto-enzymes in the form of micro-vesicles. *Biochim. Biophys. Acta, Biomembr.* 645: 63–70. https://doi.org/10.1016/0005-2736(81)90512-5.

3 Bunggulawa, E.J., Wang, W., Yin, T. et al. (2018). Recent advancements in the use of exosomes as drug delivery systems. *J. Nanobiotechnol.* 16: 1–13. https://doi.org/10.1186/s12951-018-0403-9.

4 Vader, P., Mol, E.A., Pasterkamp, G., and Schiffelers, R.M. (2016). Extracellular vesicles for drug delivery. *Adv. Drug Delivery Rev.* 106: 148–156. https://doi.org/10.1016/j.addr.2016.02.006.

5 Johnstone, R.M. (2005). Revisiting the road to the discovery of exosomes. *Blood Cells Mol. Dis.* 34: 214–219. https://doi.org/10.1016/j.bcmd.2005.03.002.

6 Denzer, K., Kleijmeer, M.J., Heijnen, H.F.G. et al. (2000). Exosome: from internal vesicle of the multivesicular body to intercellular signaling device. *J. Cell Sci.* 113: 3365–3374.

7 Colombo, M., Raposo, G., and Théry, C. (2014). Biogenesis, secretion, and intercellular interactions of exosomes and other extracellular vesicles. *Annu. Rev. Cell Dev. Biol.* 30: 255–289. https://doi.org/10.1146/annurev-cellbio-101512-122326.

8 Wang, M., Yu, F., Ding, H. et al. (2019). Emerging function and clinical values of exosomal microRNAs in cancer. *Mol. Ther. Nucleic Acids* 16: 791–804. https://doi.org/10.1016/j.omtn.2019.04.027.

9 J.M. Gudbergsson, K. Jønsson, J.B. Simonsen, K.B. Johnsen, Systematic review of targeted extracellular vesicles for drug delivery – considerations on methodological and biological heterogeneity *J. Controlled Release* (2019). 306, 108-120, https://doi.org/10.1016/j.jconrel.2019.06.006.

10 Raposo, G. and Stoorvogel, W. (2013). Extracellular vesicles: exosomes, microvesicles, and friends. *J. Cell Biol.* 200: 373–383. https://doi.org/10.1083/jcb.201211138.

11 Hessvik, N.P. and Llorente, A. (2018). Current knowledge on exosome biogenesis and release. *Cell. Mol. Life Sci.* 75: 193–208. https://doi.org/10.1007/s00018-017-2595-9.

12 Kim, D.-K., Kang, B., Kim, O.Y. et al. (2013). EVpedia: an integrated database of high-throughput data for systemic analyses of extracellular vesicles. *J. Extracell. Vesicles* 2: 20384. https://doi.org/10.3402/jev.v2i0.20384.

13 Maas, S.L.N., Breakefield, X.O., and Weaver, A.M. (2017). Extracellular vesicles: unique intercellular delivery vehicles. *Trends Cell Biol.* 27: 172–188. https://doi.org/10.1016/j.tcb.2016.11.003.

14 Lee, Y., El Andaloussi, S., and Wood, M.J.A. (2012). Exosomes and microvesicles: extracellular vesicles for genetic information transfer and gene therapy. *Hum. Mol. Genet.* 21: 1–10. https://doi.org/10.1093/hmg/dds317.

15 Marcus, M.E. and Leonard, J.N. (2013). FedExosomes: engineering therapeutic biological nanoparticles that truly deliver. *Pharmaceuticals* 6 (5): 659–680. https://doi.org/10.3390/ph6050659.

16 Nawaz, M., Fatima, F., Vallabhaneni, K.C. et al. (2016). Extracellular vesicles: evolving factors in stem cell biology. *Stem Cells Int.* 2016: 1073140–1073140. https://doi.org/10.1155/2016/1073140.

17 Greening, D.W., Gopal, S.K., Xu, R. et al. (2015). Exosomes and their roles in immune regulation and cancer. *Seminars in Cell & Developmental Biology.* 40: 72–81. https://doi.org/10.1016/j.semcdb.2015.02.009.

18 Admyre, C., Johansson, S.M., Qazi, K.R. et al. (2007). Exosomes with immune modulatory features are present in human breast milk. *J. Immunol.* 179: 1969. https://doi.org/10.4049/jimmunol.179.3.1969.

19 Liu, W., Bai, X., Zhang, A. et al. (2019). Role of exosomes in central nervous system diseases. *Front. Mol. Neurosci.* 12: 240. https://doi.org/10.3389/fnmol.2019.00240.

20 Schneider, A. and Simons, M. (2013). Exosomes: vesicular carriers for intercellular communication in neurodegenerative disorders. *Cell Tissue Res.* 352: 33–47. https://doi.org/10.1007/s00441-012-1428-2.

21 Bellin, G., Gardin, C., Ferroni, L. et al. (2019). Exosome in cardiovascular diseases: a complex world full of hope. *Cells* 8: 166. https://doi.org/10.3390/cells8020166.

22 Urbanelli, L., Buratta, S., Tancini, B. et al. (2019). The role of extracellular vesicles in viral infection and transmission. *Vaccines (Basel)* 7: 102. https://doi.org/10.3390/vaccines7030102.

23 Jan, A.T., Malik, M.A., Rahman, S. et al. (2017). Perspective insights of exosomes in neurodegenerative diseases: a critical appraisal. *Front. Aging Neurosci.* 9: 317–317. https://doi.org/10.3389/fnagi.2017.00317.

24 D'Anca, M., Fenoglio, C., Serpente, M. et al. (2019). Exosome determinants of physiological aging and age-related neurodegenerative diseases. *Front. Aging Neurosci.* 11: 232. https://doi.org/10.3389/fnagi.2019.00232.

25 Katakowski, M., Buller, B., Zheng, X. et al. (2013). Exosomes from marrow stromal cells expressing miR-146b inhibit glioma growth. *Cancer Lett.* 335: 201–204. https://doi.org/10.1016/j.canlet.2013.02.019.

26 Mizrak, A., Bolukbasi, M.F., Ozdener, G.B. et al. (2013). Genetically engineered microvesicles carrying suicide mRNA/protein inhibit schwannoma tumor growth. *Mol. Ther.* 21: 101–108. https://doi.org/10.1038/mt.2012.161.

27 Rana, S., Yue, S., Stadel, D., and Zöller, M. (2012). Toward tailored exosomes: the exosomal tetraspanin web contributes to target cell selection. *Int. J. Biochem. Cell Biol.* 44: 1574–1584. https://doi.org/10.1016/j.biocel.2012.06.018.

28 Jang, S.C., Kim, O.Y., Yoon, C.M. et al. (2013). Bioinspired exosome-mimetic nanovesicles for targeted delivery of chemotherapeutics to malignant tumors. *ACS Nano* 7: 7698–7710. https://doi.org/10.1021/nn402232g.

29 Sutaria, D.S., Badawi, M., Phelps, M.A., and Schmittgen, T.D. (2017). Achieving the promise of therapeutic extracellular vesicles: the devil is in details of therapeutic loading. *Pharm. Res.* 34: 1053–1066. https://doi.org/10.1007/s11095-017-2123-5.

30 Lunavat, T.R., Jang, S.C., Nilsson, L. et al. (2016). RNAi delivery by exosome-mimetic nanovesicles – Implications for targeting c-Myc in cancer. *Biomaterials* 102: 231–238. https://doi.org/10.1016/j.biomaterials.2016.06.024.

31 Hood, J.L. (2016). Post isolation modification of exosomes for nanomedicine applications. *Nanomedicine (Lond).* 11: 1745–1756. https://doi.org/10.2217/nnm-2016-0102.

32 Lamichhane, T.N., Raiker, R.S., and Jay, S.M. (2015). Exogenous DNA loading into extracellular vesicles via electroporation is size-dependent and enables limited gene delivery. *Mol. Pharmaceutics* 12: 3650–3657. https://doi.org/10.1021/acs.molpharmaceut.5b00364.

33 Li, S.-P., Lin, Z.-X., Jiang, X.-Y., and Yu, X.-Y. (2018). Exosomal cargo-loading and synthetic exosome-mimics as potential therapeutic tools. *Acta Pharmacol. Sin.* 39: 542–551. https://doi.org/10.1038/aps.2017.178.

34 Alvarez-Erviti, L., Seow, Y., Yin, H. et al. (2011). Delivery of siRNA to the mouse brain by systemic injection of targeted exosomes. *Nat. Biotechnol.* 29: 341–345. https://doi.org/10.1038/nbt.1807.

35 Wahlgren, J., Karlson, T.D.L., Brisslert, M. et al. (2012). Plasma exosomes can deliver exogenous short interfering RNA to monocytes and lymphocytes. *Nucleic Acids Res.* 40: e130–e130. https://doi.org/10.1093/nar/gks463.

36 Kooijmans, S.A.A., Stremersch, S., Braeckmans, K. et al. (2013). Electroporation-induced siRNA precipitation obscures the efficiency of siRNA loading into extracellular vesicles. *J. Controlled Release* 172: 229–238. https://doi.org/10.1016/j.jconrel.2013.08.014.

37 Fuhrmann, G., Serio, A., Mazo, M. et al. (2015). Active loading into extracellular vesicles significantly improves the cellular uptake and photodynamic effect of porphyrins. *J. Controlled Release* 205: 35–44. https://doi.org/10.1016/j.jconrel.2014.11.029.

38 Antimisiaris, S.G., Mourtas, S., and Marazioti, A. (2018). Exosomes and exosome-inspired vesicles for targeted drug delivery. *Pharmaceutics* 10: 218. https://doi.org/10.3390/pharmaceutics10040218.

39 Munson, P. and Shukla, A. (2015). Exosomes: potential in cancer diagnosis and therapy. *Medicines* 2 (4): 310–327. https://doi.org/10.3390/medicines2040310.

40 Joyce, J.A. and Pollard, J.W. (2009). Microenvironmental regulation of metastasis. *Nat. Rev. Cancer* 9: 239–252. https://doi.org/10.1038/nrc2618.

41 McAndrews, K.M. and Kalluri, R. (2019). Mechanisms associated with biogenesis of exosomes in cancer. *Mol. Cancer* 18: 52. https://doi.org/10.1186/s12943-019-0963-9.

42 Li, Y., Yokoyama, W., Xu, S. et al. (2017). Formation and stability of W/O microemulsion formed by food grade ingredients and its oral delivery of insulin in mice. *J. Funct. Foods* 30: 134–141. https://doi.org/10.1016/j.jff.2017.01.006.

43 Milman, N., Ginini, L., and Gil, Z. (2019). Exosomes and their role in tumorigenesis and anticancer drug resistance. *Drug Resist. Updat.* 45: 1–12. https://doi.org/10.1016/j.drup.2019.07.003.

44 Smyth, T.J., Redzic, J.S., Graner, M.W., and Anchordoquy, T.J. (2014). Examination of the specificity of tumor cell derived exosomes with tumor cells in vitro. *Biochim. Biophys. Acta, Biomembr.* 1838: 2954–2965. https://doi.org/10.1016/j.bbamem.2014.07.026.

45 Batrakova, E.V. and Kim, M.S. (2015). Using exosomes, naturally-equipped nanocarriers, for drug delivery. *J. Controlled Release* 219: 396–405. https://doi.org/10.1016/j.jconrel.2015.07.030.

46 K.B. Johnsen, J.M. Gudbergsson, M.N. Skov, L. Pilgaard, T. Moos, M. Duroux, A comprehensive overview of exosomes as drug delivery vehicles — Endogenous nanocarriers for targeted cancer therapy, *Biochim. Biophys. Acta, Rev. Cancer* (2014). 1846, 1, 75-87, https://doi.org/10.1016/j.bbcan.2014.04.005.

47 Yin, W., Ouyang, S., Li, Y. et al. (2013). Immature dendritic cell-derived exosomes: a promise subcellular vaccine for autoimmunity. *Inflammation* 36: 232–240. https://doi.org/10.1007/s10753-012-9539-1.

48 Tian, W., Liu, S., and Li, B. (2019). Potential role of exosomes in cancer metastasis. *BioMed Res. Int.* 2019: 4649705. https://doi.org/10.1155/2019/4649705.

49 Wang, J., Zheng, Y., and Zhao, M. (2017). Exosome-based cancer therapy : implication for targeting cancer. *Stem Cells* 7: 1–11. https://doi.org/10.3389/fphar.2016.00533.

50 Luketic, L., Delanghe, J., Sobol, P.T. et al. (2007). Antigen presentation by exosomes released from peptide-pulsed dendritic cells is not suppressed by the presence of active CTL. *J. Immunol.* 179: 5024–5032. https://doi.org/10.4049/jimmunol.179.8.5024.

51 Hu, G., Gong, A.-Y., Roth, A.L. et al. (2013). Release of luminal exosomes contributes to TLR4-mediated epithelial antimicrobial defense. *PLoS Pathog.* 9: e1003261. https://doi.org/10.1371/journal.ppat.1003261.

52 Aspe, J.R. and Wall, N.R. (2010). Survivin-T34A: molecular mechanism and therapeutic potential. *OncoTargets Ther.* 3: 247–254. https://doi.org/10.2147/OTT.S15293.

53 Lee, Y.S., Kim, S.H., Cho, J.A., and Kim, C.W. (2011). Introduction of the *CIITA* gene into tumor cells produces exosomes with enhanced anti-tumor effects. *Exp. Mol. Med.* 43: 281. https://doi.org/10.3858/emm.2011.43.5.029.

54 Lai, R.C., Tan, S.S., Teh, B.J. et al. (2012). Proteolytic potential of the MSC exosome proteome: implications for an exosome-mediated delivery of therapeutic proteasome. *Int. J. Proteomics* 2012: 971907–971907. https://doi.org/10.1155/2012/971907.

55 Hadla, M., Palazzolo, S., Corona, G. et al. (2016). Exosomes increase the therapeutic index of doxorubicin in breast and ovarian cancer mouse models. *Nanomedicine* 11: 2431–2441. https://doi.org/10.2217/nnm-2016-0154.

56 Kim, M.S., Haney, M.J., Zhao, Y. et al. (2016). Development of exosome-encapsulated paclitaxel to overcome MDR in cancer cells. *Nanomed. Nanotechnol. Biol. Med.* 12: 655–664. https://doi.org/10.1016/j.nano.2015.10.012.

57 Saari, H., Lázaro-Ibáñez, E., Viitala, T. et al. (2015). Microvesicle- and exosome-mediated drug delivery enhances the cytotoxicity of paclitaxel in autologous prostate cancer cells. *J. Controlled Release* 220: 727–737. https://doi.org/10.1016/j.jconrel.2015.09.031.

58 Munagala, R., Aqil, F., Jeyabalan, J., and Gupta, R.C. (2016). Bovine milk-derived exosomes for drug delivery. *Cancer Lett.* 371: 48–61. https://doi.org/10.1016/j.canlet.2015.10.020.

59 Yang, T., Fogarty, B., LaForge, B. et al. (2017). Delivery of small interfering RNA to inhibit vascular endothelial growth factor in zebrafish using natural brain endothelia cell-secreted exosome nanovesicles for the treatment of brain cancer. *AAPS J.* 19: 475–486. https://doi.org/10.1208/s12248-016-0015-y.

60 Ohno, S.I., Takanashi, M., Sudo, K. et al. (2013). Systemically injected exosomes targeted to EGFR deliver antitumor microrna to breast cancer cells. *Mol. Ther.* 21: 185–191. https://doi.org/10.1038/mt.2012.180.

61 Zhang, D., Lee, H., Wang, X. et al. (2018). Exosome-mediated small RNA delivery: a novel therapeutic approach for inflammatory lung responses. *Mol. Ther.* 26: 2119–2130. https://doi.org/10.1016/j.ymthe.2018.06.007.

62 WHO (n.d.). Global atlas on cardiovascular disease prevention and control, WHO. http://www.who.int/cardiovascular_diseases/publications/atlas_cvd/en/ (accessed 19 September 2019).

63 Gheorghe, A., Griffiths, U., Murphy, A. et al. (2018). The economic burden of cardiovascular disease and hypertension in low- and middle-income countries: a systematic review. *BMC Public Health* 18: 975. https://doi.org/10.1186/s12889-018-5806-x.

64 Gyöngyösi, M., Lukovic, D., Zlabinger, K. et al. (2017). Cardiac stem cell-based regenerative therapy for the ischemic injured heart — a short update 2017. *J. Cardiovasc. Emergencies* 3: 81–83. https://doi.org/10.1515/jce-2017-0009.

65 Ribeiro-Rodrigues, T.M., Laundos, T.L., Pereira-Carvalho, R. et al. (2017). Exosomes secreted by cardiomyocytes subjected to ischaemia promote cardiac angiogenesis. *Cardiovasc. Res.* 113: 1338–1350. https://doi.org/10.1093/cvr/cvx118.

66 Adamiak, M. and Sahoo, S. (2018). Exosomes in myocardial repair: advances and challenges in the development of next-generation therapeutics. *Mol. Ther.* 26: 1635–1643. https://doi.org/10.1016/j.ymthe.2018.04.024.

67 Davidson, S.M. and Yellon, D.M. (2018). Exosomes and cardioprotection – a critical analysis. *Mol. Aspects Med.* 60: 104–114. https://doi.org/10.1016/j.mam.2017.11.004.

68 Takahashi, Y., Nishikawa, M., Shinotsuka, H. et al. (2013). Visualization and in vivo tracking of the exosomes of murine melanoma B16-BL6 cells in mice after intravenous injection. *J. Biotechnol.* 165: 77–84. https://doi.org/10.1016/j.jbiotec.2013.03.013.

69 Gallet, R., Dawkins, J., Valle, J. et al. (2017). Exosomes secreted by cardiosphere-derived cells reduce scarring, attenuate adverse remodelling, and improve function in acute and chronic porcine myocardial infarction. *Eur. Heart J.* 38: 201–211. https://doi.org/10.1093/eurheartj/ehw240.

70 Hergenreider, E., Heydt, S., Tréguer, K. et al. (2012). Atheroprotective communication between endothelial cells and smooth muscle cells through miRNAs. *Nat. Cell Biol.* 14: 249–256. https://doi.org/10.1038/ncb2441.

71 Jansen, F., Yang, X., Hoelscher, M. et al. (2013). Endothelial microparticle–mediated transfer of microRNA-126 promotes vascular endothelial cell repair via SPRED1 and is abrogated in glucose-damaged endothelial microparticles. *Circulation* 128: 2026–2038. doi: 10.1161/CIRCULATION-AHA.113.001720.

72 Liu, J., Jiang, M., Deng, S. et al. (2018). miR-93-5p-containing exosomes treatment attenuates acute myocardial infarction-induced myocardial damage. *Mol. Ther. Nucleic Acids* 11: 103–115. https://doi.org/10.1016/j.omtn.2018.01.010.

73 Pan, J., Alimujiang, M., Chen, Q. et al. (2019). Exosomes derived from miR-146a-modified adipose-derived stem cells attenuate acute myocardial infarction–induced myocardial damage via downregulation of early growth response factor 1. *J. Cell. Biochem.* 120: 4433–4443. https://doi.org/10.1002/jcb.27731.

74 Luo, Q., Guo, D., Liu, G. et al. (2017). Exosomes from MiR-126-overexpressing ADSCS are therapeutic in relieving acute myocardial ischaemic injury. *Cell. Physiol. Biochem.* 44: 2105–2116. https://doi.org/10.1159/000485949.

75 Chen, C.C., Liu, L., Ma, F. et al. (2016). Elucidation of exosome migration across the blood-brain barrier model in vitro. *Cell. Mol. Bioeng.* 9: 509–529. https://doi.org/10.1007/s12195-016-0458-3.

76 Yang, T., Martin, P., Fogarty, B. et al. (2015). Exosome delivered anticancer drugs across the blood-brain barrier for brain cancer therapy in *Danio rerio*. *Pharm. Res.* 32: 2003–2014. https://doi.org/10.1007/s11095-014-1593-y.

77 Matsumoto, J., Stewart, T., Banks, W.A., and Zhang, J. (2017). The transport mechanism of extracellular vesicles at the blood-brain barrier. *Curr. Pharm. Des.* 23: 6206–6214. https://doi.org/10.2174/1381612823666170913164738.

78 Xin, H., Katakowski, M., Wang, F. et al. (2017). MicroRNA cluster miR-17-92 cluster in exosomes enhance neuroplasticity and functional recovery after stroke in rats. *Stroke* 48: 747–753. https://doi.org/10.1161/STROKEAHA.116.015204.

79 Banizs, A.B., Huang, T., Dryden, K. et al. (2014). In vitro evaluation of endothelial exosomes as carriers for small interfering ribonucleic acid delivery. *Int. J. Nanomed.* 9: 4223–4230. https://doi.org/10.2147/IJN.S64267.

80 Zahid, M., Phillips, B.E., Albers, S.M. et al. (2010). Identification of a cardiac specific protein transduction domain by in vivo biopanning using a M13 phage peptide display library in mice. *PLoS One* 5: e12252–e12252. https://doi.org/10.1371/journal.pone.0012252.

81 Kanki, S., Jaalouk, D.E., Lee, S. et al. (2011). Identification of targeting peptides for ischemic myocardium by in vivo phage display. *J. Mol. Cell Cardiol.* 50: 841–848. https://doi.org/10.1016/j.yjmcc.2011.02.003.

82 Vandergriff, A., Huang, K., Shen, D. et al. (2018). Targeting regenerative exosomes to myocardial infarction using cardiac homing peptide. *Theranostics* 8: 1869–1878. https://doi.org/10.7150/thno.20524.

83 Wang, X., Chen, Y., Zhao, Z. et al. (2018). Engineered exosomes with ischemic myocardium-targeting peptide for targeted therapy in myocardial infarction. *J. Am. Heart Assoc.* 7: e008737–e008737. https://doi.org/10.1161/JAHA.118.008737.

84 Mentkowski, K.I. and Lang, J.K. (2019). Exosomes engineered to express a cardiomyocyte binding peptide demonstrate improved cardiac retention in vivo. *Sci. Rep.* 9: 10041. https://doi.org/10.1038/s41598-019-46407-1.

85 Nwe, K. and Brechbiel, M.W. (2009). Growing applications of "click chemistry" for bioconjugation in contemporary biomedical research. *Cancer Biother. Radiopharm.* 24: 289–302. https://doi.org/10.1089/cbr.2008.0626.

86 Tian, T., Zhang, H.-X., He, C.-P. et al. (2018). Surface functionalized exosomes as targeted drug delivery vehicles for cerebral ischemia therapy. *Biomaterials* 150: 137–149. https://doi.org/10.1016/j.biomaterials.2017.10.012.

87 Zhang, H., Wu, J., Wu, J. et al. (2019). Exosome-mediated targeted delivery of miR-210 for angiogenic therapy after cerebral ischemia in mice. *J. Nanobiotechnol.* 17: 29. https://doi.org/10.1186/s12951-019-0461-7.

88 Jiang, M., Wang, H., Jin, M. et al. (2018). Exosomes from MiR-30d-5p-ADSCs reverse acute ischemic stroke-induced, autophagy-mediated brain injury by promoting M2 microglial/macrophage polarization. *Cell. Physiol. Biochem.* 47: 864–878. https://doi.org/10.1159/000490078.

89 Tutrone, R., Donovan, M.J., Torkler, P. et al. (2020). Clinical utility of the exosome based Exo Dx Prostate(*IntelliScore*) EPI test in men presenting for initial

Biopsy with a PSA 2–10 ng/mL. *Prostate Cancer Prostatic Dis.* 23: 607–614. https://doi.org/10.1038/s41391-020-0237-z.

90 Morse, M.A., Garst, J., Osada, T. et al. (2005). A phase I study of dexosome immunotherapy in patients with advanced non-small cell lung cancer. *J. Transl. Med.* 3: 9–9. https://doi.org/10.1186/1479-5876-3-9.

91 Escudier, B., Dorval, T., Chaput, N. et al. (2005). Vaccination of metastatic melanoma patients with autologous dendritic cell (DC) derived-exosomes: results of the first phase I clinical trial. *J. Transl. Med.* 3: 10–10. https://doi.org/10.1186/1479-5876-3-10.

92 Besse, B., Charrier, M., Lapierre, V. et al. (2015). Dendritic cell-derived exosomes as maintenance immunotherapy after first line chemotherapy in NSCLC. *Oncoimmunology* 5: e1071008–e1071008. https://doi.org/10.1080/2162402X.2015.1071008.

93 Dai, S., Wei, D., Wu, Z. et al. (2008). Phase I clinical trial of autologous ascites-derived exosomes combined with GM-CSF for colorectal cancer. *Mol. Ther.* 16: 782–790. https://doi.org/10.1038/mt.2008.1.

94 Buscail, E., Alix-Panabières, C., Quincy, P. et al. (2019). High clinical value of liquid biopsy to detect circulating tumor cells and tumor exosomes in pancreatic ductal adenocarcinoma patients eligible for up-front surgery. *Cancers (Basel).* 11: 1656. https://doi.org/10.3390/cancers11111656.

95 ClinicalTrials.gov (n.d.). Search of: exosome – List results. https://clinicaltrials.gov/ct2/results?cond=exosome&term=&cntry=&state=&city=&dist= (accessed 20 July 2020).

96 ClinicalTrials.gov (n.d.). Study investigating the ability of plant exosomes to deliver curcumin to normal and colon cancer tissue – Full text view. https://clinicaltrials.gov/ct2/show/NCT01294072 (accessed 11 September 2020).

97 Kamerkar, S., LeBleu, V.S., Sugimoto, H. et al. (2017). Exosomes facilitate therapeutic targeting of oncogenic KRAS in pancreatic cancer. *Nature* 546: 498–503. https://doi.org/10.1038/nature22341.

98 ClinicalTrials.gov (n.d.). iExosomes in treating participants with metastatic pancreas cancer with KrasG12D mutation – Full text view. https://clinicaltrials.gov/ct2/show/NCT03608631 (accessed 11 September 2020).

99 ClinicalTrials.gov (n.d.). New biomarkers and difficult-to-treat hypertension – Full text view. https://clinicaltrials.gov/ct2/show/NCT03034265 (accessed 11 September 2020).

100 ClinicalTrials.gov (n.d.). Early detection of autoimmune thyroid heart disease via urinary exosomal proteins – Full text view. https://clinicaltrials.gov/ct2/show/NCT03984006 (accessed 11 September 2020).

101 ClinicalTrials.gov (n.d.). Differential expression and analysis of peripheral plasma exosome miRNA in patients with myocardial infarction – Full text view. https://clinicaltrials.gov/ct2/show/NCT04127591 (accessed September 11, 2020).

102 ClinicalTrials.gov (n.d.). Role of exosomes derived from epicardial fat in atrial fibrillation – Full text view. https://clinicaltrials.gov/ct2/show/NCT03478410 (accessed 11 September 2020).

103 ClinicalTrials.gov (n.d.). Exercise and diet restriction on cardiovascular function in obese children and adolescents – Full text view. https://clinicaltrials.gov/ct2/show/NCT03762629 (accessed 11 September 2020).

104 ClinicalTrials.gov (n.d.). Randomized controlled trial of time-restricted feeding (TRF) in acute ischemic stroke patients – Full text view. https://clinicaltrials.gov/ct2/show/NCT04184076 (accessed 11 September 2020).

105 ClinicalTrials.gov (n.d.). Rheo-erythrocrine dysfunction as a biomarker for RIC treatment in acute ischemic stroke – Full text view. https://clinicaltrials.gov/ct2/show/NCT04266639 (accessed 11 September 2020).

106 ClinicalTrials.gov (n.d.). Allogenic mesenchymal stem cell derived exosome in patients with acute ischemic stroke – Tabular view. https://clinicaltrials.gov/ct2/show/record/NCT03384433 (accessed 12 September 2020).

107 ClinicalTrials.gov (n.d.). Exosome of mesenchymal stem cells for multiple organ dysfuntion syndrome after surgical repaire of acute type A aortic dissection – Full text view. https://clinicaltrials.gov/ct2/show/NCT04356300 (accessed 12 September 2020).

9

Delivery of Nucleic Acids, Such as siRNA and mRNA, Using Complex Formulations

*Ananya Pattnaik, Swarnaparabha Pany, A. S. Sanket, Sudiptee Das, Sanghamitra Pati, and Sangram K. Samal**

ICMR-Regional Medical Research Centre, Laboratory of Biomaterials and Regenerative Medicine for Advanced Therapies, Chandrasekharpur, Bhubaneswar 751023, Odisha, India

9.1 Introduction

Nucleic acids (NAs) are macromolecular biochemical entities majorly found in the form of deoxyribonucleic acid (DNA) and ribonucleic acid (RNA) [1, 2]. DNA is the major gene-carrying entity in the chromosome, found in eukaryotes (present inside the nucleus) and prokaryotes (scattered in the nucleoids) [3]. However, RNA is abundantly found in viruses as the genetic material than DNA [4]. The term deoxy in DNA is derived to reflect the absence of oxygen at the 2′ position in the ribose sugar moiety but not in RNA, as shown in Figure 9.1 [5]. DNA is observed in three different forms, such as A-DNA, B-DNA, and Z-DNA, based on the helicity. Among these, A and B forms of DNA coil right-handed, whereas Z form has left-handed helicity.

RNA can be majorly found in the form of messenger RNA (mRNA), ribosomal RNA (rRNA), transfer RNA (tRNA), and some other small-sized molecules, such as micro-RNA (miRNA), small-interfering RNA (siRNA), circular RNA (CircRNA), long noncoding RNA (lncRNA), noncoding RNA (ncRNA), small nucleolar RNA (snoRNA), and piwi-interacting RNA (piRNA), as shown in Figure 9.2 [6].

RNA acts as the intermediate to transfer the genetic information from DNA to proteins, and if not regulated properly at any stage, can cause various genetic disorders, [7, 8] such as Werner syndrome, ataxia–telangiectasia, Cockayne syndrome, Nijmegen breakage syndrome, Fanconi anemia, Bloom syndrome, xeroderma pigmentosum, and trichothiodystrophy [9–13].

Synthetic NA analogs called xeno nucleic acids (peptide nucleic acids, locked nucleic acids, glycol nucleic acids, threose nucleic acids, etc.) are also observed, which are chemically synthesized with different sugar puckers than those of naturally occurring NAs [14]. They exhibit similar base-pairing features as that of

* Ananya Pattnaik and Swarnaparabha Pany contributed equally. The corresponding author is Sangram K. Samal (sksamalrec@gmail.com).

Targeted Drug Delivery, First Edition. Edited by Yogeshwar Bachhav.

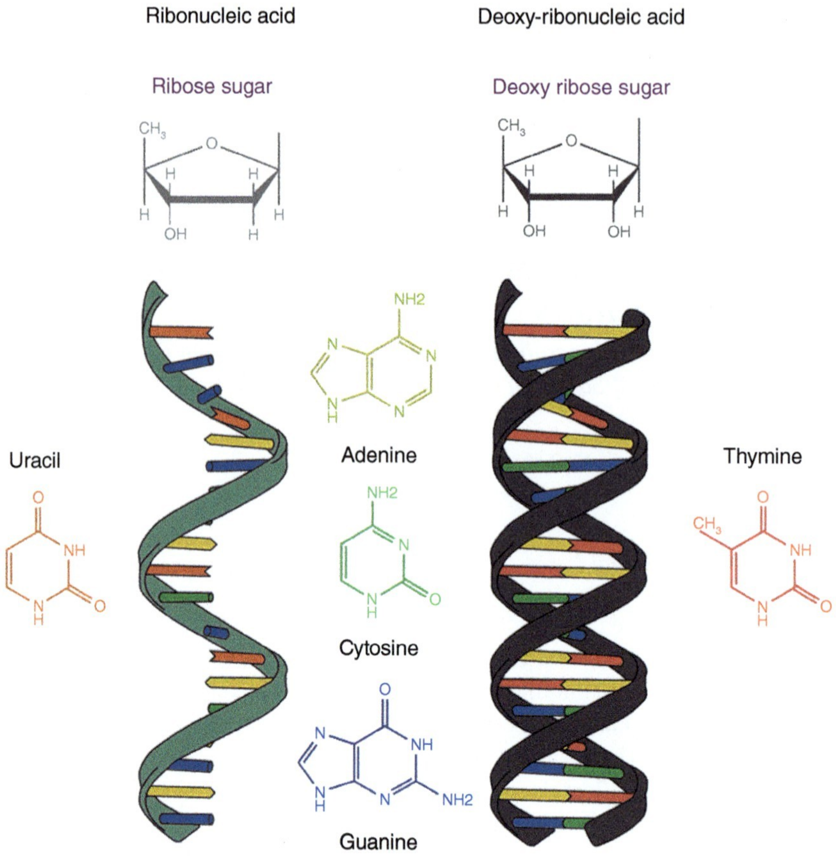

Figure 9.1 Different sugar backbone and bases of DNA and RNA. Source: Twyman and Wisden [5] / with permission of Taylor & Francis.

the natural ones, but the difference in their backbone makes them a novel genetic information storing and retrieving system [15–17]. The abundance of NAs in the cell with wide functionality leads to the exploration of new therapeutic potential. Table 9.1 summarizes the functionality of a wide variety of naturally available NAs. The delivery of the NAs is different compared to other small molecules. Despite the tremendous therapeutic potential of NAs, their delivery has not been successfully achieved in research and clinical applications [18].

During the administrations of NAs through various routes (parenteral, nasal, intramuscular, intravenous [IV], intracutaneous, and intradermal, etc.), they have to either enter the cytoplasm of the cell (siRNA, mRNA) or to the nucleus (antisense oligonucleotide, DNA) by crossing the intracellular, and endosomal barriers (Figures 9.3 and 9.4) [19–22].

In addition, NAs face several challenges in the circulatory system, such as negative charge, high-molecular weight, toxicity, low serum stability, susceptibility

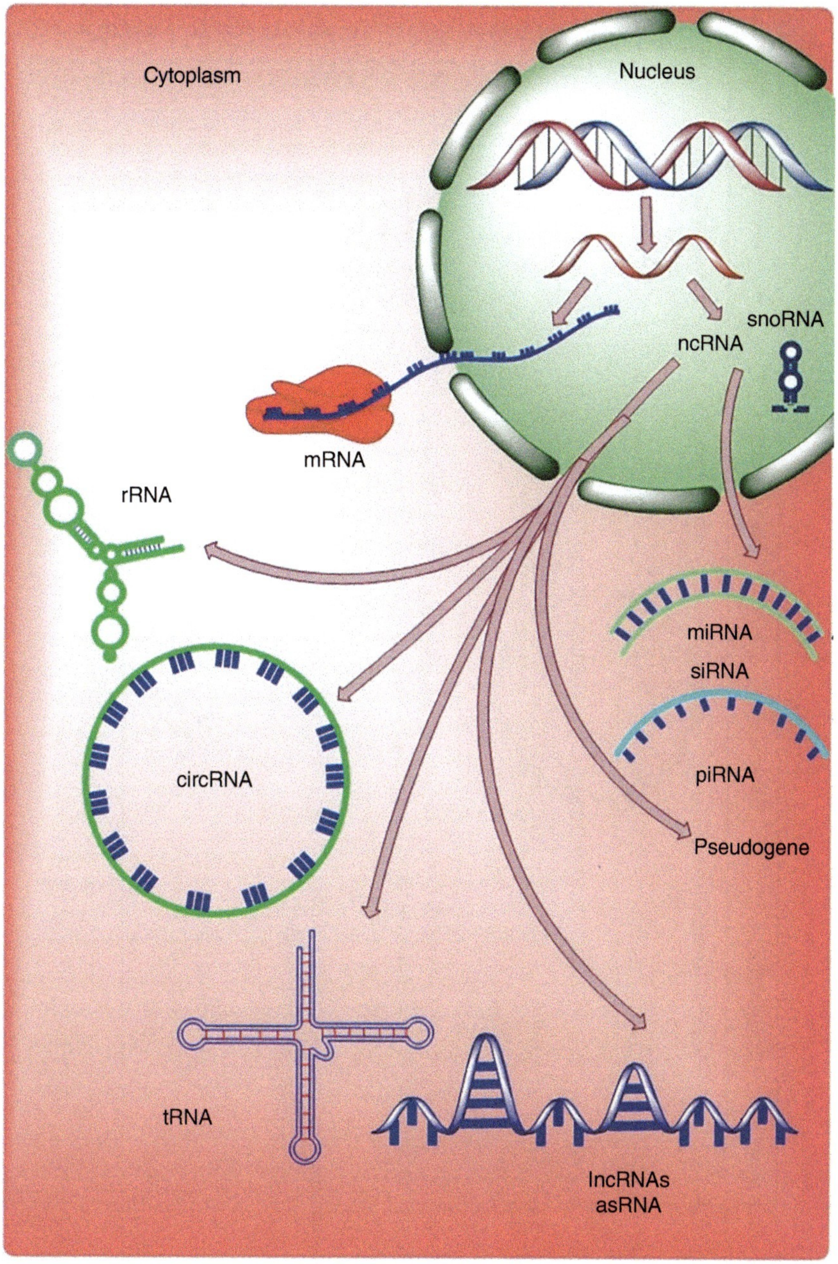

Figure 9.2 Location and types of different RNA. Source: Adapted from [6].

Table 9.1 Physical properties and mode of action of various NAs.

	Name of the cargoes	Shape and size	Structure	Intended action	Mode of action	References
DNA based therapeutics	Plasmid DNA	Double-stranded, circular (10^3–10^5 base pairs)		Gene overexpression	Transcription to mRNA in the nucleus	[41–43]
	Antisense oligonucleotides (ASO)	Single-stranded, linear (16–20 nucleotides)		(gene silencing) Inhibit single protein inhibition	Block translation of mRNA and spliceosome assembly	[44–46]
	DNA aptamers	Double-stranded (56–120 nucleotides)		Protein expression inhibition	Direct binding to extracellular and intracellular proteins	[43]
RNA based therapeutics	siRNA	Double-stranded, linear (~20–25 base pairs)		Knockdown of specific gene	RNAi pathway in the cytoplasm	[41, 45, 46]

(Continued)

Table 9.1 (Continued)

	Name of the cargoes	Shape and size	Structure	Intended action	Mode of action	References
	miRNA	Double-stranded, linear (~20–25 base pairs)		Gene regulation of gene pool	RNAi pathway in the cytoplasm	[41, 45, 46]
	mRNA	Single-stranded, linear (~1–10 kilobases)		Gene overexpression	Translated to protein in the cytoplasm	[7, 10, 11]
	RNA aptamers	Single-stranded (56–120 nucleotides)	—	Inhibition of protein expression	Directly bind to intracellular protein	[7, 10, 11]
Others	CpG oligodeoxynucleotides (ODNs)	Single-stranded, linear (~10–30 bases)		—	—	[7]
	Cyclic dinucleotides (CDNs)	Heterocyclic (2 nucleotides)	—	—	—	[7]

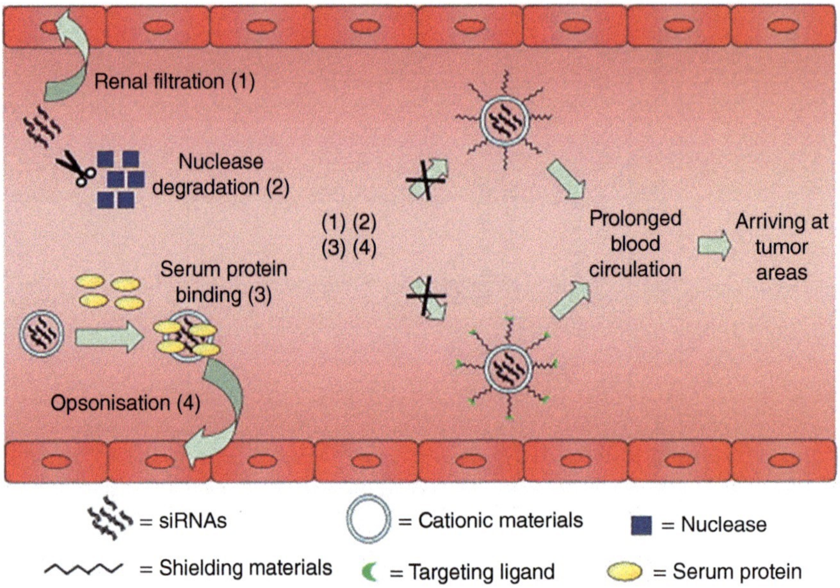

Figure 9.3 Working of siRNA and hindrances to delivery of siRNA. Source: Guo et al. [23] / with permission of Elsevier.

to enzymatic degradation, rapid renal filtration, entrapment by phagocytes, poor extravasations from blood to target tissues, and inability to penetrate target cell membrane phospholipid bilayer [25–27]. Moreover, they also have to evade the immune surveillance system as it can eliminate the exogenous NA by identifying them as foreign entities [27, 28].

To overcome all these barriers, several engineered NA-complex delivery systems have been designed in recent times to deliver NAs efficiently at specific target sites in a controlled manner to enhance their therapeutic efficacy. Different complex delivery systems influence the mode of delivery and regulation mechanisms [26]. Generally, the successful delivery of NAs into cells depends on the use of efficient viral or nonviral vectors. Most of the earlier clinical trials were based on viral vectors due to their inherent attribute to overcome cellular barriers. However, they have many limitations, such as small-scale production, triggered inflammatory reactions, high cytotoxicity, immunological response, integration at unwanted sites in the host genome, and other side effects [29–33]. This is supplemented with a short length of exogenous DNA, induced tumorigenic mutations, and the generation of active viral particles through recombination [32]. These limitations forced scientists to find an alternative approach in the form of nonviral vector-based NA-complex delivery systems [34].

The NA-complex delivery system uses polymer-based, nanoparticle (NP)-based, lipid-based, lipid-polymer hybrid, and peptide-based materials for controlled release of NA to the target site. To design an efficient targeted NA-complex delivery system, different natural and synthetic materials are being used as carriers that can form

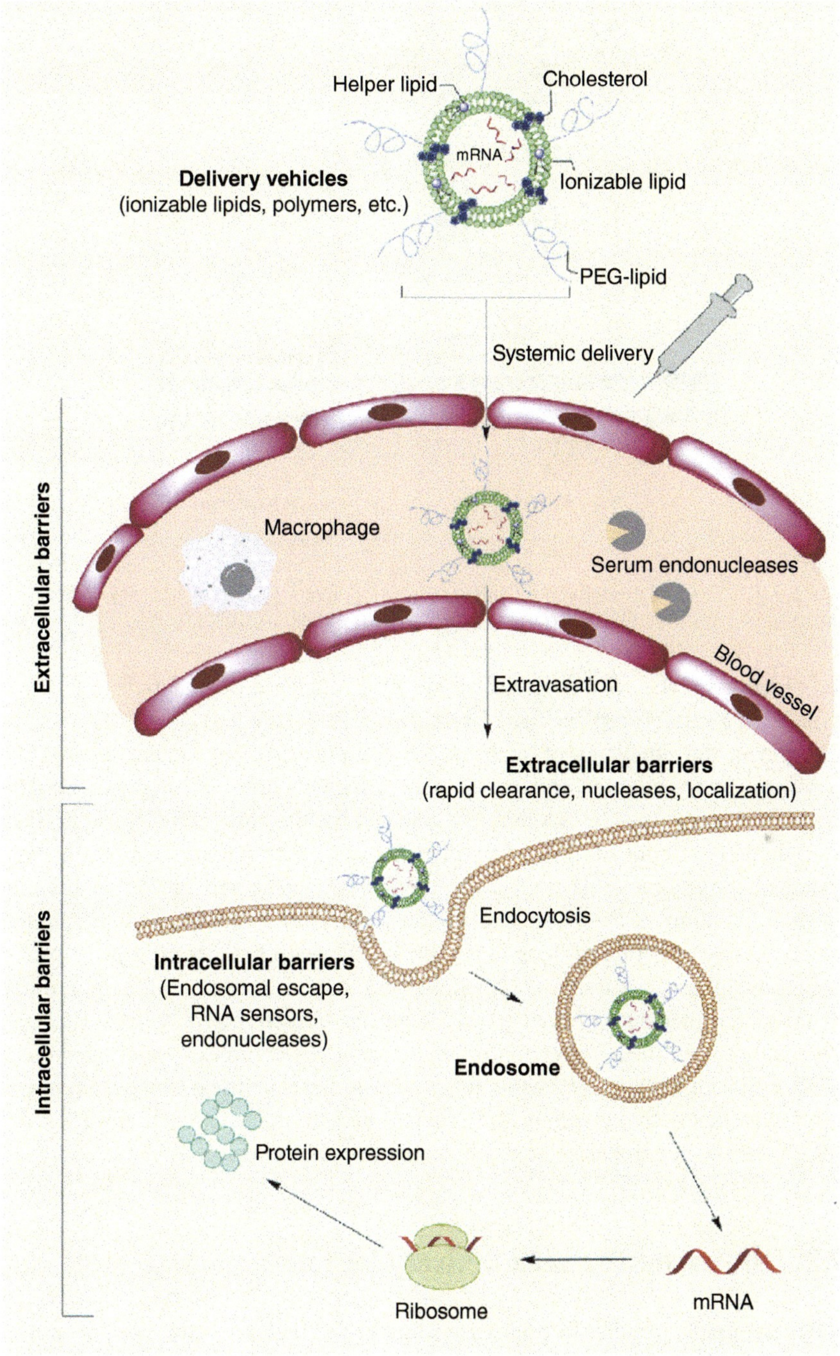

Figure 9.4 Working of mRNA and hindrances to delivery of mRNA. Source: Kowalski et al. [24] / with permission of Elsevier.

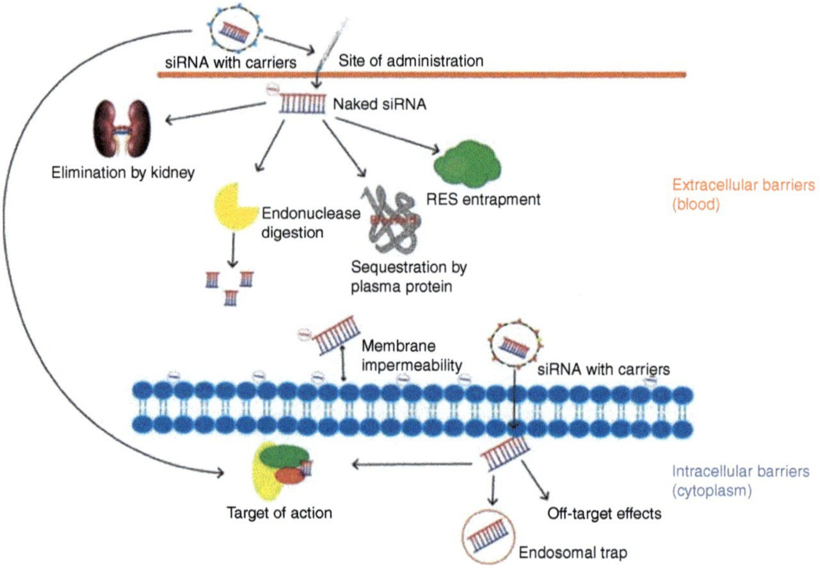

Figure 9.5 siRNA when delivered using a complex delivery system overcomes all the barriers against naked siRNA. Source: Sajid et al. [39] / MDPI / CC BY4.0.

complexes with the NAs to protect them from nuclease degradation and can be considered a suitable replacement for viral vectors (Figure 9.5) [24, 35–38].

However, the efficiency of such an NA-complex delivery system recedes due to several unprecedented reasons. It requires much effort to engineer nonviral carriers for target-specific NA delivery to overcome the various biological barriers. A fundamental understanding of modifications (for both NA and complex delivery systems), the interaction between the therapeutic agent and the target site, and the manipulation of the extracellular and intracellular environment is required for successful targeting of NAs [36]. The knowledge of underlying principles of several attributes for the NA-complex delivery system would provide the grounds for future advancement of their efficacy over the viral counterparts [40]. However, the advancement in the direction of engineering NA-complex delivery systems is still insufficient to date. In this chapter, particular focus on the recent developments of NA-complex delivery systems for efficient delivery of NAs (such as siRNA and mRNA) has been emphasized for various therapeutic applications.

9.2 NA-Based Complex Delivery System

Polymers, lipids, proteins/peptides are some of the materials that have been used for the formulation of NA-complex carriers. In the complexed delivery systems, the charge attraction and hydrophobic interaction between the biomaterials and NAs package the larger macromolecular NAs into nano-sized entities, which remains protected from degradation in biological environments. The novel formulations exhibit variability in structure and properties when transfected through different cellular internalization pathways. In this section, both the classical and advanced NA-complex delivery systems are elaborated.

9.2.1 Classical NA-Based Complex Delivery System

The classical NA-complex delivery system comprises natural, engineered, and synthetic constructs are depicted in Figure 9.6. The objective of delivery of NAs relies on the complex formation with the inclusion of polymers, lipids, glycosides, and polypeptides.

9.2.1.1 Polymer-Based NA-Complex Delivery System

Polymers, such as chitosan, hyaluronic acid, dextran, and gelatin, are some of the natural polymers widely studied as NA-complexed delivery systems [47]. The cationic polysaccharide, chitosan, is obtained by deacetylation of chitin usually obtained from fungal cell walls, crustaceans, and other sources [48, 49]. Chitosan is biocompatible and also has a low *in vivo* cytotoxic effect [50]. The cationic nature of chitosan makes it a suitable biopolymer that can interact with negatively charged NAs to form polyelectrolyte complexes and help in the efficient transfection of NAs into the cell. However, the strength of chitosan alone is low and hence scientists have tried to design formulations to increase the efficiency of NA uptake and transfection ability [49]. In the year 2007, Yu et al. have shown the increased DNA transfection ability of a modified NA-chitosan-g-poly-L-lysine complex and showed high DNA-binding ability with reduced cytotoxicity [49]. Besides, several synthetic polymers have proved to be excellent NA carriers. Some of the examples of these polymers are poly-L-Lysine (PLL), polyethylenimine (PEI), poly[(2-dimethylamino) ethyl methacrylate] (DMAEMA), poly(amidoamine), and poly(propylene imine) (PPI) [32]. Among them, PEI is the most frequently used cationic biopolymer for NA delivery; its transfection efficiency depends on the molecular weight and structure (linear or branched). PEI is a non-biodegradable and cytotoxic polymer due to its dense positive charge [31, 36]. Low molecular weight (LMW) (<10 kDa)

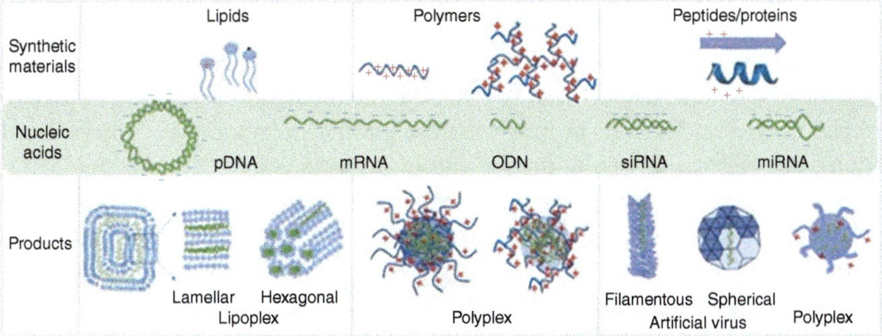

Figure 9.6 Classical strategy for condensed NAs formulations. Source: Ni et al. [21] / MDPI / CC BY4.0.

PEI shows lesser transfection capacity and low-cytotoxic effect in comparison to its high-molecular weight (25 kDa). Nyamay'Antu et al. have developed a PEI-based nonviral delivery of NA in mouse model and humans as well, which neither had inflammatory response nor any hepatotoxicity. Furthermore, it was found to be safe in human patients having pancreatic cancer [52]. Polyethylene glycol (PEG) can be useful to reduce the cytotoxic effect and extend the bloodstream clearance effect by shielding the excessive positive charge. Harguindey et al. have designed a DNA-complex NP, poly(ethylene glycol)-*b*-click NA-*b*-lactic-*co*-glycolic acid (PEG-CNA-PLGA) that successfully entrapped a larger hydrophilic dsDNA and also can co-encapsulate hydrophobic drugs at the same time [53]. Baek et al. have also studied the efficiency of PEG-engrafted graphene oxide as a biocompatible carrier for ss-peptide NA (PNA) and efficiently delivered to the targeted lung cancer cells without affecting the cell viability. The complex was also used to deliver antisense PNA and effectively downregulate the genes in the targeted cancer cells [54]. Poly-DMAEMA is a pH-responsive cationic polymer that exhibits reduced cytotoxicity and high transfection efficiency of DNA, siRNA, mRNA, and miRNA delivery [34]. Convertine et al. have designed a new complex di-block copolymer named poly(DMAEMA)-*b*-(DMAEMA-*co*-butyl methacrylate-*co*-propylacrylic acid), which was composed of a block polymer responsive to pH and a longer endosomolytic block with high hydrophobic content to induce micelle formation. The modified design of the polymer successfully delivered siRNA to the cytoplasm and also enhanced the transfection efficiency of siRNA with low cytotoxicity [55].

9.2.1.2 Lipid-Based Complex NA Delivery System

Lipid is hydrophobic organic compounds composed of hydrocarbon that include fats, oils, vitamins, and hormones. The nonpolar molecules are grouped into fatty acids and their derivatives, cholesterol and its derivatives, and lipoproteins. The complex of NA with the lipid-based formulation is extremely beneficial as it bypass the problem of hydrophobicity when endocytosed by the cell and also dodges immune surveillance at higher rates than other types of delivery systems. At physiological pH, liposomes made of ionizable lipids that act neutral, but under acidic pH, they are ionized and protonated. Cationic lipids 1,2-di-*O*-octadecenyl-3-trimethylammonium propane (DOTMA) spontaneously form the liposomes with hydrophobic tail and hydrophilic head groups connected via a linker molecule. The chemical composition and bonding pattern decide the extent of biodegradability of these systems. Felgner et al. have used DOTMA–DNA complex that was observed to fuse with the plasma membrane of cells in tissue culture. The cells were seen to uptake and express the delivered DNA [56]. While the neutral lipids (1,2-dioleoyl-*sn*-glycero-3-phosphoethanolamine) DOPE and cholesterol are the helper lipids that enhance particle stability and transfection efficiency. The fusogenic property of DOPE makes it feasible to use stimuli-responsive particles in pH-sensitive liposomal formulations [57]. Hattori et al. have reported that a mannosylated cationic liposome (Man-C4-Chol/DOPE) was able to deliver DNA to the target cells via IV-route but rapid lysosomal degradation limited the use of the complex. They also concluded that the presence of DOPE in the NA-complex delivery system increases the efficiency of gene expression in the target cells and also improves intracellular trafficking [58]. *In vivo* cytotoxicity and immune response are two main hurdles for the use of lipoplex delivery [59].

9.2.1.3 Peptide-Based Complex NA Delivery System

Peptides and proteins are biodegradable polymers that have long been used as NAs delivery systems. Some of the classical examples are the cell-penetrating peptide (CPP) and "nuclear localization signals" (NLS) that are being used to deliver NA forming polyplexes. NLS are cationic peptide sequences recognised by the importins present on the cytoplasm of the cell that directs the NA complex into the nucleus. Peptide based modification in the vector delivering NA helps enhancing the transfection ability of the complex into the cell as well as the nucleus. Hence, they are used in combination with other hydrophobic moieties. The bonding between the peptide and NAs is either by covalent linkage or by noncovalent interactions. The linkage formed between them is through charged amino acid residues. CPP are short protein sequences with 20 amino acids and a net positive charge, which help in the direct internalization of the attached NA. PLL, being a cationic polypeptide, shows strong interaction with NAs. The encapsulation capacity is also noteworthy to explore more on the compound to use as a delivery vehicle. The very initial trial of PLL as a NA delivery vector proved that it cannot be a good transfection vector due to its insufficient buffer capacity to disrupt the endosomal membrane. PLL can be modified with a hydrophobic polymer which enhances the stability of polyplexes and protects the NAs from degradation. One such example is a PEGylated PLL with enhancing *in vivo* cargo conveyance and reduced toxicity. Another example is arginine-rich protein protamine used for NA condensation [27]. These complex peptide-based NA systems are used in combination with lipids or polymers to increase polyplex stability (Table 9.2). Yamagata et al. studied the effect of the structures of gene-carrying molecules with their activities for delivery into the cells. They used dendritic PLL (KG6) and linear PLL to form a complex with DNA.

Table 9.2 Classical complex delivery systems for siRNA and mRNA.

Type of the material	Name of the cargo	Constituent of the NA-complex delivery systems	Size (nm)	References
Lipid	siRNA	DOTAP/DOPE/cholesterol	100–200	[49, 61]
		DOTAP/DOPE/cholesterol; PEI	468 ± 19; 209 ± 17	[21]
Polymer	siRNA	POCG-PEG-PEI polymer	151; 172; 245	[50, 61]
	mRNA	PEI-stearic acid copolymer	117.77 ± 3.89	[61, 62]
Peptide/ Protein	siRNA	Glucose-GSGSGSKKKKKKKKGGS GGSWKWEWKWEWKWEWG	~70	[21]
	mRNA	RALA (WEARLARALARALARHLARAL ARALRACEA)	89 (N/P = 5); 91(N/P = 10)	[27, 41]

Note: The letters represent single letter code of amino acids and their arrangement in the peptide. N/P is the Ratio of polymeric Nitrogen and Nucleic Acid Phosphate.

Both the NA-complex delivery carriers were compared for their efficiency in DNA compaction level, DNA binding and uptake amount, and gene expression level. They observed that compaction of DNA was found stronger in the case of KG6 due to neutral DNA complex formation but PLL was found more efficient (fourfold higher) in binding with DNA and its uptake by the cells than KG6. They also observed that the gene expression in the case of a complex KG6-NA delivery system was 100-fold more than that of NA delivered by only PLL [60].

9.2.2 Advanced NA-Based Complex Delivery Systems

Nowadays, many advanced NA-complex-based nanocarriers are emerging with high transfection rate, low cytotoxicity, and high biodegradability, such as liposomes, exosomes, bacteria-derived nano cell, polymeric NP, DNA and RNA nanostructure, inorganic NP, and dendrimer-based NP. These advanced complex delivery systems are built upon the previous framework of formulations with some modifications. Figure 9.7 shows the representative structures of some advanced NA-complex-based delivery systems.

9.2.2.1 Inorganic and Hybrid NPs

Unlike organic compound-based nanostructures (dendrimers, micelles, ferritin, etc.), inorganic NPs can tune their physicochemical properties and exhibit several benefits for drug-delivery systems. Few examples of the NA-complex delivery carriers are metal NPs, calcium-based NPs, and quantum dots. One classic example is the Au-NPs, which are preferred over all other metallic NPs for NA delivery because of many unique properties, such as it can form ultrasmall size and its functional diversity can easily be achieved. For pH-responsive delivery of NAs, calcium-based NPs are a good choice. At low pH values, the calcium-based NPs are highly soluble

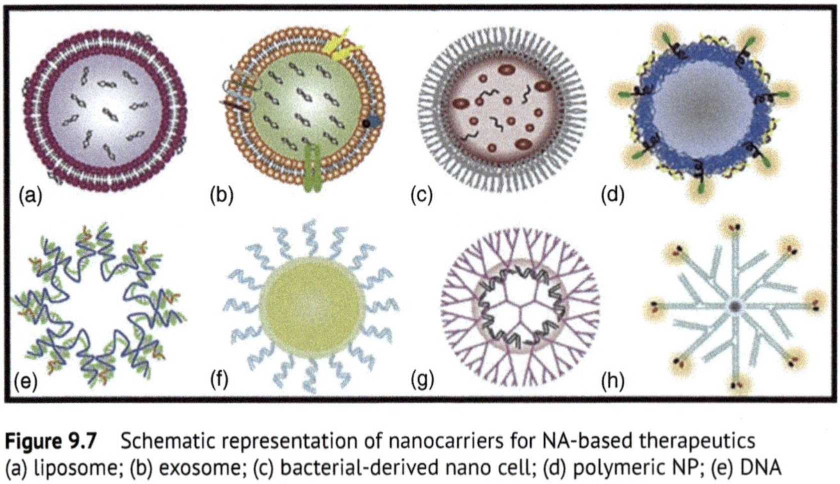

Figure 9.7 Schematic representation of nanocarriers for NA-based therapeutics (a) liposome; (b) exosome; (c) bacterial-derived nano cell; (d) polymeric NP; (e) DNA nanostructure; (f) inorganic NP; (g) dendrimer-based NP; and (h) RNA nanostructure. Source: Weng et al. [63] / with permission of Elsevier.

to form ionic substances. These calcium-based NPs release the NAs easily in an acidic environment like lysosomes. Though the transfection rate through inorganic NPs is very high, individually its efficiency is low. So, hybrid nanosystems or nanocomposites are some good alternatives as they are a combination of both inorganic NPs and polymers. Fornaguera et al. designed a novel NP-encapsulating mRNA which was further modified by oligopeptide end-modified poly(β-amino esters) and transfected to various model cell lines, including APCs. Furthermore, they were also tested *in vivo* to target the APCs through IV administration, which efficiently penetrated and transfected the cell accumulating in the spleen [64]. In another study, engineered *in vitro*-transcribed mRNA (IVT-mRNA) integrated with LNPs when injected into tumor-bearing mice, significantly decreased the volume of specific HER2-positive tumors with increased animal survivability [65].

9.2.2.2 Self-Assembled NA Nanostructures

One recent discovery is the self-assembly of the DNA macromolecules into nanostructures with many biological applications, such as NAs delivery. Transfection with such self-assembled multidimensional DNA nanostructures requires no transfection agents. DNA nanostructures are used as NA-based therapeutic agents to deliver different NAs, such as siRNA, ASOs, and guide RNA (gRNA), for clustered regularly interspaced short palindromic repeats-CRISPR-associated protein 9 (CRISPR-Cas9) system, and aptamers. Packaging RNA (pRNA) are the small RNA molecules having the ability to package DNA into procapsids. Besides pRNA, there are many natural RNA motifs, such as kink-turns, paranemic motifs, C-loops, kissing hairpin loops, multi-helix junctions, loop/loop-receptor pairs, and protein-binding motifs. The catalytic activity and self-folding DNA nanostructure assembly carrying siRNA were reported by Lee et al. [57]. Their formulation demonstrated a higher efficiency in silencing the gene and exhibited enhanced biocompatibility than the existing commercial cationic lipid transfecting agents [66, 67]. Slieman et al. synthesized DNA nanocages for delivery of siRNA for gene therapy. This made siRNA less vulnerable to nuclease degradation and thus showing a 99% release yield from the nanocages [44, 68].

9.2.2.3 Exosomes and NanoCells

Exosomes (30–150 nm) are endogenous NPs isolated from extracellular fluids or released by most of the cells in the body. Exosomes are composed of a large number of proteins and other cells which communicate active substances that are found in the parent cell. In a study, Cas9 mRNA from red blood cells integrated with exosomes was targeted to cure breast cancer both *in vitro* and *in vivo*. The induction of exosomes demonstrated a significant reduction in oncogenic cell proliferation [69]. In a recent study, exosome-encapsulated mRNA demonstrated a decreased rate of oral carcinoma cell proliferation *in vitro* [70]. In a similar study, pancreatic cancer was treated with exosome-encapsulated $Kras^{G120}$ siRNA isolated from fibroblast-like mesenchymal cells. There was a significant increase in apoptosis of Panc-1 cell *in vitro* experiments and a decrease in tumor size as observed from the *in vivo* studies [71].

9.3 Applications of NA-Complex Delivery Systems

The NA-complex-based delivery systems have brought a breakthrough in the delivery of siRNA, mRNA, miRNA, ASO, and aptamers, which have shown great potential in clinical trials. Pegaptanib (Macugen), an aptamer drug, having a PEG formulated complex carrier system to deliver the aptamer increasing its half-life time [72] was the first RNA-based therapeutics that was approved by Food and Drug Administration (FDA) in 2004 [73]. Since then, RNA studies had not progressed a lot, because of their instability in presence of ribonucleases and difficulties faced during intracellular delivery. In the last few years, advancement in NAs delivery with suitable carriers has made the first siRNA-based drug Patisiran, approved by FDA (Figure 9.8). This siRNA-based drug Partisan helped in treating polyneuropathy [45]. The siRNA in Patisiran is carried into the cells with the help of lipid vesicles that are internalized by endocytosis and get disrupted in the cell cytoplasm to release siRNAs which trigger the RNAi-based gene silencing [74]. Ever since the FDA approval of the Patisiran drug, RNA-based therapeutics has gained much success in all stages of the clinical trial for a wide range of diseases, such as various types of cancer, Hepatitis C, Duchenne muscular dystrophy, different genetic disorders, cystic fibrosis, lung diseases through protein therapy, gene therapy, translation and transcription inhibition [42, 48, 64].

A second siRNA-based drug, Givosiran was approved in late 2019 which used GalNAc-conjugated siRNA instead of LNP. Both the FDA-approved siRNA-based drugs use distinct mechanisms (ApoE vs. GalNAc), but both show strong uptake in the liver [75]. Recently, two mRNA vaccines have been proved to be effective against the SARS-Cov-2 virus. Currently, different mRNA vaccines are being used in clinical trials for COVID-19 [76]. mRNA-based vaccines against COVID-19 have been developed by many companies, such as Pfizer-BioNTech, Curevac

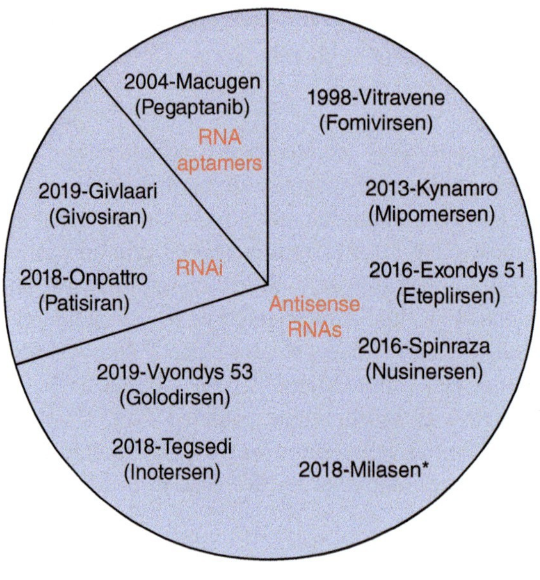

Figure 9.8 FDA-approved RNA therapy drugs. *Milasen is a special case that received approval while a single individual being in the trial.

Arcturus, and Moderna, [77]. Dagan et al. evaluated the effectiveness of an mRNA-based vaccine BNT162b2 in Israel from the data of different healthcare systems [78]. The Pfizer-BioNTech COVID-19 vaccine and BNT162b2 vaccine contain a nucleoside-modified mRNA (modRNA) encoding for a mutated form of SARS-CoV-2 virus spike protein encapsulated by LNPs. The lipid can be ALC-0315, ((4-hydroxybutyl)azanediyl)bis(hexane-6,1-diyl)bis(2-hexyldecanoate), ALC-0159, 2-[(polyethylene glycol)-2000]-*N*,ditetradecylacetamide, 1,2-distearoyl-*sn*-glycero-3-phosphocholine (DSPC), and cholesterol from which ALC-0159 is a PEGylated lipid [79]. These all make a nanocomplex with the mRNA and not only act as carriers for the mRNA to be delivered into human cells but also as adjuvants [80]. In another study, Chemaitelly et al. reported that mRNA-1273 (Moderna) vaccine showed a high efficiency against symptomatic COVID-19 (94%) while used upon B.1.1.7. α variant and the B.1.351 β variant [81]. In the case of Moderna, the complex mRNA delivery system includes encapsulated mRNA within the LNPs that constitute all the above lipids along with SM-102 [82]. The wide applications have shown much success in the field of pharmaceuticals and the biotechnological industry. With the wide spectrum availability of NA and adaptation within the secondary host, the applications are wide. Some of the applications for NA-based therapeutics are given in Table 9.3.

Applications of NA-based therapeutics delivered by various NA-complex delivery systems, including genome editing, cancer therapy, and protein therapy, are elaborated in Section 9.3.1.

9.3.1 Genome Editing

NAs can be used as therapeutics in genome-editing applications. The genome-editing technology uses the re-programmable NAs enzymes, i.e. the nucleases. The use of nucleases-mediated genome editing explores the utility of siRNA and mRNA to treat the genomic aberrations (those) causing life-threatening disorders. The extent of genome editing has reached the era of CRISPR/Cas9 technology (Figure 9.9) that relies on Cas9/sgRNA (single gRNA) ribonucleoprotein complexes (RNPs) for editing DNA at the target site. CRISPR-Cas9 technology takes the advantage of gRNA to redirect to a genetic sequence where the Cas9 enzyme can exhibit its nuclease activity [88]. Yip, reviewed the efficiency of NA-based delivery of the CRISPR-Cas9 system. It was suggested by the author that the Cas9 enzyme could be delivered at an accelerated rate by the use of mRNA which can result in faster gene editing [89]. The author also reviewed the use of the CRISPR-Cas9 system for clinical trials. The trial succeeded in treating lung cancer by reactivating T-cells [90].

A similar study by Wei et al. has reported the efficiency of modified LNPs for the delivery of RNPs to edit genes in the defective sites that includes liver, muscle, brain, and lungs. The complex LNP carrier system for RNPs shows tissue specificity while administered through the IV route. The carrier potency of the complex LNP–RNPs was used to maximize the advantage that is capable of multiplexed editing (knockout) of six genes in the mouse lungs. The authors have also used the designed complex carrier to deliver RNPs to Duchenne muscular dystrophy (DMD) mice where it restores the expression of dystrophin [92].

Table 9.3 The route of administration and targets of some RNA-based drugs.

Name of the cargo	Route of administration	Purpose	Drugs	Target	References
siRNA	Intravenous injection	Systemic delivery	15NP (QPI-1002)	P^{53}	[83]
	Subcutaneous injection	Systemic delivery	ALN-GOI (Lumarisan)	HA01	[69, 84]
			DCR-PHXC	LDHA	[85]
	Intravitreal injection	Localized delivery	Bevasiranib	VEGF	[83]
			QPI-1007	Caspase-2	[86]
	Inhalation/Intranasal/Intratracheal administration	Pulmonary delivery	ALN-RSV01	RSV nucleocapsid	[83]
	Ophthalmic drops	—	SYL1001	TRPV1	[87]
			SYL040012	ADRB2	[51]
	Intradermal	Local but little systemic delivery	BMT 101	CTGF	[83]
mRNA	Intradermal	Local but little systemic delivery	Tumor mRNA vaccine	Melan A, MAGE A1, MAGE A3, survivin, gp100,	[83]
	Subcutaneous	Systemic delivery	CV7201	Rabies virus glycoprotein	[83]
			Tumor mRNA vaccine	Melan A, MAGE A1, MAGE A3, surviving, gp100,	[83]
	Intramuscular	Systemic delivery	CV7201	Rabies virus glycoprotein	[83]

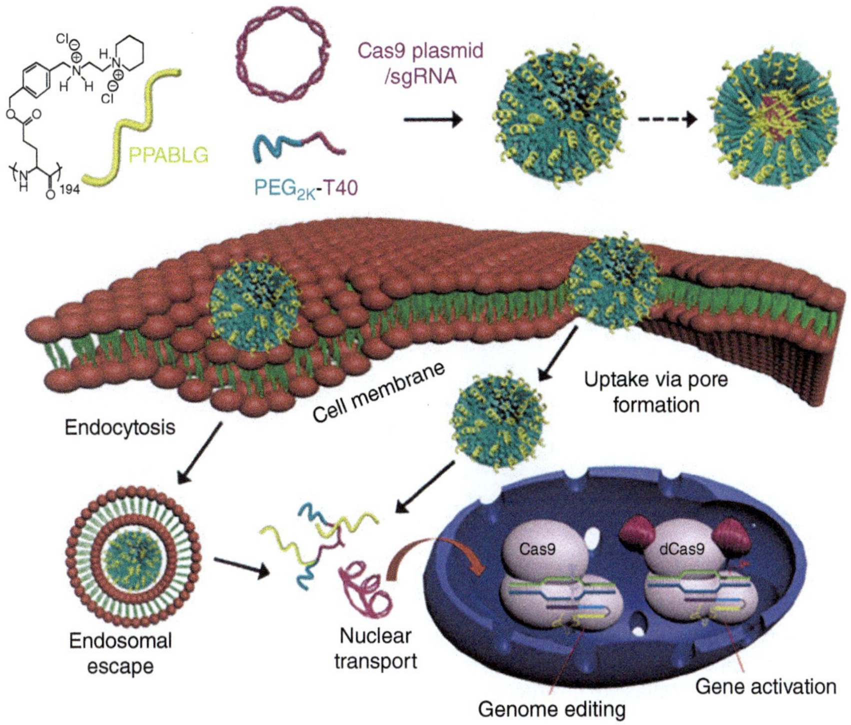

Figure 9.9 Diagrammatic explanation for NA-CRISPR-Cas9 complex delivery system. Source: Wang et al. [91].

9.3.2 Cancer Therapy

Cancer is a complex disease in which tumorigenic cells rapidly proliferate from one part to another, thereby becoming malignant. It is well understood that this accelerated drive is due to rapid angiogenesis in oncogenic cells. Thus, it becomes necessary to terminate the process which may prove beneficial to human cellular functions. The siRNA-mediated treatment of cancers is mainly driven by chemotactically engineered siRNA, complex systems mediated delivered siRNA, and composites as well. The efficacy of any of the methods is dependent on the uptake of contact inhibition independent oncogenic cells. RNAi-based treatment has been recently used in clinical trials. Atu027 is a novel LNP–siRNA complex system (Stable NA–lipid particles [SNALPs]) [93] targeting protein kinase N3 (PKN3), that is combined with Gemcitabine and used in clinical trials to treat advanced or metastatic pancreatic cancer (NCT01808638) [37]. Figure 9.10 describes the underlying principle of the complex delivery system used for siRNA delivery to its target site for gene silencing. Baek et al., in their study, have reported that the complex containing PEG-nGO nanoconstructs have shown efficient delivery of PNA in lung cancer and subsequently downregulates the targeted cancer genes [54]. The authors have observed that the complex efficiently delivered the suicidal genes to the specific target sites of cancer and regulated the gene expression of

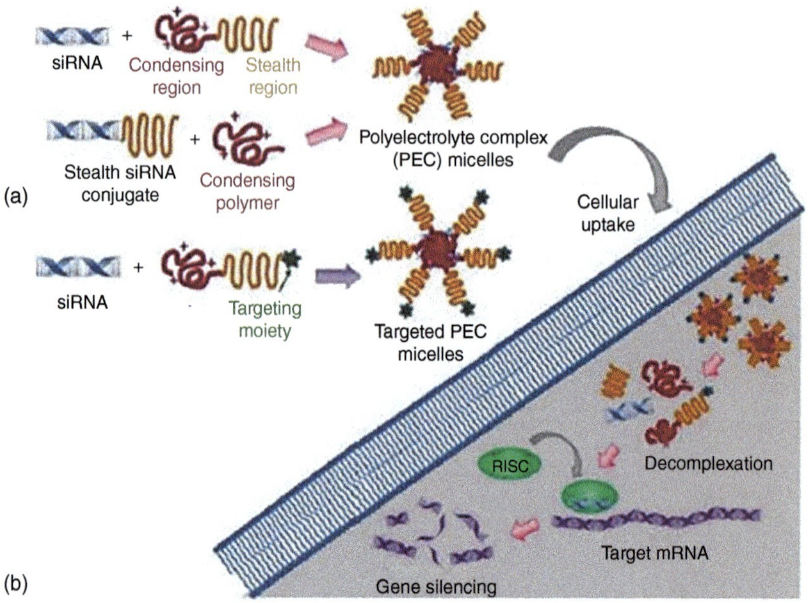

Figure 9.10 Mechanism of Polyelectrolyte complex (PEC) micelles mediated siRNA delivery. (a) Nano-sized PEC micelles are formed by condensing siRNA and a cationic material. (b) PEC micelles are incorporated with a targeting moiety for localization to target sites. Source: Bae et al. [94].

the tumor-suppressor gene and oncogenes. The pre-clinical studies on nonhuman primate models have reported the safety of the complex formulation delivering NA-based therapeutics. The complex formulation has also been tested on human participants having pancreatic cancer [94]. The participants were seen to tolerate the local and repeated intratumoral delivery of NA via *in vivo*-jetPEI® and hence reported to be safe for human use. Hence, the *in vivo*-jetPEI formulation is available in the market sold under the company name Polyplus® [52].

9.3.3 Protein Therapy

In NA-based therapeutics, it can either activate (Figure 9.11) or inhibit the protein synthesis. The delivery of the mRNA with the help of complex delivery systems expresses therapeutic proteins which can help in curing a wide array of chronic diseases. The potential therapeutic applications include replenishment of disordered protein (s), reprogrammed cellular functioning in presence of therapeutic NAs, and antibody-mediated therapies [95, 96].

Hereditary transthyretin-mediated (hATTR) amyloidosis is a rare, fatal, neurodegenerative disease in which misfolded transthyretin (TTR) protein is produced due to some mutations in the TTR gene. TTR protein stabilization rate is very limited with the help of drugs, such as difusinal and tafamidis [98]. Later inhibiting the TTR gene expression comes as an alternative strategy to correct the above problem.

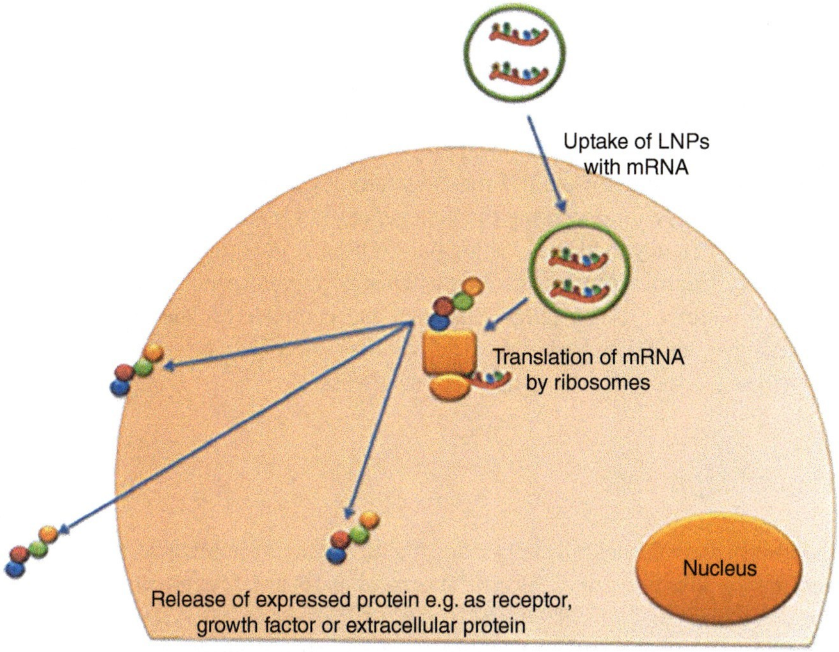

Figure 9.11 Mechanism of protein expression by an encapsulated mRNA at its target cell. Source: Michel et al. [97] / Intech open / CC BY3.0.

A siRNA drug Onpattro (Patisiran) manufactured by Alnylam pharmaceuticals is a LNP (lipid nanoparticles) formulation of siRNA that has shown expressive and promising results during clinical trials (NCT01960348) [63].

9.4 Future Prospective

The NAs-complex delivery systems have fascinated the biological community along with growing interest from both academia and pharmaceutical companies. The economy in biomedical science has risen significantly due to the increasing number of approved NAs-based therapeutics. There are few NAs-complex formulations under different phases of clinical trials, such as Curevac AG's CV7202 (Phase I), Moderna TX Inc.'s mRNA-2416 (Phase I), mRNA-3704 (Phase I), mRNA-2752 (Phase I), NIAID's mRNA-1273 (Phase I), Biontech's BNT162 (Phase I/II), Ludwig Institute for Cancer Research's CV9202 (Phase I/II) and Sylentis's SYL1001 (NCT03108664, Phase III), Quark's QPI-1007 (NCT02341560, Phase II/III), Silenseed's SIG12D (NCT01676259, Phase II), Rxi's RXT-109 (NCT02599064, Phase I/II, NCT02079168, Phase II), respectively [61, 99]. The basic understanding of the interactions between the physicochemical properties of complex delivery agents and NAs will help in designing next-generation NA-complex delivery systems. The future developments in the designing of stimuli-responsive NA-complex delivery systems will be of great significance in clinical research.

9.5 Conclusion

The utilization of NAs-complex formulations as delivery systems has shown an immense rate of success to deliver NAs for therapeutic applications. The precise understanding of the physicochemical properties (surface charge, size, morphology, stability, immunogenicity, etc.) of the NAs-complex delivery systems will provide a broad perspective for designing the next generation therapeutics ensuring safety and higher efficiency than the existing ones. These NAs-complex delivery systems significantly improve the transfection efficiency, target specificity, release time, reduced hindrances to the cellular barriers and cytotoxicity. The exploration for new NAs-complex delivery systems to deliver NAs safely into the cells of the human body will be worthwhile for treating various diseases.

Acknowledgments

Ms. Ananya Patnaik acknowledges the Indian Council for Medical Research for awarding the fellowship of ICMR-SRF (Proposal Id-15036). Ms. Swarnaprabha Pany expresses thanks to CSIR-SRF [09/547(0005)/2018-EMR-I]. Mr. A. Swaroop Sanket is thankful to Indian Council for Medical Research for awarding the fellowship ICMR-SRF (File No. 5/3/8/88/ITR-F/2020). Dr. Sangram Keshari Samal highly acknowledges the Ramanujan fellowship (SB/S2/RJN-038/2016) of the Department of Science and Technology, and Ramalingaswami Re-entry fellowship (Ref: D.O. No. BT/HRD/35/02/2006) of Department of Biotechnology, Government of India. The authors also thank the Indian Council of Medical Research-Regional Medical Research Centre, Bhubaneswar for providing a scientific platform.

References

1 Minchin, S. and Lodge, J. (2019). Understanding biochemistry: structure and function of nucleic acids. *Essays Biochem.* 63 (4): 433–456.

2 Sun, X. and Liu, H. (2020). Nucleic acid nanostructure assisted immune modulation. *ACS Appl. Bio Mater.* 3 (5): 2765–2778.

3 Jackson, M., Marks, L., May, G.H.W., and Wilson, J.B. (2018). The genetic basis of disease. *Essays Biochem.* 62 (5): 643–723.

4 Durmuş, S. and Ülgen, K. (2017). Comparative interactomics for virus–human protein–protein interactions: DNA viruses versus RNA viruses. *FEBS Open Bio* 7 (1): 96–107.

5 Twyman, R.M. and Wisden, W. (2018). Nucleic Acid Structure. *Advanced Molecular Biology* (eds. Richard M. Twyman and Wisden W.), 1e, 223–234. Taylor and Francis group publisher.

6 Lakhotia, S.C., Mallick, B., and Roy, J. (2020). Non-coding RNAs: ever-expanding diversity of types and functions. *RNA-Based Regul. Human Heal. Dis.* 19: 5–57.

7 Asha, K., Kumar, P., Sanicas, M. et al. (2018). Advancements in nucleic acid based therapeutics against respiratory viral infections. *J. Clin. Med.* 8 (1): 6.

8 Alberts, B., Johnson, A., Lewis, J. et al. (2015). *Molecular Biology of the Cell*, 6e, 1–1464. W.W. Norton & Co.

9 Asati, S. and Chaudhary, U. (2017). Prevalence of biofilm producing aerobic bacterial isolates in burn wound infections at a tertiary care hospital in northern India. *Ann. Burns Fire Disasters* 30 (1): 39–42.

10 Sharma, R., Lewis, S., and Wlodarski, M.W. (2020). DNA repair syndromes and cancer: insights into genetics and phenotype patterns. *Front. Pediatr.* 8: 683.

11 McKinnon, P.J. (2009). DNA repair deficiency and neurological disease. *Nat. Rev. Neurosci.* 10 (2): 100–112.

12 Knoch, J., Kamenisch, Y., Kubisch, C., and Berneburg, M. (2012). Rare hereditary diseases with defects in DNA-repair. *Eur. J. Dermatol.* 22 (4): 443–455.

13 Tiwari, V. and Wilson, D.M. (2019). DNA damage and associated DNA repair defects in disease and premature aging. *Am. J. Hum. Genet.* 105 (2): 237–257.

14 Duffy, K., Arangundy-Franklin, S., and Holliger, P. (2020). Modified nucleic acids: replication, evolution, and next-generation therapeutics. *BMC Biol.* 18 (1): 1–14.

15 Morihiro, K., Kasahara, Y., and Obika, S. (2017). Biological applications of xeno nucleic acids. *Mol. Biosyst.* 13 (2): 235–245.

16 Taylor, A.I., Houlihan, G., and Holliger, P. (2019). Beyond DNA and RNA: the expanding toolbox of synthetic genetics. *Cold Spring Harbor Perspect. Biol.* 11 (6): a032490.

17 Schmidt, M. (2010). Xenobiology: a new form of life as the ultimate biosafety tool. *BioEssays* 32 (4): 322–331.

18 Slivac, I., Guay, D., Mangion, M. et al. (2017). Non-viral nucleic acid delivery methods. *Expert Opin. Biol. Ther.* 17 (1): 105–118.

19 Wei, M. and Good, D. (2013). *Novel Gene Therapy Approaches*, 1–406. IntechOpen.

20 Vermeulen, L.M.P., Brans, T., Samal, S.K. et al. (2018). Endosomal size and membrane leakiness influence proton sponge-based rupture of endosomal vesicles. *ACS Nano* 12 (3): 2332–2345.

21 Ni, R., Feng, R., and Chau, Y. (2019). Synthetic approaches for nucleic acid delivery: choosing the right carriers. *Life* 9 (3): 59.

22 Abbasi, Y.F. and Bera, H. (2021). Konjac glucomannan-based nanomaterials in drug delivery and biomedical applications. In: *Biopolymer-Based Nanomaterials in Drug Delivery and Biomedical Applications* (eds. Hriday Bera, Chowdhury Mobaswar Hossain and Sudipta Saha), 119–141. Academic Press, Copyright © 2021 Elsevier Inc. All rights reserved.

23 Guo, J., Bourre, L., Soden, D.M. et al. (2011). Can non-viral technologies knockdown the barriers to siRNA delivery and achieve the next generation of cancer therapeutics? *Biotechnol. Adv.* 29 (4): 402–417.

24 Kowalski, P.S., Rudra, A., Miao, L., and Anderson, D.G. (2019). Delivering the messenger: advances in technologies for therapeutic mRNA delivery. *Mol. Ther.* 27 (4): 710–728.

25 Peeler, D.J., Sellers, D.L., and Pun, S.H. (2019). PH-sensitive polymers as dynamic mediators of barriers to nucleic acid delivery. *Bioconjugate Chem.* 30 (2): 350–365.

26 Draghici, B. and Ilies, M.A. (2015). Synthetic nucleic acid delivery systems: present and perspectives. *J. Med. Chem.* 58 (10): 4091–4130.

27 Jones, C.H., Chen, C.K., Ravikrishnan, A. et al. (2013). Overcoming non-viral gene delivery barriers: perspective and future. *Mol. Pharmaceutics* 10 (11): 4082–4098.

28 Durymanov, M. and Reineke, J. (2018). Non-viral delivery of nucleic acids: insight into mechanisms of overcoming intracellular barriers. *Front. Pharmacol.* 9: 971.

29 Raper, S.E., Chirmule, N., Lee, F.S. et al. (2003). Fatal systemic inflammatory response syndrome in a ornithine transcarbamylase deficient patient following adenoviral gene transfer. *Mol. Genet. Metab.* 80 (1–2): 148–158.

30 Manno, C.S., Arruda, V.R., Pierce, G.F. et al. (2006). Successful transduction of liver in hemophilia by AAV-factor IX and limitations imposed by the host immune response. *Nat. Med.* 12 (3): 342–347.

31 Howe, S.J., Mansour, M.R., Schwarzwaelder, K. et al. (2008). Insertional mutagenesis combined with acquired somatic mutations causes leukemogenesis following gene therapy of SCID-X1 patients. *J. Clin. Invest.* 118 (9): 3143–3150.

32 Bråve, A., Ljungberg, K., Wahren, B., and Liu, M.A. (2007). Vaccine delivery methods using viral vectors. *Mol. Pharmaceutics* 4 (1): 18–32.

33 Elsabahy, M., Nazarali, A., and Foldvari, M. (2012). Non-viral nucleic acid delivery: key challenges and future directions. *Curr. Drug Deliv.* 8 (3): 235–244.

34 Bhat, S.I., Ahmadi, Y., and Ahmad, S. (2018). Recent advances in structural modifications of hyperbranched polymers and their applications. *Ind. Eng. Chem. Res.* 57 (32): 10754–10785.

35 Dong, Y., Siegwart, D.J., and Anderson, D.G. (2019). Strategies, design, and chemistry in siRNA delivery systems. *Adv. Drug Delivery Rev.* 144: 133–147.

36 Sung, Y.K. and Kim, S.W. (2019). Recent advances in the development of gene delivery systems. *Biomater. Res.* 23 (1): 1–7.

37 Mahmoodi Chalbatani, G., Dana, H., Gharagouzloo, E. et al. (2019). Small interfering RNAs (siRNAs) in cancer therapy: a nano-based approach. *Int. J. Nanomed.* 14: 3111–3128.

38 Wadhwa, A., Aljabbari, A., Lokras, A. et al. (2020). Opportunities and challenges in the delivery of mRNA-based vaccines. *Pharmaceutics* 12 (2): 102.

39 Sajid, M.I., Moazzam, M., Kato, S. et al. (2020). Overcoming barriers for siRNA therapeutics: from bench to bedside. *Pharmaceuticals* 13 (10): 294.

40 Bujold, K.E., Hsu, J.C.C., and Sleiman, H.F. (2016). Optimized DNA "nano-suitcases" for encapsulation and conditional release of siRNA. *J. Am. Chem. Soc.* 138 (42): 14030–14038.

41 Gomes dos Reis, L., Lee, W.-H., Svolos, M. et al. (2020). Delivery of pDNA to lung epithelial cells using PLGA nanoparticles formulated with a cell-penetrating peptide: understanding the intracellular fate. *Drug Dev. Ind. Pharm.* 46 (3): 427–442.

42 Gómez-Aguado, I., Rodríguez-Castejón, J., Vicente-Pascual, M. et al. (2020). Nucleic acid delivery by solid lipid nanoparticles containing switchable lipids: plasmid DNA vs messenger RNA. *Molecules* 25 (24): 5995.

43 Sridharan, K. and Gogtay, N.J. (2016). Therapeutic nucleic acids: current clinical status. *Br. J. Clin. Pharmacol.* 659–672.

44 Sun, H., Zhu, X., Lu, P.Y. et al. (2014). Oligonucleotide aptamers: new tools for targeted cancer therapy. *Mol. Ther. Nucleic Acids* 3: e182.

45 Bajan, S. and Hutvagner, G. (2020). RNA-based therapeutics: from antisense oligonucleotides to miRNAs. *Cells* 9 (1): 137.

46 Hanna, J., Hossain, G.S., and Kocerha, J. (2019). The potential for microRNA therapeutics and clinical research. *Front. Genet.* 10: 478.

47 Samal, S.K., Dash, M., Chiellini, F. et al. (2014). Silk/chitosan biohybrid hydrogels and scaffolds via green technology. *RSC Adv.* 4 (96): 53547–53556.

48 Sahoo, R., Sanket, A.S., Pany, S. et al. (2020). Latest development of biopolymers based on polysaccharides. In: *Processing and Development of Polysaccharide-Based Biopolymers for Packaging Applications*, 281–299. Elsevier.

49 Samal, S.K., Dash, M., Van Vlierberghe, S. et al. (2012). Cationic polymers and their therapeutic potential. *Chem. Soc. Rev.* 41 (21): 7147–7194.

50 Daniel, E.-A.-K. and Hamblin Michael, R. (2016). Chitin and chitosan: production and application of versatile biomedical nanomaterials. *Int. J. Adv. Res.* 4 (3): 411–427.

51 Juliano, R.L. (2016). The delivery of therapeutic oligonucleotides. *Nucleic Acids Res.* 44 (14): 6518–6548.

52 Nyamay' Antu, A., Dumont, M., Kedinger, V., and Erbacher, P. (2019). Non-viral vector mediated gene delivery: the outsider to watch out for in gene therapy. *Cell Gene Ther. Insights* 5 (S1): 51–57.

53 Harguindey, A., Domaille, D.W., Fairbanks, B.D. et al. (2017). Synthesis and assembly of click-nucleic-acid-containing PEG–PLGA nanoparticles for DNA delivery. *Adv. Mater.* 29 (24): 1700743.

54 Baek, A., Baek, Y.M., Kim, H.M. et al. (2018). Polyethylene glycol-engrafted graphene oxide as biocompatible materials for peptide nucleic acid delivery into cells. *Bioconjugate Chem.* 29 (2): 528–537.

55 Convertine, A.J., Diab, C., Prieve, M. et al. (2010). pH-responsive polymeric micelle carriers for siRNA drugs. *Biomacromolecules* 11 (11): 2904–2911.

56 Felgner, P.L., Gadek, T.R., Holm, M. et al. (1987). Lipofection: a highly efficient, lipid-mediated DNA-transfection procedure. *Proc. Natl. Acad. Sci.* 84 (21): 7413–7417.

57 Lee, H., Lytton-Jean, A.K.R., Chen, Y. et al. (2012). Molecularly self-assembled nucleic acid nanoparticles for targeted in vivo siRNA delivery. *Nat. Nanotechnol.* 7 (6): 389–393.

58 Hattori, Y., Suzuki, S., Kawakami, S. et al. (2005). The role of dioleoylphosphatidylethanolamine (DOPE) in targeted gene delivery with mannosylated cationic liposomes via intravenous route. *J. Control. Release* 108 (2-3): 484–495.

59 Liu, C., Zhang, L., Zhu, W. et al. (2020). Barriers and strategies of cationic liposomes for cancer gene therapy. *Mol. Ther. Methods Clin. Dev.* 18: 751–764.

60 Yamagata, M., Kawano, T., Shiba, K. et al. (2007). Structural advantage of dendritic poly(L-lysine) for gene delivery into cells. *Bioorg. Med. Chem.* 15 (1): 526–532.

61 Saw, P.E. and Song, E.W. (2020). siRNA therapeutics: a clinical reality. *Sci. China Life Sci.* 63 (4): 485–500.

62 Li, Y. and Ju, D. (2017). The application, neurotoxicity, and related mechanism of cationic polymers. In: *Neurotoxicity of Nanomaterials and Nanomedicine* (ed. X. Jiang and H. Ga), 285–329. Academic Press.

63 Weng, Y., Huang, Q., Li, C. et al. (2020). Improved nucleic acid therapy with advanced nanoscale biotechnology. *Mol. Ther. Nucleic Acids* 19: 581–601.

64 Fornaguera, C., Guerra-Rebollo, M., Ángel Lázaro, M. et al. (2018). mRNA delivery system for targeting antigen-presenting cells in vivo. *Adv. Healthcare Mater.* 7 (17): 1800335.

65 Rybakova, Y., Kowalski, P.S., Huang, Y. et al. (2019). mRNA delivery for therapeutic anti-HER2 antibody expression in vivo. *Mol. Ther.* 27 (8): 1415–1423.

66 Bandaru, R., Sanket, A.S., Rekha, S. et al. (2021). Biological interaction of dendrimers. In: *Dendrimer-Based Nanotherapeutics* (ed. P. Kesharwani), 63–74. Elsevier.

67 Patil, S., Gao, Y.G., Lin, X. et al. (2019). The development of functional non-viral vectors for gene delivery. *Int. J. Mol. Sci.* 20 (21): 5491.

68 Lu, J.J., Langer, R., and Chen, J. (2009). A novel mechanism is involved in cationic lipid-mediated functional siRNA delivery. *Mol. Pharmaceutics* 6 (3): 763–771.

69 Dowdy, S.F. (2017). Overcoming cellular barriers for RNA therapeutics. *Nat. Biotechnol.* 35 (3): 222–229.

70 Zhao, X., Wu, D., Ma, X. et al. (2020). Exosomes as drug carriers for cancer therapy and challenges regarding exosome uptake. *Biomed. Pharmacother.* 128: 110237.

71 György, B., Sage, C., Indzhykulian, A.A. et al. (2017). Rescue of Hearing by gene delivery to inner-ear hair cells using exosome-associated AAV. *Mol. Ther.* 25 (2): 379–391.

72 Keefe, A.D., Pai, S., and Ellington, A. (2010). Aptamers as therapeutics. *Nat. Rev. Drug Discov.* 9 (7): 537–550.

73 Parashar, A. (2016). Aptamers in therapeutics. *J. Clin. Diagn. Res.* 10 (6): BE01–BE06.

74 Titze-de-Almeida, S.S., de Paula Brandão, P.R., Faber, I., and Titze-de-Almeida, R. (2020). Leading RNA interference therapeutics part 1: silencing hereditary transthyretin amyloidosis, with a focus on patisiran. *Mol. Diagn. Ther.* 24 (1): 49–59.

75 Yonezawa, S., Koide, H., and Asai, T. (2020). Recent advances in siRNA delivery mediated by lipid-based nanoparticles. *Adv. Drug Delivery Rev.* 154–155: 64–78.

76 Chow, M.Y.T., Qiu, Y., and Lam, J.K.W. (2020). Inhaled RNA therapy: from promise to reality. *Trends Pharmacol. Sci.* 41 (10): 715–729.

77 Gaviria, M. and Kilic, B. (2021). A network analysis of COVID-19 mRNA vaccine patents. *Nat. Biotechnol.* 39 (5): 546–548.

78 Dagan, N., Barda, N., Kepten, E. et al. (2021). BNT162b2 mRNA Covid-19 vaccine in a nationwide mass vaccination setting. *N. Engl. J. Med.* 384 (15): 1412–1423.

79 Medicines and Healthcare products Regulatory Agency (2020). Public Assessment Report Authorisation for Temporary Supply COVID-19 mRNA Vaccine BNT162b2 (BNT162b2 RNA) concentrate for solution for injection.

80 de Vrieze, J. (2020). Suspicions grow that nanoparticles in Pfizer's COVID-19 vaccine trigger rare allergic reactions. *Science Magazine. December, 21*

81 Chemaitelly, H., Yassine, H.M., Benslimane, F.M. et al. (2021). mRNA-1273 COVID-19 vaccine effectiveness against the B.1.1.7 and B.1.351 variants and severe COVID-19 disease in Qatar. *Nat. Med.* 27 (9): 1614–1621.

82 Centers for Disease Control and Prevention (2021). Moderna COVID-19 Vaccine standing orders for administering vaccine to persons 18 years and older. pp. 1–3.

83 Uludag, H., Ubeda, A., and Ansari, A. (2019). At the intersection of biomaterials and gene therapy: progress in non-viral delivery of nucleic acids. *Front. Bioeng. Biotechnol.* 7: 131.

84 Huang, J., Ma, W., Sun, H. et al. (2020). Self-assembled DNA nanostructures-based nanocarriers enabled functional nucleic acids delivery. *ACS Appl. Bio Mater.* 3 (5): 2779–2795.

85 Dindo, M., Conter, C., Oppici, E. et al. (2019). Molecular basis of primary hyperoxaluria: clues to innovative treatments. *Urolithiasis* 47 (1): 67–78.

86 Solano, E.C.R., Kornbrust, D.J., Beaudry, A. et al. (2014). Toxicological and pharmacokinetic properties of QPI-1007, a chemically modified synthetic siRNA targeting caspase 2 mRNA, following intravitreal injection. *Nucleic Acid Ther.* 24 (4): 258–266.

87 Moreno-Montañés, J., Bleau, A.M., and Jimenez, A.I. (2018). Tivanisiran, a novel siRNA for the treatment of dry eye disease. *Expert Opin. Invest. Drugs* 27 (4): 421–426.

88 Arora, L. and Narula, A. (2017). Gene editing and crop improvement using CRISPR-Cas9 system. *Front. Plant Sci.* 8: 1932.

89 Yip, B.H. (2020). Recent advances in CRISPR/Cas9 delivery strategies. *Biomolecules* 10 (6): 839.

90 Cyranoski, D. (2016). Chinese scientists to pioneer first human CRISPR trial. *Nature* 535 (7613): 476–477.

91 Wang, H.-X., Song, Z., Lao, Y.-H. et al. (2018). Nonviral gene editing via CRISPR/Cas9 delivery by membrane-disruptive and endosomolytic helical polypeptide. *Proc. Natl. Acad. Sci. U.S.A.* 115 (19): 4903–4908.

92 Wei, T., Cheng, Q., Min, Y.-L. et al. (2020). Systemic nanoparticle delivery of CRISPR-Cas9 ribonucleoproteins for effective tissue specific genome editing. *Nat. Commun.* 11 (1): 3232.

93 Schultheis, B., Strumberg, D., Santel, A. et al. (2014). First-in-human phase I study of the liposomal RNA interference therapeutic Atu027 in patients with advanced solid tumors. *J. Clin. Oncol.* 32 (36): 4141–4148.

94 Bae, K.H., Chung, H.J., and Park, T.G. (2011). Nanomaterials for cancer therapy and imaging. *Mol. Cells* 31 (4): 295–302.

95 Buscail, L., Bournet, B., Vernejoul, F. et al. (2015). First-in-man phase 1 clinical trial of gene therapy for advanced pancreatic cancer: safety, biodistribution, and preliminary clinical findings. *Mol. Ther.* 23 (4): 779–789.

96 Jackson, N.A.C., Kester, K.E., Casimiro, D. et al. (2020). The promise of mRNA vaccines: a biotech and industrial perspective. *npj Vaccines* 5 (1): 1–6.

97 Michel, T., Wendel, H.-P., and Krajewski, S. (2016). Next-generation therapeutics: mRNA as a novel therapeutic option for single-gene disorders. In: *Modern Tools for Genetic Engineering* (ed. M. Kormann), 20. InTech.

98 Rosenblum, H., Castano, A., Alvarez, J. et al. (2018). TTR (transthyretin) stabilizers are associated with improved survival in patients with TTR cardiac amyloidosis. *Circ. Heart Fail.* 11 (4): e004769.

99 Meng, C., Chen, Z., Li, G. et al. (2021). Nanoplatforms for mRNA therapeutics. *Adv. Ther.* 4 (1): 2000099.

10

Application of PROTAC Technology in Drug Development

Prashant S. Kharkar and Atul L. Jadhav

Institute of Chemical Technology, Department of Pharmaceutical Sciences, Nathalal Parekh Marg, Matunga, Mumbai, Maharashtra 400 019, India

10.1 Introduction

Design of ligands, e.g. inhibitors, antagonists, and targeting pathologically relevant biomacromolecules for modulating the disease process dominated the drug discovery and development for decades. Despite the availability of unprecedented number of macromolecular targets following completion of the human genome project, the speed of exploitation of the generated information was awfully slow. One of the major reasons was the druggability of the identified protein targets. Several of these targets involved protein–protein interactions and multifunctional protein complexes with clear lack of druggable binding sites [1]. Transcription factors, protein involved in transcription (DNA to RNA) process, present a promising class of macromolecules as drug targets, particularly in cancer. Historically, these crucial proteins presented several obstacles while being targeted by small molecules. Consistent efforts in this direction over the last decade or so have led to the identification of significant number of transcription factor inhibitors for various types of cancers [2]. Significant progress is being made in various therapeutic disciplines involving once-difficult-to-target proteins using small-molecule agents.

While the conventional medicinal chemist was busy looking out for druggable binding sites in individual proteins, the lateral thinkers came up with an interesting approach called PROteolysis-TArgeting Chimeras (PROTACs), in which the target protein degradation is promoted by matching it with E3 ubiquitin (Ub) ligases. The interesting feature of this approach requires on one side, a selective and specific interaction with the target protein, while on the other side, initiation of proteolytic mechanism by E3 ligase. Figure 10.1 depicts the philosophy of this new-generation approach, which holds promise to deliver small-molecule therapeutics for otherwise undruggable targets [3]. The PROTACs are small-molecular entities comprised of E3 ubiquitin ligase-binding end combined via a linker with the other end comprising of a high-affinity ligand of the protein-of-interest (POI), i.e. target protein to be degraded. The E3 ligase–E2 ubiquitin (Ub)-binding enzyme complex, when engaged

Targeted Drug Delivery, First Edition. Edited by Yogeshwar Bachhav.

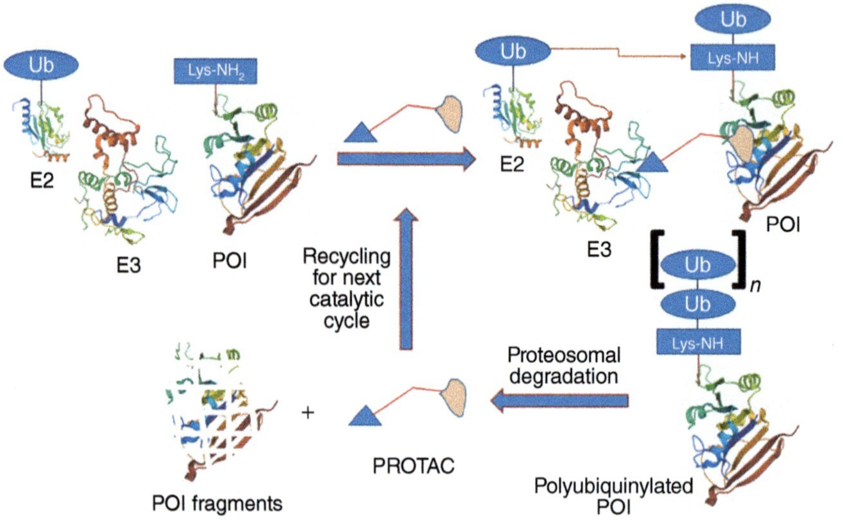

Figure 10.1 Mechanism of proteolytic degradation of protein of interest (POI) by PROTAC; E3, E3 ubiquitin ligase; E2, ubiquitin-binding enzyme; Ub, ubiquitin. The protein 3D structures are for representation only. Source: Adapted from Qi et al. [3].

by the PROTAC at the E3 ligase site (Figure 10.1), transfers Ub to the Lys-NH_2 of the POI. This further leads to polyubiquitination of the POI, which is then targeted for the proteasomal degradation by 26S proteasome. The PROTAC further participates in the next round of the catalysis. The Ub molecules dissociate from the polyubiquitinylated POI and return to the cytoplasm for recycling. The most intriguing fact of this approach is the requirement of relatively high-affinity binding site anywhere on the protein surface, which may be very different from its catalytic and/or binding site. Overall, the PROTACs use the ubiquitin-proteasomal system (UPS) to ubiquitinate and degrade the POI.

A closer look at the eukaryotic UPS reveals the precise mechanism of the ubiquitination, which involves – (i) formation of a high-energy bond between ubiquitin-activation enzyme E1-Cys-SH and the C-terminal Ub-Gly molecules in an energy (ATP)-dependent process, (ii) transfer of activated Ub from E1 to ubiquitin-binding enzyme E2, (iii) the transfer of Ub from E2, in presence of E3 ligase, to Lys-NH_2 of the POI via an isopeptide bond, followed by (iv) connection of the C-terminal of next Ub molecule with the former at the Lys48, subsequently leading to the polyubiquitinylated POI. The PROTACs, since their discovery in 2001, has revolutionized cancer chemotherapy, especially targeting of the major targets in cancer to tackle multidrug resistance problems, such as androgen receptor (AR), estrogen receptor (ER), Bruton's tyrosine kinase (BTK), bromodomain and extraterminal domain (BET), and several others [4]. A large number of PROTACs are under development, targeting various POIs and E3 ligases. While Chart 10.1 shows the representative PROTAC structures under various developmental stages, Table 10.1 lists their code names, targets (POI), and potency data wherever available. The present chapter outlines the PROTAC chemistry, the rationale behind their

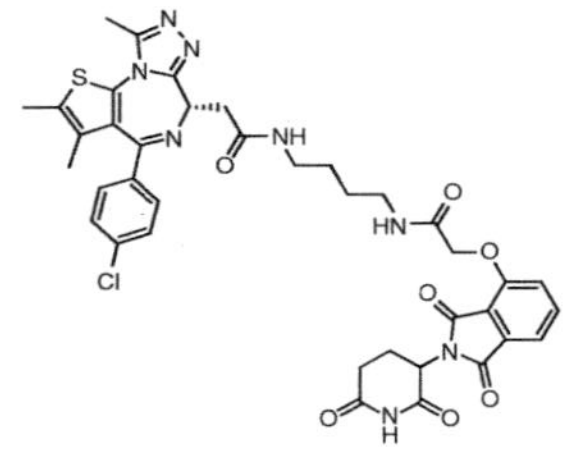

1

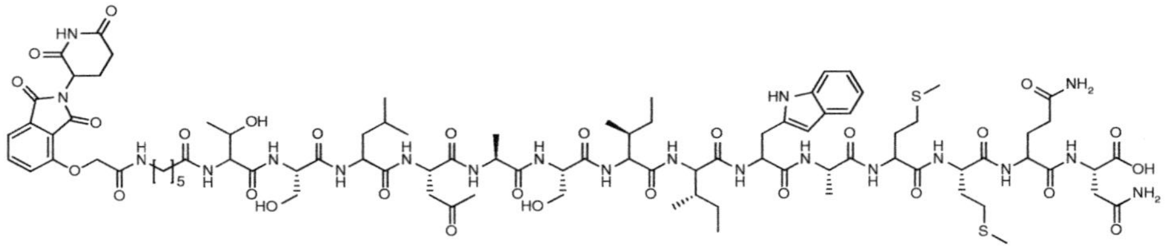

2

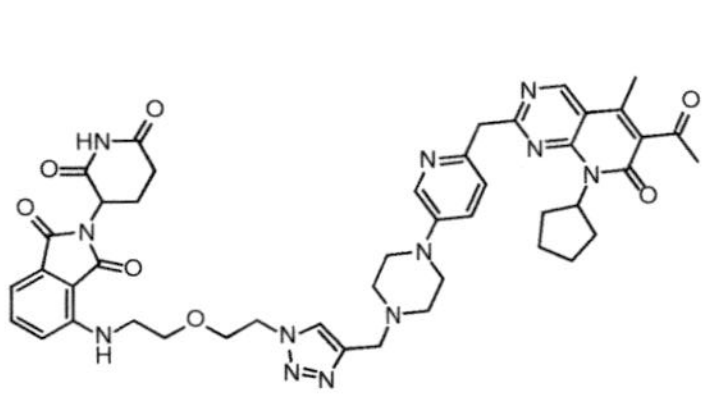

3

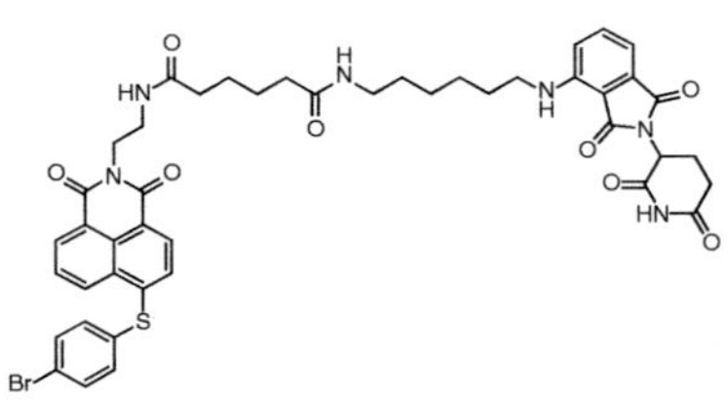

4

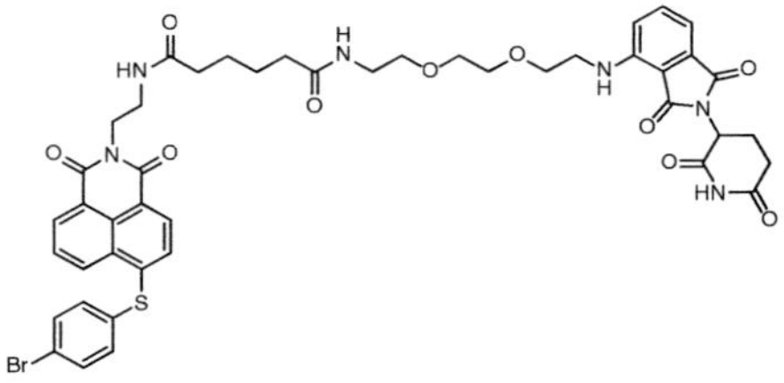

5

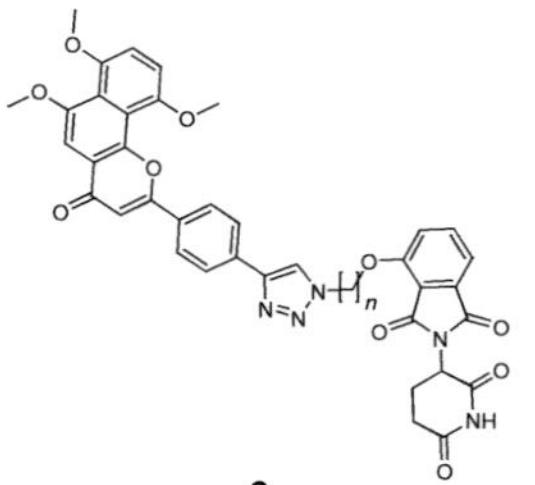

6

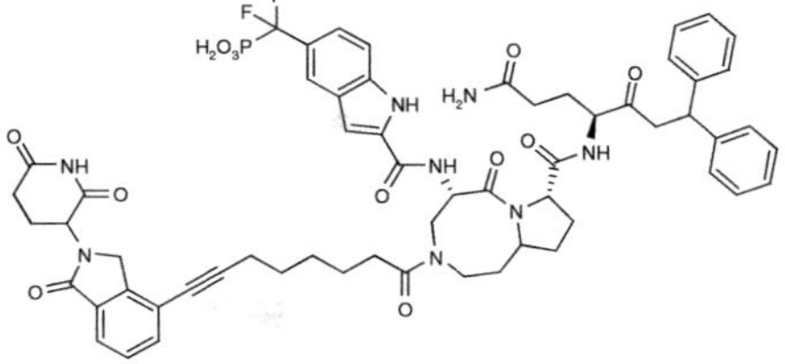

7

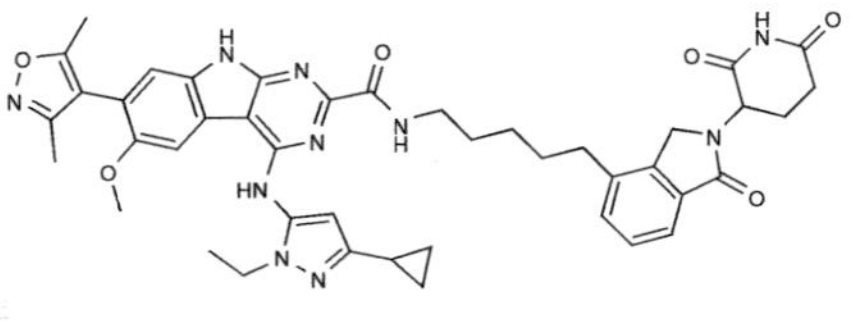

8

Chart 10.1 Molecular structures of representative PROTACs listed in Table 10.1.

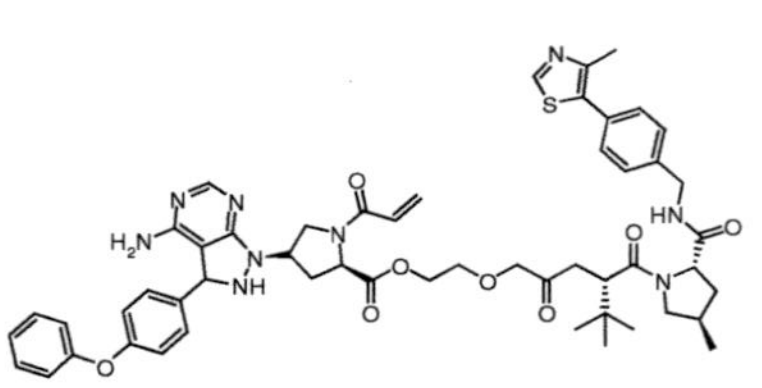

9

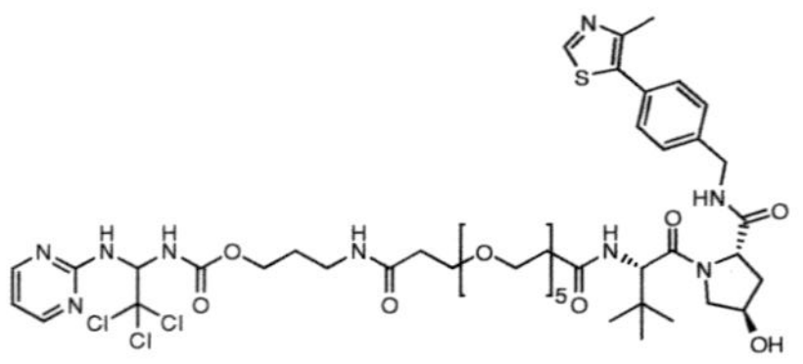

10

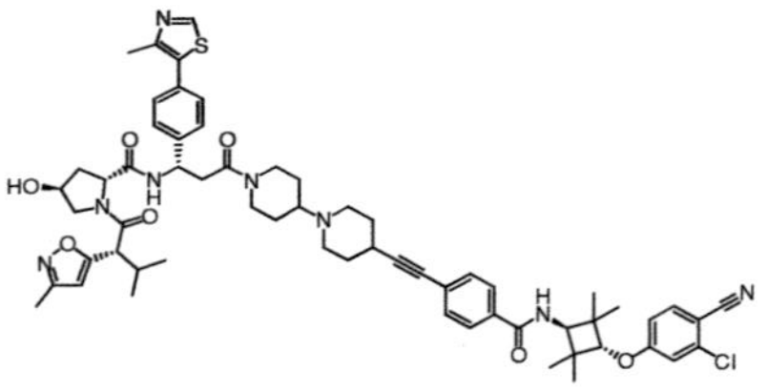

11

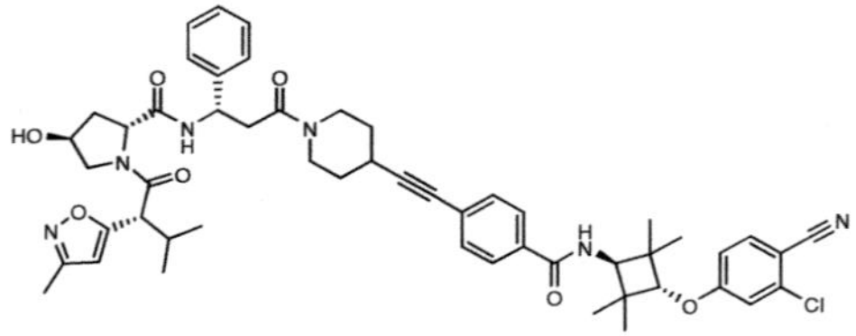

12

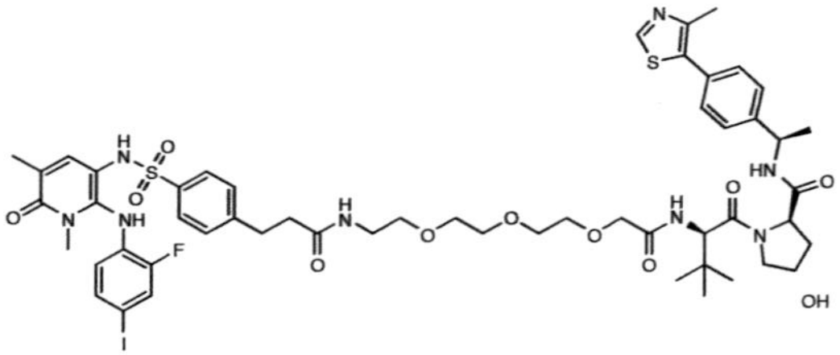

14

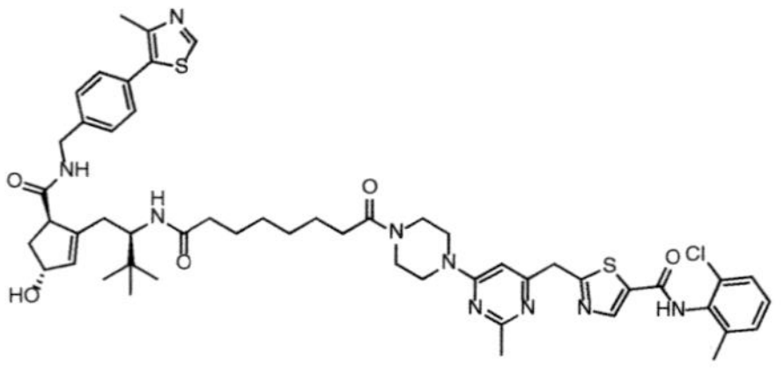

15

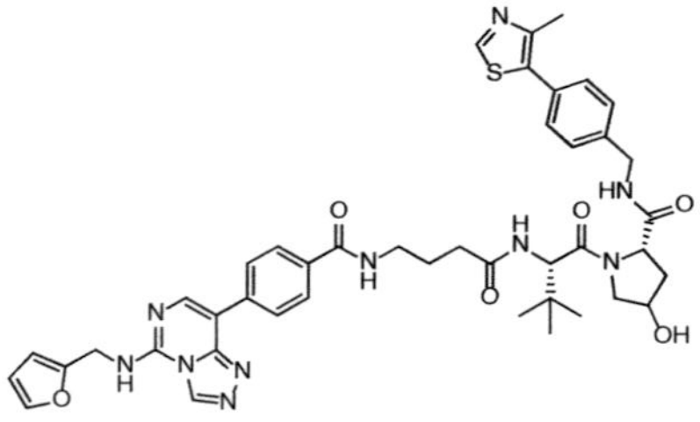

16

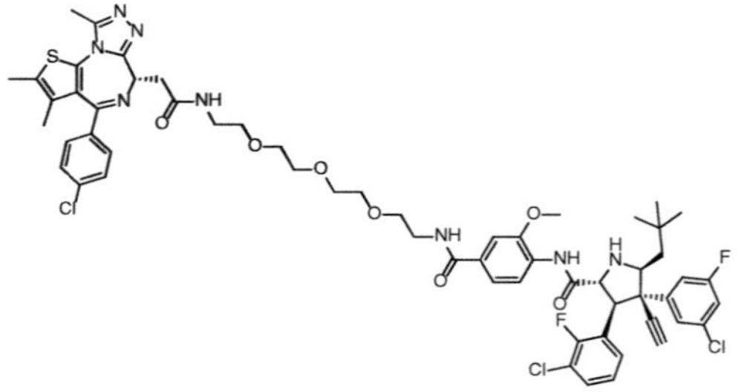

17

Chart 10.1 (*Continued*)

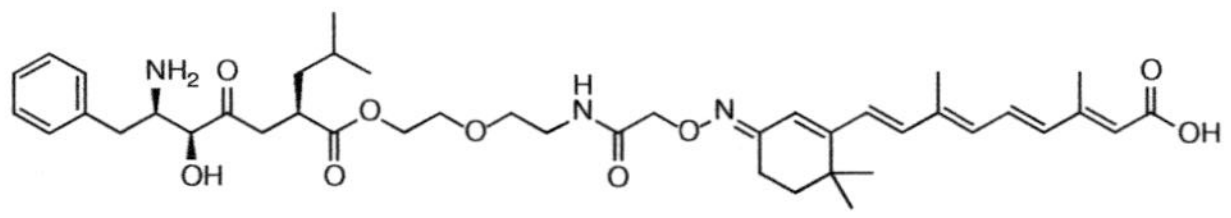

18

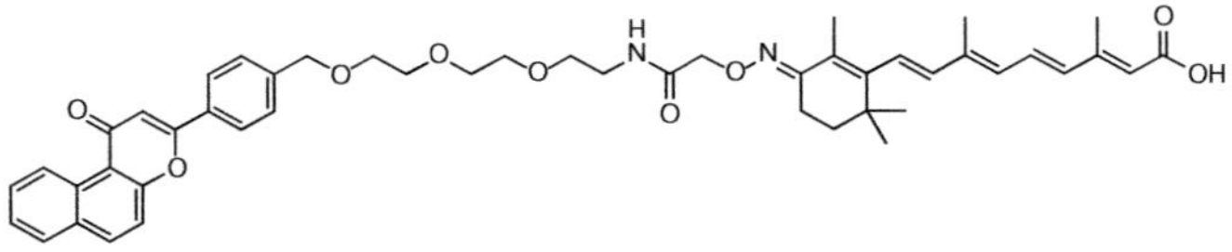

19

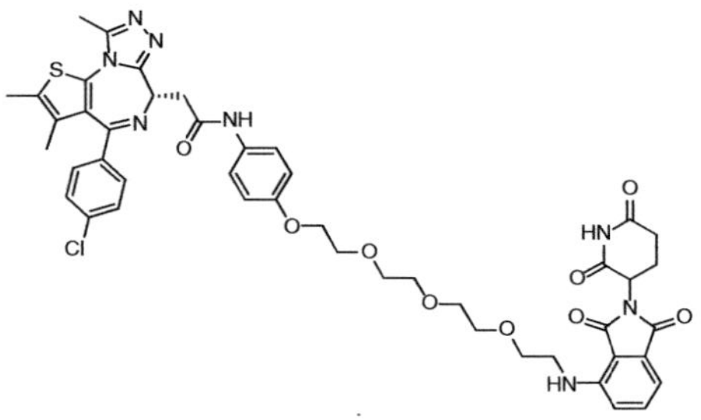

20

Chart 10.1 (*Continued*)

Table 10.1 Representative PROTACs reported in the literature.

Code	POI	E3 ligase	IC_{50} (nM)[b]/ DC_{50} (nM)[c]
dBET1 (**1**, Chart 10.1)	BRD	CRBN	20[b]
DT-6 (**2**)	TGF-β1	CRBN	—
CP-10 (**3**)	CDK6	CRBN	2.1[c]
C3 (**4**)	Mcl-1	CRBN	700[c]
C5 (**5**)	Bcl-2	CRBN	3000[c]
Compounds 6A-D (**6**)	CYP1B1	CRBN	—
SD-36 (**7**)	STAT3	CRBN	60[c]
BETd-260 (**8**)	BET	CRBN	—
PROTAC7 (**9**)	BTK BLK	VHL VHL	136[c] 220[c]
CP5V (**10**)	Cdc20	VHL	2600[b] 1600[c]
ARD-61 (**11**)	AR	VHL	2[b] 7.2[c]
ARD-266 (**12**)	AR	VHL	2[b] 0.5[c]
Compound I-6 (**13**)[a]	ERα	VHL	9700[b]
Compound 3 (**14**)	MEK	VHL	—
SIAIS178 (**15**)	BCR-ABL	VHL	24[b] 8.5[c]
UNC6852 (**16**)	PRC2	VHL	247[b]
A1874 (**17**)	BRD4	MDM2	32[c]
Compounds 4 (**18**)	CRABPs	cIAP1	—
β-NF-ATRA (**19**)	CRABPs	AhR	—
ARV-825 (**20**)	BRD4	CRBN	—

Abbreviations: BRD, bromodomain; TGF-β1, transforming growth factor beta 1; Mcl-1, myeloid cell leukemia-1; Bcl-2, B-cell lymphoma 2; CYP1B1, cytochrome P450 family 1 subfamily B member 1; BLK, B lymphoid kinase; Cdc20, cell division cycle 20; MEK, mitogen-activated protein kinase kinase; PRC2, polycomb repressive complex 2; CRABP, cellular retinoic acid-binding protein 1.

a) Peptidic PROTAC – structure not shown.

b) IC_{50} represents the concentration of a drug that is required for 50% inhibition.

c) DC_{50} represents the compound concentration at which the target is degraded by 50%.

Source: Adapted from Qi et al. [3].

design, potency, efficacy, pitfalls, and other relevant therapeutic details along with their future prospects.

10.2 Design of PROTACS: A Brief Overview

Human genomes encode for more than 600 E3 ligases; only a handful of these, such as $SCF^{\beta\text{-TrCP}}$ [5], Von Hippel–Lindau (VHL) [6], murine double minute 2 (MDM2) [7], inhibitor of apoptosis proteins (IAPs) [8], and cereblon (CRBN) [9], have been exploited for PROTAC design and targeting so far. The current pace of

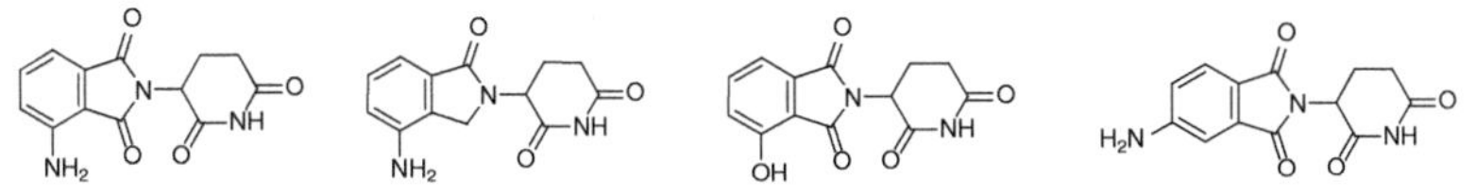

21 **22** **23** **24**

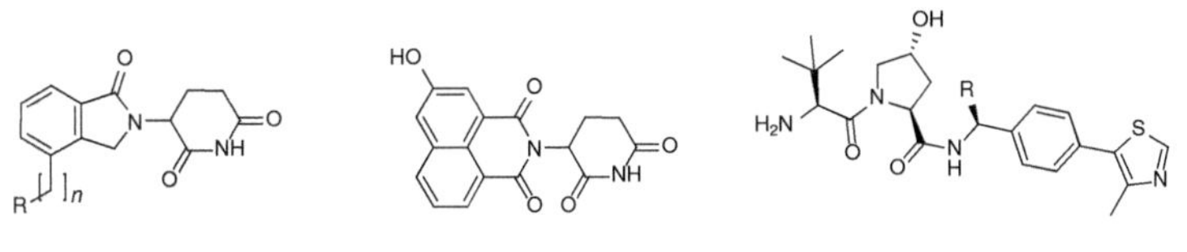

25 **26** **27**

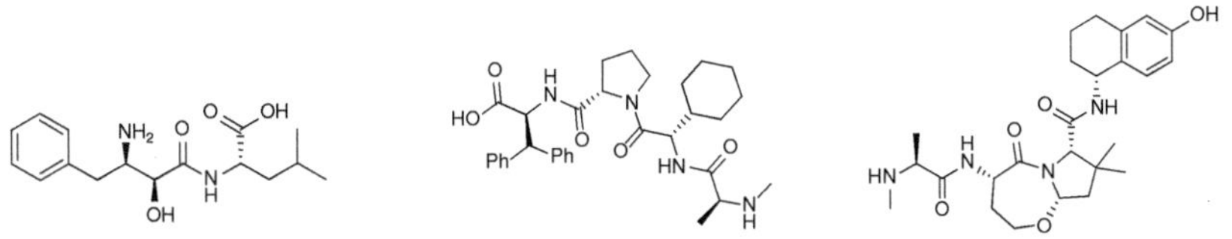

28 **29** **30**

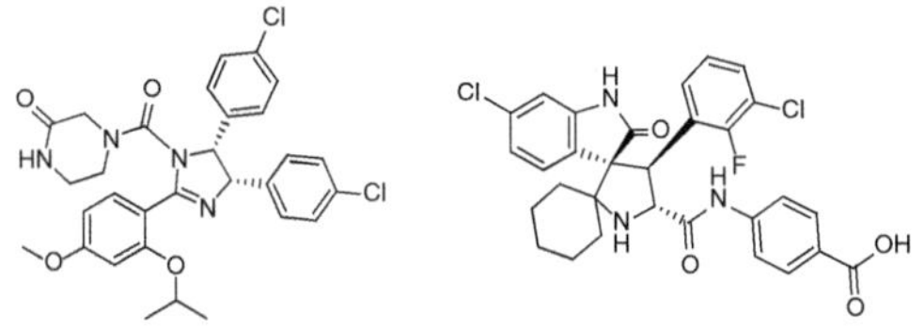

31 **32**

Chart 10.2 Representative chemotypes present in PROTACs for targeting E3 ligases.

research in this area will lead to discovery of many more E3 ligases for targeting by PROTACs. Bricelj et al. have extensively reviewed E3 ligase ligands in PROTACs under development with in-depth discussion on the synthetic strategies and linker attachment points [10]. On similar lines, Ishida and Ciulli outlined the discovery of newer E3 ligase ligands for PROTACs targeting varied POIs [11]. Several chemotypes are in use for targeting E3 ligases. Chart 10.2 lists representative examples. These include (i) CRBN ligands – pomalidomide (**21**, Chart 10.2), lenalidomide (**22**), 4-hydroxythalidomide (**23**), 5-aminothalidomide (**24**), alkyl-connected lenalidomide derivatives (**25**), tricyclic imides (**26**); (ii) VHL ligands – hydroxyproline-derived peptidic/peptidomimetic ligands (**27**); (iii) IAP ligands – bestatin (**28**), MV-1 derivative (**29**), SBP-0636457 derivative (**30**), and (iv) MDM2 ligands – Nutlin-3a derivatives (**31**), MI-1061 derivatives (**32**). Various handles on these chemotypes were tried for attaching the linker and the POI-targeting moiety.

Literature is full of reports on PROTACs targeting various POIs. Depending on the chemistry of the POI ligand involved, appropriate synthetic strategies have been employed for their synthesis. Few recent POIs targets using PROTAC technology include – (i) focal-adhesion kinase (FAK) [12]; (ii) interleukin-1 receptor-associated kinase 4 (IRAK4) [13]; (iii) IRAK1 [14]; (iv) ER [15]; (v) anaplastic lymphoma kinase (ALK) [16]; (vi) hematopoietic prostaglandin D synthase (H-PGDS) [17]; (vii) Kirsten rat sarcoma virus (KRAS) [18]; (viii) Janus kinase (JAK) [19]; (ix) BTK [20], and several others. SciFinder [21] search (conducted on 29 August 2021) using keyword "PROTAC" yielded 875 hits (journal articles: 729; patents 73; and others: 73).

Given the relatively bigger size of the PROTACs (two ligands connected with a linker), it is imperative that their physicochemical properties can be modulated to improve the permeability and *t* residence time *in vivo*. Several authors systematically explored the connection between the structural features and the physicochemical properties of the PROTACs. Scott et al. employed five AR ligands, four E3 ligase ligands, and nine linkers and investigated their permeability using parallel artificial membrane permeability assay (PAMPA), Caco-2 A2B along with human serum albumin binding, and $logD_{7.4}$ [22]. The lower permeability values could be a limiting factor for PROTAC's cellular uptake. In addition, the stability of PROTACs in cells over time and kinetics of protein degradation and clearance could be critical for overall efficacy of the therapeutic modality. A similar study by Klein et al. acknowledged the importance of membrane permeability in the overall action of PROTACs as protein degraders [23]. Ermondi et al., in their viewpoint article, reiterated the importance of the rational control on molecular properties for fully exploiting the enormous potential offered by orally active PROTACs [24]. In yet another interesting article, Atilaw et al. studied the solution conformations of the PROTACs to understand their cell permeability despite the higher molecular weight and permeability [25]. Gao et al. took a 360° view of the PROTAC technology highlighting the opportunities and challenges [26]. Knowledge gained from such studies is crucial for the design and development of better PROTACs for their rapid translation to the clinic.

10.3 Therapeutic Applications of PROTACs

Degradation of misfolded and/or pathologically relevant proteins by various approaches has proved to be a viable strategy to halt the disease process. These approaches include – (i) hydrophobic tagging (HyT), wherein the POI instability is induced by a hydrophobic fragment attached to its ligand, thereby mimicking the protein misfolding; (ii) HaloTag®, which is a modified bacterial dehalogenase enzyme binding covalently with a hexyl chloride label bound to a POI ligand. Once the POI is fused to the HaloTag protein, it can be marked for degradation by the UPS; and (iii) PROTAC technology [27]. The twenty-first century, so far, has witnessed the discovery of novel therapeutic modalities in the form of PROTACs. Over the last decade, these agents have revolutionized the drug discovery arena using protein degradation as the mechanism of action. The catalytic and non-catalytic functions of the POIs are deleted completely. Moreover, the dissociation of the PROTACs from the POIs, following their polyubiquitination, is a cherry on the cake, since the catalytic nature of this process necessitates the PROTAC action at very low doses, as long as their cellular permeability and stability are taken care of by using appropriate structural changes. In comparison to the drug target inhibition by small-molecular entities, and the ensuing problems, such as multidrug resistance, due to mutations in the drug targets, PROTACs are less prone to such issues including the increased target expression. In fact, these agents can be targeted effectively against targets with multiple mutations leading to failure of highly selective and potent target inhibitor drugs, raising hope for patients with no viable therapeutic options, e.g. triple-negative breast cancer (TNBC) [28]. The application of PROTACs is widespread, including cancer, inflammatory diseases, such as rheumatoid arthritis (RA), neurodegenerative disorders, autoimmune diseases, idiopathic pulmonary fibrosis, and many others.

10.3.1 Cancer

The most exploited therapeutic area for PROTAC development is cancer. Approximately 82% of all the reported PROTACs are in cancer areas. In this approach, a cancer-causing protein is recruited to an E3 ligase for subsequent degradation. Additionally, these wondrous molecules can be used as genetic tools of chemical origin for "knocking down" the proteins for studying their functions. Recent reports summarize the therapeutic utility of PROTACs in cancer, where they are targeted against nuclear hormone receptors, kinases, epigenetic proteins, poly(ADP-ribose) polymerases (PARPs), signal transducer and activator of transcription 3 (STAT3), E3 ligases, and others [29]. Many PROTACs are under various phases, such as "Biological Testing," "Preclinical," "Phase 1," and "Phase 2"; the most notable of these are ARV-110 (**33**, Chart 10.3; Phase 1/2 – NCT03888612; target: AR), and ARV-471 (**34**; Phase 1/2 – NCT04072952; target: ER), discovered by Arvinas LLC (www.arvinas.com), for the treatment of metastatic castration-resistant prostate cancer (mCRPC) and locally advanced or metastatic ER-positive/HER2-negative breast cancer, respectively. ARV-766 (**35**) is in Phase 1 for mCRPC and possesses different profile than ARV-110. A complete list of PROTACs in various developmental phases is given in Table 10.2.

The PROTACs targeted at BET, MZ1 (**36**), and ARV-825 were evaluated in TNBC, ovarian cancer, and BET inhibitor JQ1-resistant cell lines [28]. The results clearly demonstrated the downregulation of BRD4, a BET protein, leading to antiproliferative activity in resistant and sensitive cell lines. Additional *in vitro* and *in vivo* efficacy studies halted tumor growth in JQ1-resistant xenograft model owing to reduced BRD4 expression levels. Overall, the results provided a therapeutic option for the otherwise untreatable TNBC. Cimas et al. designed a delivery system containing MZ1 encapsulated into polymeric nanoparticles (NPs) conjugated with trastuzumab for directing the delivery of MZ1 to breast cancer cells with higher HER2 expression [30]. The NPs exhibited superior efficacy and therapeutic properties of the PROTACs. SD-36 (**37**), a STAT3 degrader, showed potent and selective activity *in vitro* in acute myeloid leukemia and anaplastic large-cell lymphoma cell lines [31]. The complete and long-lasting tumor regression in several xenograft mouse models confirmed the potential of STAT3-targeted PROTACs in several cancers.

Kinase-targeted PROTACs are one of the favorite strategies for tackling present issues with multitude of kinase inhibitors, namely target mutations and the

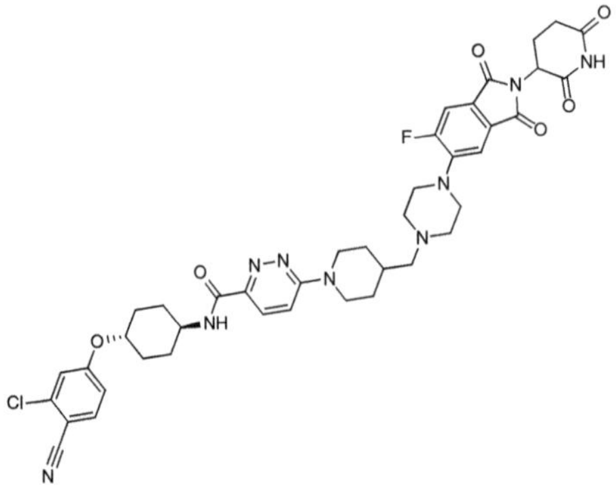

33

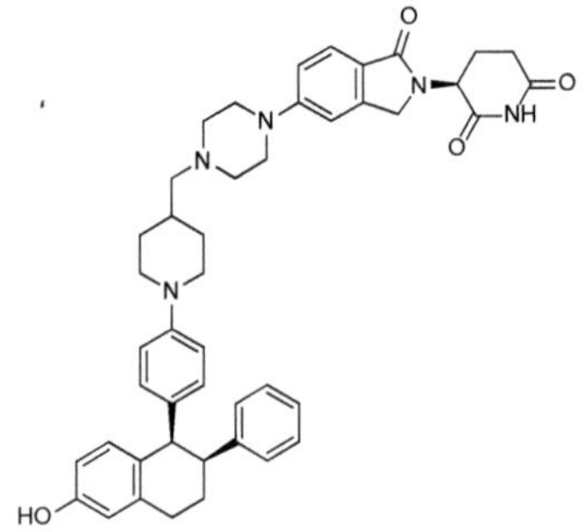

34

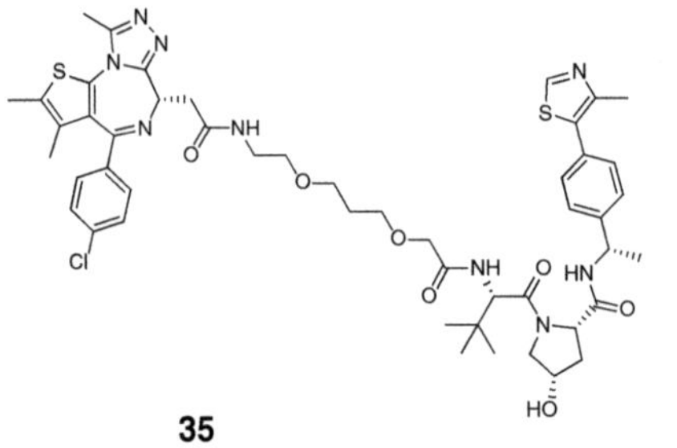

35

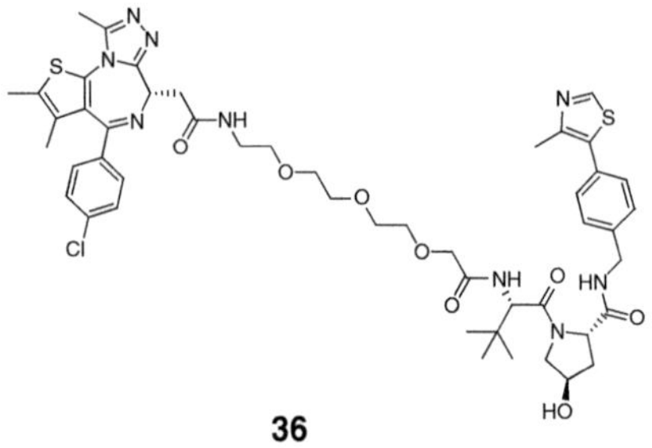

36

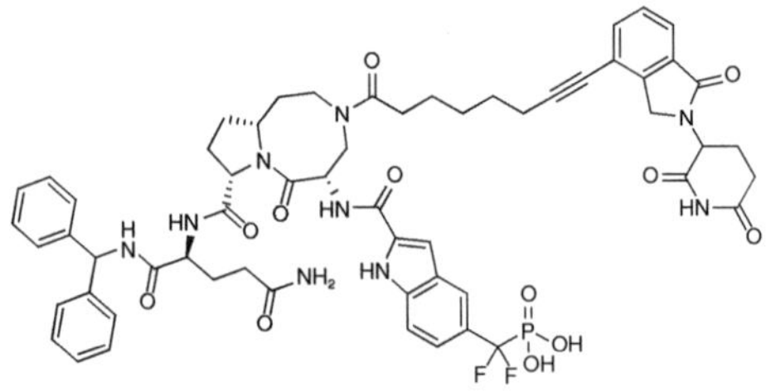

37

Chart 10.3 PROTACs under clinical development or "biological testing" phase for cancer indication.

resulting drug resistance. These molecules have shown sensitivity to drug-resistant kinases. The catalytic role played by PROTACs, which necessitates only smaller doses, and the lesser influence on aberrant kinase expression, add to the advantages offered by these new-age therapeutic agents. Additionally, kinase selectivity is another problem with the clinically used kinase inhibitors. The incorporation of a relatively selective kinase inhibitor, e.g. ibrutinib, into a PROTAC has been shown to degrade only BTK, sparing the other homologs, such as interleukin-2-inducible T-cell kinase (ITK) and TEC [32]. One of the widely explored kinase targets for PROTAC development is ALK. In past couple of years, TL13-12 (**38**, Chart 10.4, ceritinib with pomalidomide) and TL13-112 (**39**, TAE-684 with pomalidomide) targeted

Table 10.2 PROTACs in or near clinical developmental phases.

Candidate	Originator/ sponsor	Target	Indication	Clinical development phase
ARV-110	Arvinas	AR	mCRPC	Phase 2
ARV-471	Arvinas	ER	Breast cancer	Phase 2
ARV-766	Arvinas	AR	mCRPC	Phase 1
AR-LDD	Bristol Myers Squibb	AR	PC	Phase 1
DT2216	Dialectic	BCL-XL	Liquid and solid cancers	Phase 1
KT-474	Kymera/Sanofi	IRAK4	Autoimmune including AD, HS, and RA	Phase 1
KT-413	Kymera	IRAK4[a)]	MYD88-mutant DLBCL	Phase 1
KT-333	Kymera	STAT3	Liquid and solid cancers	Phase 1
NX-2127	Nurix	BTK[a)]	B-cell malignancies	Phase 1
NX-5948	Nurix	BTK	B-cell malignancies and autoimmune	Phase1
CG001419	Cullgen	TRK	Cancer and other diseases	IND
CFT8634	C4 Therapeutics	BRD9	Synovial sarcoma	IND
FHD-609	Foghorn	BRD9	Synovial sarcoma	IND

Abbreviations: mCRPC, metastatic castration resistant prostate cancer; PC, prostate cancer; AD, atopic dermatitis; HS, hidradenitis suppurativa; DLBCL, diffuse large B-cell lymphoma; RA, rheumatoid arthritis.
a) With immunomodulatory activity.
Source: Adapted from Wang et al. [27].

at ALK appeared in literature [33]. Few other ALK PROTACs MS4077, MS4088, and TD-004 (structures not shown) demonstrated excellent ALK degradation and cell growth inhibition, as well as *in vivo* efficacy. These studies provided conclusive evidence that ALK was a promising target for exploitation.

DAS-VHL and DAS-CRBN (**40**), derived from dasatinib (BCR-ABL tyrosine kinase inhibitor), caused degradation of c-ABL (65% at 1 μM) and both c-ABL (>85% at 1 μM) and BCR-ABL (>60% at 1 μM), respectively [33]. The cellular EC_{50} was 4.4 nM (K562 cell line). DAS-IAP was yet another BCR-ABL degrader exhibiting sustained antiproliferative effects. On similar line, wild-type (degradation concentration, DC_{50} = 9.2 nM) and ibrutinib-resistant C481S BTK (DC_{50} = 30 nM) were targeted with P13I (**41**) [34]. The PROTAC could effectively downregulate self-phosphorylation of C481S BTK, while ibrutinib could not. The improved version L18I (structure not shown) showed promising *in vitro* activity and *in vivo* efficacy in mouse xenograft models inoculated with HBL-1 cells harboring C481S mutation.

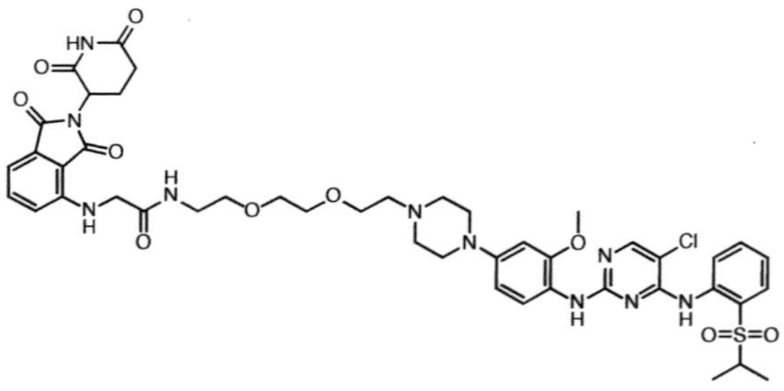

38 (ALK)

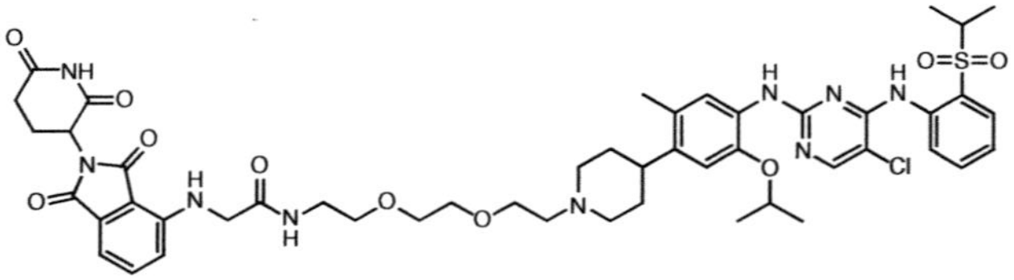

39 (ALK)

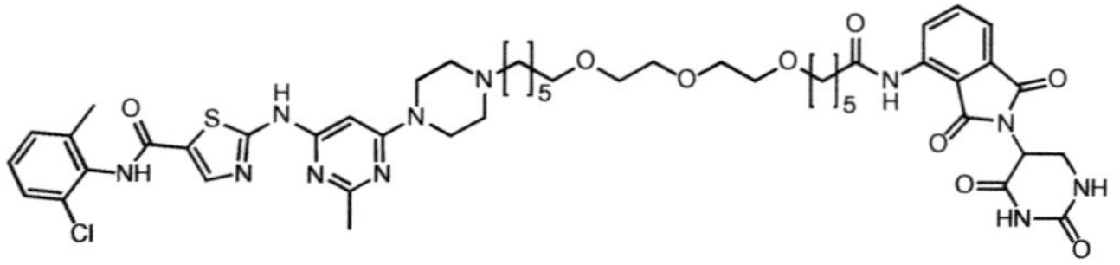

40 (Bcr-Abl)

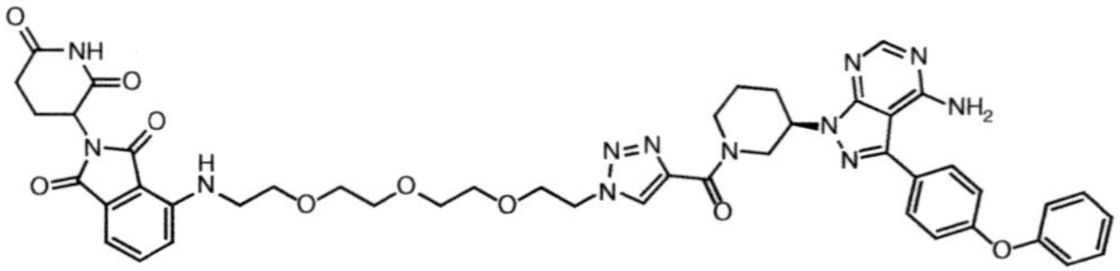

41 (BTK)

Chart 10.4 Kinase-targeting PROTACs. The kinase target is given in the parentheses.

Multikinase degraders containing highly promiscuous kinase inhibitor conjugated with CRBN ligand were reported; they could degrade many kinases in addition to BTK. Subsequently, many groups reported BTK degraders containing CRBN ligand pomalidomide (cross-refer [32]). Such efforts were then replicated for cell-cycle regulating cyclin-dependent kinases CDK4/6 [35]. This was followed by the dual and selective CDK4 and CDK6 degraders. Even though highly potent degraders were developed, the challenges related to PROTACs applied to cyclin-dependent kinases (CDK)-resistant cells still remain. Several other kinases, such as CDK8, CDK9, casein kinase 2, epidermal growth factor receptor (EGFR), HER2, FAK, FMS-like tyrosine kinase 3 (FLT-3), extracellular signal-regulated kinase 1 (ERK1) and extracellular signal-regulated kinase 2 (ERK2), p38 mitogen-activated protein kinases (MAPKs), phosphoinositide 3-kinases (P13Ks), serum/glucocorticoid-inducible protein kinase (SGK), TANK-binding kinase 1 (TBK1), and others, have been successfully targeted by PROTACs [32]. Overall, the availability of selective and preferentially selective kinase inhibitors for various cancer types provides greater opportunities for PROTAC development, obviously for addressing the kinase-inhibitor resistance problem.

The AR- and ER-targeting PROTACs are abundantly reported in the literature. ARCC-4 (**42**, Chart 10.5; target: AR; $DC_{50} = 5\,nM$) exhibited highly effective antiproliferative activity on prostate tumor cells and degraded clinically relevant AR-mutant F876L in LNCaP cells (LNCaP/F876L) along with other point-mutated versions [36]. This was very encouraging in light of the fact that the standard-care therapy consisting of enzalutamide failed completely. In addition to previously described clinical candidate ARV-471, a potent ER degrader ERD-308 (**43**; target: ER; $DC_{50} = 0.43\,nM$ in T47D ER+ cell line) was reported to degrade ER completely and inhibited cell growth completely. These hormone receptor-targeting PROTACs are important for targeting mutant receptors, thereby providing therapeutic alternatives when the conventional drugs acting on these receptors fail [37].

Eukaryotic translation initiation factor 4E (eIF4E), a protein crucial for cell proliferation, differentiation, and metastasis, has been explored as a potential target for developing PROTACs in several cancers. Its overexpression in many malignant cell lines and primary tumors derived from animals and humans makes it an attractive target. Degradation of eIF4E by PROTACs (GMP/guanosine diphosphate (GDP) derivatives conjugated to lenalidomide and VHL ligand) was attempted by Kaur et al. [38]. The developed agents, particularly the GDP derivatives (exemplified by **44**), were moderately potent in degrading eIF4E, while the GMP derivatives were inactive. Cellular permeability was the major constraint for the demonstration of intracellular activity. Despite the limited success, the approach has great potential in further development and optimization of the eIF4E degraders.

Lee et al. designed PROTACs targeted at ligand-activated transcription factor aryl hydrocarbon receptor (AhR) that acts as a ligand-dependent cullin-based E3 ligase [39]. The authors developed their designs starting from natural product apigenin, which is known to interact with AhR. The conjugation of apigenin with a pentapeptide VHL ligand along with other structural modifications led to Api-Protac-II, which degraded AhR effectively. In a similar report, Ohoka et al.

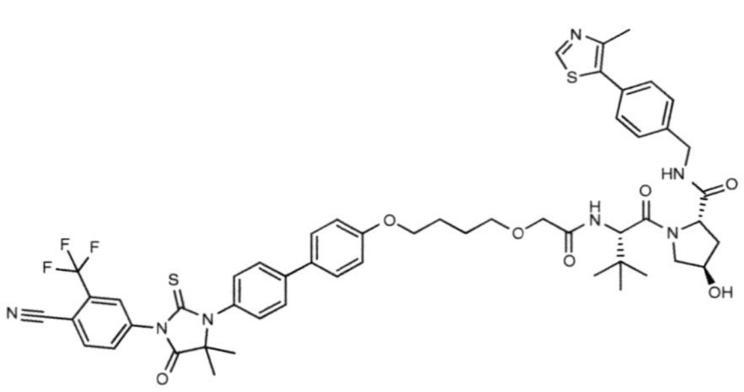

42 (AR)

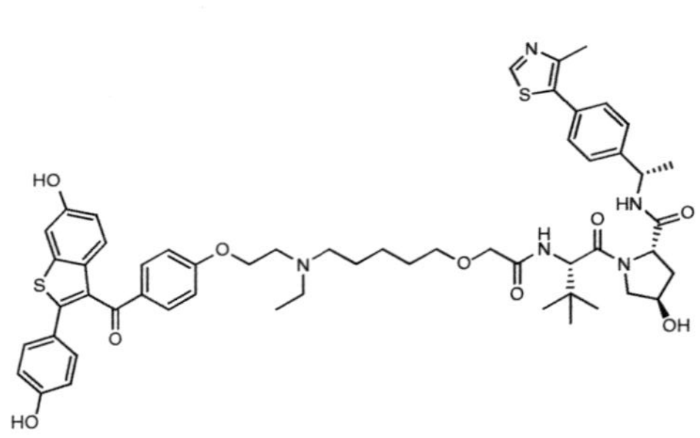

43 (ER)

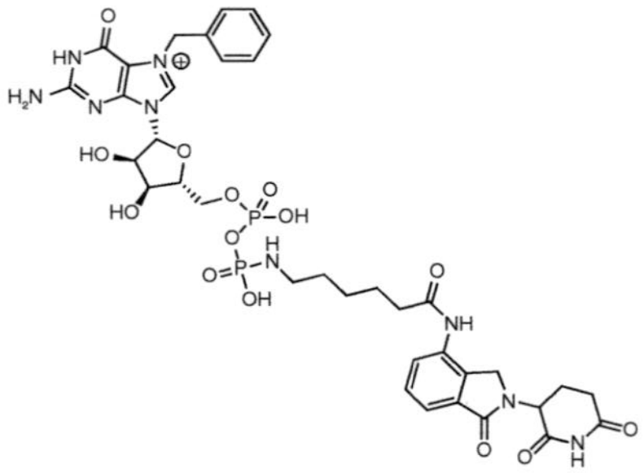

44 (eIF4E)

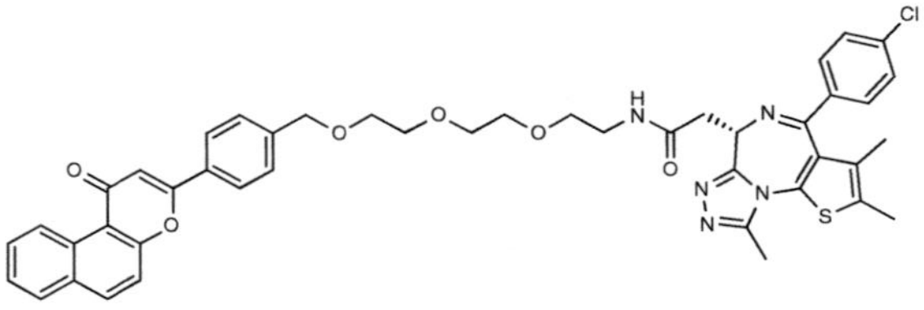

45 (AhR)

Chart 10.5 Miscellaneous anticancer PROTACs. The macromolecular target is mentioned in the parentheses.

developed small-molecule chimeras that recruited AhR E3 ligase complex [40]. β-NF-JQ1 (**45**) was directed against bromodomain (BRD)-containing proteins with β-NF serving as the AhR ligand, resulting in the degradation of the target proteins. The study opened newer avenues for targeting POIs via AhR E3 ligase.

The design and development of PROTACs, aimed at cancer targets is a happening field. Several groups are working together to address the biochemical evaluation of these agents in relevant assays for drug-resistant cell lines. These assays are critical for validating the hypothesis pertaining to their therapeutic applications. The availability of numerous POI ligands, as well as the slowly-growing number of E3 ligases for recruiting the POIs to the UPS, yields countless possibilities for novel chemical matter. The scientific community is hopeful that the number of clinical candidates for various types of cancers based on PROTAC philosophy will increase in near future and bring back the euphoric state once experienced by the industry when several kinase inhibitors made it to the clinic.

10.3.2 Neurodegenerative Disorders

PROTACs have found wide applications in neurodegenerative disorders due to the presence of aberrant proteins in the disease pathology, e.g. mutant huntingtin (mHTT) protein in Huntington's disease and tau protein in Alzheimer's disease (AD). One of the earlier tau degraders was a peptide consisting of tau-recognition motif, linker, E3 ligase-binding motif, and cell-penetrating motif [41]. Systematic attempts to optimize the motifs led to the identification of TH006, a 32 amino acid containing cell-penetrable peptide, which induces tau degradation effectively. It reduced the neurotoxicity of amyloid Aβ in AD transgenic mouse model. Lu et al. used Kelch-like ECH-associated protein 1 (Keap 1) E3 ligase in peptidic PROTAC for targeting tau [42]. The design exhibited submicromolar K_d values for Keap 1 and tau proteins. The cellular tau protein was degraded in dose- and time-dependent fashion. Interestingly, the study added Keap 1 E3 ligase in the kitty of handful, widely used members. The major breakthrough came in 2019 when first small-molecule tau degrader was reported in a patent application [43]. Typical design included pyridoindole moiety as tau-binding motif conjugated to CRBN or VHL-binders via polyethylene glycol (PEG)-based linker (**46**, Chart 10.6). The molecules exhibited favorable pharmacokinetic profile in mice. QC-01-75 (**46** with $n = 1$) demonstrated tau degradation ability in frontotemporal dementia neuronal cell model. The improved *in vivo* efficacy and pharmacokinetic profile were indeed the good news.

On similar lines, PROTACs for mHTT were shown to induce its degradation in fibroblasts from HD patients. 6-Methyl-2-phenylbenzo[d]thiazole was the POI-binding moiety, and the dipeptidyl bestatin analog of cellular inhibitor of apoptosis protein 1 (cIAP1) served as the E3 ligase-binding motif (**47**) [44]. The degrader was effective in reducing the levels of mHTT at 10 μM concentration with no apparent cytotoxicity. Based on the results, further structural modifications were done in the PROTAC structures, including the E3 ligase binding motifs, which yielded potent mHTT degraders with dose- and time-dependent degradation patterns. α-Synuclein protein in Lewy bodies, which is the hallmark of Parkinson's disease (PD),

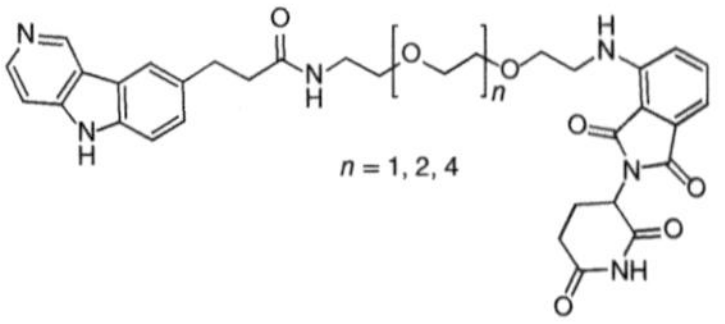

46 (tau)

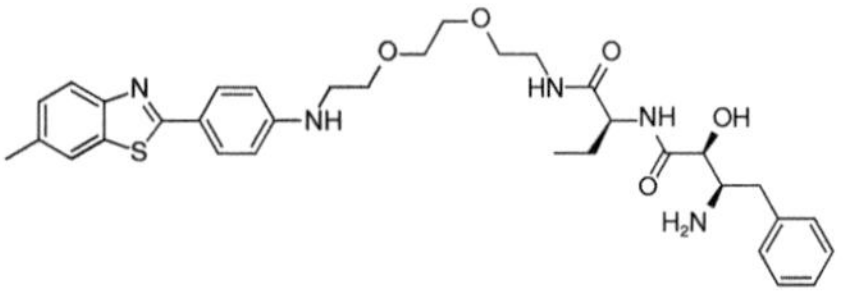

47 (mHTT)

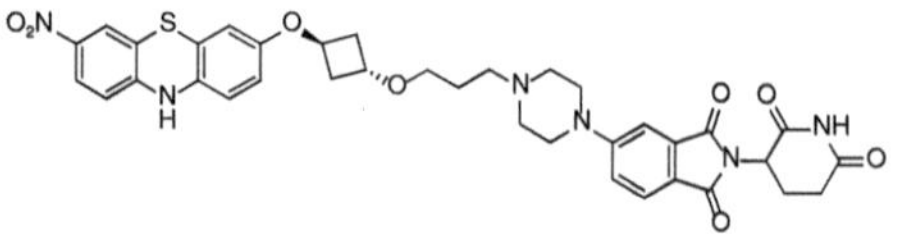

48 (α-synuclein)

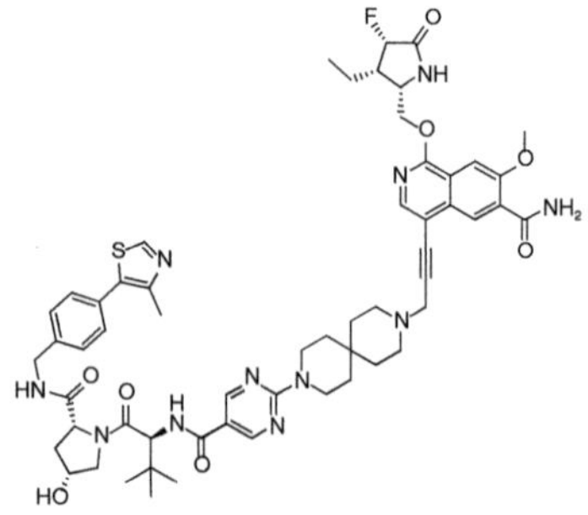

49 (IRAK4)

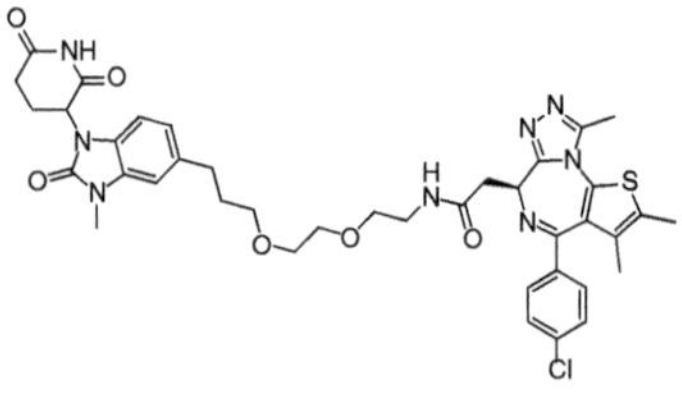

50 (IRAK4)

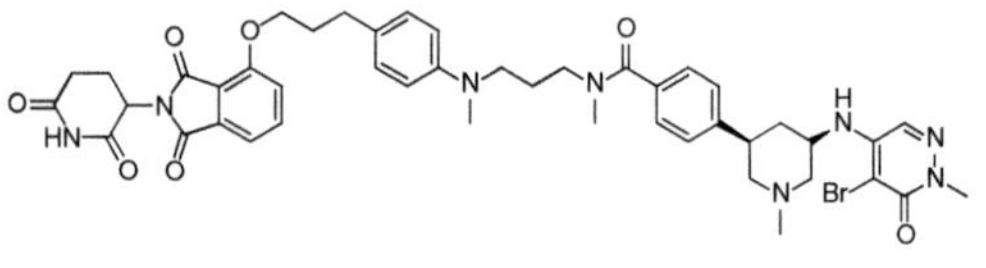

51 (PCAF/GCN5)

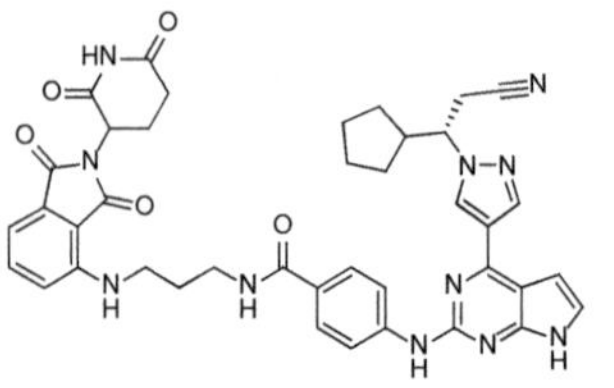

52 (JAK2)

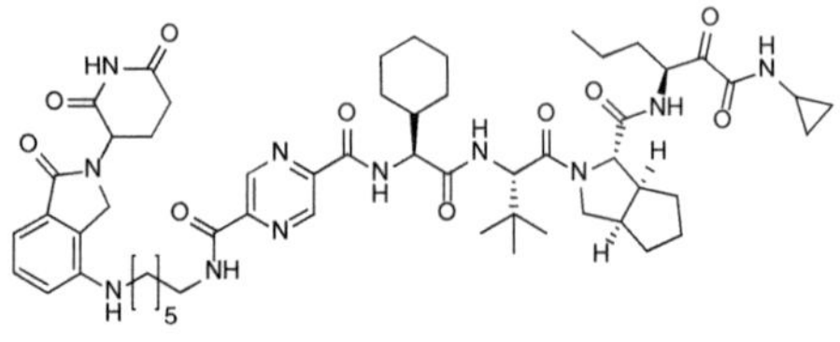

53 (HCV NS3/4A)

Chart 10.6 PROTACs for neurodegenerative diseases (**46–48**), immunological diseases (**49–52**), and viral infections (**53**). The macromolecular targets are listed in parentheses.

was targeted by the peptidic and small-molecule degraders. Initial candidates were peptidic in nature. Recent reports on small-molecule agents included 2-(4-*N*-methyl-phenyl)benzo-thiazole, 1-benzyl-3-(3-(4-nitrophenyl)allylidene)indolin-2-one, and 3-nitro-10*H*-phe-nothiazine substructures as α-synuclein-binding motifs, and CRBN or VHL-ligands as E3 ligase-binding moieties. Cyclic amine or PEG served as linkers; structure **48** exemplifies the design [45]. The DC_{50} values in HEK293 T-Rex expressing A53T α-syn were less than 1 μM. These molecules contained structural features appropriate for penetration through blood–brain barrier, a requirement for their *in vivo* efficacy.

Transactive response DNA-binding protein (TDP-43) is a culprit in a progressive neurodegenerative disorder, amyotrophic lateral sclerosis (ALS). The ubiquitinated and hyperphosphorylated cytosolic aggregates of TDP-43 is a distinctive feature in ALS patients. Although HyT has been explored for degradation of TDP-43 [46], there are no reports on PROTACs targeting this already ubiquitinylated protein. It is just a matter of time that such molecules appear in the literature. Overall, neurodegenerative diseases are an attractive area for PROTAC development for obvious reasons. The newer agents capable of halting, if not reversing, the disease progression will become a reality soon.

10.3.3 Immunological Diseases

Several key proteins play an important role in immune responses of both innate and acquired immunity. Lack of or overactivation of these proteins can lead to either increased susceptibility to pathogens or autoimmune diseases, respectively. The IRAK4, a member of the serine/threonine kinase family, IRAK, is crucial for innate immunity. It is a component of an intracellular multiprotein complex, termed myddosome, that mediates signaling through toll-like receptors (TLRs) and interleukin-1 receptors (IL-1Rs). The nonkinase or scaffolding functions of IRAK4 are more important in certain cell types; inhibition of its kinase function did not yield desired efficacy in autoimmune diseases.

A team of researchers at GlaxoSmithKline synthesized IRAK4 degrader starting from IRAK4 inhibitor, PF-06650833. Systematic structure–activity relationship (SAR) studies led to a potent, appreciably soluble, and metabolically stable PROTAC (**49**, human peripheral blood mononuclear cells (*h*PBMC) DC_{50} = 151 nM, dermal fibroblast DC_{50} = 36 nM) [47]. Surprisingly, it did not possess a different pharmacological profile compared to the IRAK4 inhibitor PF-06650833. In a related study, Zhang et al. tried to assess IRAK4 functions in activated B-cell-like diffuse large B-cell lymphoma by its kinase inhibition and degradation modalities, none of which led to growth inhibition or cell-killing effects [48]. Nonetheless, the developed PROTACs shed some light on its protein scaffolding function. KT-474 (**50**), the most advanced candidate (Phase 1) from Kymera Therapeutics (www.kymeratx.com), targets IRAK4. It is a potent, highly selective, orally bioavailable agent for the treatment of IL-1R/TLR-driven conditions and diseases, such as hidradenitis suppurativa (HS), AD, RA, and other diseases. It is the only non-oncology PROTAC in the clinic for conditions with unmet medical needs.

Novel PROTAC, **51**, for epigenetic proteins, P300/CBP-associated factor (PCAF, DC_{50} = 1.5 nM) and general control nonderepressible 5 (GCN5, DC_{50} = 3 nM), was developed [49]. Both proteins are crucial for cell proliferation, cell differentiation DNA damage repair, and metabolic regulation. The degradation of both proteins compromised the ability of dendritic cells and macrophages to respond to lipopolysaccharide and lowered the production of inflammatory cytokines.

The distinctive biological profile of the degrader proposed a novel approach for the treatment of inflammation. Recently, anti-RA JAK2-targeting PROTACs were reported in a recent patent application [19]. The degraders (exemplified by **52**) consisted of ruxolitinib and baricitinib conjugated via diamino linker to either pomalidomide, lenalidomide, or 5-amino thalidomide motifs. The compounds were evaluated for cytotoxicity in MHH-CALL-4 cell line and exhibited submicro/nanomolar potency. Several such targets at the intersection of cancer and inflammatory pathways can be targeted using PROTAC approach. The findings can be applied in cross-disciplines to generate a good starting point for basic biological and/or pharmacological investigations. The use of protein degraders as tool molecules/probes offers an immense potential to advance biology.

10.3.4 Viral Infections

The current pandemic has changed the world completely. Viral infections have become commonplace in last four decades, majorly starting from AIDS, Ebola, Dengue, H1N1, Zika, Chikungunya, SARS-CoV-1, middle-east respiratory syndrome (MERS), and now, SARS-CoV-2, in addition to the age-old Hepatitis C and others. Efficacious drugs have been developed against many of these pathogens so far. In a nutshell, antiviral drug discovery is difficult with several issues to address, such as drug resistance, altered efficacy of drugs against newer variants, and inability to translate findings from one virus to the other. Drugs acting against viral targets put selective pressure on the pathogens and the ensuing mutations make the life of clinicians and drug discovery researchers quite difficult. Newer antiviral drugs are often needed for the well-known viral infections and the newly-found ones. Targeted protein degradation strategy using PROTACs has been successfully used for the design and discovery of novel small-molecule antiviral agents; Hepatitis C virus (HCV) has been used as a model case.

de Wispelaere et al. chose HCV NS3/4A protease as the starting point for novel PROTAC design, wherein antiviral drug telaprevir was incorporated [50]. The attachment point for the linker was selected following molecular modeling studies of the protease:telaprevir complex. Three degraders (represented by **53**, Chart 10.6) were designed, synthesized, and evaluated for antiviral activity. Compound 53 exhibited NS3/4A protease IC_{50} = 385 nM and DC_{50} = 669 nM. The most potent compound in the series (structure not given) showed DC_{50} = 50 nM. It was conclusively proved that the antiviral activity of the designed compounds was due to the target degradation. Interestingly, the degraders were active against the HCV clones harboring the mutations in the protease. In summary, the developed PROTACs were truly superior as antivirals compared to the viral protease inhibitors. More research is needed in this direction to address the challenges associated with antiviral drug discovery and the unmet medical need for existing and emerging viral infections. PROTACs have a potential to deliver the much needed solutions.

10.4 Challenges and Limitations in the Development PROTACs

The PROTACs present various challenges, right from the design, through synthesis, all the way up to cellular permeability and *in vivo* pharmacokinetics and delivery. The larger structural motifs comprising of two ligands connected via an optimal linker are responsible for moving out of drug-like chemical space; the obvious problems related to solubility, permeability, and metabolic liability kick in. The orally bioavailable PROTACs seem to be a distant dream, at least with majority of the reported ones [26]. Nonetheless, researchers have developed molecules that are orally bioavailable and metabolically stable. Reports on such molecules are important for fine-tuning the physicochemical and pharmacokinetic properties of the new-generation agents. Additionally, the synthetic approaches pose several hurdles due to multistep, low-yielding, and difficult-to-scale-up reactions. Optimization of linkers is challenging as well. With every new E3 ligase, the possibilities of designing newer PROTACs open up.

The precise molecular mechanism of action of PROTACs is debated over similar modality called "molecular glue" [51]. Recent reports confirm the "molecular glue" mode along with PROTAC one or only the former in the molecular mechanism of their action. Their optimal combination, if at all, for the target protein degradation is still uncertain. In fact, identifying these modes itself can be difficult. Due to the basic premise of the catalytic mode opted by the PROTACs, the pharmacokinetic–pharmacodynamic evaluation in early phases of discovery is quite challenging. Successful transition of the PROTACs in the clinic to higher phases may throw light on this aspect. On another note, since PROTACs do not necessarily bind at the binding/catalytic site, candidate ligand selection for POIs can be pragmatic, i.e. there could be many more ligands that could potentially qualify as POI ligands. Medicinal chemist's intuition may not really help in the design process, at least by today's PROTAC philosophy.

Delivery of these degraders at the site of action is a daunting task. Earlier peptidic PROTACs were fused with cell-penetrating peptides (CPPs). The present-day all small-molecule agents can be designed in NP systems, such as liposomes, for targeted delivery to the site of action, e.g. intratumoral delivery. Chen et al. attempted the intracellular delivery of pre-fused PROTACs with the help of lipid-like NPs [52]. The authors pre-fused PROTACs with E3 ligases related proteins and delivered the two-component system by lipid NPs, which led to rapid and dramatic degradation of the POI. This novel approach needs to be replicated for various E3 ligases related proteins to generalize its utility.

Lastly, the development of resistance to PROTACs is an issue as seen from some reports. This is due to the genomic alterations in the core E3 ligase complexes. The cell survival mechanism is responsible for making the E3 ligase components redundant [53]. More reports on the underlying causes of drug resistance to these protein degraders are required for fully comprehending the issue.

10.5 Future Perspectives

The PROTACs and related approaches present a novel strategy to degrade the "pathological" versions of the POIs. The ability to charter the undruggable genome is of particular interest. More number of successful applications of this technology for the POIs, difficult to be perturbed by the small-molecules and/or other macromolecular approaches, will affirm its broader applicability. The rapid "chemical knockdown" process catalyzed by the PROTACs is interesting and rewarding, for the potential to discover tool molecules for advancing biology as well as discovering therapeutic alternatives for targets and diseases where drug resistance problems surface. Despite the difficulties, limitations, and challenges, design and development or PROTACs for targets, such as G-protein coupled receptors, e.g. α_{1A}-adrenergic receptors [54], will bring newer therapeutic avenues to the table. Such approaches can be extrapolated to channelopathies and other rare genetic diseases, where faulty protein is the main culprit. A recent report on defective RNA-binding proteins in many diseases has prompted a great deal of efforts centered around RNA-PROTACs [55]. This is not the end, in fact it just a beginning of endless possibilities and opportunities. The prioritization of these approaches for unmet medical need is required. Also, the microbial proteins may be targeted with PROTACs for developing antimicrobial PROTACs. More efforts are definitely needed in view of the growing strength of the "superbugs." Creative thoughts processes coupled with lateral thinking will improve the therapeutic landscape in near and distant future.

References

1 Poso, A. (2021). The future of medicinal chemistry, PROTAC, and undruggable drug targets. *J. Med. Chem.* 64 (15): 10680–10681.

2 Bushweller, J.H. (2019). Targeting transcription factors in cancer – from undruggable to reality. *Nat. Rev. Cancer* 19 (11): 611–624.

3 Qi, S.M., Dong, J., Xu, Z.Y. et al. (2021). PROTAC: an effective targeted protein degradation strategy for cancer therapy. *Front. Pharmacol.* 12: 1–13.

4 Sun, X. and Rao, Y. (2020). PROTACs as potential therapeutic agents for cancer drug resistance. *Biochemistry* 59 (3): 240–249.

5 Ci, Y., Li, X., Chen, M. et al. (2018). SCFβ-TRCP E3 ubiquitin ligase targets the tumor suppressor ZNRF3 for ubiquitination and degradation. *Protein Cell* 9 (10): 879–889.

6 Yang, K., Wu, H., Zhang, Z. et al. (2020). Development of selective histone deacetylase 6 (HDAC6) degraders recruiting von Hippel–Lindau (VHL) E3 ubiquitin ligase. *ACS Med. Chem. Lett.* 11 (4): 575–581.

7 Li, W., Peng, X., Lang, J., and Xu, C. (2020). Targeting mouse double minute 2: current concepts in DNA damage repair and therapeutic approaches in cancer. *Front. Pharmacol.* 11: 1–18.

8 Hrdinka, M. and Yabal, M. (2019). Inhibitor of apoptosis proteins in human health and disease. *Genes Immun.* 20 (8): 641–650.

9 Girardini, M., Maniaci, C., Hughes, S.J. et al. (2019). Cereblon versus VHL: hijacking E3 ligases against each other using PROTACs. *Bioorg. Med. Chem.* 27 (12): 2466–2479.

10 Bricelj, A., Steinebach, C., Kuchta, R. et al. (2021). E3 ligase ligands in successful PROTACs: an overview of syntheses and linker attachment points. *Front. Chem.* 9: 1–46.

11 Ishida, T. and Ciulli, A. (2021). E3 ligase ligands for PROTACs: how they were found and how to discover new ones. *SLAS Discovery* 26 (4): 484–502.

12 Gao, H., Wu, Y., Sun, Y. et al. (2020). Design, synthesis, and evaluation of highly potent FAK-targeting PROTACs. *ACS Med. Chem. Lett.* 11 (10): 1855–1862.

13 Nunes, J., McGonagle, G.A., Eden, J. et al. (2019). Targeting IRAK4 for degradation with PROTACs. *ACS Med. Chem. Lett.* 10 (7): 1081–1085.

14 Kargbo, R.B. (2021). Targeting IRAK1 for degradation with PROTACs. *ACS Med. Chem. Lett.* 12 (6): 943–944.

15 Kargbo, R.B. (2020). Estrogen receptor degrading PROTACS for the treatment of breast cancer. *ACS Med. Chem. Lett.* 11 (12): 2361–2363.

16 Kargbo, R.B. (2019). PROTACs and targeted protein degradation for treating ALK-mediated cancers. *ACS Med. Chem. Lett.* 10 (8): 1102–1103.

17 Yokoo, H., Shibata, N., Naganuma, M. et al. (2021). Development of a hematopoietic prostaglandin D synthase-degradation inducer. *ACS Med. Chem. Lett.* 12 (2): 236–241.

18 Kargbo, R.B. (2020). PROTAC-mediated degradation of KRAS protein for anticancer therapeutics. *ACS Med. Chem. Lett.* 11 (1): 5–6.

19 Kargbo, R.B. (2021). PROTAC-mediated degradation of Janus kinase as a therapeutic strategy for cancer and rheumatoid arthritis. *ACS Med. Chem. Lett.* 12 (6): 945–946.

20 Kargbo, R.B. (2021). PROTAC-mediated degradation of Bruton's tyrosine kinase as a therapeutic strategy for cancer. *ACS Med. Chem. Lett.* 12: 688–689.

21 Chemical Abstracts Service (2021). SciFinder, version 2021. Chemical Abstracts Service (CAS), Columbus, OH. cas.org (accessed 29 August 2021).

22 Scott, D.E., Rooney, T.P.C., Bayle, E.D. et al. (2020). Systematic investigation of the permeability of androgen receptor PROTACs. *ACS Med. Chem. Lett.* 11 (8): 1539–1547.

23 Klein, V.G., Townsend, C.E., Testa, A. et al. (2020). Understanding and improving the membrane permeability of VH032-based PROTACs. *ACS Med. Chem. Lett.* 11 (9): 1732–1738.

24 Ermondi, G., Garcia Jimenez, D., Rossi Sebastiano, M., and Caron, G. (2021). Rational control of molecular properties is mandatory to exploit the potential of PROTACs as oral drugs. *ACS Med. Chem. Lett.* 12 (7): 1056–1060.

25 Atilaw, Y., Poongavanam, V., Svensson Nilsson, C. et al. (2021). Solution conformations shed light on PROTAC cell permeability. *ACS Med. Chem. Lett.* 12 (1): 107–114.

26 Gao, H., Sun, X., and Rao, Y. (2020). PROTAC technology: opportunities and challenges. *ACS Med. Chem. Lett.* 11 (3): 237–240.

27 Wang, Y., Jiang, X., Feng, F. et al. (2020). Degradation of proteins by PROTACs and other strategies. *Acta Pharm. Sin. B* 10 (2): 207–238.

28 Noblejas-López, M.D.M., Nieto-Jimenez, C., Burgos, M. et al. (2019). Activity of BET-proteolysis targeting chimeric (PROTAC) compounds in triple negative breast cancer. *J. Exp. Clin. Cancer Res.* 38 (1): 1–9.

29 (a) Ocaña, A. and Pandiella, A. (2020). Proteolysis targeting chimeras (PROTACs) in cancer therapy. *J. Exp. Clin. Cancer Res.* 39 (1): 1–9. (b) Zhou, X., Dong, R., Zhang, J.Y. et al. (2020). PROTAC: a promising technology for cancer treatment. *Eur. J. Med. Chem.* 203: 112539. (c) Wan, Y., Yan, C., Gao, H., and Liu, T. (2020). Small-molecule PROTACs: novel agents for cancer therapy. *Future Med. Chem.* 12 (10): 915–938. (d) Li, X. and Song, Y. (2020). Proteolysis-targeting chimera (PROTAC) for targeted protein degradation and cancer therapy. *J. Hematol. Oncol.* 13 (1): 1–14.

30 Cimas, F.J., Niza, E., Juan, A. et al. (2020). Controlled delivery of BET-PROTACS: in vitro evaluation of MZ1-loaded polymeric antibody conjugated nanoparticles in breast cancer. *Pharmaceutics* 12 (10): 1–11.

31 Bai, L., Zhou, H., Xu, R. et al. (2019). A potent and selective small-molecule degrader of STAT3 achieves complete tumor regression *in vivo*. *Cancer Cell* 36 (5): 498–511.

32 Sun, X., Gao, H., Yang, Y. et al. (2019). PROTACs: great opportunities for academia and industry. *Signal Transduction Targeted Ther.* 4 (1): 64.

33 Lai, A.C., Toure, M., Hellerschmied, D. et al. (2016). Modular PROTAC design for the degradation of oncogenic BCR-ABL. *Angew. Chem. Int. Ed.* 55 (2): 807–810.

34 (a) Sun, Y., Ding, N., Song, Y. et al. (2019). Degradation of Bruton's tyrosine kinase mutants by PROTACs for potential treatment of ibrutinib-resistant non-Hodgkin lymphomas. *Leukemia* 33 (8): 2105–2110. https://doi.org/10.1038/s41375-019-0440-x. (b) Buhimschi, A.D., Armstrong, H.A., Toure, M. et al. (2018). Targeting the C481S ibrutinib-resistance mutation in Bruton's tyrosine kinase using PROTAC-mediated degradation. *Biochemistry* 57 (26): 3564–3575.

35 Zhao, B. and Burgess, K. (2019). PROTACs suppression of CDK4/6, crucial kinases for cell cycle regulation in cancer. *Chem. Commun.* 55 (18): 2704–2707.

36 Salami, J., Alabi, S., Willard, R.R. et al. (2018). Androgen receptor degradation by the proteolysis-targeting chimera ARCC-4 outperforms enzalutamide in cellular models of prostate cancer drug resistance. *Commun. Biol.* 1 (1): 1–9.

37 Hu, J., Hu, B., Wang, M. et al. (2019). Discovery of ERD-308 as a highly potent proteolysis targeting chimera (PROTAC) degrader of estrogen receptor (ER). *J. Med. Chem.* 62 (3): 1420–1442.

38 Kaur, T., Menon, A., and Garner, A.L. (2019). Synthesis of 7-benzylguanosine cap-analogue conjugates for eIF4E targeted degradation. *Eur. J. Med. Chem.* 166: 339–350.

39 Lee, H., Puppala, D., Choi, E.Y. et al. (2007). Targeted degradation of the aryl hydrocarbon receptor by the PROTAC approach: a useful chemical genetic tool. *ChemBioChem* 8 (17): 2058–2062.

40 Ohoka, N., Tsuji, G., Shoda, T. et al. (2019). Development of small molecule chimeras that recruit AhR E3 ligase to target proteins. *ACS Chem. Biol.* 14 (12): 2822–2832.

41 Chu, T.T., Gao, N., Li, Q.Q. et al. (2016). Specific knockdown of endogenous tau protein by peptide-directed ubiquitin-proteasome degradation. *Cell Chem. Biol.* 23 (4): 453–461. https://doi.org/10.1016/j.chembiol.2016.02.016.

42 Lu, M., Liu, T., Jiao, Q. et al. (2018). Discovery of a Keap1-dependent peptide PROTAC to knockdown Tau by ubiquitination-proteasome degradation pathway. *Eur. J. Med. Chem.* 146: 251–259. https://doi.org/10.1016/j.ejmech.2018.01.063.

43 Kargbo, R.B. (2019). Treatment of Alzheimer's by PROTAC-tau protein degradation. *ACS Med. Chem. Lett.* 10 (5): 699–700.

44 (a) Yamashita, H., Tomoshige, S., Nomura, S. et al. (2020). Application of protein knockdown strategy targeting β-sheet structure to multiple disease-associated polyglutamine proteins. *Bioorg. Med. Chem.* 28 (1): 115175. (b) Tomoshige, S., Nomura, S., Ohgane, K. et al. (2017). Discovery of small molecules that induce the degradation of huntingtin. *Angew. Chem. Int. Ed.* 56 (38): 11530–11533.

45 Kargbo, R.B. (2020). PROTAC compounds targeting α-synuclein protein for treating neurogenerative disorders: Alzheimer's and Parkinson's diseases. *ACS Med. Chem. Lett.* 11 (6): 1086–1087.

46 Gao, N., Huang, Y.P., Chu, T.T. et al. (2019). TDP-43 specific reduction induced by di-hydrophobic tags conjugated peptides. *Bioorg. Chem.* 84 (2018): 254–259.

47 Chen, Y., Ning, Y., Bai, G. et al. (2021). Design, synthesis, and biological evaluation of IRAK4-targeting PROTACs. *ACS Med. Chem. Lett.* 12 (1): 82–87.

48 Zhang, J., Fu, L., Shen, B. et al. (2020). Assessing IRAK4 functions in ABC DLBCL by IRAK4 kinase inhibition and protein degradation. *Cell Chem. Biol.* 27 (12): 1500–1509.

49 Bassi, Z.I., Fillmore, M.C., Miah, A.H. et al. (2018). Modulating PCAF/GCN5 immune cell function through a PROTAC approach. *ACS Chem. Biol.* 13 (10): 2862–2867.

50 de Wispelaere, M., Du, G., Donovan, K.A. et al. (2019). Small molecule degraders of the hepatitis C virus protease reduce susceptibility to resistance mutations. *Nat. Commun.* 10 (1): 3468.

51 Thiel, P., Kaiser, M., and Ottmann, C. (2012). Small-molecule stabilization of protein-protein interactions: an underestimated concept in drug discovery? *Angew. Chem. Int. Ed.* 51 (9): 2012–2018.

52 Chen, J., Qiu, M., Ma, F. et al. (2021). Enhanced protein degradation by intracellular delivery of pre-fused PROTACs using lipid-like nanoparticles. *J. Controlled Release* 330: 1244–1249.

53 Zhang, L., Riley-Gillis, B., Vijay, P., and Shen, Y. (2019). Acquired resistance to BET-PROTACs (proteolysis-targeting chimeras) caused by genomic alterations in core components of E3 ligase complexes. *Mol. Cancer Ther.* 18 (7): 1302–1311.

54 Li, Z., Lin, Y., Song, H. et al. (2020). First small-molecule PROTACs for G protein-coupled receptors: inducing α_{1A}-adrenergic receptor degradation. *Acta Pharm. Sin. B* 10 (9): 1669–1679.

55 Ghidini, A., Cléry, A., Halloy, F. et al. (2021). RNA-PROTACs: Degraders of RNA-binding proteins. *Angew. Chem. Int. Ed.* 60 (6): 3163–3169.

11

Metal Complexes as the Means or the End of Targeted Delivery for Unmet Needs

Trevor W. Hambley

The University of Sydney, School of Chemistry, NSW 2006, Australia

11.1 Introduction

Targeted delivery of drugs, particularly for cancer, can reduce side effects, enhance efficacy, and enable personalized approaches to treatment. Metal complexes have the potential to make an enormous contribution to targeted delivery, including the chaperoning and caging of other active agents and the selective delivery of active metal compounds, minimizing off-target effects, and increasing activity [1, 2]. At least 20 different metals might be considered for use in these ways and many of these have multiple oxidation states each of which has different reactivities. The reactivity can be further tuned by variation of the other groups bound to the metal giving rise to an extraordinary range of properties. The complexes can also respond to their chemical environment, enabling selective activation in target tissues. Despite this extraordinary flexibility and responsiveness, there are few examples of the use of metal complexes in targeted drug delivery. Therefore, in this chapter, I will focus first on the principles for selecting metal and for designing a complex to enable drug targeting in the context of the role of the metal (Sections 11.2–11.4) and then describe the work that has been done to date to develop and test such compounds in the context of the targeting strategies being employed (Sections 11.5–11.8).

Three classes of complexes will be considered – those in which one of the attached ligands is the active agent and the role of the metal is to deliver it to the target site, those in which the metal complex itself is active, and those in which the metal does both. In the first case, the focus will be on cobalt complexes and in the latter two, platinum complexes, since these have been the subject of the largest amount of work. However, the principles developed and tested can be applied in the context of many other metals. Radioactive metals have been used in diagnosis and therapy for many decades and increasingly, targeted versions of these are being developed. Since the application of these metal complexes is based on their radioactivity rather than their chemical properties, they are not discussed extensively in this chapter but are referenced where they provide exemplars of what targeted delivery of metal complexes might achieve.

Targeted Drug Delivery, First Edition. Edited by Yogeshwar Bachhav.

The work done to date will primarily be discussed in terms of unmet needs in cancer treatment, though the principles for targeted delivery can be applied to other diseases and illnesses. Cancer treatment is at a turning point because checkpoint inhibitors have created new options for previously difficult to treat cancers, such as melanomas and non-small-cell lung cancer [3]. It has been suggested that as many as 25% of cancer patients may benefit from such inhibitors [4], but in many cases, they are most effective when coupled with a cytotoxic agent, such as carboplatin [5]. At the same time, molecularly targeted agents, despite spectacular successes, such as Gleevec and Herceptin and their importance in precision medicine, have reached a point where progress has slowed. For both of these reasons, there is renewed interest in cytotoxic agents, such as platinum. Indeed, use of the platinum has continued over the past 40 years and in Australia, nearly 50% of those treated with chemotherapy for a solid tumor have platinum in their regimen [6, 7]. Similar or higher levels of usage are likely in many other countries [8]. However, severe side effects continue to plague their use with most patients having to have their dosage reduced [7, 9]. Therefore, targeted delivery to reduce toxicity and increase efficacy is being actively pursued with some promising developments as outlined below. Equally, many other cytotoxic agents would benefit from targeted delivery as would most molecularly targeted agents. An impediment to progress in this area is the synthetic challenge of generating targeted versions of these agents and in this respect, metal complexes have a role to play because they can deliver unmodified forms of many drugs, increasing their efficacy and minimizing off-target effects.

A major cause of failure of current chemotherapy and radiotherapy for cancer is the cells that escape destruction because (i) they are quiescent and therefore not as susceptible to drugs that rely on rapid cellular cycling, (ii) they are in hypoxic environments and therefore not as susceptible to radiation or chemotherapy that relies on the presence of oxygen, or (iii) they lie too far from the vasculature to receive lethal doses of the chemotherapeutic agent. In many cases, all three factors apply to the same population of cells. Targeting these groups of cells remains an unmet need and metal complexes can make a major contribution. Indeed, it may be that cisplatin in particular already does by virtue of its relatively poor cellular uptake which results in better diffusion throughout the tumor [10] and its relative effectiveness irrespective of the presence of hypoxia [11]. This example shows how understanding of the properties of metal complexes might be exploited to enable the development of agents to help address unmet needs in targeting and this is the focus of this chapter.

11.2 Class 1: Chaperones

All drugs are subject to biotransformation which in most cases results in a loss of activity and/or removal from cells and/or from the body. Also, some drugs are designed to bind to metal centers in proteins and most drugs have heteroatoms that can bind to metals. Consequently, sequestration by metals in the body is one form of biotransformation that can contribute to deactivation or removal. For example, drugs designed to target zinc-containing enzymes often incorporate a

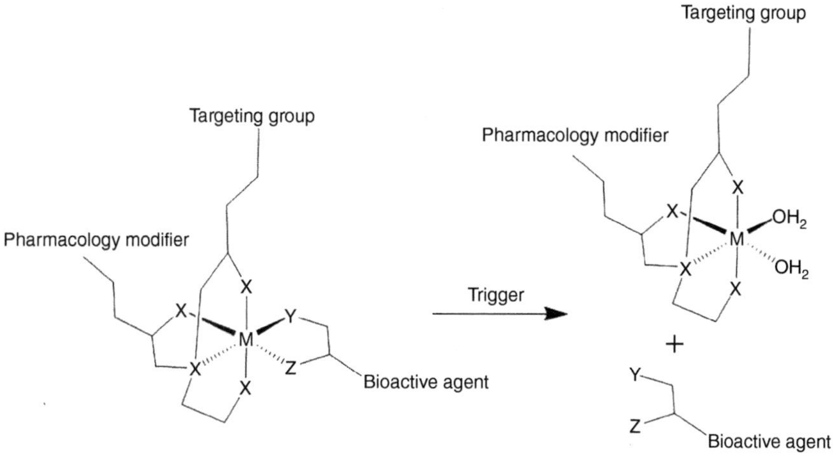

Figure 11.1 Schematic showing the principle of a chaperone complex carrying a drug molecule that accumulates at the target site and releases the drug as a result of trigger event.

hydroxamate group for binding to the zinc. However, hydroxamates also bind avidly to iron(III) [12, 13] which will reduce the bioavailability of the drug. Coordination of the drug to a metal prior to administration to form a stable entity that protects the metal-binding moiety of the drug from adventitious metals is one means of reducing these unwanted side reactions. It can also reduce the rate of other types of reaction that contribute to deactivation, such as the conversion of hydroxamates to carboxylates with the associated loss of NO [14]. Also, binding to a metal can greatly reduce the toxicity of the drug by caging it and thereby reduce side effects as has been reported for nitrogen-mustards [15–17] and matrix metalloproteinase (MMP) inhibitors [18]. This approach, known as chaperoning, depends on the drug being released where and when required, and indeed, this release process can be the basis of the mechanisms used to target particular cells or cellular environments. The chaperone can also be decorated with groups that increase accumulation in the cells or environments to be targeted and groups that tune the pharmacology (Figure 11.1). A chaperone constructed in this way can potentially be used to deliver a range of drugs, greatly decreasing the cost and time taken to generate targeted versions of each of them. In this section, we discuss the principles for choosing a chaperone to (i) protect the drug to be delivered, (ii) deliver it to the cells or environments to be targeted, and (iii) release it where and when required.

11.2.1 Chaperones that Protect Drugs

For a metal to be an effective chaperone of a drug, it must form a complex that is stable enough to withstand the environment it encounters on route to the target, typically the bloodstream or the intraperitoneal cavity. These environments are populated by a wide variety of compounds that can bind to the metal, potentially displacing the drug from the chaperone. They can also contain reducing agents or reducing environments that lead to destabilization of the entire chaperone complex and loss of the drug. This is particularly true of the intracellular environment in red blood cells and minimizing uptake by these cells is likely to be crucial to the success of a chaperone [19, 20]. The principles for generating stable metal complexes and understanding the environmental factors that can lead to destabilization have been the subject of extensive study for more than a century. However, investigations of the stability of metal complexes in the environments found in the human body have been less frequent. Much work has been done over the past 50 years in the context of platinum anticancer agents and more recently a few other metals, such as ruthenium, but in most cases, these investigations have been focused on the inherent activity of the metal or its complexes, not their use as chaperones.

The stability of a metal complex is influenced by the metal, its oxidation state, and the nature of the ligands bound to the metal. The compatibility between the electronic environment these create at the metal center and the nature of the heteroatoms of the drug that interact with the metal will determine how tightly it is bound. The stability of the bonds between the metal and the drug being transported are influenced by the other groups bound to the metal, resulting in the ability to finely tune this stability, often with relative synthetic ease. These other groups are referred to as "spectator ligands" or "co-ligands" and can generally be connected to target moieties, such as glucose or folate without affecting the stability of the attachment to the drug being chaperoned (the targeting group in Figure 11.1). The chemical and physical properties of the complex can also be tuned by the addition of one or more moieties that function as pharmacology modifiers (Figure 11.1).

Among the most stable complexes are those of cobalt(III), ruthenium(II), and platinum(IV), and consequently, these have been the subject of many studies. They also have the benefit of being redox-active in that they have another accessible oxidation state which has quite different stability properties providing a mechanism for drug release under redox control. For example, cobalt(III) complexes with nitrogen donors are particularly stable, while those with oxygen donors are somewhat less so. On reduction to cobalt(II), which can occur intracellularly, drugs attached through oxygen donors are rapidly released and those with nitrogen donors less rapidly. Indeed, we have shown that it is possible by varying the nature of the nitrogen donors of the spectator ligands to develop cobalt(III) complexes which range from those that are either completely resistant to reduction inside cells to those that rapidly release the compound, curcumin in this example, under redox control [21].

Curcumin is an excellent example of the potential for stabilization and pharmacology modification by coordination. There have been numerous reports of curcumin's antioxidant properties and cancer-cell-killing abilities and its use as a supplement are widespread. However, it is highly unstable in the body, decomposing rapidly, resulting in it having very low bioavailability [22]. However, formation of its cobalt complexes stabilizes the curcumin and enables it to demonstrate its biological potential [23].

Many other metal ions have been investigated as chaperones for drugs or bioactive agents, though in some cases without appropriate consideration of stability issues. Copper is probably the most widely investigated example, both as copper(II) and copper(I), and both oxidation states can give rise to complexes that are sufficiently stable to modify the pharmacological properties of bioactive agents. Generally, copper(II) favors donors such as nitrogen and oxygen, while copper(I) favors sulfur and phosphorus. In both cases, protection of the metal from the environment by steric crowding can enhance the stability and potentially generate complexes that are stable enough for targeted delivery. For example, the highly cytotoxic gold, silver, and copper complexes of 1,2-bis(diphenylphosphino)ethane, $[M(dppe)_2]^+$ give rise to almost spherical molecules with only the phenyl groups exposed at the surface, generating lipophilic cations that are highly effective mitochondrial poisons. These complexes almost certainly function as delivery vehicles for the highly toxic 1,2-bis(diphenylphosphino)ethane molecules and other similar bidentate phosphines and this is confirmed by the similarity of the biological properties of the complexes of the three metals, showing that they play little role in inducing cell death [24–26].

11.2.2 Delivery to the Cells or Environments to Be Targeted

Given a drug bound sufficiently well to metal to protect it en route to its target, the next task is to deliver it to the desired location. Delivery, or more accurately, accumulation in the desired location, can be based on proteins expressed on the surface of cells of interest, or on the features of the extracellular environment, including active enzymes found in that environment, and these mechanisms are mentioned briefly here and are discussed in more detail in Sections 11.5–11.7.

Exploiting the chemical or physical features of the extracellular environment to activate prodrugs is unlikely to be a rate-limiting step in cellular uptake because the environment will not be changed significantly by the reactions with the prodrug and these reactions are rapid. Any extracellular proteases exploited in the activation of the prodrug will also be unchanged, but the rate of cleavage may limit the amount of drug available for cellular uptake and this can be negatively impacted by the attachment of the bioactive agent being delivered. Also, activation in the extracellular space enables diffusion and uptake by healthy cells in the vicinity; the so-called "bystander effect." Conversely, intracellular activation, unless followed by extensive efflux of the active agent from the cell, should not give rise to bystander effects. However, as discussed below, cellular uptake via nutrient transporters depends on the number of transporters expressed and activated at the cell surface, the rate of uptake, the mechanism of uptake, and the associated intracellular disposition.

The physical features of the extracellular environment: The extracellular environment of most tumors, for example, is typified by multiple microenvironments, including hypoxic and unusually acidic regions. These features can be exploited to trigger release of the drug from the metal, or to change the properties of the metal/drug conjugate so that it is more readily taken up by cells. Our research group has reported examples of both, exploiting hypoxia to facilitate reduction of cobalt(III) and release of the drug, and exploiting the lower pH to promote protonation, a change in charge from negative to neutral, and more rapid cellular uptake by passive diffusion [27–30]. Others have investigated the use of acid-sensitive linkers that could function with or without a metal center [31].

Exploiting nutrient uptake pathways: Tumor cells have very high demands for nutrients and in many cases, this results in upregulation of the transporters for a specific nutrient, such as glucose, folate, or glutamine. These transport mechanisms can be exploited for the targeted delivery of a chaperone and its load by attaching the nutrient to the co-ligand of the chaperone.

Exploiting enzymes in the extracellular environment: The extracellular environments of tumors and inflamed regions, such as those found in arthritic joints, are characterized by much higher than usual levels of enzymes, such as MMPs. These enzymes can be exploited to cleave peptidic groups to remove moieties that prevent cellular uptake (see Section 11.7).

11.2.3 Release from the Metal Where and When Required

Having delivered the agent to a target cell, it is essential that the metal releases it where required and does so rapidly. It is plausible that this release could occur extracellularly, but there are few mechanisms that would trigger this with the selectivity desired and the risk of substantial and undesirable bystander effects would be high. Consequently, there are few examples of extracellular release and this mostly occurs intracellularly following reduction of the complex or reaction with one or more of the many strong nucleophiles found in the intracellular environment.

11.3 Class 2: Active Metal Complexes

Metals and metal complexes have been used as medicines for thousands of years, starting with gold, and moving on to highly toxic metals and semimetals, such as mercury and arsenic [32]. More recently, lower toxicity elements, such as bismuth, have found a role [33] and many metal salts and complexes are used to treat trace-element deficiencies. However, there is no doubt that the metal complex that has had the greatest impact in medicine is cisplatin, *cis*-diamminedichlorido-platinum(II), first developed by Barnett Rosenberg and colleagues in the 1960s. Cisplatin and the related complexes, carboplatin and oxaliplatin (Figure 11.2), are together one of the most widely used groups of anticancer agents with 40–50% of patients treated with chemotherapy for a solid tumor receiving one of the three compounds [7, 9]. Also in use are compounds, such as arsenic trioxide (Trisenox)

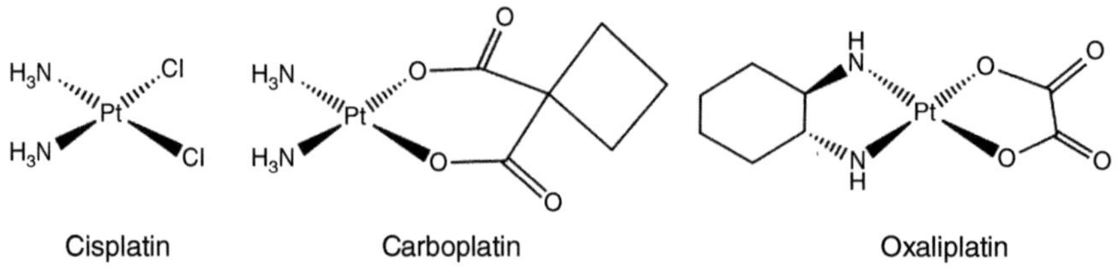

Figure 11.2 The three platinum agents approved worldwide for the treatment of a variety of cancers.

for treating acute promyelocytic leukemia and auranofin for treating rheumatoid arthritis [34]. Many of these compounds are highly toxic and their use is associated with substantial and dose-limiting side effects. Consequently, there has been extensive interest over the last 20 years, in particular, in developing targeted versions of these agents. The great bulk of the work done to date has been focused on the platinum agents and the platinum(IV) oxidation state in particular because of the unique opportunities it provides for the attachment of targeting groups and fine control of activation in the tumor environment [1, 35].

11.3.1 Targeted Platinum Agents

The three existing platinum anticancer agents (Figure 11.2) are all in the 2+ oxidation state (platinum(II)) and as such, are moderately reactive. All undergo reaction with water on a time scale of minutes to hours and while this is a critical part of their mechanism of action because it enables reactions with targets, such as DNA; it also enables undesirable side reactions that can result in deactivation en route to the target and side effects from the damage done in these reactions. Platinum(IV) (4+ oxidation state) complexes are much more inert and have an additional two binding sites. They must be reduced to platinum(II) to be effective, but this has the additional advantage of releasing the targeting groups so that they do not interfere with action of the platinum(II) product (Figure 11.3). The ideal platinum agent will be stable in the bloodstream and en route to the tumor, but be rapidly reduced and activated once inside the cancer cell and much progress has been made toward achieving these goals in recent years [20, 36–38]. Targeting groups can be added to either platinum(II) or platinum(IV) complexes and examples of each are known. They have been reviewed recently [1] and therefore we will focus here on the platinum(IV) complexes since these are the focus of most of the current interest.

The optimal mode of action of a targeted platinum(IV) complex is outlined in Figure 11.4. Since platinum agents are generally administered intravenously, the

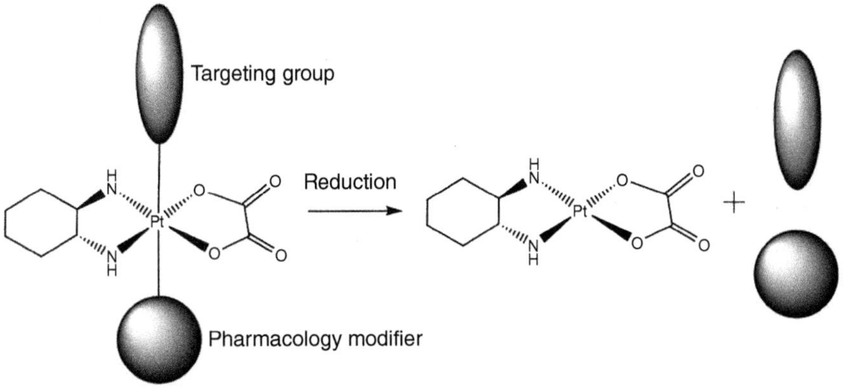

Figure 11.3 A platinum(IV) analog of oxaliplatin with targeting and pharmacology-modifying groups in the axial positions showing how reduction leads to loss of these groups and release of the unmodified drug.

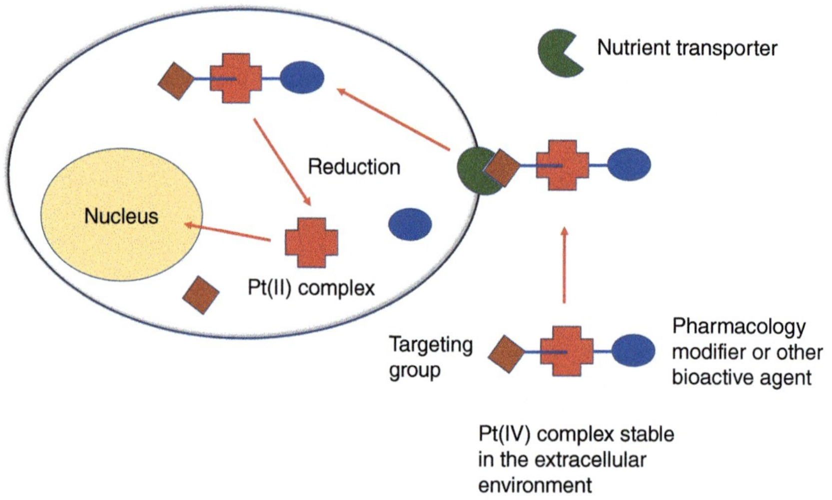

Figure 11.4 Schematic showing the mode of action of a platinum(IV) prodrug taken up via a nutrient transporter.

complex should have good water solubility and high stability in the bloodstream and the extracellular environment of organs and tumors. Generally, stability translates to how resistant the complex is to reduction and in these environments that corresponds to reduction by thiols, ascorbic acid, or other simple reductants. More than 25 years ago, a correlation between the reduction potential of a platinum(IV) complex and its stability in the presence of such simple reductants had been identified, and this structure–activity relationship formed the basis of our drug design strategy for many years [39]. However, resistance to reduction by simple reductants also corresponded to slow reduction and activation in the intracellular environment, limiting the level of activity seen in two-dimensional cell culture. Fortunately, a set of "rule-breakers" that do not follow the relationship between reduction potential and reactivity with simple reductants have been identified [36]. Specifically, complexes with four carboxylate donors and two amine donors have reduction potentials that are in the middle of the range observed but are, remarkably, the most resistant to reduction by ascorbate. The explanation for this unexpected behavior derives from the role of intermolecular contacts in the reduction process and these are less effective for carboxylate groups than for other types of donor groups. Equally fortuitously, the same is not true for reduction inside the cell which appears to follow the relationship with the reduction potentials such that these complexes are reduced and activated relatively rapidly. Intracellular reduction of platinum(IV) complexes is believed to be primarily due to reductases rather than simple reductants, such as ascorbic acid and cysteine [40, 41] and this is probably responsible for the difference between the rates of extracellular and intracellular reduction. It has been shown, for example, that such complexes are relatively stable in blood plasma, but are rapidly reduced in whole blood, presumably as a result of cellular uptake and intracellular reduction [20]. Subsequently, it has been shown that some other donor groups, such as carbamates, can give rise to platinum(IV) complexes with similarly good or better properties [37, 42] and this has expanded the ranges of synthetic options for such complexes.

Platinum(IV) complexes with anticancer potential typically have two nitrogen donors, two anionic donors trans to these (the leaving groups), and two anionic groups in the axial sites (Figure 11.3). Targeting groups can be attached via any of these groups, but most of the focus has been via those in the axial sites since these are lost on reduction and therefore the targeting group will not interfere with the activity of the platinum(II) complex that is released. Indeed, in most cases, the complex released is one of the three well-established platinum anticancer agents (Figure 11.2) and this is considered to be an advantage since these are agents of known activity and toxicity. The chemistry for the attachment of targeting groups is now relatively well established and typically follows oxidation of the platinum(II) complex with hydrogen peroxide, which results in hydroxide occupying one or both axial sites depending on the solvent used [43, 44]. The hydroxides can then be reacted with anhydrides or coupled to carboxylates using standard coupling chemistry. For example, reaction with succinic anhydride generates a free carboxylate group which can then be coupled to an amine using standard peptide-coupling methods.

Platinum(IV) complexes with glutamine [45], glucose [46, 47], or folate [48] in axial sites have been investigated with encouraging results as outlined below. Platinum(II) complexes with folate or various sugars in the axial sites have also been investigated with similarly encouraging results [49–54].

11.4 Class 3: Dual-Threat Metal Complexes

Platinum(IV) complexes can be used to deliver a bioactive agent in addition to the platinum(II) moiety released on reduction and such complexes are known as dual-threat reagents. Such agents have features of both Classes 1 and 2 described above in that the bioactive agent is chaperoned by the platinum. If a targeting group is attached via one of the axial sites, the other is available for binding of a bioactive agent (for example, in place of the pharmacology-modifying group in Figure 11.3), or both are available if the targeting group is attached to the amines or the equatorially bound leaving group(s). Attaching a bioactive agent in this way protects it en route to the tumor, delivers it via any targeting mechanism in use, and can greatly increase its cellular uptake. For example, aspirin (salicylic acid) has a negative charge which slows its cellular uptake, but it has been attached to platinum(IV) enabling enhanced uptake [55]. The idea of attaching an active agent to platinum(II), either as part of the amine ligands or the leaving group, has been investigated extensively, but none have reached the clinic. An aspect of this approach that needs to be considered is that the concentrations of the platinum moiety and the other bioactive are constrained to be the same and it is unlikely that this would enable delivery of the optimal dose of both. However, the use of such an agent could be complemented by co-admission with either a standard platinum agent or the other bioactive, and the benefits of improved uptake could then be exploited.

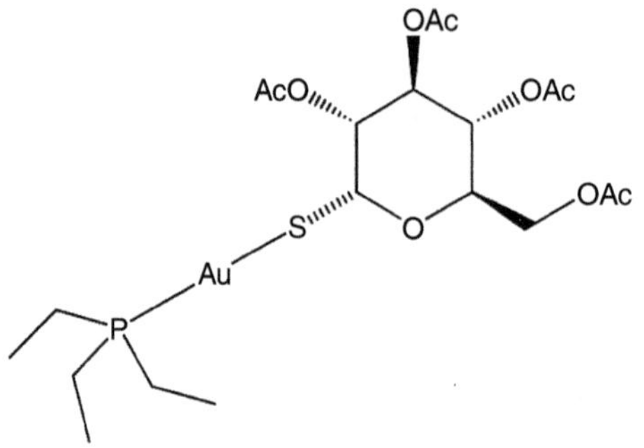

Figure 11.5 Chemical structure of auranofin.

To date, while there have been a number of studies of dual-threat platinum(IV) reagents [55–61], none that we are aware of which have also exploited an active-targeting mechanism, but it is an approach that has significant potential. In the context of the platinum(II) complexes mentioned above, the approach has downsides in that the synthesis may be challenging and the platinum/bioactive conjugate may have suboptimal pharmacological properties. These are less of an issue for platinum(IV) complexes because attachment of the bioactive can be via a site that does not compromise activity and there remains the possibility of tuning the pharmacology using one or more of the other sites on the platinum.

In theory, complexes of other metals that have inherent activity could function in the same way and this particularly applies to complexes of second- and third-row metals, such as ruthenium, osmium, silver, and gold. Gold is an interesting and widely studied example in that it functions as a thioredoxin inhibitor [62]. In many cases, the ligands are expected to have separate biological activity and the gold will modify the pharmacology of these ligands, protecting them from deactivation and possibly increasing their intracellular accumulation. For example, auranofin (Figure 11.5) has a triethylphosphine ligand that is likely to be responsible for at least some of the activity seen against cancer cells. However, there has not been as much effort directed toward fine-tuning the properties of such complexes so that they can both release any chaperoned bioactive and have inherent activity and it is not yet clear that other metals can achieve both without compromising activity.

11.5 Targeting Strategies: The Chemical and Physical Environment

Tumors exhibit both extracellular and intracellular environments that differ from those of normal tissues. Both sets of differences arise primarily because of either an unusual metabolism and/or the rapid growth, poor vascularization, and poor or nonexistent lymphatic system [63]. As a result, the extracellular environment is characterized by lower than usual pH (5.5–7.2), low oxygenation (hypoxia), and low levels of other nutrients [16, 64, 65]. Intracellularly, the pH is typically higher than normal (7.4–7.6) and oxygen levels are low. These abnormalities in the extracellular tumor environment can be exploited to activate prodrugs and thereby target cells in or near such environments [66].

11.5.1 Hypoxia

The development of hypoxia-activated prodrugs has been pursued for more than three decades, but with limited clinical application to date [67–69]. The principle is that the prodrug must be reduced to be activated, but in the presence of normal levels of oxygen, any reduction is followed by rapid reoxidation. One of the most promising agents, tirapazamine, has undergone phase III clinical trials but has not received approval for clinical use [70, 71]. Transition-metal complexes have great potential as hypoxia-activated agents because many of them have multiple oxidation states, with the higher oxidation state typically being the more inert. Consequently, reduction of an inert prodrug can result in an activated agent or release of a caged drug. Platinum(IV) complexes must be reduced to platinum(II) to be active, and while this is not a reversible process and therefore is not directly dependent on the concentration of oxygen, it is expected to occur more rapidly in cells in hypoxic environments because of the accumulation of greater reductive capacity [72]. However, cobalt complexes that function as chaperones for active agents have received the most attention [73].

Cobalt(III) forms some of the most stable complexes, particularly among the less toxic first-row transition metals, but on reduction to cobalt(II), the stability decreases greatly and a drug bound to the cobalt will be released much more readily [73]. The first to propose using cobalt to deliver drugs to hypoxic regions were Ware et al. who investigated the chaperoning of highly toxic nitrogen–mustard compounds [17, 74, 75]. Coordination of these agents to cobalt(III) effectively caged the nitrogen–mustards and greatly diminished their toxicity, but this was recovered on reduction to cobalt(II). For optimal selectivity, it was suggested that the cobalt(III) complexes should have a reduction potential in the range 200–400 mV vs. NHE because all cells have intracellular reductases with the potentials in that range. However, we have subsequently found that complexes with potentials substantially more negative than this range, and therefore more difficult to reduce, appear to function at least as effectively [18, 28].

Both the stability and the ease of reduction of cobalt complexes can be tuned by variation of the other ligands bound to the metal (the co-ligands) and there have been numerous studies aimed at determining the optimal properties for selective delivery to hypoxic regions [73]. Ascorbate and cysteine have been widely used to typify the simple reductants that a carrier complex would encounter in the bloodstream. If challenge experiments carried out in the presence and absence of oxygen show differences in the rate of the release of the drug model then hypoxia selectivity can be anticipated.

We have made extensive use of fluorescent analogs of carboxylate and hydroxamate-based drugs to monitor release of the drug, both under challenge from simple reductants and in cellular systems. Anthraquinone-2-carboxylic acid (AQ2CH), coumarin-343 (C343H), and their hydroxamic acid analog (C343-haH), for example, bind strongly to cobalt(III) and in doing so, lose their fluorescence [27, 28, 30]. Release of the fluorophore from the cobalt, most likely following reduction to cobalt(II), results in recovery of the fluorescence (Figure 11.6) enabling

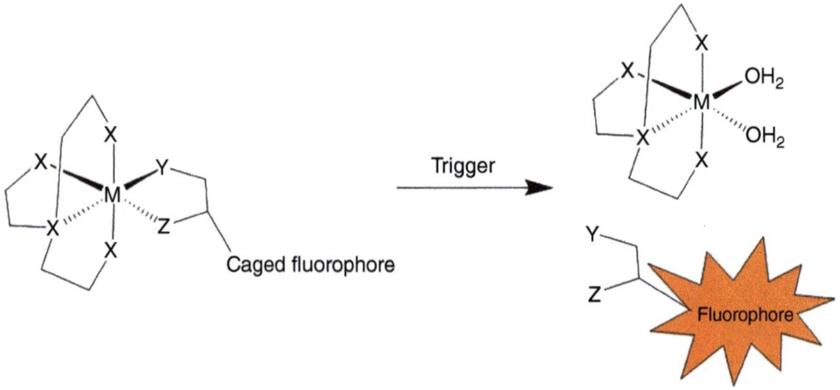

Figure 11.6 The caging of a fluorophore by attachment to a metal complex.

this release to be tracked, both in solution and by fluorescence mapping in 2D- and 3D-cellular models. These model systems have enabled us to investigate the effects of the co-ligand and of the presence of oxygen on the rate release of the hydroxamic acid-containing moiety, confirming the potential of the anthraquinone system in particular as a hypoxia-activated chaperone [30]. Modification of the co-ligands in the cobalt complexes of C343-ha increased the oxygen dependence of the rate of reduction of those complexes, demonstrating the fine-tuning that is possible in these systems [27, 76]. These systems have also enabled the spatial location of the release to be observed. In 2D-cellular models, extensive intracellular fluorescence is observed [27, 28] and particularly encouraging is the observation of selective release in the hypoxic region of tumor cell spheroids [30]. An advantage of metals, particularly those found at very low or zero levels in cells, is that the distribution of the chaperoning metal can be mapped by using techniques, such as synchrotron-induced X-ray emission [77] and laser-ablation inductively coupled plasma-mass spectrometry (ICPMS) [78]. This coupled with fluorescence maps where fluorescent analogs of the drug have been used, enables any similarities or differences in the distributions to be identified enabling an understanding of the processes of activation and subsequent diffusion of the products.

There have been a small number of studies of other metals as hypoxia-activated chaperones, including chromium(III) [74] and iron(III) [12], but with much less evidence of the efficacy seen for cobalt(III) complexes. Other metals, such as platinum(IV), could also be used as reductively activated chaperones [56, 79, 80], but have not been investigated for their hypoxia selectivity in a chaperone role.

11.5.2 pH-Based Targeting

The low extracellular pH in tumor environments promotes protonation of weak bases which will result in an increased population of neutral compounds and better uptake if the prodrug is initially negatively charged [64, 81]. The converse, of poorer uptake resulting from protonation of a neutral compound to give a cation, is well known for doxorubicin, a weak base with a pK_a of 7.6 which has a twofold lower

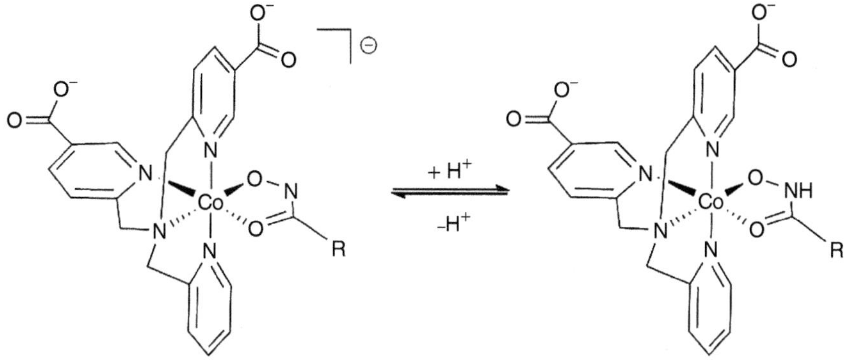

Figure 11.7 Protonation of the complex on the left in the acidic extracellular environment of tumors results in a net-neutral complex and facilitates cellular uptake. Source: Based on Yamamoto et al. [27].

activity at a pH of 6.8 than at 7.4 [81]. Since negatively charged compounds are even less well taken up, a greater impact can be anticipated for protonation of such drugs and indeed this is observed for chlorambucil, a negatively charged drug with a pK_a of 5.8 [64]. Critical to the success of this approach is that the protonatable group should have a pK_a close to the range of 6–7 so that small changes in pH lead to a substantial difference in the population of the charged and uncharged forms. Fortunately, hydroxamate-based drugs exhibit a pK_a of about 6.5 when bound to cobalt(III) and the excellent sensitivity to pH of the cellular accumulation of such complexes has been reported on (Figure 11.7) [27].

Many metal complexes undergo ligand-exchange reactions with water, either extracellularly or intracellularly, and the coordinated water molecule, which results often, has a pK_a in the range of interest. For example, cisplatin reacts with water to produce $[PtCl(NH_3)_2OH_2]^+$, which will accumulate in cells because of the positive charge. However, in cancer cells with a higher pH, more of this complex will exist as the hydroxide $[PtCl(NH_3)_2OH]$ which will be able to escape, diminishing selectivity for cancer cells. This complex has a pK_a of 6.4 and therefore will exist primarily in the neutral deprotonated form so the effect is expected to be minor [82]. This effect might be exploited for preferential accumulation of negatively charged complexes, but is an approach that has received little attention.

11.5.3 The EPR Effect

The enhanced permeability and retention (EPR) effect has been widely reported as a basis for targeting tumors [83]. The rationale is that the blood vessels in a tumor are leaky and larger particles are able to escape through the vessel walls into the tumor than through the walls of blood vessels in healthy tissues. Particles up to 400 μm are believed to be able to extravasate in this way and numerous studies have been conducted with metal-containing nanoparticles to attempt to exploit this [84].

Albumin binding is believed to provide for an *in situ*-generated form of the EPR-targeting effect [83] and this is part of the rationale for the attachment of drugs to albumin, either pre- or post-administration. However, a recent study of a platinum(IV) prodrug that binds to albumin has shown that it was equally effective against leukemia where the EPR effect was not expected to be operational [42]. Also, the EPR effect is expected to give rise to only a modest advantage in uptake [85] and in many cases, this will not make up for the additional challenges large particles face following extravasation because of the high-interstitial pressure in tumors.

11.6 Targeting Strategies: Transporters

Cancer cells have a high demand for nutrients because of either rapid growth and division or their unusual metabolic properties and a consequent need for a particular nutrient in very high amounts. Therefore, exploiting nutrient uptake pathways, in general, is a mechanism for targeted delivery of active moieties to cancer cells [86], and exploiting specific pathways based on the metabolic profile of an individual cancer is a basis for precision oncology [1]. While it is often stated that cancer cells grow and divide rapidly, this is only true for the subset of cancer cells that have access to oxygen and other nutrients at the levels needed to sustain such activity. Those cells that are more than about 70 μm from the vasculature are typically in a hypoxic environment, unable to access these nutrients at such levels, and consequently enter a quiescent state. Despite this, they often have aberrant metabolic properties, and therefore targeting is still possible. Finding a means of targeting, these quiescent cells is particularly important because they are resistant to treatments that depend on either rapid cell division or normal oxygen levels to be effective.

The potential of nutrient-based targeting is amply demonstrated by the use of substrates of prostate-specific membrane antigen (PSMA) to deliver radionuclides to prostate cancer (PCa) cells. PSMA is a carboxypeptidase and its preferred substrate is glutamate [87]. A urea-based substrate has been developed [88] that has proven to be a very effective agent for delivering radiolabeled compounds to PSMA-expressing cells and this has revolutionized the diagnosis and treatment of PCa [89, 90]. PSMA-targeting groups have been used to deliver technetium, rhenium, gallium, yttrium, actinium, and lutetium complexes for radio imaging and/or therapy and a number are in clinical use [91–96]. It remains to be seen whether the PSMA-targeting approach could be used to deliver a sufficient amount of a nonradioactive drug for it to function more broadly as a targeting mechanism, but the principle is clearly established by this example.

The nutrients that have received the most attention as targeting vectors include folate and glucose, while other nutrients, such as glutamine, have substantial potential because of the reliance of many prostate and breast cancers on it as a primary nutrient, but have only received limited attention. To be effective, a targeting vector must have the following features – a substantially higher rate of accumulation in cancer cells than in healthy cells (selectivity), a rate of uptake and accumulation that enables delivery of sufficient compound to result in cell death, and an uptake pathway that delivers the compound to the regions of the cell where it can exert its action.

To achieve selectivity, uptake via other pathways and via passive diffusion, in particular, must be low to minimize uptake by healthy cells. Attachment of a targeting vector to the drug moiety to be delivered can interfere with this selectivity, but this is an aspect that has received limited attention. Also, the level of nutrient uptake by the cells being targeted must be much higher than uptake by healthy cells. In this respect, folate is an excellent targeting vector because most cells in the body do not express folate receptors whereas many cancer cells express them at a high level [97]. Healthy cells typically take up reduced folates, such as tetrahydrofolate, via a different pathway, but kidney and lung cells do take up folate and it is converted to tetrahydrofolate in the liver. Consequently, toxicity toward these organs needs to be considered when developing folate-receptor-targeted agents.

The rate of uptake and accumulation needed for an agent being delivered via a nutrient transporter depends on the potency of that compound. Generally, this has been described in terms of the cytotoxicity of the compound, and it has been suggested that the IC_{50} needs to be of the order of 10 nM for delivery by the folate and other pathways to be effective [86, 98]. However, cytotoxicity determined in the usual way is not necessarily a good indicator of the potential of targeted delivery. Cisplatin, for example, has modest cytotoxicity with values against most cancer cell lines being in the 1–10 μM range. However, this is greatly influenced by its relatively poor cellular uptake and the addition of lipophilic groups can give rise to much higher cytotoxicities as a result of better cellular uptake, despite the active agent remaining the same [99]. Therefore, a better indicator of the potential for targeted delivery would be the intracellular levels of the compound needed to kill a cell. Fortunately, for metal complexes, particularly those of metals that are not present naturally, this is relatively easy to determine by using methods, such as ICPMS.

The pathway associated with nutrient uptake can have a substantial effect on the ultimate location of the agent being delivered, particularly if it involves endocytosis. Uptake via folate receptors does involve endocytosis and the endosome generated has unusual features, such as low pH that might interfere with the functioning of the agent being delivered. Also, escape from the endosome will generally be necessary and it has been suggested that this is one of the greatest remaining impediments to successful targeting via the folate uptake pathway [86].

While not usually thought of as a nutrient, albumin is taken up by some cancer cells to be used as a source of amino acids and consequently can be used for the targeted delivery of drugs. Because of its size, it may also exploit the EPR effect (see above) and attachment to albumin is known to greatly increase the plasma half-life of the attached group. Albumin targeting is used successfully in the agent Abraxane, which is albumin-delivered paclitaxel [100] and albumin-delivered forms of doxorubicin, and other agents are being developed [101, 102]. The drug to be delivered can be attached to albumin *ex vivo* or can be attached to a group that will bind to albumin in the bloodstream. Platinum(IV) complexes with maleimide in the axial site are showing great promise and are expected to enter clinical trials in the near future. In this case, the maleimide reacts with a cysteine thiol on albumin, creating a covalent link. The active platinum will be released from the albumin on reduction, ideally within a cancer cell and accumulation as a nutrient will promote this. Albumin has also been investigated for the delivery of triethylphosphinegold(I) for the treatment of rheumatoid arthritis [103].

Hormone receptors, such as HER2, have also been investigated as targeting mechanisms for platinum(IV) complexes with encouraging results [104] as have a number of peptides taken up selectively by cancer cells. However, the ability of such mechanisms to deliver sufficient cytotoxin remains unproven.

Nanoparticles, such as carbon nanotubes and aptamers that have targeting vectors, such as folate or PSMA attached along with a prodrug form of a therapeutic such as platinum(IV), have been investigated [48, 105, 106]. This approach has the potential to exploit both the benefits of the nutrient transporter pathways and the EPR effect.

11.7 Targeting Strategies: Enzyme Activation

The extracellular environment of tumors and inflamed tissues is typified by the presence of high concentrations of enzymes, and particularly of proteases. MMPs are involved in the extensive tissue remodeling associated with rapid tumor growth and can be produced by either the tumor cells or the healthy cells surrounding the tumor [107]. In PCa, the protease kallikrein-3, more widely known as prostate-specific antigen (PSA), is greatly overexpressed and the level of expression is used as a marker of PCa progression [108]. MMPs and PSA can be used to cleave peptides that link a drug to a group that greatly slows uptake by cells, removing the blocking group and enabling uptake by cells in the region (Figure 11.8). This is an approach that can be exploited for the delivery of drugs bound to metal by attaching the complex to a peptide that is cleaved by the target enzyme.

The principle underlying protease activation is that the intact peptide with the drug moiety attached is unable to readily enter cells, either because the peptide is sufficient to slow uptake or because charged groups, particularly the negatively

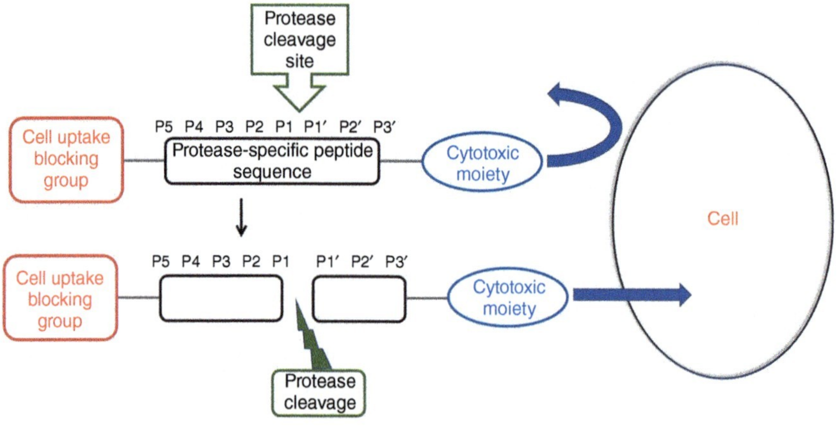

Figure 11.8 Schematic showing the cleavage of a peptide sequence by a protease to release the biologically active moiety.

charged amino acids, glutamic acid, and aspartic acid, have been incorporated in the peptide sequence. Cleavage of the peptide by a protease removes sufficient of the amino acids, including any added charged amino acids, enabling the remainder of the peptide along with the cytotoxin to enter cells. Rapid uptake at this point is important to minimize diffusion from the site and collateral damage known as "bystander effects." Also, excellent selectivity is needed to minimize cleavage by proteases found elsewhere in the body, but rapid cleavage is desirable to enable high concentrations of the active agent to be released and these two aspects can be in conflict. Protease activation can have significant advantages over other strategies in that the proteases are generated in some cases by healthy cells in the vicinity as part of the body's response to the presence of the tumor. Consequently, the rapid emergence of resistance arising from mutations in the cancer is much less likely.

We and others have generated proof-of-principle entities that demonstrate the ability of this approach to control cellular uptake. However, in the process, we have observed that the cellular models reported to express high levels of the proteases being targeted appear to produce less active or lower amounts of protease than expected [109]. Similarly, there have been numerous attempts to develop protease-activated targeted agents for clinical application, and significant promise has been shown for both MMP and PSA-activated agents [110–112]. However, no such compound has reached the clinic and a contributor to this may be the inadequacy of many preclinical models. Both MMPs and PSA exist in inactive forms and must be activated. In real tumors, more than 90% of the PSA enzyme exists in the active form, but in 2D-cell culture and xenografts most of it is inactive [108]. Consequently, modest preclinical results may be due in part to the amount of active enzyme present being much less than would be found in a clinical situation resulting in agents with significant potential being abandoned.

11.8 Other Targeting Strategies

Molecular targeting of kinases: Kinase inhibitors have been at the forefront in the developments of new anticancer agents for the past 25 years and have made major contributions to the management of many forms of cancer. Metal complexes can be designed to have the stereochemical features needed to function as kinase inhibitors and modular construction techniques and greater range of geometries available can be exploited to generate these features quickly and cheaply. Meggers has demonstrated the potential of this approach, producing kinase inhibitors with very potency and selectivity [113]. However, he also noted that the number of isomers that can be generated can make it difficult to produce a single pure compound.

Mitochondrial targeting: Generally, cancer cells have a much higher mitochondrial membrane potential than healthy cells and therefore are unusually sensitive to poisons that are taken up by the mitochondria. Lipophilic cations, in particular, have been used in this respect [114], ranging from the cationic phosphine complexes mentioned above [24] to gadolinium complexes attached to arylphosphonium cations [115]. However, it is not yet clear whether this approach will give rise to the desired level of targeting because the compounds will be taken up by all cells, and toxicity outside the mitochondria must, therefore, be minimal.

Antibody targeting: Many of the most successful molecularly targeted pharmaceuticals are based on antibodies, such Herceptin (trastuzumab) and the use of antibodies to deliver metal complexes as antibody–drug conjugates (ADCs) has been investigated. For example, trastuzumab has been investigated for the selective delivery of platinum(II) and platinum(IV) complexes to HER2-positive cells [116, 117]. Also, oxaliplatin has been selectively targeted to hepatocellular carcinoma using anti-c-Met IgG [118].

Platinum has also been used as a linker to create a "plug-and-play" ADC motif in which the platinum is attached to an antibody and the payload of choice is then bound to that platinum, an approach that has more in common with the chaperones described above than the active platinum complexes [119–121]. Much more widespread is the use of antibodies for delivering radio imaging agents and radiotherapeutic agents [122, 123].

Bioorthogonal coupling: The administration of a targeting vector attached to a coupling group followed, after an appropriate washout time, of the cytotoxin attached a coupling partner is an approach with substantial promise because of the ability to minimize toxicity and maximize the rate and selectivity of localization of the cytotoxin in the vicinity of the target [124]. There are examples of it being used in the delivery of metal-based radiopharmaceuticals [125] and this is an approach worthy of further investigation. Bioorthogonal coupling could be used for delivery of metal-based chaperones, but to the best of our knowledge is not an approach that has been investigated in this regard.

11.9 Conclusions

As the foregoing demonstrates, metal complexes have much to offer in the targeting of disease, both as the means to deliver toxic agents or the end by functioning as toxic agents themselves. The synthetic versatility of coordination chemistry, coupled with the ability to attach unmodified drugs and/or targeting groups, makes it possible to investigate a wide range of options quickly and cheaply. There is a perception that metals are too toxic to use in these ways, but metals such as cobalt and copper that are naturally present and therefore can be managed by the body can be used. Also, targeted delivery of platinum anticancer agents has the potential to greatly reduce the toxicity of these important drugs. Finally, the ability of metal complexes to respond to their environments creates additional means of targeting that are less readily achievable using other types of compounds.

At present, we know of no active clinical trials of targeted metal complexes, though one or two trials are anticipated for platinum(IV) complexes that couple *in vivo* to albumin. Given the potential outlined above, it is important to consider why more clinical trials are not underway or being planned. Srinivasarao and Low have noted that the intracellular fate of targeted drugs and in particular their possible entrapment in subcellular organelles is one of the remaining challenges [86] and this applies to all targeted drugs, including metal complexes. However, two other issues are likely to be the most significant for metal complexes – developing or using compounds that are sufficiently toxic to kill cells at the concentrations achievable using the uptake mechanisms described above, and developing prodrugs that are sufficiently stable to survive *in vivo* for long enough for the delivery mechanisms to take effect. It has been suggested that a cytotoxin with an IC_{50} of 10 nM or less is needed for targeted delivery to be effective [98]. Most metal complexes have cytotoxicities of 1 mM or greater, though there are examples of platinum and osmium complexes among others that are likely to be sufficiently cytotoxic [126–128]. As discussed above, cytotoxicities may not be an adequate indicator because delivery of cisplatin as a lipophilic platinum(IV) complex can result in cytotoxicity that is in the range needed. Therefore, a better understanding of the effect of metal-based drugs delivered into cells is needed and those that are effective at intracellular concentration achievable by targeted delivery need to be identified. Finally, the prodrug form of the complex must survive in the bloodstream long enough to reach the tumor and be accumulated by the tumor cells at sufficient concentrations. This is a major challenge because there are many reductants and alternative ligands in the bloodstream and uptake by red blood cells is likely to contribute to the loss of prodrug from the pool available for the delivery to the tumor. Much work has gone into finding platinum(IV) and cobalt(III) complexes that meet these criteria, but very little for any other metals. The limited stability of most of the metal complexes that have been investigated as anticancer agents is likely to mean that they are unsuitable for targeted delivery and a greater focus is needed on identifying complexes that are sufficiently stable but are activated rapidly once in the cell.

References

1 Hambley, T.W. (2019). Transporter and protease mediated delivery of platinum complexes for precision oncology. *J. Biol. Inorg. Chem.* 24: 457.

2 Mc Carron, P., Crowley, A., O'Shea, D. et al. (2018). Targeting the folate receptor: improving efficacy in inorganic medicinal chemistry. *Curr. Med. Chem.* 25 (23): 2675.

3 O'Donnell, J.S., Long, G.V., Scolyer, R.A. et al. (2017). Resistance to PD1/PDL1 checkpoint inhibition. *Cancer Treat. Rev.* 52: 71.

4 Mullard, A. (2019). Roger Perlmutter. *Nat. Rev. Drug Discovery* 18 (11): 818.

5 Mathew, M., Enzler, T., Shu, C.A., and Rizvi, N.A. (2018). Combining chemotherapy with PD-1 blockade in NSCLC. *Pharmacol. Ther.* 186: 130.

6 Wheate, N.J., Walker, S., Craig, G.E., and Oun, R. (2010). The status of platinum anticancer drugs in the clinic and in clinical trials. *Dalton Trans.* 39 (35): 8113.

7 Um, I.S., Armstrong-Gordon, E., Moussa, Y.E. et al. (2019). Platinum drugs in the Australian cancer chemotherapy healthcare setting: is it worthwhile for chemists to continue to develop platinums? *Inorg. Chim. Acta* 492: 177.

8 Kelland, L.R. (2007). The resurgence of platinum-based cancer chemotherapy. *Nat. Rev. Cancer* 7: 573.

9 Oun, R., Moussa, Y.E., and Wheate, N.J. (2018). The side effects of platinum-based chemotherapy drugs: a review for chemists (vol 47, pg 6645, 2018). *Dalton Trans.* 47 (23): 7848.

10 Bryce, N.S., Zhang, J.Z., Whan, R.M. et al. (2009). Accumulation of an anthraquinone and its platinum complexes in cancer cell spheroids: the effect of charge on drug distribution in solid tumour models. *Chem. Commun.*, 2673

11 Mellor, H.R., Snelling, S., Hall, M.D. et al. (2005). The influence of tumour microenvironmental factors on the efficacy of cisplatin and novel platinum(IV) complexes. *Biochem. Pharmacol.* 70: 1137.

12 Failes, T.W. and Hambley, T.W. (2007). Towards bioreductively activated prodrugs: Fe(III) complexes of hydroxamic acids and the MMP inhibitor marimastat. *J. Inorg. Biochem.* 101: 396.

13 Raymond, K.N., Allred, B.E., and Sia, A.K. (2015). Coordination chemistry of microbial iron transport. *Acc. Chem. Res.* 48 (9): 2496.

14 Samuni, Y., Samuni, U., and Goldstein, S. (2012). The mechanism underlying nitroxyl and nitric oxide formation from hydroxamic acids. *Biochim. Biophys. Acta-Gen. Subj.* 1820 (10): 1560.

15 Ware, D.C., Brothers, P.J., Clark, G.R. et al. (2000). Synthesis, structures and hypoxia-selective cytotoxicity of cobalt(III) complexes containing tridentate amine and nitrogen mustard ligands. *J. Chem. Soc., Dalton Trans.* 925.

16 Wilson, W.R., Moselen, J.W., Cliffe, S. et al. (1994). Exploiting tumour hypoxia through bioreductive release of diffusible cytotoxins: the cobalt(III)-nitrogen mustard complex SN 24771. *Int. J. Radiat. Oncol. Biol. Phys.* 29 (2): 323.

17 Ware, D.C., Palmer, B.D., Wilson, W.R., and Denny, W.A. (1993). Hypoxia-selective antitumor agents. 7. Metal complexes of aliphatic mustards as a new class of hypoxia-selective cytotoxins. Synthesis and evaluation of cobalt(III) complexes of bidentate mustards. *J. Med. Chem.* 36: 1839.

18 Failes, T.W., Cullinane, C., Diakos, C.I. et al. (2007). Studies of a Co(III) complex of the MMP inhibitor marimastat: a potential hypoxia activated pro-drug. *Chem. Eur. J.* 13: 2974.

19 Chen, C.K.J., Kappen, P., Gibson, D., and Hambley, T.W. (2020). *trans*-Platinum(IV) pro-drugs that exhibit unusual resistance to reduction by endogenous reductants and blood serum but are rapidly activated inside cells: ^{1}H NMR and XANES spectroscopy study. *Dalton Trans.* 48: 7722.

20 Chen, C.K.J., Kappen, P., and Hambley, T.W. (2019). The reduction of *cis*-platinum(IV) complexes by ascorbate and in whole human blood models using ^{1}H NMR and XANES spectroscopy. *Metallomics* 11 (3): 686.

21 Renfrew, A.K., Bryce, N.S., and Hambley, T. (2015). Cobalt(III) chaperone complexes of curcumin: photoreduction, cellular accumulation and light-selective toxicity towards tumour cells. *Chem. Eur. J.* 21 (43): 15224.

22 Bansal, S.S., Goel, M., Aqil, F. et al. (2011). Advanced drug delivery systems of curcumin for cancer chemoprevention. *Cancer Prev. Res.* 4 (8): 1158.

23 Renfrew, A.K., Bryce, N.S., and Hambley, T.W. (2013). Delivery and release of curcumin by a hypoxia-activated cobalt chaperone: a XANES and FLIM study. *Chem. Sci.* 4 (9): 3731.

24 Berners-Price, S.J. and Sadler, P.J. (1987). Phosphines in medicine. *Chem. Br.* 541.

25 Bowen, R.J., Navarro, M., Shearwood, A.M.J. et al. (2009). 1 : 2 Adducts of copper(I) halides with 1,2-bis(di-2-pyridylphosphino)ethane: solid state and solution structural studies and antitumour activity. *Dalton Trans.* (48): 10861. https://doi.org/10.1039/b912281h

26 Berners-Price, S.J., Bowen, R.J., Galettis, P. et al. (1999). Structural and solution chemistry of gold(I) and silver(I) complexes of bidentate pyridyl phosphines: selective antitumour agents. *Coord. Chem. Rev.* 186: 823.

27 Yamamoto, N., Renfrew, A.K., Kim, B.J. et al. (2012). Dual targeting of hypoxic and acidic tumor environments with a cobalt(III) chaperone complex. *J. Med. Chem.* 55 (24): 11013.

28 Yamamoto, N., Danos, S., Bonnitcha, P.D. et al. (2008). Cellular uptake and distribution of cobalt complexes of fluorescent ligands. *J. Biol. Inorg. Chem.* 13: 861.

29 Hall, M.D., Failes, T.W., Yamamoto, N., and Hambley, T.W. (2007). Bioreductive activation and drug chaperoning in cobalt pharmaceuticals. *Dalton Trans.* (36): 3983. https://doi.org/10.1039/b707121c

30 Kim, B.J., Hambley, T.W., and Bryce, N.S. (2011). Visualising the hypoxia selectivity of cobalt(III) prodrugs. *Chem. Sci.* 2 (11): 2135.

31 Liu, S.Q., Ono, R.J., Yang, C. et al. (2018). Dual pH-responsive shell-cleavable polycarbonate micellar nanoparticles for in vivo anticancer drug delivery. *ACS Appl. Mater. Interfaces* 10 (23): 19355.

32 Sadler, P.J. and Guo, Z. (1998). Metal complexes in medicine: design and mechanism of action. *Pure and Appl. Chem.* 70 (4): 863.

33 Sadler, P.J., Li, H., and Sun, H. (1999). Coordination chemistry of metals in medicine: target sites for bismuth. *Coord. Chem. Rev.* 185–186: 689.

34 Finkelstein, A.E., Walz, D.T., Batista, V. et al. (1976). Auranofin – new oral gold compound for treatment of rheumatoid arthritis. *Ann. Rheum. Dis.* 35 (3): 251.

35 Johnstone, T.C., Suntharalingam, K., and Lippard, S.J. (2016). The next generation of platinum drugs: targeted Pt(II) agents, nanoparticle delivery, and Pt(IV) prodrugs. *Chem. Rev.* 116 (5): 3436.

36 Zhang, J.Z., Wexselblatt, E., Hambley, T.W., and Gibson, D. (2012). Pt(IV) analogs of oxaliplatin that do not follow the expected correlation between electrochemical reduction potential and rate of reduction by ascorbate. *Chem. Commun.* 48 (6): 847.

37 Wilson, J.J. and Lippard, S.J. (2011). Synthesis, characterization, and cytotoxicity of platinum(IV) carbamate complexes. *Inorg. Chem.* 50 (7): 3103.

38 Hofer, D., Varbanov, H.P., Hejl, M. et al. (2017). Impact of the equatorial coordination sphere on the rate of reduction, lipophilicity and cytotoxic activity of platinum(IV) complexes. *J. Inorg. Biochem.* 174: 119.

39 Ellis, L.T., Er, H.M., and Hambley, T.W. (1995). The influence of the axial ligands of a series of Pt(IV) anti-cancer complexes on their reduction to Pt(II) and reaction with DNA. *Aust. J. Chem.* 48: 793.

40 Wexselblatt, E. and Gibson, D. (2012). What do we know about the reduction of Pt(IV) pro-drugs? *J. Inorg. Biochem.* 117: 220.

41 Nemirovski, A., Kasherman, Y., Tzaraf, Y., and Gibson, D. (2007). Reduction of *cis,trans,cis*-$[PtCl_2(OCOCH_3)_2(NH_3)_2]$ by aqueous extracts of cancer cells. *J. Med. Chem.* 50 (23): 5554.

42 Kowol, C. R., Heffeter, P., Berger, W. et al. (2016). Monomaleimide-functionalized platinum compounds for cancer therapy. WIPO, Ed. Austria, filed 09 December 2016 and published 30 July 2018.

43 Zhang, J.Z., Bonnitcha, P., Wexselblatt, E. et al. (2013). Facile preparation of mono-, di- and mixed-carboxylato platinum(IV) complexes for versatile anti-cancer prodrug design. *Chem. Eur. J.* 19 (5): 1672.

44 Giandomenico, C.M., Abrams, M.J., Murrer, B.A. et al. (1995). Carboxylation of kinetically inert platinum(IV) hydroxy complexes. An entree into orally active platinum(IV) antitumor agents. *Inorg. Chem.* 34: 1015.

45 Ravera, M., Gabano, E., Tinello, S. et al. (2017). May glutamine addiction drive the delivery of antitumor cisplatin-based Pt(IV) prodrugs? *J. Inorg. Biochem.* 167: 27.

46 Ma, J., Liu, H.F., Xi, Z.Q. et al. (2018). Protected and de-protected platinum(IV) glycoconjugates with GLUT1 and OCT2-mediated selective cancer targeting: demonstrated enhanced transporter-mediated cytotoxic properties *in vitro* and *in vivo*. *Front. Chem.* 6: 15.

47 Ma, J., Wang, Q.P., Huang, Z.L. et al. (2017). Glycosylated platinum(IV) complexes as substrates for glucose transporters (GLUTs) and organic cation transporters (OCTs) exhibited cancer targeting and human serum albumin binding properties for drug delivery. *J. Med. Chem.* 60 (13): 5736.

48 Dhar, S., Liu, Z., Thomale, J. et al. (2008). Targeted single-wall carbon nanotube-mediated Pt(IV) prodrug delivery using folate as a homing device. *J. Am. Chem. Soc.* 130 (34): 11467.

49 Aronov, O., Horowitz, A.T., Gabizon, A., and Gibson, D. (2003). Folate-targeted PEG as a potential carrier for carboplatin analogs. Synthesis and in vitro studies. *Bioconjugate Chem.* 14 (3): 563.

50 Patra, M., Johnstone, T.C., Suntharalingam, K., and Lippard, S.J. (2016). A potent glucose-platinum conjugate exploits glucose transporters and preferentially accumulates in cancer cells. *Angew. Chem. Int. Ed.* 55 (7): 2550.

51 Patra, M., Awuah, S.G., and Lippard, S.J. (2016). Chemical approach to positional isomers of glucose–platinum conjugates reveals specific cancer targeting through glucose-transporter-mediated uptake *in vitro* and *in vivo*. *J. Am. Chem. Soc.* 138 (38): 12541.

52 Chen, Y., Heeg, M.J., Braunschweiger, P.G. et al. (1999). A carbohydrate linked cisplatin analogue having antitumor activity. *Angew. Chem. Int. Ed.* 38 (12): 1768.

53 Li, T.L., Gao, X.Q., Yang, L. et al. (2016). Methyl 6-amino-6-deoxy-D-pyranoside-conjugated platinum(II) complexes for glucose transporter (GLUT)-mediated tumor targeting: synthesis, cytotoxicity, and cellular uptake mechanism. *ChemMedChem* 11 (10): 1069.

54 Liu, R., Fu, Z., Zhao, M. et al. (2017). GLUT1-mediated selective tumor targeting with fluorine containing platinum(II) glycoconjugates. *Oncotarget* 8 (24): 39476.

55 Cheng, Q.Q., Shi, H.D., Wang, H.X. et al. (2016). Asplatin enhances drug efficacy by altering the cellular response. *Metallomics* 8 (7): 672.

56 Dhar, S. and Lippard, S.J. (2009). Mitaplatin, a potent fusion of cisplatin and the orphan drug dichloroacetate. *Proc. Natl. Acad. Sci. U.S.A.* 106 (52): 22199.

57 Petruzzella, E., Braude, J.P., Aldrich-Wright, J.R. et al. (2017). A quadruple-action platinum(IV) prodrug with anticancer activity against KRAS mutated cancer cell lines. *Angew. Chem. Int. Ed.* 56 (38): 11539.

58 Babak, M.V., Zhi, Y., Czarny, B. et al. (2019). Dual-targeting dual-action platinum(IV) platform for enhanced anticancer activity and reduced nephrotoxicity. *Angew. Chem. Int. Ed.* 58 (24): 8109.

59 Lee, K.G.Z., Babak, M.V., Weiss, A. et al. (2018). Development of an efficient dual-action GST-inhibiting anticancer platinum(IV) prodrug. *ChemMedChem* 13 (12): 1210.

60 Novohradsky, V., Zerzankova, L., Stepankova, J. et al. (2014). Antitumor platinum(IV) derivatives of oxaliplatin with axial valproato ligands. *J. Inorg. Biochem.* 140: 72.

61 Karmakar, S., Poetsch, I., Kowol, C.R. et al. (2019). Synthesis and cytotoxicity of water-soluble dual- and triple-action satraplatin derivatives: replacement of equatorial chlorides of satraplatin by acetates. *Inorg. Chem.* 58 (24): 16676.

62 Zhang, X.N., Selvaraju, K., Saei, A.A. et al. (2019). Repurposing of auranofin: thioredoxin reductase remains a primary target of the drug. *Biochimie* 162: 46.

63 Jain, R.K. (1998). The next frontier of molecular medicine: delivery of therapeutics. *Nat. Med.* 4 (6): 655.

64 Tannock, I.F. and Rotin, D. (1989). Acid pH in tumors and its potential for therapeutic exploitation. *Cancer Res.* 49 (16): 4373.

65 Minchinton, A.I. and Tannock, I.F. (2006). Drug penetration in solid tumours. *Nat. Rev. Cancer* 6 (8): 583.

66 Hambley, T.W. (2008). Physiological targeting to improve anticancer drug selectivity. *Aust. J. Chem.* 61 (9): 647.

67 Phillips, R.M. (2016). Targeting the hypoxic fraction of tumours using hypoxia-activated prodrugs. *Cancer Chemother. Pharmacol.* 77 (3): 441.

68 Wilson, W.R. and Hay, M.P. (2011). Targeting hypoxia in cancer therapy. *Nat. Rev. Cancer* 11 (6): 393.

69 Mao, X.J., McManaway, S., Jaiswal, J.K. et al. (2019). Schedule-dependent potentiation of chemotherapy drugs by the hypoxia-activated prodrug SN30000. *Cancer Biol. Ther.* 20 (9): 1258.

70 Rischin, D., Peters, L.J., O'Sullivan, B. et al. (2010). Tirapazamine, cisplatin, and radiation versus cisplatin and radiation for advanced squamous cell carcinoma

of the head and neck (TROG 02.02, HeadSTART): a phase III trial of the Trans-Tasman Radiation Oncology Group. *J. Clin. Oncol.* 28 (18): 2989.
71 DiSilvestro, P.A., Ali, S., Craighead, P.S. et al. (2014). Phase III randomiz ed trial of weekly cisplatin and irradiation versus cisplatin and tirapazamine and irradiation in stages IB2, IIA, IIB, IIIB, and IVA cervical carcinoma limited to the pelvis: a Gynecologic Oncology Group Study. *J. Clin. Oncol.* 32 (5): 458.
72 Schreiber-Brynzak, E., Pichler, V., Heffeter, P. et al. (2016). Behavior of platinum(IV) complexes in models of tumor hypoxia: cytotoxicity, compound distribution and accumulation. *Metallomics* 8 (4): 422.
73 Renfrew, A.K., O'Neill, E.S., Hambley, T.W., and New, E.J. (2018). Harnessing the properties of cobalt coordination complexes for biological application. *Coord. Chem. Rev.* 375: 221.
74 Ware, D.C., Siim, B.G., Robinson, K.G. et al. (1991). Synthesis and characterization of aziridine complexes of cobalt(III) and chromium(III) designed as hypoxia-selective cytotoxins. X-ray crystal structure of *trans*-$[co(Az)_4(NO_2)_2]$ $Br{\cdot}2H_2O{\cdot}LiBr$. *Inorg. Chem.* 30: 3750.
75 Ware, D.C., Wilson, W.R., Denny, W.A., and Rickard, C.E.F. (1991). Design and synthesis of cobalt(III) nitrogen mustard complexes as hypoxia selective cytotoxins. The X-ray crystal structure of bis(3-chloropentane-2,4-dionato) (*RS-N,N'*-bis(2-chloroethyl)ethylenediamine)cobalt(III) perchlorate, $[Co(Clacac)_2(bce)]ClO_4$. *J. Chem. Soc., Chem. Commun.* (17): 1171. https://doi.org/10.1039/c39910001171
76 Green, B.P., Renfrew, A.K., Glenister, A. et al. (2017). The influence of the ancillary ligand on the potential of cobalt(III) complexes to act as chaperones for hydroxamic acid-based drugs. *Dalton Trans.* 46 (45): 15897.
77 Glenister, A., Chen, C.K.J., Tondl, E.M. et al. (2017). Targeting curcumin to specific tumour cell environments: the influence of ancillary ligands. *Metallomics* 9 (6): 699.
78 O'Neill, E.S., Kaur, A., Bishop, D.P. et al. (2017). Hypoxia-responsive cobalt complexes in tumor spheroids: laser ablation inductively coupled plasma mass spectrometry and magnetic resonance imaging studies. *Inorg. Chem.* 56 (16): 9860.
79 Renfrew, A.K. (2014). Transition metal complexes with bioactive ligands: mechanisms for selective ligand release and applications for drug delivery. *Metallomics* 6 (8): 1324.
80 Ang, W.H., Khalaila, I., Allardyce, C.S. et al. (2005). Rational design of platinum(IV) compounds to overcome glutathione-S-transferase mediated drug resistance. *J. Am. Chem. Soc.* 127 (5): 1382.
81 Raghunand, N., He, X., van Sluis, R. et al. (1999). Enhancement of chemotherapy by manipulation of tumour pH. *Br. J. Cancer* 80 (7): 1005.
82 Berners-Price, S.J., Frenkiel, T.A., Frey, U. et al. (1992). Hydrolysis products of cisplatin: pK_a determinations via $[^1H, ^{15}H]$ NMR spectroscopy. *J. Chem. Soc., Chem. Commun.* 789.

83 Schilling, U., Friedrich, E.A., Sinn, H. et al. (1992). Design of compounds having enhanced tumor uptake, using serum albumin as a carrier—part II. *In vivo* studies. *Int. J. Radiat. Appl. Instrum. Part B* 19 (6): 685.

84 Bort, G., Lux, F., Dufort, S. et al. (2020). EPR-mediated tumor targeting using ultrasmall-hybrid nanoparticles: from animal to human with theranostic AGuIX nanoparticles. *Theranostics* 10 (3): 1319.

85 Youn, Y.S. and Bae, Y.H. (2018). Perspectives on the past, present, and future of cancer nanomedicine. *Adv. Drug Delivery Rev.* 130: 3.

86 Srinivasarao, M. and Low, P.S. (2017). Ligand-targeted drug delivery. *Chem. Rev.* 117 (19): 12133.

87 Chang, S.S. (2004). Overview of prostate-specific membrane antigen. *Rev. Urol.* 6: S13.

88 Kularatne, S.A., Zhou, Z.G., Yang, J. et al. (2009). Design, synthesis, and preclinical evaluation of prostate-specific membrane antigen targeted ^{99m}Tc-radioimaging agents. *Mol. Pharmaceutics* 6 (3): 790.

89 Chen, Y., Foss, C.A., Byun, Y. et al. (2008). Radiohalogenated prostate-specific membrane antigen (PSMA)-based ureas as imaging agents for prostate cancer. *J. Med. Chem.* 51 (24): 7933.

90 Kularatne, S.A., Wang, K., Santhapuram, H.K.R., and Low, P.S. (2009). Prostate-specific membrane antigen targeted imaging and therapy of prostate cancer using a PSMA inhibitor as a homing ligand. *Mol. Pharmaceutics* 6 (3): 780.

91 Banerjee, S.R., Foss, C.A., Castanares, M. et al. (2008). Synthesis and evaluation of technetium-99m- and rhenium-labeled inhibitors of the prostate-specific membrane antigen (PSMA). *J. Med. Chem.* 51 (15): 4504.

92 Banerjee, S.R., Pullambhatla, M., Byun, Y. et al. (2010). ^{68}Ga-labeled inhibitors of prostate-specific membrane antigen (PSMA) for imaging prostate cancer. *J. Med. Chem.* 53 (14): 5333.

93 Frei, A., Fischer, E., Childs, B.C. et al. (2019). Two is better than one: difunctional high-affinity PSMA probes based on a [CpM(CO)$_3$] (M = Re/^{99m}Tc) scaffold. *Dalton Trans.* 48 (39): 14600.

94 Weineisen, M., Schottelius, M., Simecek, J. et al. (2015). ^{68}Ga- and ^{177}Lu-labeled PSMA I&T: optimization of a PSMA-targeted theranostic concept and first proof-of-concept human studies. *J. Nucl. Med.* 56 (8): 1169.

95 Sathekge, M., Bruchertseifer, F., Knoesen, O. et al. (2019). ^{225}Ac-PSMA-617 in chemotherapy-naive patients with advanced prostate cancer: a pilot study. *Eur. J. Nucl. Med. Mol. Imaging* 46 (1): 129.

96 Rathke, H., Flechsig, P., Mier, W. et al. (2019). Dosimetry estimate and initial clinical experience with ^{90}Y-PSMA-617. *J. Nucl. Med.* 60 (6): 806.

97 Graybill, W.S. and Coleman, R.L. (2016). Folate receptor-targeted therapeutics for ovarian cancer. *Drugs Future* 41 (2): 137.

98 Srinivasarao, M., Galliford, C.V., and Low, P.S. (2015). Principles in the design of ligand-targeted cancer therapeutics and imaging agents. *Nat. Rev. Drug Discovery* 14 (3): 203.

99 Kelland, L.R., Murrer, B.A., Abel, G. et al. (1992). Ammine amine platinum(IV) dicarboxylates – a novel class of platinum complex exhibiting selective cytotoxicity to intrinsically cisplatin-resistant human ovarian-carcinoma cell-lines. *Cancer Res.* 52: 822.

100 Sofias, A.M., Dunne, M., Storm, G., and Allen, C. (2017). The battle of "nano" paclitaxel. *Adv. Drug Delivery Rev.* 122: 20.

101 Gong, J., Yan, J., Forscher, C., and Hendifar, A. (2018). Aldoxorubicin: a tumor-targeted doxorubicin conjugate for relapsed or refractory soft tissue sarcomas. *Drug Des. Dev. Ther.* 12: 777.

102 Seetharam, M., Kolla, K.R., and Ganjoo, K.N. (2018). Aldoxorubicin therapy for the treatment of patients with advanced soft tissue sarcoma. *Future Oncol.* 14 (23): 2323.

103 Dean, T.C., Yang, M., Liu, M.Y. et al. (2017). Human serum albumin-delivered $[Au(PEt_3)]^+$ is a potent inhibitor of T cell proliferation. *ACS Med. Chem. Lett.* 8 (5): 572.

104 Wong, D.Y.Q., Lim, J.H., and Ang, W.H. (2015). Induction of targeted necrosis with HER2-targeted platinum(IV) anticancer prodrugs. *Chem. Sci.* 6 (5): 3051.

105 Dhar, S., Gu, F.X., Langer, R. et al. (2008). Targeted delivery of cisplatin to prostate cancer cells by aptamer functionalized Pt(IV) prodrug-PLGA-PEG nanoparticles. *Proc. Natl. Acad. Sci. U.S.A.* 105 (45): 17356.

106 Feazell, R.P., Nakayama-Ratchford, N., Dai, H., and Lippard, S.J. (2007). Soluble single-walled carbon nanotubes as longboat delivery systems for platinum(IV) anticancer drug design. *J. Am. Chem. Soc.* 129 (27): 8438.

107 Vandooren, J., Opdenakker, G., Loadman, P.M., and Edwards, D.R. (2016). Proteases in cancer drug delivery. *Adv. Drug Delivery Rev.* 97: 144.

108 Denmeade, S.R., Sokoll, L.J., Chan, D.W. et al. (2001). Concentration of enzymatically active prostate-specific antigen (PSA) in the extracellular fluid of primary human prostate cancers and human prostate cancer xenograft models. *Prostate* 48 (1): 1.

109 Di Marco, L., Zhang, J.Z., Doan, J. et al. (2019). Modulating the cellular uptake of fluorescently tagged substrates of prostate-specific antigen before and after enzymatic activation. *Bioconjugate Chem.* 30 (1): 124.

110 LeBeau, A.M., Kostova, M., Craik, C.S., and Denmeade, S.R. (2010). Prostate-specific antigen: an overlooked candidate for the targeted treatment and selective imaging of prostate cancer. *Biol. Chem.* 391 (4): 333.

111 Atkinson, J.M., Siller, C.S., and Gill, J.H. (2008). Tumour endoproteases: the cutting edge of cancer drug delivery? *Br. J. Pharmacol.* 153 (7): 1344.

112 Sun, I., Yoon, H., Lim, D., and Kim, K. (2020). Recent trends in *in situ* enzyme-activatable prodrugs for targeted cancer therapy. *Bioconjugate Chem.* 31: 1012.

113 Meggers, E. (2017). Exploiting octahedral stereocenters: from enzyme inhibition to asymmetric photoredox catalysis. *Angew. Chem. Int. Ed.* 56 (21): 5668.

114 Modica-Napolitano, J.S. and Aprille, J.R. (2001). Delocalized lipophilic cations selectively target the mitochondria of carcinoma cells. *Adv. Drug Delivery Rev.* 49 (1–2): 63.

115 Morrison, D.E., Aitken, J.B., de Jonge, M.D. et al. (2014). High mitochondrial accumulation of new gadolinium(III) agents within tumour cells. *Chem. Commun.* 50 (18): 2252.

116 Huang, R., Sun, Y., Gao, Q.H. et al. (2015). Trastuzumab-mediated selective delivery for platinum drug to HER2-positive breast cancer cells. *Anti-Cancer Drugs* 26 (9): 957.

117 Huang, R., Sun, Y., Zhang, X.Y. et al. (2015). Biological evaluation of a novel Herceptin-platinum (II) conjugate for efficient and cancer cell specific delivery. *Biomed. Pharmacother.* 73: 116.

118 Ma, Y.L., Zhang, M.J., Wang, J.Y. et al. (2019). High-affinity human anti-c-Met IgG conjugated to oxaliplatin as targeted chemotherapy for hepatocellular carcinoma. *Front. Oncol.* 9: 17.

119 Merkul, E., Sijbrandi, N.J., Aydin, I. et al. (2020). A successful search for new, efficient, and silver-free manufacturing processes for key platinum(II) intermediates applied in antibody-drug conjugate (ADC) production. *Green Chem.* 22 (7): 2203.

120 Merkul, E., Sijbrandi, N.J., Muns, J.A. et al. (2019). First platinum(II)-based metal-organic linker technology (Lx®) for a plug-and-play development of antibody-drug conjugates (ADCs). *Expert Opin. Drug Delivery* 16 (8): 783.

121 Waalboer, D.C.J., Muns, J.A., Sijbrandi, N.J. et al. (2015). Platinum(II) as bifunctional linker in antibody-drug conjugate formation: coupling of a 4-nitrobenzo-2-oxa-1,3-diazole fluorophore to trastuzumab as a model. *ChemMedChem* 10 (5): 797.

122 Tafreshi, N.K., Doligalski, M.L., Tichacek, C.J. et al. (2019). Development of targeted alpha particle therapy for solid tumors. *Molecules* 24 (23): 48.

123 Bailly, C., Clery, P.F., Faivre-Chauvet, A. et al. (2017). Immuno-PET for clinical theranostic approaches. *Int. J. Mol. Sci.* 18 (1): 12.

124 Devaraj, N.K. and Weissleder, R. (2011). Biomedical applications of tetrazine cycloadditions. *Acc. Chem. Res.* 44 (9): 816.

125 Vito, A., Alarabi, H., Czorny, S. et al. (2017). A^{99m}Tc-labelled tetrazine for bioorthogonal chemistry. Synthesis and biodistribution studies with small molecule *trans*-cyclooctene derivatives (vol. 11, e0167425, 2016). *Plos One* 12 (2): 1.

126 Manzotti, C., Pratesi, G., Menta, E. et al. (2000). BBR 3464: a novel triplatinum complex, exhibiting a preclinical profile of antitumor efficacy different from cisplatin. *Clin. Cancer Res.* 6: 2626.

127 Smyre, C.L., Saluta, G., Kute, T.E. et al. (2011). Inhibition of DNA synthesis by a platinum-acridine hybrid agent leads to potent cell kill in nonsmall cell lung cancer. *ACS Med. Chem. Lett.* 2 (11): 870.

128 van Rijt, S.H., Romero-Canelon, I., Fu, Y. et al. (2014). Potent organometallic osmium compounds induce mitochondria-mediated apoptosis and S-phase cell cycle arrest in A549 non-small cell lung cancer cells. *Metallomics* 6 (5): 1014.

12

Formulation of Peptides for Targeted Delivery

Pankti Ganatra[1], Karen Saiswani[2], Nikita Nair[2], Avinash Gunjal[2], Ratnesh Jain[1], and Prajakta Dandekar[2]

[1]Institute of Chemical Technology, Department of Chemical Engineering, Nathalal Parekh Marg, Matunga, Mumbai 400 019, India
[2]Institute of Chemical Technology, Department of Pharmaceutical Sciences & Technology, Nathalal Parekh Marg, Matunga, Mumbai 400 019, India

12.1 Introduction

Peptides are the most versatile biomolecules that perform many critical roles in the human body. They act as hormones, neurotransmitters, growth factors, ion-channel ligands, anti-infective agents, or as components of the innate immune system [1]. Progress in science and technology has revealed around 7000 naturally occurring peptides, to date [2]. They are highly selective in action and act by attaching to specific cell-surface receptors, which initiate a cascade of intracellular enzymatic reactions and facilitate their pharmacological action. Considering their pharmacological advantages, peptides are regarded as revolutionary therapeutic molecules that provide high specificity, biocompatibility, efficacy, and tolerability in the human body. These characteristic features differentiate peptide molecules from conventional drug moieties. The boom in the field of biotechnology has provided simple and economic methods for the synthesis of therapeutic peptides. Also, their production cost is much lower than that of the protein molecules. Peptides have been considered for pharmacotherapy ever since the commercial availability of biomolecules, such as insulin and thyroxine. [2]. Since then, extensive research has happened in the area of peptide-based therapeutics. As a result, today we have around 60 United States Food and Drug Administration (US FDA)-approved, therapeutic peptide formulations in the market, some of which have been stated in Table 12.1. Apart from that, ~140 peptide molecules are currently being evaluated in clinical trials, while 500 peptide molecules are being evaluated in preclinical trials, respectively [1]. Despite having tremendous pharmacological potential, drug delivery of peptide-based therapeutics remains a great challenge because of their complex pharmacokinetic profile. Poor bioavailability of these molecules, contributed by their large size, hydrophilicity, and enzymatic degradation, is the biggest concern in developing oral drug-delivery systems for therapeutic peptides. Other complexities

Targeted Drug Delivery, First Edition. Edited by Yogeshwar Bachhav.

Table 12.1 Major therapeutic peptides approved by the US FDA.

Sr. no	Peptide	Company name	Therapeutic use	References
1	Plecanatide	Synergy Pharmaceuticals, Inc.	Gastrointestinal laxative	[3]
2	Etelcalcetide	KAI Pharmaceuticals, Inc.	Secondary hyperparathyroidism in adult patients with chronic kidney disease on hemodialysis	[3]
3	Abaloparatide	Radius Health, Inc.	Osteoporosis	[3]
4	Semaglutide	Novo Nordisk, Inc.	Treatment of type 2 diabetes mellitus	[3]
5	Macimorelin	Aeterna Zentaris, Inc.	For the diagnosis of adult growth hormone deficiency	[3]
6	Angiotensin II	La Jolla Pharm Co.	Control of blood pressure in adults with sepsis or other critical conditions	[3]
7	Insulin glargine	Sanofi	Treatment of diabetes	[4]
8	Calcitonin	Nastech Pharmaceutical Company Inc.	Osteoporosis	[4]
9	Leuprorelin	Abbott Laboratories	Prostate cancer	[4]
10	Liraglutide	Novo Nordisk	Treatment of diabetes	[4]
11	Colistin	Profile Pharma Ltd.	Acute or chronic infections due to multi drug resistant (MDR) Gram-negative bacilli	[5]
12	Dactinomycin	Recordati rare disease, Inc.	Wilms' tumor, childhood rhabdomyosarcoma, Ewing's sarcoma, and metastatic	[5]

faced during development of peptide-containing parenteral drug-delivery systems include immunogenicity, short plasma half-life, and inadequate cellular uptake, which also contribute to their poor bioavailability [2, 6]. An ideal delivery strategy to overcome these hurdles would be to design targeted peptide therapies, which have been primarily accomplished through nanoparticulate carriers. Such systems can protect the therapeutic peptides against degradation, prolong their half-life, and also increase their intracellular permeation at the target site [7]. Targeted peptide delivery systems usually comprise of three components, viz the pharmacologically active peptide, a nanocarrier system, and a target-identifying moiety that ensures stability and efficacy of the peptides [8].

Nanocarriers, having a size range between 1 and 1000 nm, can be fabricated using polymers, lipids, or metal ions [7]. The primary advantage of these materials is their ability to be conjugated or loaded with a variety of therapeutic substances, such as drugs, peptides, proteins, and nucleic acids, as well as with other targeting molecules, which may enable their selective delivery to specific locations within the body. Nanocarrier-based targeted peptide delivery systems are designed in such a way that they are capable of protecting the therapeutic molecules from enzymatic degradation and immune responses, prolonging the circulation of their cargo/es, and delivering them at appropriate sites. This increases the localized concentration of these substances, thereby minimizing the required dose and improving their bioavailability [6, 8]. The most commonly employed drug-delivery nanocarriers are the liposomes, the polymeric and lipidic nanoparticles (NPs), and the nanoparticles composed of metal oxides [8]. In many cases, the surface of these nanocarriers is coated with hydrophilic polymers, such as polyethylene glycol (PEG), which creates a protective layer around them and prevents their uptake by the mononuclear phagocytes. The coated nanocarriers suppress the immune response, which allows their longer circulation in the plasma [8]. Targeted nanocarriers are designed for specific delivery to the desired sites of action, which prevents their accumulation in nonspecific organs or tissues and reduces their systemic side effects, especially in case of delivering anticancer molecules [6, 8]. In this chapter, we will be focusing on nanocarriers that have been utilized for targeted delivery of therapeutic peptides to various organs and those meant for different diseased conditions, specifically different types of cancers.

Targeted drug-delivery systems can be formulated to deliver their payload via two different mechanisms, viz active targeting or passive targeting. In some diseases, such as cancers, inflammatory or infectious diseases, the blood vessels become more permeable, allowing the particles to cross the epithelial barrier and enter into the interstitial spaces. Nanocarriers having sizes of up to 200 nm can permeate through these barriers. Based on their size, charge, and composition, these particles can bypass the immune and reticuloendothelial systems and accumulate in the target spaces. Site-specific delivery is also governed by the peculiar characteristics of the target site. For example, the inflamed or neoplastic tissues exhibit lower pH and higher temperature, as compared to the normal tissues. Thus, stimuli-responsive nanocarriers can identify these tissue differences, based on their composition, and release the drug at that particular site. This forms the basis of the passive-targeting systems. In case of tumors, differences in the microenvironment and lack of lymphatic clearance facilitate passive targeting of nanocarriers, also known as enhanced permeability and retention (EPR) effect [8]. In case of active targeting, the surfaces of nanocarriers are modified with specific ligands, such as proteins, vitamins, aptamers, lectins, saccharides, hormones, glycoproteins, or peptides [6, 8]. These modified ligands specifically interact with the target site, via antibody–antigen interaction or ligand–receptor interaction. The specificity achieved with active targeting greatly reduces the systemic side effects, especially in case of anticancer therapy. The success of targeted delivery systems depends on the selection of the targeting moiety. The first step in this process involves recognition of the molecular targets that are overexpressed in disease conditions. The selected ligand should have high affinity and selectivity for the specific molecular target. It should also be stable during chemical conjugation with the nanocarriers or therapeutic peptides [9]. Peptides are one of the most widely used targeting molecules as they act as suitable ligands for cell-surface receptors. Some peptides may enhance penetration of cargoes attached to them or loaded in them, enabling their intracellular delivery. These are known as cell-penetrating peptides (CPPs) or protein transduction domains (PTDs).

The concept of CPP was proposed 20 years ago, based on the observation that some proteins, mainly transcription factors, could shuttle within the cell and from one cell to another. Generally, CPPs are cationic peptides composed of sequences of 20–30 amino acids. Most commonly used CPPs are transactivator of transcription (TAT) peptide, antennapadia (Antp), arginine, and lysine-rich peptides, which comprise of positively charged amino acids and transportan [9, 10]. The mechanism of CPP permeation is thought to be based on micropinocytosis and is not associated with receptor-mediated endocytosis [9]. The first step of internalization involves electrostatic interaction of CPPs with cell-membrane proteoglycans. This binding leads to a sequence of changes in the actin protein conformation, which is an indication of the start of the internalization process. The modulation of actin hampers membrane fluidity, permitting entry of cargo inside the cells through micropinocytosis [9]. CPPs are attached to their respective cargoes and deliver them either via covalent or non-covalent interactions. In non-covalent conjugation, short amphiphilic peptides are attached to the nanocarriers by electrostatic interaction or van der Waals forces. This is advantageous as it does not require any cross-linking or chemical modification, making it easy to release the cargo inside the cells [11]. In covalent interaction, cleavable disulfide or thioesters linkages are formed. However, reports have revealed a reduction in biological efficacy of therapeutic peptides formulated by this method. Thus, non-covalent conjugations are preferred over covalent bonding [9]. The activity of CPPs depends on their secondary structure, their interaction with the cell-surface receptors and lipid molecules, the nature and concentration of the cargoes attached to CPPs, the cell type, and the membrane composition [11]. CPPs have attracted tremendous interest as ligands for targeted drug delivery because of their ability to be conjugated with various therapeutic molecules and nanocarriers, their low cytotoxicity, high efficiency, rapid penetration into the cells, and their stability in physiological buffers.

12.2 Peptides Used in Cancer Therapy

Therapeutic peptides have exhibited high efficacy as anticancer agents due to their selectivity and targetability. Treatment of cancer using peptide targeting is a commonly used strategy that is governed by the site of action, the microenvironment, and the type of ligand expressed on cancer cells. These peptides are broadly categorized into three main groups, viz (i) antimicrobial/pore-forming peptides, (ii) cell-permeable peptides, and (iii) tumor-targeting peptides [12]. Peptides consist of specific amino-acid sequences, which mimic or interfere with important interactions during tumor progression, and they have the potential to specifically target and inhibit tumor growth. The development of a suitable delivery system is essential for delivering such peptides effectively and also ensuring their stability. Here, we have described the various targeting strategies, as well as the formulations that have been explored for the treatment of different types of cancers.

12.2.1 Lung Cancer

Lung cancer is one of the leading causes of cancer-related deaths as it is usually detected at an advanced stage when limited therapeutic options are available. Thus, novel and effective strategies are required for treating lung tumors. The use of intracellular-targeting peptides is advantageous for inhibition of cytoplasmic proteins that mediate progression of lung tumors. However, peptides must penetrate through the cancer cell membrane, which is difficult to accomplish and hence requires development of suitable formulations. The study conducted by Kim et al. involved the production of lipid calcium carbonate (LCC) nanoparticles for the pH-sensitive targeted delivery of intracellularly acting therapeutic peptide, EEEEpYFELV (EV), to the lung cancer cells, as shown in Figure 12.1. Here, the nanoparticles were conjugated with anisamide (AA), a ligand targeting the sigma receptor expressed on the lung cancer cells. The nanoparticle formulation was composed of calcium carbonate core, which was coated with 1,2-dioleoyl-*sn*-glycero-3-phosphoethanolamine-*N*-(glutaryl) (DOPE-glu), an additional layer of 1,2-dioleoyl-3-trimethylammonium-propane chloride salt (DOTAP/cholesterol), and was further modified using 1,2-distearoryl-*sn*-glycero-3-phosphoethanolamine-*N*-[methoxy (polyethyleneglycol-2000)] ammonium salt-anisamide (DSPE-PEG-AA). These components together aided in the endosomal release of the peptides. Researchers postulated that DOTAP formed ion pairs with endosomal lipids, disrupting the endosome. Additionally, the conversion of LCC to carbon dioxide was anticipated to develop stress and eventually disrupt the endosome, thereby delivering the peptide into the cytoplasm. The LCC core, prepared without the external liposomal outer layer, was tested for its ability to mediate pH-dependent release of fluorescently labeled EV peptide. This was performed to evaluate drug release in the acidic environment of the endosome. The breakdown of LCC core was rapid at a pH of 5.5 and negligible at a relatively higher pH, thus portraying a pH-dependent release. Confocal microscopy was employed to quantify the cellular

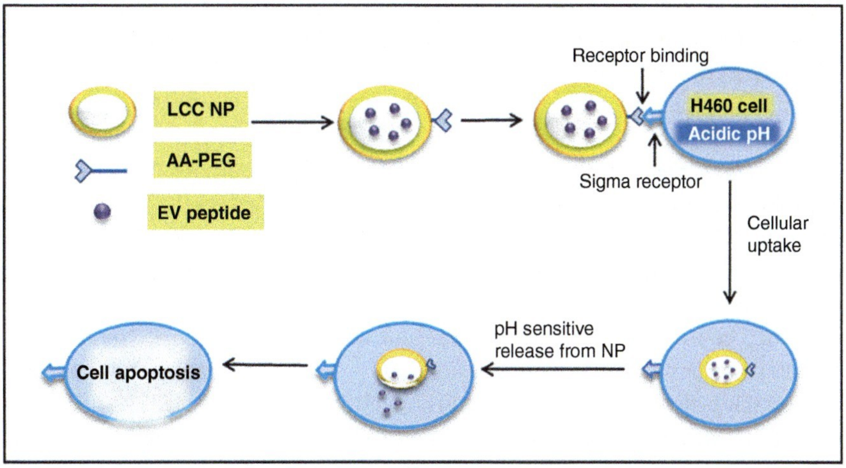

Figure 12.1 Lipid-coated calcium carbonate nanoparticle for targeting lung cancer cells.

uptake of the peptide from the nanoparticles, and suggested a higher cellular uptake of the LCC NP's, in comparison to the control NPs (free fluorescently labeled EV peptide). This confirmed the role of LCC NP's in enabling the transportation of the EV peptide into the cytoplasm, where it could interact with its molecular target, the kinase domain of epidermal growth factor receptor (EGFR). Further, the LCC NPs exhibited a 40% reduction in cell viability, using the 3-(4,5-dimethylthiazol-2-yl)-2,5-diphenyltetrazolium bromide (MTT) assay. The extent of cell apoptosis, analyzed by the flow cytometry, showed that the cells underwent a whopping 70–80% apoptosis after treatment with LCC NPs. Additionally, treatment with LCC NPs, lacking the PEG AA component, resulted in absence of cell targeting. Thus, PEG AA was shown to be essential for cellular targeting of the therapeutic peptide. *In vivo* tissue distribution of the LCC NPs revealed a higher accumulation of the fluorescently labeled EV peptide in the tumor. At the end of the treatment, the serological parameters were tested for evaluating any systemic toxicity. These results confirmed that the drug was better tolerated using such formulation. Several such therapies are being currently explored for the treatment of lung cancer [13].

12.2.2 Melanoma

Programmed cell death-ligand 1 (PD-L1), which is overexpressed by tumor cells, interacts with programmed cell death receptors (PD-1), overexpressed on exhausted T cells, and causes a decline in T-cell proliferation. This eventually leads to an increase in melanoma growth and justifies the use of PD-1/PD-L1 as a potential target for anticancer moieties. The therapeutic agents used against this target, however, have not been quite useful because of the ineffectiveness and lack of sensitivity toward the patient, which necessitated the design of combination therapy to improve the response rate. It involved targeting PD-1/PD-L1, along with indoleamine 2,3-dioxygenase (IDO), an enzyme also overexpressed by several tumors. This enzyme is responsible for the degradation of essential amino acid L-tryptophan to L-kynurenine, resulting in depletion of tryptophan levels eventually inhibiting the proliferation of T cells. An agent NLG919, a highly selective inhibitor of IDO, was used at a higher dose due to its low water solubility. Hence, a delivery system to increase the bioavailability of NLG919 and have a dual action against PD-1/PD-L1 and IDO was formulated [14].

A study by Cheng et al. involved the combination therapy of ^{D}PPA-1, a short D-peptide antagonist of PD-L1, and NLG919, an inhibitor of IDO, as depicted in Figure 12.2. The two major hallmarks of tumor microenvironment are a slightly acidic pH and a higher expression of matrix metalloproteinase-2 (MMP-2). An amphiphilic peptide, consisting of hydrophilic and hydrophobic portions, was synthesized. The hydrophobic part consisted of 3-diethylaminopropyl isothiocyanate (DEAP) and a peptide with sequence PLGLAG, which was the substrate of MMP-2, whereas the hydrophilic part was comprised of the ^{D}PPA-1 peptide. The hydrophobic portion further consisted of one lysine, two leucine molecules, and two glycine residues, which acted as a linker. The lysine residues ensured conjugation with DEAP, while the leucine residues added to the hydrophobicity. The amphiphilic

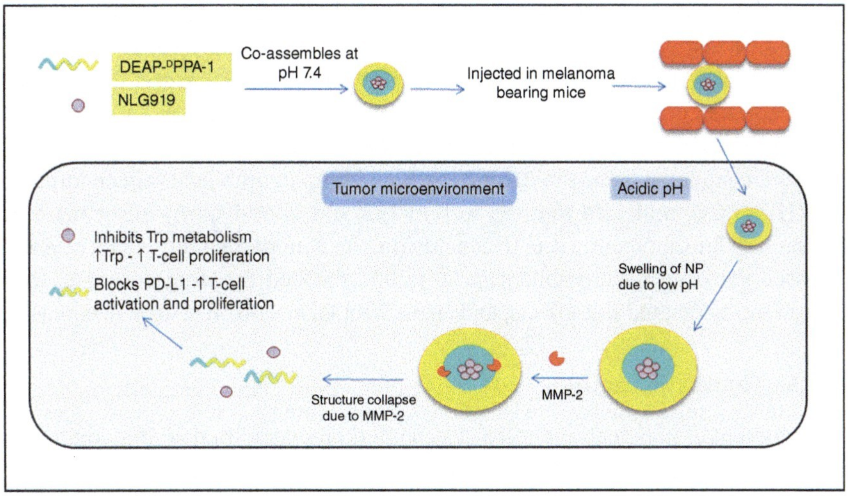

Figure 12.2 Sequentially responsive therapeutic peptide assembling nanoparticles for dual-targeted cancer immunotherapy.

peptide was co-assembled with NLG919 at the physiological pH, which resulted in the formation of nanoparticles. In a mildly acidic environment, like that of the tumor site, the DEAP molecules present in the hydrophobic part of the amphiphilic peptide were protonated, which resulted in the swelling of the nanoparticles. It allowed the MMP-2 to gain entry into the nanostructure and cleave its substrate peptide, resulting in the burst release of encapsulated NLG919 and ^{D}PPA-1. This timely release of ^{D}PPA-1 and NLG919 increased the survival rate of cytotoxic T cells and hence decreased tumor growth. Further, the particle size and surface charge of the nanoparticles were analyzed. Results indicated an increase in size of the particles due to their swelling at the acidic pH. Additionally, an increase in the surface charge, associated with protonation of DEAP, was also observed at the lower pH. The nanoparticles failed to disassemble at the pH of 7.4, even in the presence of MMP-2, thus confirming sequential response of DEAP-^{D}PPA-1 nanostructure, initiated at a lower pH, and further upon exposure to MMP-2. The pH of 6.8, along with the overexpression of MMP-2, contributed to the collapse of peptide-loaded nanoparticles, which released the drug and the bioactive peptide. The release of NLG919 was analyzed over a period of six hours. A release of 20.3% of the NLG919 was observed at the physiological pH of 7.4, while a 73% release was obtained at the pH of 6.8. The inhibitory activity of NLG919 was evaluated in B16-F10 cells by the measurement of the levels of kynurenine, indicating the extent of tryptophan metabolism. The inhibitory potential of free NLG919 was concentration dependent, while its release from the nano-formulation was pH dependent. Transmission electron microscopy (TEM) analysis depicted the complete disruption of nanoparticles at the low pH, in presence of MMP-2. Further, fluorescence imaging studies revealed specific localization of fluorescence within the tumor. Additionally, flow cytometry and immunohistochemistry analysis revealed an increase in the density of cytotoxic T lymphocytes and NK cells within tumor tissues. These results indicated that the formulation could be effectively used to treat tumors. The safety of the nanoparticles, analyzed by histological examination in tumor-bearing mice, revealed that there were no observable abnormalities in the heart, liver, spleen, lung, or kidney, in any of the treatment groups. In addition to this, a gradual increase in the body weight of tumor-bearing mice was observed. The results of the toxicity studies and cell viability assays, conducted using the B16-F10 cells and human umbilical vein endothelial cells (HUVECs), indicated that the formulation did not adversely affect the cells and was safe for therapeutic use. In conclusion, the nanoparticulate delivery system increased the drug bioavailability and stability, reduced the drug dose, and thus lowered the associated side effects, and also exhibited a controlled drug release [14].

12.2.3 Pancreatic Cancer

The treatment of pancreatic cancer with chemotherapy and radiation is commonly associated with severe adverse effects and toxicity. Hormones, in particular, somatostatin (SST) is known to be useful in treating pancreatic tumors. A study conducted by Dubey et al. involved synthesis of an SST analog, octreotide, along with the evaluation of octreotide-loaded nanoparticles. The resulting nanoparticles were passively targeted to the cancer cells, wherein they were retained via the EPR effect, as explained in Figure 12.3. Biodegradable polymers, such as PEG and poly(ε-caprolactone) (PCL), were used due to their high biocompatibility and ease of being formulated into nanoparticles. Octreotide nanoparticles were prepared using double emulsion solvent evaporation method. The primary oil in water (O/W) emulsion consisting of octreotide, PEG, and PCL was re-emulsified using Pluronic® F-68 solution to obtain water in oil in water (W/O/W) double emulsion. Thus, the targeted release of octreotide achieved was due to the surface modification and stabilization of PCL nanoparticles by Pluronic F-68. Particle-size analysis of the nanoparticles revealed that the size of the particles increased from 130 to 195 nm

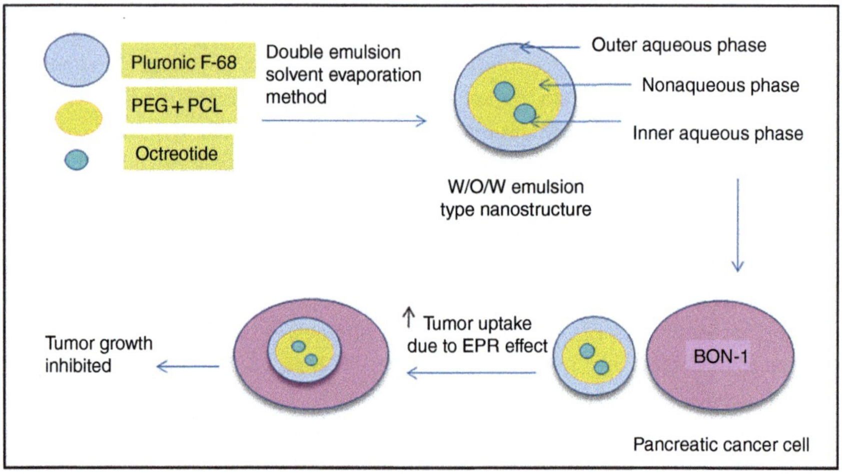

Figure 12.3 Biodegradable PCL/PEG nanoparticles for targeted delivery of octreotide to the neuroendocrine tumor.

with an increase in the amount of PCL and PEG. On the other hand, Pluronic F-68 resulted in decrease in particle size and an increase in the entrapment efficiency, the latter was analyzed in presence of sonication. The use of Pluronic F-68 allowed formation of nanoparticles having the size suitable for tumor targeting (200–300 nm). It also prevented leakage of the peptide, thus improving the drug-loading efficiency. The zeta potential of the nanoparticles was -18.66 ± 1.75 mV, which confirmed the stability of nanoparticles. Monolayer cultures of human pancreatic BON-1 cells, expressing the SSTR2 receptors, were used for *in vitro* and *in vivo* evaluation of octreotide nanoparticles. 99mTc-labeled octreotide was employed during *in vitro* receptor-binding assay and the results were analyzed using γ-scintillation counting. Cytotoxicity of both, the native drug and the nanoparticles, was similar up to the concentration of 1 $\mu g\,ml^{-1}$. However, varying effects were observed with an increase in the drug concentration. In case of octreotide nanoparticles, dose-dependent cytotoxicity was observed. The inhibition was 54.03% with the nanoparticle containing 25 $\mu g\,ml^{-1}$ of drug, as against 33.06% with the native octreotide at the same concentration. MTT assay successfully proved the target-specific cytotoxicity of the octreotide nanoparticles. *In vivo* tumor regression studies performed in BON-1 tumor-bearing mice revealed tumor inhibition extent of $8.42 \pm 0.3\%$ and $14.6 \pm 0.3\%$ with octreotide nanoparticles and free drug, respectively. The biodistribution studies depicted a higher concentration of octreotide inside the tumors when it was encapsulated within the PCL/PEG nanoparticles. Thus, biodegradable PCL/PEG nanoparticles acted as an effective alternative to chemotherapy and radiation in treating pancreatic cancer [15].

12.2.4 Brain Cancer

Glioblastoma is the most common and aggressive form of brain tumor that is derived from glial cells, such as astrocytes and oligodendrocytes, and is classified as grade IV by the World Health Organization [16]. The two major strategies that have been widely explored by various researchers for the treatment of glioblastoma include (i) the stem cells associated with glial cells correspond to brain tumor-initiating cells (BTICs), which are responsible for tumor initiation, progression, and recurrence due to the resistance that they acquire against the current therapies. Therefore, newer strategies have been developed to treat and target these aggressive cancer cells. (ii) Further, when the tumor grows, the blood vessels associated with it also grow, which takes place through angiogenesis, and thus this process has been targeted to avoid further growth of the tumor. Tumor blood vessels are markedly different from normal vessels due to the presence of various cell surface and extracellular matrix proteins, which encourage binding of ligands, such as peptides and antibodies. Thus, these proteins (receptors) can be used for targeting various drugs that may be attached to ligands for these receptors, thereby reducing their exposure to the normal tissues. In a study, the effect of NFL-TBS.40–63 peptide on BTICs isolated from human glioblastoma was examined in detail. The NFL-TBS.40–63 peptide (YSSYSAPVSSSLSVRRSYSSSSGS), a tubulin-binding agent derived from neurofilaments, was biotinylated or coupled with 5-carboxyfluorescein (5-FAM) at the N-terminal domain to allow its distribution. Additionally, amidation was conducted at the C-terminal domain to protect it from proteolysis. The peptide, by itself or conjugated with nanoparticles, could extensively enter the BTICs. Thereafter, it exhibited its antitumor effect by inhibiting the proliferation of BTICs and inducing death via alteration of their microtubule network and cell–cell adhesion, thereby causing decrease in the self-renewal ability of these cancer cells. The peptide-conjugated nanoparticle system preferentially targeted the GBM cells *in vitro*, in the BTIC cell line, and *in vivo*, in adult rats. However, their uptake was very low in the astrocytes and neurons due to the peptide-specific effect. The peptide was specifically attached to the C-terminal domain of the βIII-tubulin isotype, which was overexpressed in GBM cells. Also, lipid nanocapsules (LNCs) were prepared by the same research group, as shown in Figure 12.4, and characterized. The systems prepared were LNCs coupled with the biotinylated peptide (LNC-NFL), LNCs containing the lipophilic tracer DiD (1,10-dioctadecyl-3,3,30,30-tetramethylindodicarbocyanine, 4-chlorobenzenesulfonate) (LNCs-DiD) and LNC-NFL-DiD. The uptake of these LNCs by viable cells, namely BTIC12, BTIC25, and BTIC53, was analyzed by flow cytometry, which revealed a dose-dependent, massive and rapid uptake of the peptide in these BTICs. Thus, the NFL-TBS.40–63 peptide proved to be a promising therapeutic agent for the treatment of glioblastoma, by targeting as well as killing the glioblastoma cancer and glioblastoma stem cells (BTICs), through destruction of their microtubule network. This also minimized the risk of recurrence of this type of cancer [17].

In another study, a nanosystem was formulated to enhance the delivery of a pro-apoptotic peptide for the treatment of glioblastoma via antiangiogenic therapy, as explained in Figure 12.5. The nanosystem was made up of the following three components, viz a tumor-homing peptide, i.e. the CGKRK (Cys–Gly–Lys–Arg–Lys) peptide, a pro-apoptotic peptide, i.e. the $_{\mathrm{D}}$[KLAKLAK]$_2$, and iron oxide. Also, a tumor-penetrating peptide was added to this nanosystem to enhance its therapeutic effect via improved penetration into the tumor tissue. The tumor-homing peptide was particularly delivered to the mitochondria of the tumor cells, where it conjugated with the proapoptotic peptide and acted on the mitochondria.

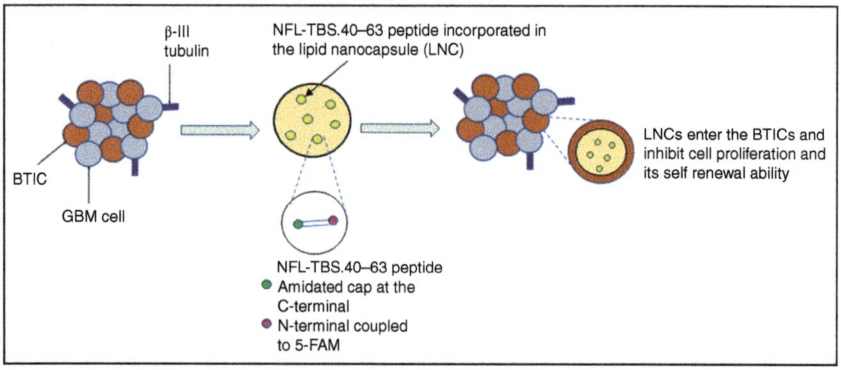

Figure 12.4 Therapeutic peptide incorporated in LNCs for targeting GBM stem cells.

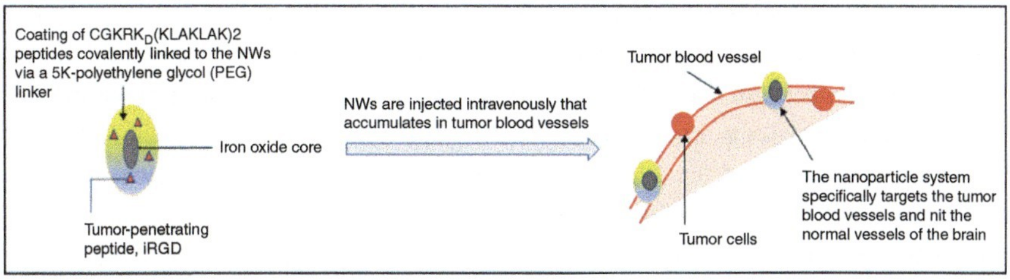

Figure 12.5 Proapoptotic peptide-coated iron oxide NWs (nanoworms) for targeting GBM cells.

The CGKRK peptide was employed for targeting the nanoparticles to the mitochondria of the tumor cells. The rhodamine (Rd)-labeled CGKRK peptide was largely accumulated in the GBM tumors but not in the normal brain tissues, which improved the delivery of the proapoptotic peptide or the drug to the mitochondria of the tumor cells. $_D$[KLAKLAK]$_2$, an α-helical amphipathic peptide, was originally designed to disrupt the bacterial cell membrane. In eukaryotic cells, it disrupted the mitochondrial membrane, and thus initiated apoptotic cell death. It was a highly toxic compound, even when particularly targeted to the tumor site [18]. Therefore, nanoparticles were formulated for their administration to reduce their toxicity due to their unique affinity and selectivity toward the malignant cells [19]. Further, iron oxide was included to enhance the pro-apoptotic activity and to allow imaging of the nanoparticles inside the tumors. Lastly, the nanoparticles were amalgamated with the tumor-penetrating peptide, iRGD. This nanosystem was named as "nanoworms" (NWs), in which CGKRK$_D$(KLAKLAK)$_2$ was coated around iron oxide, because of the elongated shape of the nanoparticles, which resulted in more efficient targeting as compared to the spherical nanoparticles. The two peptides (CGKRK and $_D$[KLAKLAK]$_2$) were synthesized as a chimeric peptide, which was then covalently linked to the NWs via a 5K-PEG linker. NWs coated with the chimeric CGKRK$_D$[KLAKLAK]$_2$ peptides were injected intravenously and accumulated in the tumor vessels of the GBM models, but not in the normal vessels of the brain, signifying the target specificity of the CGKRK peptide. Approximately 80% of the total tumor vessels could engulf both, CGKRK-NWs and CGKRK$_D$[KLAKLAK]$_2$-NWs, whereas $_D$[KLAKLAK]$_2$-NWs were internalized in only 4% of the vessels. After an i.v. injection of CGKRK$_D$[KLAKLAK]$_2$-NWs, magnetic resonance imaging (MRI) of the tumors exhibited hypointense signals throughout the tumor due to the action of NWs. The nanoparticle-coupled peptide was 100-fold more effective than the free peptide, on a molar basis. Thus, this nanosystem acted as a suitable theranostic agent, which was rendered even more effective due to the incorporation of a tumor penetrating peptide, iRGD [18].

12.2.5 Breast Cancer

Resistance to chemotherapeutic agents remains a major barrier in the effective treatment of metastatic breast cancer [12]. Gold nanoparticles (AuNPs) have emerged as highly successful candidates for targeted delivery of therapeutic peptides or drugs.

The gold core is essentially inert, nontoxic, and possesses favorable physicochemical properties, which allows the drug release even in remote areas. Binding of ligands to the receptors present on the cell surfaces triggers intracellular signaling, which aids in transportation of drugs into the cells via receptor-mediated endocytosis. In a study conducted by Kumar et al., two peptides, namely PMI (p12; TSFAEYWNLLSP), which is a therapeutic peptide, and a peptide with sequence of CRGDK, which is a targeting peptide, were conjugated on the surface of ultrasmall (2 nm) AuNPs for selective binding to the neuropilin-1 (Nrp-1) receptors. These receptors are present in abundance in various types of cancer cells, which also regulate the process of membrane receptor-mediated internalization. The binding of the therapeutic peptide to the Nrp-1 receptors present on the MDA-MB-321 (human breast cancer cell line) cell surface was elaborately examined for the treatment of breast cancer. PMI, commonly known as p12, is a widely known potent inhibitor that binds to MDM2 (murine double minute 2) and MDMX (murine double minute X) complexes. MDM2 is a master regulator, whereas MDMX is its homolog that plays a crucial role in controlling and maintaining the integrity of the p53 pathway. The binding of p12 to these complexes resulted in regulated activity and stability of the tumor-suppressive protein, p53, a well-known protein that transcriptionally modulates the expression of various target genes in response to cellular stress, indirectly resulting in cell-cycle arrest, DNA damage, or apoptosis. Due to the presence of the targeting peptide sequence, such as CRGDK, the intracellular uptake of AuNPs was found to be highly increased. Besides, it also resulted in maximum binding of the AuNPs to the targeted receptor, which was overexpressed on the MDA-MB-321 cell surface. This led to an improvement in the delivery of the therapeutic p12 peptide to the breast cancer cells. Such ultrasmall AuNPs were found to be effective for the treatment of breast cancer types that overexpressed Nrp-1 receptors depicted in Figure 12.6 [20].

In another study, CPPs were examined. These peptides contained a C-terminal Cend Rule (CendR) sequence motif (R/K)XX(R/K), which was responsible for tissue penetration and cell internalization, while X denoted any amino acid. The tumor-specific CendR peptides contained both, a tumor-homing moiety and a cryptic CendR moiety, which was proteolytically unmasked in the tumor cells. LyP-1 (sequence: CGNKRTRGC), a cyclic tumor-homing nonapeptide, containing a CendR element, was identified by phage display and exhibited the potential for

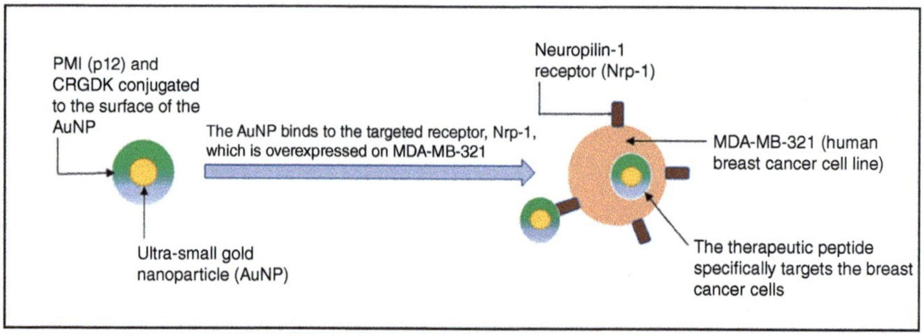

Figure 12.6 Peptide-coated AuNPs for targeting tumor cells in breast cancer.

tissue penetration. LyP-1 could attach to the tumor cells, tumor macrophages, and tumor lymphatics by binding to their p32 receptor, which is a mitochondrial protein present on the surface of these cells, followed by internalization (MDA-MB-435 cells in this case). In this study, truncated form of LyP-1 (CGNKRTR; tLyP-1) was selected, wherein the CendR motif was exposed. Further, a saturation assay was performed with the help of fluorescein-labeled peptides (FAM-peptides), which resulted in low binding of tLyP-1 to p32. Moreover, it was shown that tLyP-1 internalized into the tumor cells through the Nrp-1-dependent CendR internalization pathway and could also bind to neuropilin-2, which resulted in the activation of the CendR pathway. Both fluorescein-labeled tLyP-1 peptide and tLyP-1-conjugated nanoparticles exhibited a strong and selective homing to tumors by penetrating from the blood vessels into the tumor parenchyma. The truncated form of LyP-1, tLyP-1, was more potent as compared to LyP-1 and also enhanced the extravasation of a co-injected nanosystem into the tumor tissue. The binding and internalization of tLyP-1 occurred in MDA-MB-435 breast carcinoma cells that expressed only NRP2 and not NRP1, and thus tLyP-1 could co-localize with NRP2.

The tissue distribution of intravenously injected tLyP-1 was examined in mice bearing orthotopic breast cancers, where the peptide was found to internalize within the 4T1 cells, *in vitro*. These cells overexpressed both, NRP1 and NRP2, and exhibited a very high sequence homology between mice and humans, corresponding to 93% and 95%, respectively. The ability of tLyP-1 to deliver nanoparticles into the tumor cells was evaluated by conjugating FAM-tLyP-1 to elongated iron oxide nanoparticles-dubbed NWs (tLyP-1-NWs), as shown in Figure 12.7, wherein their biodistribution indicated selective homing to 4T1 (epithelial breast cancer cell line) tumor cells. These NWs were also specifically homed to the tumors located in another breast cancer model, i.e. human MDA-MB-231 xenograft model that expressed both, NRP1 and NRP2. In addition, they showed a significantly wider distribution in the tumor tissue, which was attributed to the exposed CendR motif. It was also found that tLyP-1 could induce penetration of a co-administered compound by the activation of the CendR pathway when it was injected together with NWs coated with CGKRK, the tumor-homing peptide. Confocal microscopy analysis depicted enhanced tumor penetration of CGKRK, when administered along with tLyP-1. The tLyP-1 peptide, alone or in conjugation with nanoparticles,

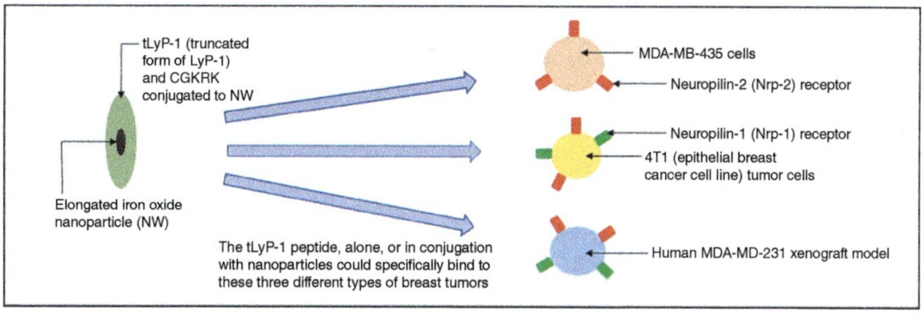

Figure 12.7 Peptide-coated iron oxide NWs for targeting various tumor cell lines in breast cancer.

could specifically bind to three different types of breast tumors and could spread widely within the tumor tissue as well [21].

A class of peptides known as cytolytic peptides were used in *in vivo* cancer models but were limited due to their toxicity, non-specificity, and degradation. These barriers were overcome via a nanoscale delivery vehicle that was synthesized by incorporating a nonspecific amphipathic cytolytic peptide, melittin, into the outer lipid layer of a perfluorocarbon nanoparticle. It is a water-soluble, cationic, 26-aa α-helical peptide, which is obtained from the venom of the honeybee *Apis mellifera*. Being a cytolytic peptide, it is highly nonspecific and leads to toxicity when injected via the i.v. route, causing physical and chemical disruption of the membrane structures. It enters into the cell membrane as a monomer and oligomerizes into toroidal or barrel stave structures, thereby facilitating pore formation and leading to cell death. In a study by Soman et al., the authors demonstrated the desirable pharmacokinetics of a nanoparticulate system containing melittin, which allowed its accumulation in murine tumors *in vivo,* followed by a drastic reduction in the tumor growth, without any apparent signs of toxicity. Direct assay methods revealed the selectivity of the nanocarriers to multiple tumor sites through a hemi-fusion mechanism, which triggered apoptosis. These nanocarriers were formulated as an oil-in-water emulsion, composed of a liquid perfluorooctyl bromide (PFOB) core that contained a monolayer of phospholipid, thereby stabilizing their interface with the aqueous media, followed by incorporation of melittin into the phospholipid monolayer. The incorporation of melittin was confirmed by techniques, such as surface plasmon resonance, fluorescence, and circular dichroism spectroscopy. Therefore, formulating a potent cytolytic peptide into a nano-vehicle, with the use of passive or active targeting, proved to be a promising strategy for chemotherapy at multiple stages. Significant payload of melittin was delivered safely via the i.v. route, in this preclinical study, to target the xenograft (MDA-MB-435 human breast cancer) [22].

12.2.6 Leukemia

Leukemia is a type of cancer that is difficult to cure and remarkably threatens human health. During the pathogenesis of acute myeloid leukemia (AML), immature leukemic cells grow rapidly, whereas the production of healthy blood cells is faulty. High relapse rate of AML is still a crucial problem despite considerable advances in anticancer therapies. One crucial cause of relapse is the existence of leukemia stem cells (LSCs) with self-renewal ability, which contributes to repeated treatment resistance and recurrence [23].

Recent studies have shown the efficacy of lanthanide-tagged nanoparticles (LDNp) as an imaging tool because of their excellent photoluminescence property and biocompatibility. In a study conducted at Xi'an Jiaotong University, China, a fluorescent lanthanide oxyfluoride nanoparticle (LONp)-based multifunctional peptide drug-delivery vehicle was formulated for the potential treatment of AML, as explained in Figure 12.8. The AML cells are commonly composed of wild-type p53, high levels of MDM2 and/or MDMX that are the functional inhibitors of p53, and an overexpressed cell-surface receptor, CD33. p53 is a tumor-suppressor protein

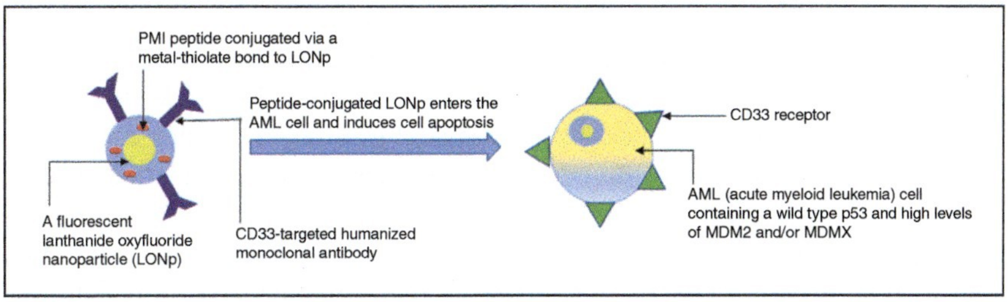

Figure 12.8 LONp-based PMI delivery system for the treatment of AML.

that induces strong growth-inhibitory responses, followed by apoptosis and cellular stress. It plays a crucial role in preventing the conversion of damaged cells into cancerous cells. The LONp was conjugated to a p53-activating dodecameric peptide antagonist of both MDM2 and MDMX, namely the PMI (TSFAEYWALLSP), via a metal–thiolate bond and was also bound to a CD33-targeted humanized monoclonal antibody, to allow AML-specific intracellular delivery of a stabilized PMI. The resultant nanoparticle antiCD33-LONp-PMI was found to be nontoxic to the normal cells of the body, whereas it induced apoptosis of AML cells and primary leukemic cells isolated from AML patients, by antagonizing MDM2 and/or MDMX and activating the p53 pathway. Conjugation of peptides to nanosystems highly enhanced their resistance to proteolysis, membrane permeability, and bioavailability. Being fluorescent in nature, it also allowed real-time visualization of a series of apoptotic events associated with the AML cells, thereby proving itself as a useful tool for tracking and treatment response monitoring. Thus, the data collected on the therapeutic efficacy and safety profile of antiCD33-LONp-PMI rendered it as a novel class of antitumor agents for the potential treatment of AML [24]. This further needs to be validated via clinical trials.

In another study, bovine lactoferricin (LfcinB), a cationic antimicrobial peptide (AMP) exhibiting cytotoxic activity against microorganisms and human cancer cells was used. It could kill Jurkat T-leukemia cells via the mitochondrial pathway of apoptosis, but the process by which LfcinB triggered the pathway was not well understood. A biotinylated form of LfcinB was also used to monitor the ability of LfcinB to bind to Jurkat T-leukemia cells, along with an unmodified form of LfcinB, wherein both showed equivalent apoptosis-inducing activity. The apoptosis that was preceded by the binding of LfcinB to the cell membrane, was followed by its progressive permeabilization, mediating cytotoxicity. The effect of LfcinB on the cell membrane integrity was examined by measuring propidium iodide uptake and there was dose-dependent permeabilization of the cell membrane within five minutes. LfcinB first entered the cytoplasm of Jurkat T-leukemia cells and then triggered the onset of mitochondrial depolarization, which was examined by colloidal-gold electron microscopy and flow cytometry. However, the co-localization of intracellular LfcinB with the mitochondria was monitored by confocal microscopy. Furthermore, the attack of LfcinB on the purified mitochondria led to rapid loss of transmembrane potential, thereby releasing cytochrome C (Cyt C). LfcinB did not internalize via endocytosis since endocytosis inhibitors did not prevent LfcinB-induced cytotoxicity. Therefore, a fusogenic liposomal delivery system, based on the formulation of p14 fusion-associated small transmembrane (FAST) protein, was prepared for the intracellular delivery of LfcinB, which caused the death of Jurkat T-leukemia cells. It mediated the fusion of liposomes to the target cell plasma membranes into cationic multilamellar vesicles. LfcinB-induced apoptosis was observed in Jurkat T-leukemia cells due to the damage to the cell membrane, the subsequent disruption of mitochondrial membranes, and DNA damage by internalized LfcinB. In conclusion, these findings and results showed that LfcinB-induced apoptosis in Jurkat T-leukemia cells was triggered by a series of events ranging from LfcinB-mediated permeabilization of the cell membrane and its uptake across the damaged cell membrane, followed by its co-localization with mitochondria to LfcinB-mediated depolarization of mitochondria, thereby resulting in Cyt C release and initiation of the intrinsic pathway of apoptosis [25].

Permeability transition pore (PTP) responsible for mitochondrial permeability transition (PT) is one of the potential target sites for the treatment of tumors and to reduce resistance to cancer treatment. Mastoparan (MP), an amphipathic peptide obtained from wasp venom, has been reported to promote mitochondrial PT and hence, it can be used as an antitumor agent. However, it must be selectively delivered to the cancer cells to avoid damage to noncancerous cells. A study conducted by Yamada et al. reported the use of transferrin-modified liposomes (Tf-L) for encapsulating MP, along with a pH-sensitive fusogenic peptide (GALA), as depicted in Figure 12.9. The liposomes were prepared using reverse-phase evaporation. MP, encapsulated in Tf-L equipped cholesteryl GALA (Chol-GALA), can be internalized by tumor cells via receptor-mediated endocytosis. In the endosomes, GALA enhances the fusion of liposomes and endosomes at a lower pH causing the release of MP to the cytosol. MP that escapes from the endosomes can attack the mitochondria, PT is induced, and Cyt C is released from the mitochondria.

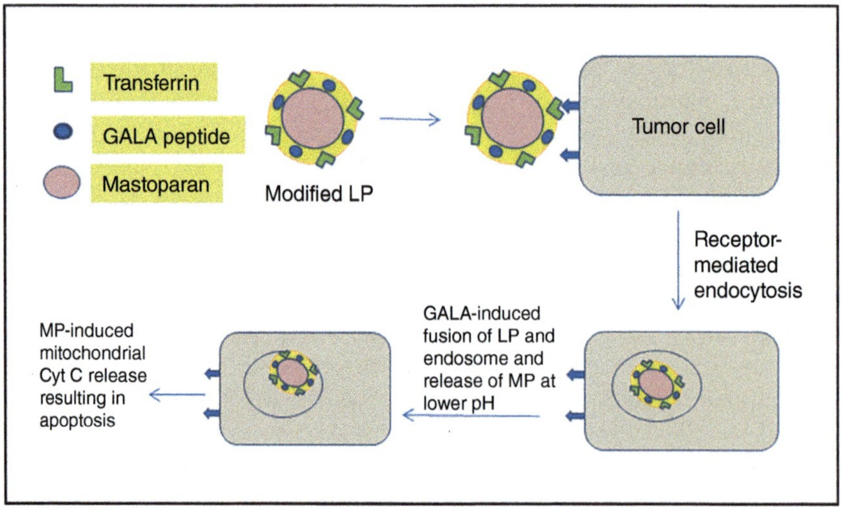

Figure 12.9 Mitochondrial delivery of mastoparan (MP) with transferrin liposomes equipped with a pH-sensitive fusogenic peptide for selective cancer therapy.

Hence, Chol-GALA was incorporated to promote the selective delivery of MP into the cytosol. K562 cells were cultured and the mitochondrial fraction was isolated. The concentration of Cyt C was determined in the isolated fraction of mitochondria, in the presence of phosphate, using western blotting assay. Based on the results of western blotting, MP was found to be potent as a cytotoxic agent at a concentration above 25 μM. Sulforhodamine B (Rho), a commonly used aqueous marker for liposomes, was encapsulated into the liposomal structure to ensure targeted delivery of Cyt C into the cytosol. In presence of Chol-GALA, the fluorescence marker Rho, was effectively delivered into the cytosol of K562 cells, which confirmed the targeting potency of Chol-GALA. The cytotoxic effect of MP in free as well as in the encapsulated form was tested. The results depicted a selective cytosolic release of Cyt C from the mitochondria, in case of the formulation consisting of Tf-L and Chol-GALA encapsulating MP. The surface modification of liposomes with pH-sensitive fusogenic peptide, GALA, contributed to the endosomal escape of liposomes, followed by cytosolic release of MP. In absence of GALA, MP could not induce PT, which confirmed the importance of GALA. Besides, the treatment of K562 cells with the modified Tf-L resulted in higher Cyt C release from mitochondria. The concentration of MP in the cytosol was estimated to be around 50 μM, which was sufficient to exhibit cytotoxic effect. Hence, transferrin liposomes modified with a pH-sensitive fusogenic peptide were successfully utilized as a delivery system for the selective delivery of therapeutic peptide MP into the mitochondria [26].

12.3 Peptide-Targeting Based on Site of Action

12.3.1 Topical Delivery of Peptides

Skin is the largest organ of the body and is an important site for drug administration. The topical route being noninvasive is highly convenient for the delivery of biomacromolecules or macromolecules to the patients. However, the stratum corneum (SC) is the main barrier that interrupts drug delivery. Due to the unique properties of the SC, only small molecules that are lipophilic can cross this barrier. Biomolecules, such as therapeutic peptides, require an appropriate formulation system for topical delivery [27]. This section emphasizes on some of the formulation strategies that have been utilized to enhance peptide delivery via the topical route.

Cyclosporine A (CsA), one of the non-ribosomal peptides, is topically used as a potent anti-inflammatory drug against a broad range of inflammatory diseases, including psoriasis and atopic dermatitis. However, topically used peptides exhibit lower effectiveness because of lower absorption. To enable effective permeation of CsA, it was conjugated with the heptamer of arginine (biotin r7), which acted as a CPP. The conjugation was established by a linker that was unstable at the physiological pH. At a pH higher than 7, the conformation of the linker was altered such that the conjugation between CsA and biotin r7 was broken, which intracellularly released the CsA to exert its action. This conjugated peptide formulation was dispensed in the form of a lotion. The formulation was tested on nude mice skin, where contact dermatitis was induced experimentally. It was observed that biotin r7-CsA

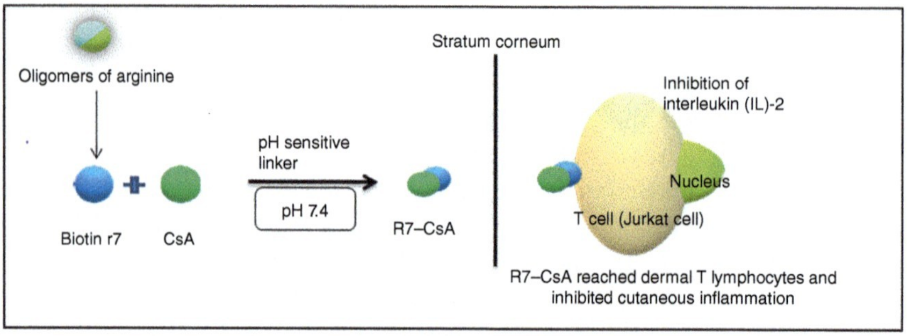

Figure 12.10 Topical delivery of CsA conjugated with arginine oligomer for inhibition of inflammation.

exhibited better penetration, as compared to native CsA. Similar results were also observed in human skin, wherein the epidermal layer is thicker than that of the mice. Moreover, biotin r7-CsA was also observed in CD3 T cells in the epidermis and other inflammatory cells, such as macrophages and fibroblasts. Thus, it was concluded that the CsA peptide, in combination with biotin r7, could successfully target the T lymphocytes and could be used for the treatment of inflammatory diseases of skin [28]. The formulation strategy is depicted in Figure 12.10.

Antimicrobial resistance is the most critical challenge that the healthcare professionals of the twenty-first century are facing. Commonly, skin infections are caused by *Staphylococcus aureus* and its methicillin-resistant analog, *Pseudomonas aeruginosa,* and multi-drug resistant (MDR) variants, *Staphylococcus epidermidis* and *Escherichia coli.* Treatment with AMPs may be a promising therapy of future, especially when treating MDR microorganisms. This is because AMPs are less susceptible for developing antimicrobial resistance. AMPs act rapidly and possess a broad spectrum of bactericidal activity. A study conducted by Boge et al. involved the development of lipidic cubosomes, loaded with LL-37 AMP, as depicted in Figure 12.11, for the effective treatment of *S. aureus.* LL-37 is bactericidal, immunomodulatory, and it promotes healing of chronic wounds. However, peptide molecules are prone to proteolytic degradation and have poor penetration through the skin. Therefore, cubosomes were developed as suitable carriers to enhance the penetration and protect the encapsulated peptide against degradation. They are liquid crystalline nanoparticles produced using lipid bilayer, in a 3D cube shape [29]. In this study, glyceryl monooleate-based cubosomes were synthesized. Peptide was loaded into the cubosomes by three different methods, viz (i) preloading, where the peptide was mixed with the liquid crystalline gel that was then loaded into nanoparticles, (ii) post-loading, where peptide molecules were allowed to adsorb on to previously synthesized cubosomes, and (iii) hydrotrope loading, in which LL-37 was simultaneously incorporated during the formation of cubosomes. Among the three approaches, preloading resulted in the smallest particle size (130 nm), with a narrow particle-size distribution (polydispersity index, PDI 0.15). Peptide loading resulted in the formation of lamellar phase and later formed vesicles. In an *ex vivo* study conducted using pig skin, it was observed that LL-37 solution, alone, exhibited

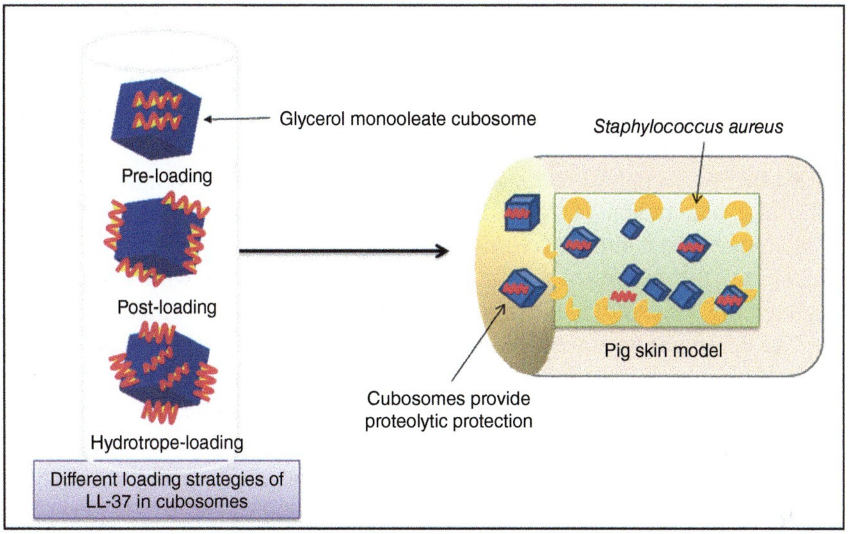

Figure 12.11 Cubosomes loaded with the antimicrobial peptide LL-37.

a greater antimicrobial activity as compared to the cubosomes. This was thought to be due to complete encapsulation of peptides inside the cubosome structure, which hindered its release. However, out of above-mentioned approaches, the preloaded cubosomes exhibited better action due to their small particle size and large surface area. Cubosomes resulted in efficient protection of the LL-37 peptide against proteolytic degradation but exhibited limitations with respect to their antimicrobial properties. Development of a stimuli-responsive carrier was thus, envisaged as a viable vehicle for improving the delivery of AMP [29].

12.3.2 Ocular Delivery of Peptides

Peptides are being increasingly used as promising therapeutics for the treatment of acute- and chronic-eye conditions that have affected the population worldwide, which include age-related macular degeneration (AMD), cataracts, diabetic retinopathy (DR), retinal inflammation, dry-eye conditions, and glaucoma. Even though eye is the most accessible organ for the delivery of small and large therapeutic molecules, ocular drug delivery has been very challenging due to the anatomy and physiology of the eye. Barriers that affect the bioavailability of macromolecules are the anterior and posterior segment structures, such as the conjunctiva, sclera, retinal pigment epithelium, and cornea. These barriers restrict the delivery of therapeutic agents, particularly to the back of the eye [30, 31]. This section highlights the formulation strategies to deliver peptides that have been investigated to enhance the ocular administration to both, anterior segment (cornea) and posterior segment (retina) of the eye.

The pharmacological barriers against topical administration in the ocular region are the tear turnover and the rapid drainage of the tear film, which creates a narrow contact time window between the drug and cornea leading to poor corneal permeation. Nanocarriers have been developed for delivering therapeutic peptides to overcome these barriers. Diebold and coworkers developed a liposome-based ocular delivery system to encapsulate thrombospondin-1-derived peptide (KRFK). The therapeutic effect of KRFK prevents chronic ocular surface inflammation. The liposomes provided protection to this peptide against the tear enzymes and released the encapsulated peptide in a controlled manner. Moreover, the excipients used in the preparation of liposomes, such as phosphatidylcholine, cholesterol, and vitamin E, are naturally present in the tear film. They also act as surfactants and improve the uptake of KRFK by connecting the lipids with the aqueous components of the tear film and promote the replenishment of the lipids in the tear film. The liposomes were prepared by the solvent evaporation technique. *In vitro* models of conjunctiva and cornea were utilized to study their permeability. There was no improvement in the permeability of KRFK peptide, when delivered using the liposomes, through the conjunctival epithelium. This was anticipated due to the high permeability of the epithelial layer, which easily allows the entry of macromolecules, such as peptides. However, the liposomes improved the corneal permeation of KRFK, as compared to the native peptide [32].

Retinal ischemia is a pathological condition that occurs when there is insufficient blood supply to the retina, resulting in altered metabolic functions due to the lack of oxygen supply. In this study, connexin 43 mimetic peptide (Cx43 MP) was selected due to its potential to prevent secondary damage associated with retinal ischemia and other inflammatory disorders. It is known to reduce edema, inflammation, neuronal death, and vascular leakage after retinal injury by obstructing the pathological opening of gap junction hemichannels. Connexins (Cx) are transmembrane proteins that play a crucial role in cellular communication, as subunits of hemichannels and gap junctions. Cx43 isoform is the most extensively expressed member of the Cx family, wherein the number 43 represents its molecular weight, i.e. 43 kDa. Attempts were made to enhance the retinal delivery of this peptide by loading it into hyaluronic acid (HA)-coated nanoparticles of serum albumin (HSA NPs), as depicted in Figure 12.12. This was hypothesized to prolong the activity of Cx43 MP and enable its targeted delivery to the retina. HSA NPs displayed number of advantages, such as high drug-loading capacity, possibility for surface modification, as well as non-immunogenicity, and thus were considered as an ideal delivery system. Blank HSA NPs were prepared by desolvation technique and two different methods were chosen to load the peptide into the nanoparticles, viz adsorption and incorporation. HA was coated over two types of fluorescein isothiocyanate-labeled

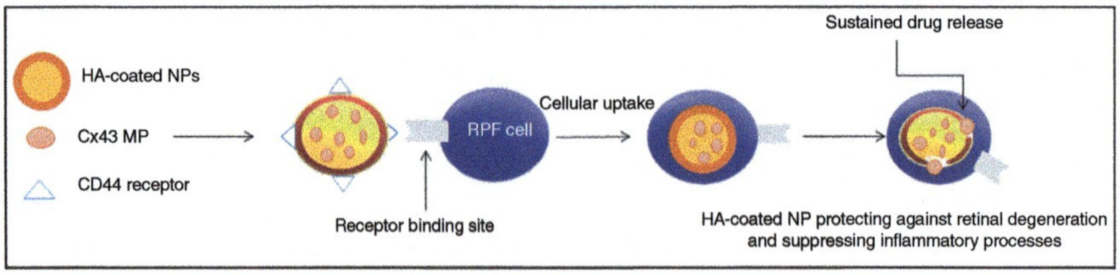

Figure 12.12 HA-coated albumin nanoparticle for targeted delivery of Cx43 MP for the treatment of retinal ischemia.

Cx43 MP (FITC-Cx43 MP)-HSA NPs, via electrostatic interaction and chemical modification, respectively. It was observed that the peptide was slowly released from the nanoparticles for four months. NPs administered intravitreally diffused through the vitreous and then penetrated the neural retina and retinal pigment epithelial (RPE) cells. Enhanced *in vitro* cellular uptake as well as *ex vivo* retinal penetration were observed for the HA-coated NPs, as compared to the uncoated particles, via HA-CD44 receptor-mediated interactions. *In vitro* cell viability, cellular uptake, and Cx43 MP hemichannel blocking were investigated using human RPE cells. The penetration ability of the nanoparticles across the neural retina was evaluated in an *ex vivo* model, using fresh bovine eyes. CD44 is a widely expressed transmembrane glycoprotein in the retina and a cell-surface receptor for HA. Results of cell viability and Cx43 MP functionality assays revealed the protective nature of the nanoparticles toward Cx43 MP, as they prevented the degradation of this peptide and sustained its release in the retina. This was anticipated due to the slow degradation and hydration of HSA, which prolonged its action, without reducing the cell viability, at concentrations that were used for Cx43 hemichannel blocking. Therefore, HA-coated HSA NPs exhibited a high potential for sustained and targeted delivery of Cx43 MP to treat various types of retinal inflammatory disorders. Furthermore, HA-coated HSA NPs released the Cx43 MP in a sustained manner and exhibited cellular biocompatibility, which in turn reduced the need for frequent intravitreal dosing of this system during the clinical studies [33].

12.3.3 Brain Delivery of Peptides

Targeted delivery of small and large molecules to the central nervous system (CNS) is quite challenging. Access to the brain is highly controlled due to the presence of the blood–brain barrier (BBB). The main component of the BBB is the endothelial cell layer that lines the blood vessels and creates a barrier against the transport of substances due to the presence of tight junctions. The capillaries in the brain are highly specialized and exhibit low permeability for water-soluble drugs. Other supportive cells of the BBB include the pericytes and astrocytes, which are present in the supporting tissue at the base of the endothelial membrane and form the solid envelope around the brain capillaries. The BBB allows only lipid-soluble compounds, water, nutrients, and small molecules (400–600 Da) to move in and out of the brain via mechanisms, such as passive diffusion, active transport, carrier-mediated transport, and endocytosis [34]. One of the challenges in the drug discovery process is reaching and maintaining effective CNS permeation and drug concentrations. The brain uses efflux pumps at the luminal side of the BBB to recognize and remove foreign substances. Anatomical features, such as the presence of multi-drug resistance proteins, the changes in the morphology and physiology of BBB during pathological conditions, and the blood–brain tumor barrier (BBTB), may also prevent drug accumulation inside the brain. Specific therapeutic peptides play an essential role in several neurological diseases, such as neurodegeneration, psychiatric disorders, pain, stroke, and brain cancer therapies. Advanced techniques are being explored for the targeted delivery of peptides to the CNS, which include the liposomal carriers, lipidic and polymeric nanoparticles [35, 36].

Alzheimer's disease (AD) is one of the most common forms of dementia prevalent among the aging population. The challenges in delivery of drugs to the CNS, as stated earlier, have necessitated the development of newer, effective therapies. Neurotrophic agents, such as neuropeptides, can be efficiently used for treating AD. Previously conducted *in vitro* and *in vivo* studies have proven the efficacy of NAP (NAPVSIPQ), an octapeptide derived from activity-dependent neuroprotective protein (ADNP), in the treatment of AD. However, owing to its low stability and short half-life, the development of an effective delivery system is essential. A study conducted by Liu et al. involved the production of B6 (CGHKAKGPRK) peptide-modified poly(ethylene glycol)–poly(lactic acid) (PEG–PLA) nanoparticles for the targeted delivery of NAP, as shown in Figure 12.13. The polymer used in this study, which is the PEG–PLA block copolymer, aided in prolonged drug release and longer circulation time of NAP. It helped to overcome the problems associated with its short half-life. Also, prolonged drug release was found to be suitable for the treatment of AD. The nanoparticles were prepared using the emulsion/solvent evaporation technique. The conjugation efficiency of B6 peptide was confirmed using X-ray photoelectron spectroscopy (XPS). *In vitro* release of NAP from the unmodified NPs and the B6 modified NAP NPs (B6 NP) were found to be similar. Cellular uptake of these NPs was analyzed using fluorescent microscopy. The cell-associated fluorescence intensity of B6-NPs was much higher than that of the unmodified NPs, which was further confirmed by quantitative cellular uptake analysis. There was no significant cytotoxicity observed when NAP NPs and B6-NAP-NPs were compared using *in vitro* cell viability studies. The brain-targeting ability of the B6 peptide was analyzed by real-time *in vivo* imaging. The fluorescence intensity of B6 NPs in the brain was significantly higher than that of unmodified NPs. The Morris water maze (MWM) behavioral experiment, along with the tests for biochemical indicators, were used to analyze the neuroprotective potential of NAP in the AD mouse model. The data obtained indicated better potential of B6 NPs than the unmodified NPs for the treatment of AD. The drug dose in case of B6 NPs was 0.02 μg, whereas that of unmodified NPs was 0.08 μg. The animals receiving NAP solutions at the dose of 0.02 and 0.08 μg did not show any significant improvement in their behavioral or biochemical indicators due to the rapid degradation of NAP after *in vivo* injection. Hematoxylin and eosin (HE) staining was performed to analyze the possible damage associated with the use of the nanoparticles. The B6 NPs did not cause any visible damage. Hence it was concluded that the B6 NPs could be efficiently used as a vehicle for delivering neuropeptides for the treatment of AD [37].

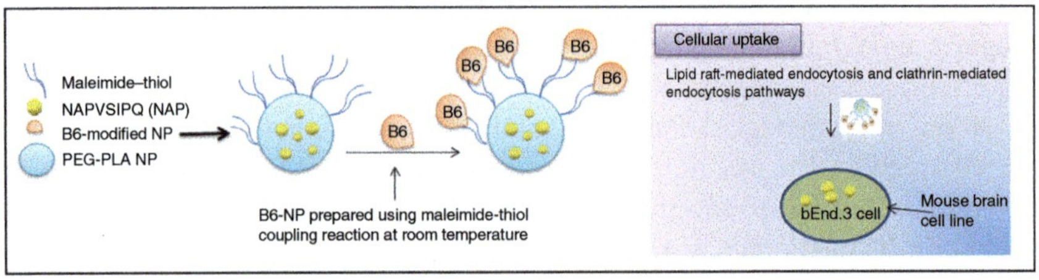

Figure 12.13 B6-modified PEG–PLA nanoparticle for targeted delivery of NAP.

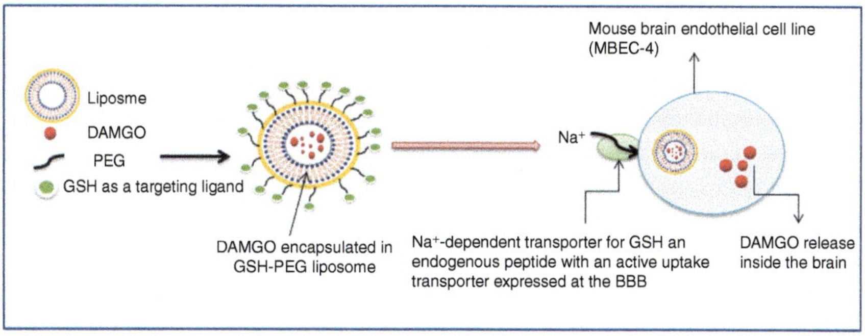

Figure 12.14 Brain-targeted delivery of DAMGO peptide using GSH-PEGylated liposomal formulation.

Drug delivery to the brain can be enhanced using glutathione (GSH)-conjugated PEG-coated liposomes because of their biocompatibility. These liposomes have been already proven to be efficacious as drug carriers for diseases, such as cancer and fungal infections. Subsequently, an addition of GSH, which is an endogenous peptide with an active uptake transporter expressed by the BBB, can potentially translocate the liposomal formulation to the brain and enhance CNS delivery. Therefore, Lindqvist et al. formulated glutathione-PEGylated (GSH-PEG) liposomes, as depicted in Figure 12.14, and evaluated their ability to increase and prolong blood-to-brain delivery of the opioid peptide DAMGO (H-TyrD-Ala-Gly-MePhe-Gly-ol). This was followed by evaluation of the pharmacokinetics-pharmacodynamics (PK-PD) profile of the liposomal formulation. DAMGO is an enkephalin analog with high selectivity as well as affinity for μ-opioid receptors. An intravenous loading dose of DAMGO was administered to rats, followed by GSH-PEG liposomal formulation of DAMGO. The concentration of free DAMGO in the brain and plasma was measured by using microdialysis, whereas GSH-PEG liposome-encapsulated DAMGO was measured using regular plasma sampling. The anti-nociceptive effect of DAMGO was examined with the tail-flick method. The short infusion of only DAMGO resulted in a fast decline of the peptide concentration in plasma, whereas its encapsulation in GSH-PEG liposomes prolonged the half-life by more than 40 folds (several minutes to seven hours). Thus, by calculating and monitoring the released peptide concentration from the GSH-PEG liposomal formulation in, blood and brain interstitial fluid over time, it was demonstrated that these liposomes were successful in enhancing and prolonging the delivery of the peptide to the brain [38].

12.3.4 Lung-Targeted Delivery of Peptides

Delivery to the lungs is essential and important to cure chronic diseases, such as pulmonary fibrosis, asthma, chronic obstructive pulmonary disease (COPD), cystic fibrosis (CF), and lung cancer. Pulmonary delivery describes various systems, devices, formulations, and methods of delivery of the active molecules to the respiratory tract and for systemic delivery through the lungs. The importance of pulmonary delivery is due to the availability of extensive surface area of the air–blood barrier, which permits rapid absorption by passive diffusion. Pulmonary delivery is the most attractive target in healthcare research as the lungs may be used for local, systemic as well as targeted delivery of therapeutic molecules [39].

Peptides, such as vasoactive intestinal peptide (VIP), exert efficient therapeutic effects in various lung diseases. When VIP interacts with its membrane receptor, it undergoes rapid internalization, which leads to reduction in availability of surface for the attachment of other VIP molecules. Thus, the peptide molecules need to wait for the receptor to recycle itself for further binding. The peptides undergo degradation during this period, and thus may face problems of short half-life and low stability. Thus, liposomes were designed to improve the effectiveness of this therapeutic peptide. Stabilized liposomes were fabricated to provide a depot effect. A study conducted by Hajos et al. involved the production of an inhalable liposomal formulation for the targeted delivery of VIP to the lungs, which acted as an alternative to IV administration. The liposomal formulation was evaluated for its suitability to be administered via the commercial mouthpiece ventilation inhaler. The inhaler used was micro-drop master jet (MPV). The droplet size generated by this device was suitable for bronchiolar deposition. The inhalable, vasoactive intestinal peptide-loaded liposomes (VLLs) were formulated using a phospholipid. The association of VIP to the negatively charged phospholipids induced a change in its conformation, from random coil to alpha-helical structure, which is the preferred form for ligand–receptor interaction. It also prevented the degradation of VIP due to enzymes present in the lungs. Further, a PEG shell was employed as a protective shield for the surface-attached VIP, which would prevent its aggregation and degradation. The impact of nebulization on liposomes was tested. No major changes in the particle size and PDI were observed. The spray pattern of the MPV inhaler was analyzed, which displayed a droplet size distribution ranging from 1 to 20 μ. The release of VIP was evaluated using strips of arteries that contained more muscle cells having the VIP receptors. The analysis was performed in the presence and absence of the target muscle cells. It was concluded that VIP was released only in the presence of target cells, which confirmed the ability of the liposomes to facilitate the targeted release of VIP. Sustained release of VIP was confirmed by testing the relaxation response. The formulation was further tested for its relaxation activity, upon storage. It was observed that the relaxation activity was stable over a month and thereafter declined. Thus, it was concluded that the inhalable liposomes could be successfully used for the targeted delivery of VIP to the lungs [40].

Tuberculosis (TB) is a life-threatening disease caused by *Mycobacterium tuberculosis*. Its transmission occurs mostly via the respiratory tract. The current treatments for TB witness low patient compliance due to factors, such as the high cost, prolonged duration of administration, and multi-drug therapies, which often result in treatment failure and development of drug resistance. Thus, AMPs are being increasingly used as promising candidates for the treatment of TB [41]. In a study by Silva et al., an exogenous cationic AMP LLKKK18 was selected due to its ability to efficiently kill the mycobacteria. The activity of this peptide was enhanced by incorporating it into self-assembling HA nanogels, which not only increased the stability and the targeting ability of the peptide but also exhibited reduced cytotoxicity and degradability. Additionally, this formulation also permitted targeted delivery of the peptide to the macrophages. The mycobacterial infection alters the host cell membrane. This results in the exposure of the negatively charged molecules and thereby results in high binding of the cationic peptides. This, in turn, leads to the permeation of AMPs in the mycobacteria, through the mycobacterial cell envelope [42]. Researchers produced self-assembling HA-based nanogels, involving a cationic exchange resin (AG 50W) to exchange the cations of sodium hyaluronate, with the lipophilic tetrabutylammonium (TBA) ion. This enabled solubilization of HA in dimethyl sulfoxide (DMSO). Further, the hydrophobic 11-amino-1-undecanethiol (AT) was cross-linked to form HA-AT. After preparing the nanogel, the peptide was encapsulated by mixing it with the nanogel in phosphate-buffered saline (PBS), under stirring, for 24 hours, at 25 °C. The encapsulation efficiency of the peptide-loaded nanogels was around 70%. The two properties of the peptide that may have encouraged its encapsulation within the hydrophobic core of the negatively charged nanogel include its cationic charge and hydrophobicity. The intracellular presence of the peptide within the host cells and its co-localization with mycobacteria were also confirmed, which resulted in a significant reduction in the mycobacterial load in the infected macrophages. Intratracheal administration of these peptide-loaded nanogels drastically reduced the infection levels in mice infected with *Mycobacterium avium* or *M. tuberculosis,* after 5–10 administrations every other day. Therefore, due to the low probability of cellular resistance to this peptide, the results confirmed the potential of LLKKK18-loaded nanogels for treatment of TB, especially to kill the opportunistic *M. avium* and the most infectious human pathogen, the *M. tuberculosis.* Two strains of the opportunistic *M. avium*, 2447 and 2291, which formed smooth transparent (SmT) colonies, and *M. tuberculosis* H37Rv were selected to perform further studies. *In vitro* incubation of macrophages with LLKKK18-loaded nanogels significantly reduced the intracellular levels of both organisms. Such systems, loaded with AMPs, proved to be a novel and promising strategy for the treatment of mycobacterial infections [41].

12.4 Conclusion and Future Prospects

In recent years, efforts in the synthesis of therapeutic peptides and advanced developments in the design of suitable systems for their targeted delivery have gained much attention and importance. It holds immense potential in the case of cancer, where targeted therapy can overcome primary challenges, such as cytotoxicity. We have described various approaches that were found to give promising results and may be translated into market one day. Passive targeting of therapeutic peptides to the respective tumor cells was possible because of nanocarriers, which either utilized enhanced permeability and retention (ERP) effect of cancer cells or exhibited stimuli-responsive nature. A tumor-homing peptide was the most commonly employed for active targeting. These developments were possible only due to advanced research in the field of nanotechnology and by identification of molecular targets and specific ligands for such targets. A few hundred peptides have been already evaluated in preclinical and clinical trials. Accelerated development in characterization of peptides, understanding the penetration and intracellular signaling of peptide molecules, and development of methods for establishing better correlation between *in vitro* and *in vivo* experimental data, is anticipated to hasten the development of peptide-based formulations, for better therapy of various grave health conditions [43].

References

1 Fosgerau, K. and Hoffmann, T. (2015). Peptide therapeutics: current status and future directions. *Drug Discovery Today* 20 (1): 122–128. https://doi.org/10.1016/j.drudis.2014.10.003.

2 Lu, Y., Yang, J., and Sega, E. (2006). Issues related to targeted delivery of proteins and peptides. *AAPS J.* 8 (3): 466–478.

3 Al Musaimi, O., Al Shaer, D., de la Torre, B.G., and Albericio, F. (2018). 2017 FDA peptide harvest. *Pharmaceuticals* 11 (2): 1–10.

4 Lau, J.L. and Dunn, M.K. (2018). Therapeutic peptides: historical perspectives, current development trends, and future directions. *Bioorg. Med. Chem.* 26 (10): 2700–2707. https://doi.org/10.1016/j.bmc.2017.06.052.

5 Kaur, K., Singh, I., Kaur, P., and Kaur, R. (2015). Food and Drug Administration (FDA) approved peptide drugs. *Asian J. Res. Biol. Pharm. Sci.* 3 (3): 75–88. https://pdfs.semanticscholar.org/a57d/dfa8165391c0aba179850297c37b68a399e8.pdf.

6 Andrieu, J., Re, F., Russo, L., and Nicotra, F. (2019). Phage-displayed peptides targeting specific tissues and organs. *J. Drug Targeting* 27 (5–6): 555–565. https://doi.org/10.1080/1061186X.2018.1531419.

7 Pudlarz, A. and Szemraj, J. (2018). Nanoparticles as carriers of proteins, peptides and other therapeutic molecules. *Open Life Sci.* 13 (1): 285–298.

8 Solaro, R., Chiellini, F., and Battisti, A. (2010). Targeted delivery of protein drugs by nanocarriers. *Materials* 3: 1928–1980.

9 Santos, H.A., Bimbo, L.M., Das Neves, J., and Sarmento, B. (2012). Nanoparticulate targeted drug delivery using peptides and proteins. In: *Nanomedicine: Technologies and Applications* (ed. T.J. Webster), 236–301. Woodhead Publishing Limited https://doi.org/10.1533/9780857096449.2.236.

10 Crombez, L., Morris, M., Deshayes, S. et al. (2008). Peptide-based nanoparticle for ex vivo and in vivo dug delivery. *Curr. Pharm. Des.* 14 (34): 3656–3665.

11 Deshayes, S., Konate, K., Aldrian, G. et al. (2010). Structural polymorphism of non-covalent peptide-based delivery systems: highway to cellular uptake. *Biochim. Biophys. Acta, Biomembr.* 1798 (12): 2304–2314. https://doi.org/10.1016/j.bbamem.2010.06.005.

12 Marqus, S., Pirogova, E., and Piva, T.J. (2017). Evaluation of the use of therapeutic peptides for cancer treatment. *J. Biomed. Sci.* 24 (1): 1–15.

13 Kim, S.K., Foote, M.B., and Huang, L. (2013). Targeted delivery of EV peptide to tumor cell cytoplasm using lipid coated calcium carbonate nanoparticles. *Cancer Lett.* 334 (2): 311–318. https://doi.org/10.1016/j.canlet.2012.07.011.

14 Cheng, K., Ding, Y., Zhao, Y. et al. (2018). Sequentially responsive therapeutic peptide assembling nanoparticles for dual-targeted cancer immunotherapy. *Nano Lett.* 18 (5): 3250–3258.

15 Dubey, N., Varshney, R., Shukla, J. et al. (2012). Synthesis and evaluation of biodegradable PCL/PEG nanoparticles for neuroendocrine tumor targeted delivery of somatostatin analog. *Drug Delivery* 19 (3): 132–142.

16 Ostrom, Q.T., Gittleman, H., Farah, P. et al. (2013). CBTRUS statistical report: primary brain and central nervous system tumors diagnosed in the United States in 2006–2010. *Neuro Oncol.* 15 (Suppl. 2): 15:ii1–ii56.

17 Lépinoux-Chambaud, C. and Eyer, J. (2019). The NFL-TBS.40–63 peptide targets and kills glioblastoma stem cells derived from human patients and also targets nanocapsules into these cells. *Int. J. Pharm.* 566: 218–228. https://doi.org/10.1016/j.ijpharm.2019.05.060.

18 Agemy, L., Friedmann-Morvinski, D., Kotamraju, V.R. et al. (2011). Targeted nanoparticle enhanced proapoptotic peptide as potential therapy for glioblastoma. *Proc. Natl. Acad. Sci. U.S.A.* 108 (42): 17450–17455.

19 Lee, W.H., Loo, C.Y., Young, P.M. et al. (2014). Recent advances in curcumin nanoformulation for cancer therapy. *Expert Opin. Drug Delivery* 11 (8): 1183–1201.

20 Kumar, A., Ma, H., Zhang, X. et al. (2012). Gold nanoparticles functionalized with therapeutic and targeted peptides for cancer treatment. *Biomaterials* 33 (4): 1180–1189. https://doi.org/10.1016/j.biomaterials.2011.10.058.

21 Roth, L., Agemy, L., Kotamraju, V.R. et al. (2012). Transtumoral targeting enabled by a novel neuropilin-binding peptide. *Oncogene* 31 (33): 3754–3763. https://doi.org/10.1038/onc.2011.537.

22 Soman, N.R., Baldwin, S.L., Hu, G. et al. (2009). Molecularly targeted nanocarriers deliver the cytolytic peptide melittin specifically to tumor cells in mice, reducing tumor growth. *J. Clin. Invest.* 119 (9): 2830–2842.

23 Tan, Y., Wu, Q., and Zhou, F. (2020). Targeting acute myeloid leukemia stem cells: current therapies in development and potential strategies with new dimensions. *Crit. Rev. Oncol. Hematol.* 152: 102993. https://doi.org/10.1016/j.critrevonc.2020.102993.

24 Niu, F., Yan, J., Ma, B. et al. (2018). Lanthanide-doped nanoparticles conjugated with an anti-CD33 antibody and a p53-activating peptide for acute myeloid leukemia therapy. *Biomaterials* 167: 132–142. https://doi.org/10.1016/j.biomaterials.2018.03.025.

25 Mader, J.S., Richardson, A., Salsman, J. et al. (2007). Bovine lactoferricin causes apoptosis in Jurkat T-leukemia cells by sequential permeabilization of the cell membrane and targeting of mitochondria. *Exp. Cell. Res.* 313 (12): 2634–2650.

26 Yamada, Y., Shinohara, Y., Kakudo, T. et al. (2005). Mitochondrial delivery of mastoparan with transferrin liposomes equipped with a pH-sensitive fusogenic peptide for selective cancer therapy. *Int. J. Pharm.* 303 (1–2): 1–7.

27 Witting, M., Obst, K., Friess, W., and Hedtrich, S. (2015). Recent advances in topical delivery of proteins and peptides mediated by soft matter nanocarriers. *Biotechnol. Adv.* 33 (6): 1355–1369.

28 Rothbard, J.B., Garlington, S., Lin, Q. et al. (2000). Conjugation of arginine oligomers to cyclosporin A facilitates topical delivery and inhibition of inflammation. *Nat. Med.* 6 (11): 1253–1257.

29 Boge, L., Hallstensson, K., Ringstad, L. et al. (2019). Cubosomes for topical delivery of the antimicrobial peptide LL-37. *Eur. J. Pharm. Biopharm.* 134: 60–67. https://doi.org/10.1016/j.ejpb.2018.11.009.

30 Mandal, A., Pal, D., Agrahari, V. et al. (2018). Ocular delivery of proteins and peptides: challenges and novel formulation approaches. *Adv. Drug Delivery Rev.* 126: 67–95. https://doi.org/10.1016/j.addr.2018.01.008.

31 Kim, Y.C., Chiang, B., Wu, X., and Prausnitz, M.R. (2014). Ocular delivery of macromolecules. *J. Controlled Release* 190: 172–181. https://doi.org/10.1016/j.jconrel.2014.06.043.

32 Soriano-Romaní, L., Álvarez-Trabado, J., López-García, A. et al. (2018). Improved *in vitro* corneal delivery of a thrombospondin-1-derived peptide using a liposomal formulation. *Exp. Eye Res.* 167: 118–121. https://doi.org/10.1016/j.exer.2017.12.002.

33 Huang, D., Chen, Y.S., Green, C.R., and Rupenthal, I.D. (2018). Hyaluronic acid coated albumin nanoparticles for targeted peptide delivery in the treatment of retinal ischaemia. *Biomaterials* 168: 10–23. https://doi.org/10.1016/j.biomaterials.2018.03.034.

34 Bellettato, C.M. and Scarpa, M. (2018). Possible strategies to cross the blood–brain barrier. *Ital. J. Pediatr.* 44 (S2): 127–133.

35 Dréan, A., Goldwirt, L., Verreault, M. et al. (2016). Blood–brain barrier, cytotoxic chemotherapies and glioblastoma. *Expert Rev. Neurother.* 16: 1285–1300.

36 Lalatsa, A., Schatzlein, A.G., and Uchegbu, I.F. (2014). Strategies to deliver peptide drugs to the brain. *Mol. Pharmaceutics* 11 (4): 1081–1093.

37 Liu, Z., Gao, X., Kang, T. et al. (2013). B6 peptide-modified PEG-PLA nanoparticles for enhanced brain delivery of neuroprotective peptide. *Bioconjugate Chem.* 24 (6): 997–1007.

38 Lindqvist, A., Rip, J., Gaillard, P.J. et al. (2013). Enhanced brain delivery of the opioid peptide DAMGO in glutathione PEGylated liposomes: a microdialysis study. *Mol. Pharmaceutics* 10 (5): 1533–1541.

39 Patil, J.S. and Sarasija, S. (2012). Pulmonary drug delivery strategies: a concise, systematic review. *Lung India* 29 (1): 44–49.

40 Hajos, F., Stark, B., Hensler, S. et al. (2008). Inhalable liposomal formulation for vasoactive intestinal peptide. *Int. J. Pharm.* 357 (1–2): 286–294.

41 Silva, J.P., Gonçalves, C., Costa, C. et al. (2016). Delivery of LLKKK18 loaded into self-assembling hyaluronic acid nanogel for tuberculosis treatment. *J. Controlled Release* 235: 112–124. https://doi.org/10.1016/j.jconrel.2016.05.064.

42 Gutsmann, T. (2016). Interaction between antimicrobial peptides and mycobacteria. *Biochim. Biophys. Acta, Biomembr.* 1858 (5): 1034–1043. https://doi.org/10.1016/j.bbamem.2016.01.031.

43 Kaspar, A.A. and Reichert, J.M. (2013). Future directions for peptide therapeutics development. *Drug Discovery Today* 18 (17–18): 807–817. https://doi.org/10.1016/j.drudis.2013.05.011.

13

Antibody-Based Targeted T-Cell Therapies

Manoj Bansode[1], Kaushik Deb[2], and Sarmistha Deb[2]

[1] *Saiseva Biotech Pvt. Ltd., CureCells™ Cord Blood Bank, Kant Helix, Bhoircolony, Chinchwad, Pune 411033, Maharashtra, India*
[2] *DiponEd BioIntelligence LLP, 60/A, 2nd Floor, Karnataka Bank Building, Jigani Link Road, Bomasandra Industrial Area, Bangalore 560099, Karnataka, India*

13.1 Introduction

Adaptive immunity in mammals is comprised of cell-mediated and humoral immunity. The adaptive immune system consists of cells that originate from the lymphoid system. These are the T and B lymphocytes that trigger an acquired or antigen-specific immune response. The B lymphocytes can mature into plasmocytes that produce B-cell antigen receptors, immunoglobulin (Ig), and thus initiate the host humoral immune response. On the other hand, T cells are responsible predominantly for cell-mediated immunity and also moderate humoral immunity [1, 2]. T cells are matured in the thymus through the process of positive and negative selections. The most commonly known T cells are CD4+ T cells (helper T cells) and CD8+ T cells (cytotoxic T cells or killer T cells). Both T cells express T-cell receptors (TCRs) and CD3 co-receptors on their surface [1]. The TCRs are capable of recognizing the major histocompatibility complex (MHC-1 and MHC-2)-bound processed antigens as expressed by the antigen-processing cells (APCs). CD4+ T cells recognize the MHC-2-bound antigens, whereas CD8+ T cells recognize MHC-1-bound antigens.

Cancer is the major cause of death worldwide. In 2019, 1 762 450 new cancer cases and 606 880 cancer deaths were projected to occur in the United States [3]. In India, the burden of cancer patients is increased to 2.25 million [4, 5]. The conventional cytotoxic approach is still a major part of the treatment of cancer. However, the conventional treatment has limitations and severe side effects, which is, therefore, responsible for the constant research in therapeutic approaches for cancer. The new era in the treatment of cancer utilizes immunotherapy for the treatment that means utilization of the patient's immune system [6]. Cancer immunotherapy has emerged as a safe alternative for cytotoxic treatments. The goal of cancer immunotherapy is to trigger immune responses during immunosurveillance and the elimination of cancer cells.

Targeted Drug Delivery, First Edition. Edited by Yogeshwar Bachhav.

The main cells involved in this process are cytotoxic lymphocytes, cytotoxic T cells, and natural killer (NK) cells. Cell death is triggered via different receptors, such as perforin (PRF1), granule-associated enzymes (GZM), and death-ligand/death receptor system.

13.2 Immune-Directed Cancer Cell Death

Tumor cells express the antigens on the surface that provokes the activation of immunocyte's consequential production of antigen-specific antibodies, which results in specific and efficient damages of tumor cells. For example, lysis of the cells occurs through binding of the antibodies to the tumor antigen, this leads to different antibody-mediated cell cytotoxic reactions. This process includes lysis of the cell by complement activation that occurred through binding of the antibodies to the tumor antigen [7], resulting in the generation of C3a and C3b fragments that are responsible for activation of neutrophils and macrophages, respectively, which is followed by cytotoxicity toward cancer cells through signaling of these receptors [8].

13.3 Immunotherapy Strategies in Cancer

Currently, there are several immunotherapies present in the market, majority of which are natural biological products used to activate the immune systems through genetic engineering, such as chimeric antigen receptor (CAR)-T cells. Immunotherapy against cancer is divided into three main categories, i.e. monoclonal antibodies, immune-response modifiers, and vaccines. The antibody-based therapies are passive from immunotherapy, which means the molecules are introduced into the body rather than the immune response is inherited by the body itself [9, 10].

The approach of monoclonal antibodies includes the interaction between Fc receptors that are expressed on tumor cells, which act as the target antigen. That means the targets of these therapies are protooncogenes and oncogenes, growth factor receptors of signaling, and antigenic differences between normal and cancerous cells [10].

Currently, there are approved monoclonal antibodies for the treatment of breast cancer (trastuzumab) and non-Hodgkin's lymphoma (NHL) (rituximab) and diagnosis of certain cancers [10]. Other than these, Alemtuzumab, Tositumomab, Cetuximab, Bevacizumab, Panitumumab, Catumaxomab, Ofatumumab, Ipilimumab, Pertuzumab, and Denosumab are other conjugate-based monoclonal antibodies approved for use [11]. The monoclonal antibodies have a limitation over conventional antibodies in that they have a shorter plasma half-life [10].

Immune-response modifiers are of two types, namely extrinsic molecules and intrinsic molecules. Extrinsic molecules include BCG, *Cryptosporidium parvum*, and endotoxin, which are microbes or microbial products and are known as immune potentiators, while, intrinsic factors include IL-1, IL-2, interferons, TNF,

B cells, growth factors, colony-stimulating factors. These agents exert their effects at different stages of the immune response [10].

13.4 T-Cell Therapy

The goal of T-cell therapy is to generate an immune-modulated antitumor response in the host system. The therapy is divided into two parts. The first one includes naturally occurring T-cells that are isolated from umbilical cord blood and the second type of T-cell therapy is the genetically modified blood-derived T cells or CAR-T cells that are acting at specific recognition sites of tumor cells [12].

Cancer is progressed through a carcinogenic cascade that involves the different steps, i.e. proliferation, cell-cycle progression, DNA replication, escaping apoptosis, angiogenesis, and metastasis [10]. The T-lymphocytes play an important role in the destruction of tumor cells of mammals [13]. Activation of T cells leads to the generation of sensitized and cytotoxic subset T-cell clones ensuring the release of lymphokines. These released lymphokines mobilize and activate the B cells through growth and differentiation factors. The production of lymphokines activates phagocytes from a reticuloendothelial system that digest and destroy the tumor cells. Additionally, the lymphokines (specifically interferons) enhance cytotoxic activity toward tumor cells through NK cells and macrophages. Cancer cells have unique tumor markers that can be recognized by the NK cells. Additionally, cytotoxic factors that are released due to cytolysis, engender additional unique and specific receptors on the surface of targeted cancer cells through a series of biological events. These unique markers are also recognized by NK cells resulting in selective and specific devastation of cancer cells [14]. Different *in vitro* studies reported previously have demonstrated the involvement of macrophages and NK cells to eradicate tumor cells through the mechanism of cytolysis and phagocytosis [15, 16]. Literature reports suggest that the efficiency of cytolysis is enhanced in the presence of lymphocytes and lymphokines [17–19].

13.5 Naturally Occurring T Cells

Naturally occurring T cells (tumor-infiltrating lymphocytes [TILs]) are a heterogeneous population of lymphocytes that are primarily comprised of T cells and NK cells. These cells naturally migrate into a tumor and exert their actions. Activation of T cells releases lymphokines that mobilize and activate the B cells through growth and differentiation factors. The production of lymphokines activates phagocytes that digest and destroy the tumor cells. Additionally, the lymphokines kill tumor cells through NK cells and macrophages. Cancer cells have unique tumor markers that can be recognized by the NK cells. In addition to that, cytotoxic factors generate additional unique and specific receptors on the surface of targeted cancer cells that are also recognized by NK cells, ensuring selective and specific devastation of cancer cells [14]. Literature reports have demonstrated the contribution of macrophages and NK cells to eradicate tumor cells through the mechanism of cytolysis and phagocytosis [15, 16]. The first report of T-cell therapy was published in 1972, where the patient with gastric cancer was treated with T-cell therapy and the outcome of that treatment was a total reversion of liver metastasis in absence of other cytotoxic treatments [20]. The TILs positively exhibited their presence in the prognosis of various cancer [21–24].

TILs are capable of recognizing tumor cells consisting of antigens through their endogenous TCRs. T-cell growth factors, such as IL-2, facilitate the development of large-scale *in vitro* expansion of TILs isolated from umbilical cord blood [25, 26].

TILs are applied for the treatment of late-stage metastatic melanoma cases. Additionally, TILs have illustrated durable responses, achieved a reduction in tumors size, and disease-free condition after the treatment. In the case of T-cell therapy, mortality rate, relapse of cancer cases, and refractory cancer are considerably less in consideration with other cancer therapy [27–32].

13.6 Genetically Modified Occurring T Cells

Genetically engineered T cells enhance the immune function of the immune cells when the natural tumor-specific responses have failed. The modification of T cells is designed through manipulating antigen specificity [12]. The genetic modification can be achieved by transferring genetic material encoding either a cloned TCR or by synthetic CAR targeting tumor-specific antigens toward T cells. TCRs are the natural T receptors that recognize the antigen located on tumor cells, and the process is achieved by MHC and HLA systems. In these types of treatments, T cells are obtained from peripheral blood and activated before genetic alteration. After genetic alterations, the cells are expanded and reperfusion is done. Gene transfer methods are commonly used to genetically engineer T cells, including transient mRNA transfection, retroviral vectors, lentiviral vectors, transposons, or, most recently, homologous recombination after gene editing [33–37].

Apart from having efficacy against cancer treatment, toxicities were observed in case of TILs treatments. These toxicities are mainly related to the preconditioning and high-dose administration of IL-2 [27–29]. Also, there have been reports of the mutational burden of tumor or tumor neoepitope burden as the response toward TILs [38].

13.7 Clinical Implication of T-Cell and CAR-T-Cell Therapy:

The use of TILs, peptide-induced T cells, and engineered T cells have shown the radical potency in the treatment of cancers that leads to durable and complete response (CR) against the late-stage and refractory disease stage patients [39–43].

Table 13.1 indicates the clinically available data for naturally occurring T cells and genetically modified T cells.

Table 13.1 Clinical trials of adoptive cell therapy with reported responses.

Diseases	Patient population	Antigen target	Conditioning therapy	CR	ORR	References
Melanoma	20	Various	Cy	1 (5%)	11 (55%)	[44]
Melanoma	43	Various	Cy + Flu	5 (12%)	15 (48%)	[45]
Melanoma	15	MART-1 (aa27-35, HLA-A2)	Cy + Flu	—	2 (13%)	[46]
Melanoma	16	gp100 (aa154-162, HLA-A2)	Cy + Flu	1 (6%)	3 (19%)	[47]
Metastatic melanoma	20	MART-1TCR (DMF5)	Cy + Flu	—	6 (30%)	[48]
Metastatic melanoma	16	gp100TCR (gp154)	Cy + Flu	—	3 (19%)	[48]
Synovial sarcoma Melanoma	17	NY-ESO-1 (aa157165, HLA-A2)	Cy + Flu	2 (12%)	9 (53%)	[49]
Melanoma	31	Various	Cy + Flu	4 (13%)	15 (48%)	[31]
Melanoma	31	Various	Cy + Flu	2 (6%)	13 (42%)	[32]
ALL (child/ young adult)	75	CD19	Cy + Flu Other	61 (81%)	61 (81%)	[50]
ALL (adult)	16	CD19	Cy	14 (88%)	14 (88%)	[51]
ALL (child/ young adult)	30	CD19	Cy + Flu	27 (90%)	27 (90%)	[50]
B-ALL	38	—	Cy/Flu or FLAG or IE	23 (61%)	—	[52]
CLL	14	CD19	CD19 Cy + Flu	4 (29%)	8 (58%)	[53]
Multiple myeloma	20	NY-ESO-1 (aa157-165, HLA-A2)	Other	16 (80%)	18 (90%)	[54]
NHL/CLL	15	CD19	Cy + Flu	8 (53%)	12 (80%)	[55]
B-ALL	30	—	Cy ± etoposide or Cy/Flu	27 (93%)	—	[56]
Melanoma	25	Various	Cy + Flu	3 (13%)	10 (42%)	[30]
B-ALL	59	—	Investigator's choice	55 (93%)	—	[57]
NHL	101	CD19	Cy + Flu	54%	54%	[58]
CLL	24	CD19	Cy + Flu	4 (17%)	16 (67%)	[59]
B-ALL	33	—	Cy or Cy/Flu	29 (91%)	—	[60]
ALL (adult) (CD28)	53	CD19	Cy	44 (83%)	44 (83%)	[61]
MM	16	BCMA	Cy + Flu	10 (63%)	13 (81%)	[62]

ALL, acute lymphocytic leukemia; B-ALL, B-cell acute lymphoblastic leukemia; BCMA, B-cell maturation antigen; CLL, chronic lymphocytic leukemia; CR, complete response; Cy, cyclophosphamide; FLAG, fludarabine + Ara-c + G-CSF; Flu, fludarabine; IE, ifosfamide/etoposide; MM, multiple myeloma; NHL, non-Hodgkin lymphoma; ORR, objective response rate.

13.8 Antibody-Induced T-Cell Therapy

T cells play a significant role to regulate predominantly cell-mediated immunity and moderately humoral immunity [1, 2]. Cancer cells modulate the immune system, T-cell function, and convert T cells into T-regulatory cells that suppress the immune system [63–65], resulting in enormous cell growth without any inhibition by the immune system. In addition to regular T-cell therapy, the new approach for redirecting T cells against cancer cells is the introduction of bispecific antibodies (BsAbs) that can bind with antigens expressed on the surface of tumor cells as well as T cells so that it can act as a bridging agent for two different cell types [66, 67]. Such BsAbs activate and redirect T cells toward vicinity of the cancer cells [68].

13.9 A Bispecific Antibody (BsAbs)-Induced T-Cell Therapy

This type of cancer immunotherapy seeks an approach to engross the cancer cells with effector T cells. These antibodies have two specific ends, one is having specificity toward T cells and the other with the tumor antigen. This association leads to the build immunological synapse and induces tumor cell death [69]. BsAbs are the antibodies that can bind two unique and separate antigens or epitopes on the same antigen to initiate the required response.

BsAbs are majorly divided into two categories, i.e. immunoglobulin G (IgG)-like molecules and non-IgG-like molecules. IgG-like BsAbs retain Fc-mediated effector functions, such as antibody-induced cytotoxicity, complement-dependent cytotoxicity, and antibody-dependent cellular phagocytosis [70]. Non-IgG-like BsAbs are smaller in size and have higher tissue penetration [71].

Ig-like molecules are of different types, such as quadromas [72], knobs-into-holes [73], dual-variable domains Ig (DVD-Ig) [74], IgG-single-chain Fv (scFv), two-in-one or dual action Fab (DAF) antibodies [75], half-molecule exchange [76], and κλ-bodies [77].

Non-IgG-like antibodies are scFv-based BsAbs [78], tandem scFvs [79], diabody format [80], dual-affinity retargeting (DART) molecules [80], nanobodies [81], and dock-and-lock (DNL) method [82].

The T-cell-activating antibodies (TABs) are composed of two single-chain variable fragments that are having their specific binding sites for tumor antigens and T cells, as indicated in Figure 13.1. Each fragment is having H- and L-chain fragments that are modified as per the desired targets [83]. A similar type of linker holds these fragments together. These linkers provide flexibility for the binding to the specific cellular sites of T cells and cancer cells.

These fused specificity fragments exclude other additional elements, such as Fc regions, and therefore are dependent on effector–tumor synapse formation and T-cell expansion. Currently, Fc-retaining antibodies have tumor cell-dependent

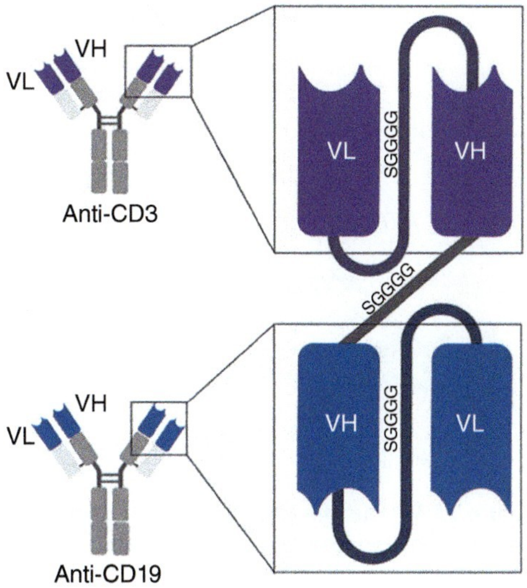

Figure 13.1 Structure of TABs: VH, variable H chain; VL, variable L chain.

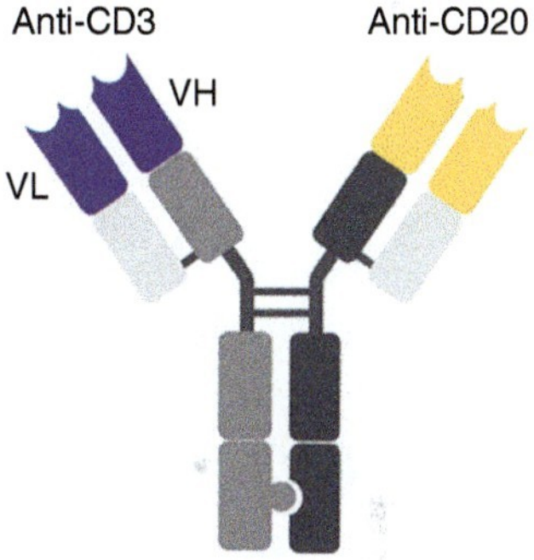

Figure 13.2 Full-length Ab retaining its Fc tail domain. Fc regions.

Fc regions, such as CD19, CD20, CD123, CD20, epithelial cell adhesion molecule (EPCAM), and T-cell-specific anti-CD3 Fc regions, as indicated in Figure 13.2.

The Fc regions are tumor-cell dependent, such as breast cancer, which has HER2; gastrointestinal tumors have the carcinoembryonic Ag (CEA); stomach, colon, and breast tumors have EPCAM; and B cells have CD19 and CD20 [70]. The main mechanism involved in these types of treatment is to initiate immune synapse formation or bridge formation between T cells and tumor cells. This immune synapse formation then endorses the release of perforins and granzymes, which get entered into tumor cells and promote apoptosis (Figure 13.3) [84].

The therapy is more advantageous as the dual-binding sites maintain more specificity, and therefore there is no random activation of T cells [85]. The cell-death rate is dependent on the ability of the molecule to induce proliferation and expansion of T cells that are unbound to the tumor (Figure 13.4) and also determine the sustained antitumor effects of these therapies [86, 87]. Currently, the investigation is ongoing to use this principle with other immune effector systems, such as NK cells [88, 89].

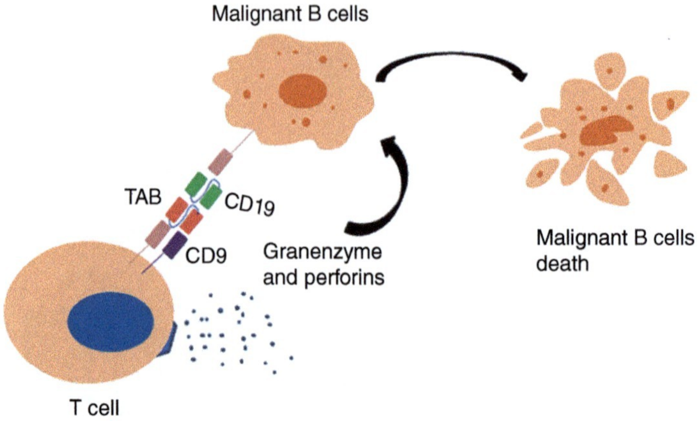

Figure 13.3 The immune synapse formation between T cells and malignant B cells. This immune synapse formation then endorses the release of perforins and granzymes, which get entered into tumor cells and promote malignant cell apoptosis.

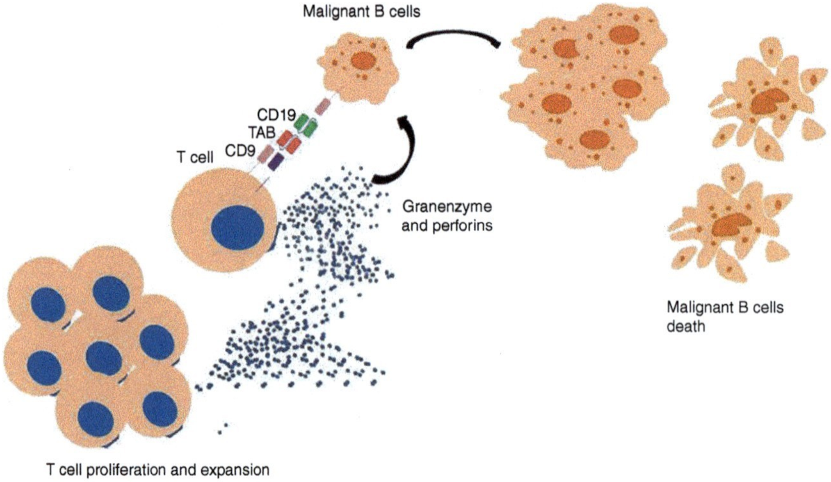

Figure 13.4 T-cell proliferation and expansion to promote the malignant cell apoptosis. Source: [86, 87].

Cytotoxic T cells play an important role in the immune response that is involved in cancer [90]. Tumor-specific T cell responses have limitations because of the immune escape mechanism induced by cancerous cells during the process of immunomodulation. This challenge could be overcome through the progress of immunotherapy. One of the strategies involved for the same is to harness the immune cells that will act as cytotoxic agents against the cancerous cells. The cells are designed in a way to redirect T cells against tumor cells [91]. The process is followed by a cytolytic synapse between T cells and targeted tumor cells. This leads to cell lysis through the activation and proliferation of T cells [92]. Apart from T cells, other immune cells that are also involved in the process are macrophages, monocytes, granulocytes, and NK cells.

13.10 Formats of BsAbs

There are several formats of BsAbs that are used in the treatment of cancer; for example Triomab, BiTE, DART, and FynomAb along with the IgG-dependent and independent BsAbs [93].

13.11 Triomab Antibodies in T-Cell Therapy

These antibodies are produced by mouse–rat hybridomas. These antibodies are trifunctional with one arm that is binding with tumor-associated antigen, second arm binding to CD3 receptors on T cells, and the chimeric Fc region ha that recognizes the necessary accessory present on macrophages, dendritic cells, and NK cells (these include CD64, CD32, CD16, and FcγR). T cells on activation cause the release of cytokines, such as TNF-α and IFN-γ. The FcγR-positive accessory cells release the pro-inflammatory markers at high levels, which include IL-6, Il-12, GM-CSF, and DC-CK1 [94]. Catumaxomab is a member of the Triomab family (Figure 13.5).

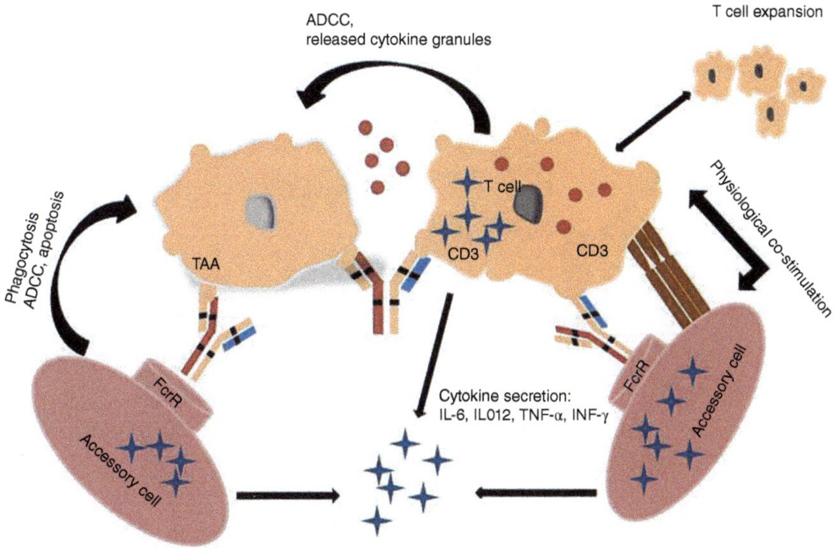

Figure 13.5 Tumor cells are killed through T-cell-mediated lysis and ADCC, as well as through phagocytosis by activated accessory cells. ADCC, antibody-dependent cell-mediated cytotoxicity.

13.12 Bispecific Antibodies in T-Cell Therapy

Bispecific antibodies (BsAbs) are those that bind two distinct epitopes to cancer. Amivantamab and Blinatumomab are example of the BiTE platform. Amivantamab (Rybrevant™): a bispecific antibody that targets EGFR and MET receptors on tumor cells; it is approved for lung cancer. Blinatumomab (Blincyto®): a bispecific antibody that targets CD19 on tumor cells as well as CD3 on T cells; it is approved for subsets of patients with leukemia [95]. BiTE type BsAbs have potent tumor cytotoxic actions by redirecting T cells to tumor cells. The cytolytic synapse between T cells and target cells results in influx of granzyme proteases during proliferation of T cells resulting in release of the inflammatory cytokines that are responsible for cell death [96, 97] (Figure 13.6).

This technology is used to produce more stable BsAbs that will reduce immunogenicity. DART involves covalent linkage between the two chains to improve stability over BiTEs. Dual-targeting strategies using bispecific antibodies directly act on cell-surface receptors, soluble factors, effector molecules, and effector cells that can act as an alternative therapy with respect to combination therapy.

Advantages of bispecific antibodies: Bispecific antibodies provide superior potency due to their novel mechanism of action. This type of therapy can be used for the treatment of solid tumor malignancies and highlights the potential they hold for future therapies to come. The major factor involved in such situations is the ability of antibodies to reach their target [66, 98]. Bispecific antibodies (BsAbs) can assist as mediators to redirect immune effector cells, such as NK

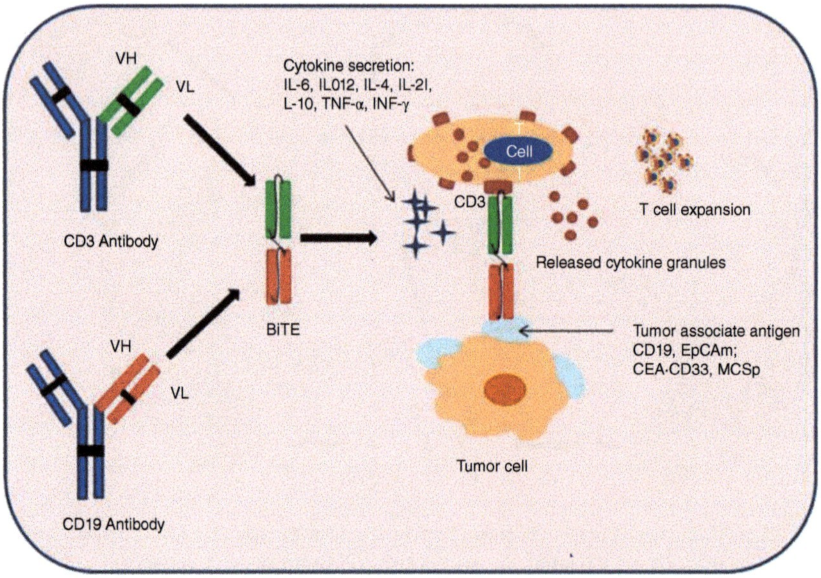

Figure 13.6 BiTE® Blinatumomab binds to CD3 on T cells and CD19-expressing B-cell malignancies.

cells and T cells, toward tumor cells to trigger tumor destruction. Additionally, by targeting two different receptors in combination on the same cell, these antibodies can induce modifications of a cell-signaling pathway.

Disadvantages of bispecific antibodies: Many limitations are faced to reach the target for drug delivery, such as poorly organized vasculature, lower accessibility for synovial joints and the kidney. Due to such limitations, there is not any successful clinical trial reported for T-cell-activating bispecific antibodies therapy in solid tumors. Due to very short half-life and dose-limiting toxicity, bispecific T-cell engagers are prohibitive in tumors. It has also been observed that tumor-associated antigen is not exclusively expressed only on transformed cells, which leads to off-tumor toxicities.

Another problem faced in the use of T-cell-activating bispecific antibodies is tumor mutation and succeeding treatment escape. In acute lymphoblastic leukemia, CD19-relapses remain, which has been found to be a major challenge in about 10–20% of patients due to disrupted CD19 membrane export [99].

13.13 Clinically Approved T-Cell-Activating Antibodies

Catumaxomab, the first clinically approved T-cell-activated bispecific antibody, was used to target CD3 and EPCAM for the treatment of malignant ascites. This antibody was trifunctional, consisting of mouse IgG2a (EPCAM) and rat IgG2b (CD3) [100, 101]. But, Catumaxomab was withdrawn in 2014 from European market due to commercial reasons.

The clinically approved bispecific T-cell engager so far is Blinatumomab (Blincyto, Amgen). One single-chain variable fragment of Blinatumomab is specific for CD3 and the other for CD19 [102–105]. Blinatumomab is used for the immunotherapy of lymphoma and leukemia. It has been shown to elevate the production of inflammatory cytokines, such as IL-2, IFN-γ, TNF-α, IL-4, IL-6, and IL-10 [106]. Blinatumomab can bridge cancerous B cells directly to CD3 positive T cells and therefore bypass TCR and MHC-1 molecules [102, 107, 108].

The hypothetical mechanism in Blinatumomab treatment was reported as increased frequencies of regulatory T cells and increased levels of PD-L1 expression on B-precursor acute lymphocytic leukemia (ALL) cells [109–111]. A fatal cytokine-release syndrome has also been observed as a very critical situation in polyclonal activation of T cells by T-cell-activating bispecific antibodies [112, 113] (Table 13.2).

13.14 Prospects

Antitumor-specific T lymphocytes isolated from tumors seemed to be an ideal candidate for killing tumor cells. TILs cells kill infected targets with great precision. Unfortunately, further challenges to treat various types of cancers using TILs find several difficulties, such as of immunosuppressive tumor environment, and also a lack of mode of treatment. The exploration for using autologous T cells with more predictable characteristics of tumor cell recognition has started.

Table 13.2 Clinical stage bispecific antibodies

BsAb	Formats	Targets	Clinical trial × identifier	Diseases
Catumaxomab	Triomab	EpCAM × CD3	Approved in EU	EpCAM-positive tumor, malignant ascites
			Completed phase IIa	Platinum refractory epithelial ovarian cancer
			Phase II	Gastric adenocarcinomas
			Phase II	Ovarian cancer
Ertumaxomab	Triomab	HER2 × CD3	Phase I/II	Her2/Neu-positive advanced solid tumors
FBTA05	TrioMab	CD20 × CD3	Phase I/II	Leukemia
Blinatumomab	BiTE	CD3 × CD19	Approved in USA	ALL
			Phase I	Relapsed NHL
			Phase II	B-cell ALL
			Phase II	Relapsed/refractory ALL
Solitomab (MT110, AMG 110)	BiTE	CD3 × EpCAM	Completed phase I	Solid tumors
AMG 330	BiTE	CD33 × CD3	Phase I	Relapsed/refractory AML
MT112 (BAY2010112)	BiTE	PSMA × CD3	Phase I	Prostatic neoplasms
MT111 (MEDI-565)	BiTE	CEA × CD3	Completed phase I	Gastrointestinal adenocarcinomas
BAY2010112	BiTE	CD3 × PSMA	Phase I	Prostatic neoplasms
MEDI-565	BiTE	CEA × CD3	Completed phase I	Gastrointestinal adenocarcinomas
HER2Bi-aATC	T cells preloaded with BsAbs	CD3 × HER2	Phase I	Ovarian, fallopian tube, or primary peritoneal cancer
GD2Bi-aATC	T cells preloaded with BsAbs	CD3 × GD2	Phase I/II	Children and young adults with neuroblastoma and osteosarcoma
			Completed phase I	Multiple myeloma and plasma cell neoplasm
MGD006	DART	CD123 × CD3	Phase I	Relapsed/refractory AML
MGD007	DART	gpA33 × CD3	Phase I	Colorectal carcinoma
IMCgp100	ImmTAC	CD3 × gp100	Phase I	Malignant melanoma
RG7802	CrossMab	CEA × CD3	Phase I	Solid cancers
MOR209/ES414	scFv-IgG	PSMA × CD3	Phase I	Prostate cancer

Source: Based on Fan et al. [93].

The efficacy of autologous CAR-T cells therapy is much more effective in the treatment of cancer. A CAR-T cell is being tested for solid tumors, such as lung, colon, prostate, and breast cancer. This CAR is a van for T cells transferred cytokine-mediated killing. At present, three CAR-T cells therapies are on the market. These therapies are approved for patients under age 25 with ALL and adults with DLBCL, and approved for adults with relapsed or refractory large B-cell NHL (RR DLBCL).

CAR-T cells therapies also have several challenges, including lack of migration of CAR-T cells from blood vessels to the tumor site, and within solid tumors. Further research on CARs modifications is essential to construct cytokines in the intracellular domain. Hence, novel approaches need to be developed in cancer to arrest antigen-negative relapse, to progress efficacy of tumor killing, to improve CAR-T cell perseverance, and to increase overall efficacy.

13.15 Conclusion

The current standard treatments of cancer, including surgery, radiotherapy, and chemotherapy, have severe side effects due to their nonspecific toxicity toward normal cells. Thus, alternative safe treatments are essential for cancer patients. In this regard, immunotherapy, such as T-cell or CAR-T cells in combination with antibodies, is effective with minimal side effects and toxicity. Furthermore, clinical trials using T-cell therapies have promising therapeutic outcomes in cancer patients. The scope to further improvement of T-cell and CAR-T cell design and delivery enhances the hope of a cure for several patients with malignancies that can sign an exciting new era in cancer treatment.

References

1 Sauls, R.S. and Taylor, B.N. (2019). *Histology, T-Cell Lymphocyte*. StatPearls Publishing.

2 Cano, R. L. E.; Lopera, H. D. E. (2013). *Introduction to T and B Lymphocytes*. Bogota (Colombia), El Rosario University Press.

3 Siegel, R.L., Miller, K.D., and Jemal, A. (2019). Cancer statistics, 2019. *CA. Cancer J. Clin.* 69 (1): 7–34. https://doi.org/10.3322/caac.21551.

4 Mathew, A., George, P.S., Jagathnath Krishna, K.M. et al. (2019). Transition of cancer in populations in India. *Cancer Epidemiol.* 58: 111–120. https://doi.org/10.1016/j.canep.2018.12.003.

5 National Institute of Cancer Prevention and Research (NICPR). (2018). Cancer Statistics – India Against Cancer. http://cancerindia.org.in/cancer-statistics (accessed 12 March 2020).

6 Miliotou, A.N. and Papadopoulou, L.C. (2018). CAR T-cell therapy: a new era in cancer immunotherapy. *Curr. Pharm. Biotechnol.* 19 (1): 5–18. https://doi.org/10.2174/1389201019666180418095526.

7 Kulcsár, G. (1997). Apoptosis of tumor cells induced by substances of the circulatory system. *Cancer Biother. Radiopharm.* 12 (1): 19–26. https://doi.org/10.1089/cbr.1997.12.19.

8 Peipp, M. and Valerius, T. (2002). Bispecific antibodies targeting cancer cells. *Biochem. Soc. Trans.* 30 (4): 507–511. https://doi.org/10.1042/bst0300507.

9 Kulcsár, G. (1997). Theoretical and literary evidence for the existence of the passive antitumor defence system. *Cancer Biother. Radiopharm.* 12 (4): 281–286.

10 Adam, J.K., Odhav, B., and Bhoola, K.D. (2003). Immune responses in cancer. *Pharmacol. Ther.* 99 (1): 113–132. https://doi.org/10.1016/S0163-7258(03)00056-1.

11 Coulson, A., Levy, A., and Gossell-Williams, M. (2014). Monoclonal antibodies in cancer therapy: mechanisms, successes and limitations. *West Indian Med. J.* 63 (6): 650–654. https://doi.org/10.7727/wimj.2013.241.

12 Met, Ö., Jensen, K.M., Chamberlain, C.A. et al. (2019). Principles of adoptive T cell therapy in cancer. *Semin. Immunopathol.* 41 (1): 49–58. https://doi.org/10.1007/s00281-018-0703-z.

13 Scaffidi, C., Kirchhoff, S., Krammer, P.H., and Peter, M.E. (1999). Apoptosis signaling in lymphocytes. *Curr. Opin. Immunol.* 11 (3): 277–285. https://doi.org/10.1016/s0952-7915(99)80045-4.

14 Paul, W.E. (1993). *Fundamental Immunology*. New York: Raven Press.

15 Gough, M.J., Melcher, A.A., Ahmed, A. et al. (2001). Macrophages orchestrate the immune response to tumor cell death. *Cancer Res.* 61 (19): 7240–7247.

16 Bingle, L., Brown, N.J., and Lewis, C.E. (2002). The role of tumour-associated macrophages in tumour progression: implications for new anticancer therapies. *J. Pathol.* 196 (3): 254–265. https://doi.org/10.1002/path.1027.

17 Paulnock, D. (1992). Macrophage activation by T cells. *Curr. Opin. Immunol.* 4 (3): 344–349. https://doi.org/10.1016/0952-7915(92)90087-U.

18 Embleton, M.J. (2003). The macrophage (2nd Edn). *Br. J. Cancer* 89 (2): 421. https://doi.org/10.1038/sj.bjc.6601103.

19 Adams, D.O. and Johnson, S.P. (1992). Molecular bases of macrophage activation: regulation of class II MHC genes in tissue macrophages. In: *Mononuclear Phagocytes*, 425–436. Dordrecht: Springer Netherlands https://doi.org/10.1007/978-94-015-8070-0_56.

20 Rosenberg, S.A., Fox, E., and Churchill, W.H. (1972). Spontaneous regression of hepatic metastases from gastric carcinoma. *Cancer* 29 (2): 472–474. https://doi.org/10.1002/1097-0142(197202)29:2<472::aid-cncr2820290235>3.0.co;2-u.

21 Clemente, C.G., Mihm, M.C., Bufalino, R. et al. (1996). Prognostic value of tumor infiltrating lymphocytes in the vertical growth phase of primary cutaneous melanoma. *Cancer* 77 (7): 1303–1310. https://doi.org/10.1002/(SICI)1097-0142(19960401)77:7<1303::AID-CNCR12>3.0.CO;2-5.

22 Galon, J., Costes, A., Sanchez-Cabo, F. et al. (2006). Type, density, and location of immune cells within human colorectal tumors predict clinical outcome. *Science* 313 (5795): 1960–1964. https://doi.org/10.1126/science.1129139.

23 Sato, E., Olson, S.H., Ahn, J. et al. (2005). Intraepithelial $CD8^+$ tumor-infiltrating lymphocytes and a high $CD8^+$/regulatory T cell ratio are associated

with favorable prognosis in ovarian cancer. *Proc. Natl. Acad. Sci. U. S. A.* 102 (51): 18538–18543. https://doi.org/10.1073/pnas.0509182102.

24 Loi, S., Sirtaine, N., Piette, F. et al. (2013). Prognostic and predictive value of tumor-infiltrating lymphocytes in a phase III randomized adjuvant breast cancer trial in node-positive breast cancer comparing the addition of docetaxel to doxorubicin with doxorubicin-based chemotherapy: BIG 02-98. *J. Clin. Oncol.* 31 (7): 860–867. https://doi.org/10.1200/JCO.2011.41.0902.

25 Smith, K.A., Gilbride, K.J., and Favata, M.F. (1980). Lymphocyte activating factor promotes T-cell growth factor production by cloned marine lymphoma cells. *Nature* 287 (5785): 853–855. https://doi.org/10.1038/287853a0.

26 Rosenberg, S.A., Packard, B.S., Aebersold, P.M. et al. (1988). Use of tumor-infiltrating lymphocytes and interleukin-2 in the immunotherapy of patients with metastatic melanoma. A preliminary report. *N. Engl. J. Med.* 319 (25): 1676–1680. https://doi.org/10.1056/NEJM198812223192527.

27 Rosenberg, S.A., Yang, J.C., Sherry, R.M. et al. (2011). Durable complete responses in heavily pretreated patients with metastatic melanoma using T-cell transfer immunotherapy. *Clin. Cancer Res.* 17 (13): 4550–4557. https://doi.org/10.1158/1078-0432.CCR-11-0116.

28 Besser, M.J., Shapira-Frommer, R., Itzhaki, O. et al. (2013). Adoptive transfer of tumor-infiltrating lymphocytes in patients with metastatic melanoma: intent-to-treat analysis and efficacy after failure to prior immunotherapies. *Clin. Cancer Res.* 19 (17): 4792–4800. https://doi.org/10.1158/1078-0432.CCR-13-0380.

29 Pilon-Thomas, S., Kuhn, L., Ellwanger, S. et al. (2012). Efficacy of adoptive cell transfer of tumor-infiltrating lymphocytes after lymphopenia induction for metastatic melanoma. *J. Immunother.* 35 (8): 615–620. https://doi.org/10.1097/CJI.0b013e31826e8f5f.

30 Andersen, R., Donia, M., Ellebaek, E. et al. (2016). Long-lasting complete responses in patients with metastatic melanoma after adoptive cell therapy with tumor-infiltrating lymphocytes and an attenuated IL2 regimen. *Clin. Cancer Res.* 22 (15): 3734–3745. https://doi.org/10.1158/1078-0432.CCR-15-1879.

31 Itzhaki, O., Hovav, E., Ziporen, Y. et al. (2011). Establishment and large-scale expansion of minimally cultured "young" tumor infiltrating lymphocytes for adoptive transfer therapy. *J. Immunother.* 34 (2): 212–220. https://doi.org/10.1097/CJI.0b013e318209c94c.

32 Radvanyi, L.G., Bernatchez, C., Zhang, M. et al. (2012). Specific lymphocyte subsets predict response to adoptive cell therapy using expanded autologous tumor-infiltrating lymphocytes in metastatic melanoma patients. *Clin. Cancer Res.* 18 (24): 6758–6770. https://doi.org/10.1158/1078-0432.CCR-12-1177.

33 Eyquem, J., Mansilla-Soto, J., Giavridis, T. et al. (2017). Targeting a CAR to the TRAC locus with CRISPR/Cas9 enhances tumour rejection. *Nature* 543 (7643): 113–117. https://doi.org/10.1038/nature21405.

34 Peng, P.D., Cohen, C.J., Yang, S. et al. (2009). Efficient nonviral sleeping beauty transposon-based TCR gene transfer to peripheral blood lymphocytes confers antigen-specific antitumor reactivity. *Gene Ther.* 16 (8): 1042–1049. https://doi.org/10.1038/gt.2009.54.

35 Tsuji, T., Yasukawa, M., Matsuzaki, J. et al. (2005). Generation of tumor-specific, HLA class I-restricted human Th1 and Tc1 cells by cell engineering with tumor peptide-specific T-cell receptor genes. *Blood* 106 (2): 470–476. https://doi.org/10.1182/blood-2004-09-3663.

36 Zhao, Y., Zheng, Z., Cohen, C.J. et al. (2006). High-efficiency transfection of primary human and mouse T lymphocytes using RNA electroporation. *Mol. Ther.* 13 (1): 151–159. https://doi.org/10.1016/j.ymthe.2005.07.688.

37 Clay, T.M., Custer, M.C., Sachs, J. et al. (1999). Efficient transfer of a tumor antigen-reactive TCR to human peripheral blood lymphocytes confers anti-tumor reactivity. *J. Immunol.* 163 (1): 507–513.

38 Lauss, M., Donia, M., Harbst, K. et al. (2017). Mutational and putative neoantigen load predict clinical benefit of adoptive T cell therapy in melanoma. *Nat. Commun.* 8 (1): 1738. https://doi.org/10.1038/s41467-017-01460-0.

39 Ruella, M. and Kalos, M. (2014). Adoptive immunotherapy for cancer. *Immunol. Rev.* 257 (1): 14–38. https://doi.org/10.1111/imr.12136.

40 Jensen, M.C. and Riddell, S.R. (2014). Design and implementation of adoptive therapy with chimeric antigen receptor-modified T cells. *Immunol. Rev.* 257 (1): 127–144. https://doi.org/10.1111/imr.12139.

41 Hinrichs, C.S. and Rosenberg, S.A. (2014). Exploiting the curative potential of adoptive T-cell therapy for cancer. *Immunol. Rev.* 257 (1): 56–71. https://doi.org/10.1111/imr.12132.

42 Dotti, G., Gottschalk, S., Savoldo, B., and Brenner, M.K. (2014). Design and development of therapies using chimeric antigen receptor-expressing T cells. *Immunol. Rev.* 257 (1): 107–126. https://doi.org/10.1111/imr.12131.

43 Yee, C. (2014). The use of endogenous T cells for adoptive transfer. *Immunol. Rev.* 257 (1): 250–263. https://doi.org/10.1111/imr.12134.

44 Rosenberg, S.A., Packard, B.S., Aebersold, P.M. et al. (1988). Use of tumor-infiltrating lymphocytes and interleukin-2 in the immunotherapy of patients with metastatic melanoma. *N. Engl. J. Med.* 319 (25): 1676–1680. https://doi.org/10.1056/NEJM198812223192527.

45 Dudley, M.E., Wunderlich, J.R., Yang, J.C. et al. (2005). Adoptive cell transfer therapy following non-myeloablative but lymphodepleting chemotherapy for the treatment of patients with refractory metastatic melanoma. *J. Clin. Oncol.* 23 (10): 2346–2357. https://doi.org/10.1200/JCO.2005.00.240.

46 Morgan, R.A., Dudley, M.E., Wunderlich, J.R. et al. (2006). Cancer regression in patients after transfer of genetically engineered lymphocytes. *Science* 314 (5796): 126–129. https://doi.org/10.1126/science.1129003.

47 Johnson, L.A., Morgan, R.A., Dudley, M.E. et al. (2009). Gene therapy with human and mouse T-cell receptors mediates cancer regression and targets normal tissues expressing cognate antigen. *Blood* 114 (3): 535–546. https://doi.org/10.1182/blood-2009-03-211714.

48 Morgan, R.A., Dudley, M.E., and Rosenberg, S.A. (2010). Adoptive cell therapy. *Cancer J.* 16 (4): 336–341. https://doi.org/10.1097/PPO.0b013e3181eb3879.

49 Robbins, P.F., Morgan, R.A., Feldman, S.A. et al. (2011). Tumor regression in patients with metastatic synovial cell sarcoma and melanoma using genetically engineered lymphocytes reactive with NY-ESO-1. *J. Clin. Oncol.* 29 (7): 917–924. https://doi.org/10.1200/JCO.2010.32.2537.

50 Maude, S.L., Frey, N., Shaw, P.A. et al. (2014). Chimeric antigen receptor T cells for sustained remissions in leukemia. *N. Engl. J. Med.* 371 (16): 1507–1517. https://doi.org/10.1056/NEJMoa1407222.

51 Davila, M.L., Riviere, I., Wang, X. et al. (2014). Efficacy and toxicity management of 19-28z CAR T cell therapy in B cell acute lymphoblastic leukemia. *Sci. Transl. Med.* 6 (224): 224ra25. https://doi.org/10.1126/scitranslmed.3008226.

52 Lee, D.W., Stetler-Stevenson, M., Yuan, C.M. et al. (2015). Safety and response of incorporating CD19 chimeric antigen receptor T cell therapy in typical salvage regimens for children and young adults with acute lymphoblastic leukemia. *Blood* 126 (23): 684. https://doi.org/10.1182/blood.V126.23.684.684.

53 Porter, D.L., Hwang, W.-T., Frey, N.V. et al. (2015). Chimeric antigen receptor T cells persist and induce sustained remissions in relapsed refractory chronic lymphocytic leukemia. *Sci. Transl. Med.* 7 (303): 303ra139. https://doi.org/10.1126/scitranslmed.aac5415.

54 Rapoport, A.P., Stadtmauer, E.A., Binder-Scholl, G.K. et al. (2015). NY-ESO-1–specific TCR–engineered T cells mediate sustained antigen-specific antitumor effects in myeloma. *Nat. Med.* 21 (8): 914–921. https://doi.org/10.1038/nm.3910.

55 Kochenderfer, J.N., Dudley, M.E., Kassim, S.H. et al. (2015). Chemotherapy-refractory diffuse large B-cell lymphoma and indolent B-cell malignancies can be effectively treated with autologous T cells expressing an anti-CD19 chimeric antigen receptor. *J. Clin. Oncol.* 33 (6): 540–549. https://doi.org/10.1200/JCO.2014.56.2025.

56 Turtle, C.J., Hanafi, L.-A., Berger, C. et al. (2016). CD19 CAR–T cells of defined $CD4^+$:$CD8^+$ composition in adult B cell ALL patients. *J. Clin. Invest.* 126 (6): 2123–2138. https://doi.org/10.1172/JCI85309.

57 Wang, Z., Wu, Z., Liu, Y., and Han, W. (2017). New development in CAR-T cell therapy. *J. Hematol. Oncol.* 10 (1): 53. https://doi.org/10.1186/s13045-017-0423-1.

58 Neelapu, S.S., Locke, F.L., Bartlett, N.L. et al. (2017). Axicabtagene ciloleucel CAR T-cell therapy in refractory large B-cell lymphoma. *N. Engl. J. Med.* 377 (26): 2531–2544. https://doi.org/10.1056/NEJMoa1707447.

59 Turtle, C.J., Hay, K.A., Hanafi, L.-A. et al. (2017). Durable molecular remissions in chronic lymphocytic leukemia treated with CD19-specific chimeric antigen receptor-modified T cells after failure of ibrutinib. *J. Clin. Oncol.* 35 (26): 3010–3020. https://doi.org/10.1200/JCO.2017.72.8519.

60 Geyer, M.B., Rivière, I., Sénéchal, B. et al. (2018). Autologous CD19-targeted CAR T cells in patients with residual CLL following initial purine analog-based therapy. *Mol. Ther.* 26 (8): 1896–1905. https://doi.org/10.1016/j.ymthe.2018.05.018.

61 Maude, S.L., Laetsch, T.W., Buechner, J. et al. (2018). Tisagenlecleucel in children and young adults with B-cell lymphoblastic leukemia. *N. Engl. J. Med.* 378 (5): 439–448. https://doi.org/10.1056/NEJMoa1709866.

62 Brudno, J.N., Maric, I., Hartman, S.D. et al. (2018). T cells genetically modified to express an anti–B-cell maturation antigen chimeric antigen receptor cause remissions of poor-prognosis relapsed multiple myeloma. *J. Clin. Oncol.* 36 (22): 2267–2280. https://doi.org/10.1200/JCO.2018.77.8084.
63 Ribas, A. and Wolchok, J.D. (2018). Cancer immunotherapy using checkpoint blockade. *Science* 359 (6382): 1350–1355. https://doi.org/10.1126/science.aar4060.
64 Kobold, S., Duewell, P., Schnurr, M. et al. (2015). Immunotherapy in tumors. *Dtsch. Aerztebl. Int.* https://doi.org/10.3238/arztebl.2015.0809.
65 Dunn, G.P., Bruce, A.T., Ikeda, H. et al. (2002). Cancer immunoediting: from immunosurveillance to tumor escape. *Nat. Immunol.* 3 (11): 991–998. https://doi.org/10.1038/ni1102-991.
66 Carter, P.J. and Lazar, G.A. (2018). Next generation antibody drugs: pursuit of the "high-hanging fruit.". *Nat. Rev. Drug Discov.* 17 (3): 197–223. https://doi.org/10.1038/nrd.2017.227.
67 Kobold, S., Pantelyushin, S., Rataj, F., and vom Berg, J. (2018). Rationale for combining bispecific T cell activating antibodies with checkpoint blockade for cancer therapy. *Front. Oncol.* 8: https://doi.org/10.3389/fonc.2018.00285.
68 Baeuerle, P.A. and Reinhardt, C. (2009). Bispecific T-cell engaging antibodies for cancer therapy. *Cancer Res.* 69 (12): 4941–4944. https://doi.org/10.1158/0008-5472.CAN-09-0547.
69 Trabolsi, A., Arumov, A., and Schatz, J.H. (2019). T cell-activating bispecific antibodies in cancer therapy. *J. Immunol.* 203 (3): 585–592. https://doi.org/10.4049/jimmunol.1900496.
70 Spiess, C., Zhai, Q., and Carter, P.J. (2015). Alternative molecular formats and therapeutic applications for bispecific antibodies. *Mol. Immunol.* 67 (2): 95–106. https://doi.org/10.1016/j.molimm.2015.01.003.
71 Kontermann, R.E. and Brinkmann, U. (2015). Bispecific antibodies. *Drug Discov. Today* 20 (7): 838–847. https://doi.org/10.1016/j.drudis.2015.02.008.
72 Seimetz, D., Lindhofer, H., and Bokemeyer, C. (2010). Development and approval of the trifunctional antibody catumaxomab (anti-EpCAM × anti-CD3) as a targeted cancer immunotherapy. *Cancer Treat. Rev.* 36 (6): 458–467. https://doi.org/10.1016/j.ctrv.2010.03.001.
73 Shatz, W., Chung, S., Li, B. et al. Knobs-into-holes antibody production in mammalian cell lines reveals that asymmetric afucosylation is sufficient for full antibody-dependent cellular cytotoxicity. *MAbs* 5 (6): 872–881. https://doi.org/10.4161/mabs.26307.
74 Wu, C., Ying, H., Bose, S. et al. Molecular construction and optimization of anti-human IL-1α/β dual variable domain immunoglobulin (DVD-Ig™) molecules. *MAbs* 1 (4): 339–347. https://doi.org/10.4161/mabs.1.4.8755.
75 Eigenbrot, C. and Fuh, G. (2013). Two-in-one antibodies with dual action Fabs. *Curr. Opin. Chem. Biol.* 17 (3): 400–405. https://doi.org/10.1016/j.cbpa.2013.04.015.

76 Labrijn, A.F., Meesters, J.I., Priem, P. et al. (2014). Controlled Fab-arm exchange for the generation of stable bispecific IgG1. *Nat. Protoc.* 9 (10): 2450–2463. https://doi.org/10.1038/nprot.2014.169.

77 Fischer, N., Elson, G., Magistrelli, G. et al. (2015). Exploiting light chains for the scalable generation and platform purification of native human bispecific IgG. *Nat. Commun.* 6: 6113. https://doi.org/10.1038/ncomms7113.

78 Le Gall, F., Kipriyanov, S.M., Moldenhauer, G., and Little, M. (1999). Di-, tri- and tetrameric single chain Fv antibody fragments against human CD19: effect of valency on cell binding. *FEBS Lett.* 453 (1–2): 164–168. https://doi.org/10.1016/s0014-5793(99)00713-9.

79 Chames, P. and Baty, D. Bispecific antibodies for cancer therapy: the light at the end of the tunnel? *MAbs* 1 (6): 539–547. https://doi.org/10.4161/mabs.1.6.10015.

80 Garber, K. (2014). Bispecific antibodies rise again. *Nat. Rev. Drug Discov.* 13 (11): 799–801. https://doi.org/10.1038/nrd4478.

81 Revets, H., De Baetselier, P., and Muyldermans, S. (2005). Nanobodies as novel agents for cancer therapy. *Expert Opin. Biol. Ther.* 5 (1): 111–124. https://doi.org/10.1517/14712598.5.1.111.

82 Goldenberg, D.M., Rossi, E.A., Sharkey, R.M. et al. (2008). Multifunctional antibodies by the dock-and-lock method for improved cancer imaging and therapy by pretargeting. *J. Nucl. Med.* 49 (1): 158–163. https://doi.org/10.2967/jnumed.107.046185.

83 Mack, M., Riethmuller, G., and Kufer, P. (1995). A small bispecific antibody construct expressed as a functional single-chain molecule with high tumor cell cytotoxicity. *Proc. Natl. Acad. Sci.* 92 (15): 7021–7025. https://doi.org/10.1073/pnas.92.15.7021.

84 Stinchcombe, J.C., Bossi, G., Booth, S., and Griffiths, G.M. (2001). The immunological synapse of CTL contains a secretory domain and membrane bridges. *Immunity* 15 (5): 751–761. https://doi.org/10.1016/S1074-7613(01)00234-5.

85 Huehls, A.M., Coupet, T.A., and Sentman, C.L. (2015). Bispecific T-cell engagers for cancer immunotherapy. *Immunol. Cell Biol.* 93 (3): 290–296. https://doi.org/10.1038/icb.2014.93.

86 Hoffmann, P., Hofmeister, R., Brischwein, K. et al. (2005). Serial killing of tumor cells by cytotoxic T cells redirected with a CD19-/CD3-bispecific single-chain antibody construct. *Int. J. Cancer* 115 (1): 98–104. https://doi.org/10.1002/ijc.20908.

87 Klinger, M., Brandl, C., Zugmaier, G. et al. (2012). Immunopharmacologic response of patients with B-lineage acute lymphoblastic leukemia to continuous infusion of T cell–engaging CD19/CD3-bispecific BiTE antibody blinatumomab. *Blood* 119 (26): 6226–6233. https://doi.org/10.1182/blood-2012-01-400515.

88 Wang, T., Sun, F., Xie, W. et al. (2016). A bispecific protein RG7S-MICA recruits natural killer cells and enhances NKG2D-mediated immunosurveillance against hepatocellular carcinoma. *Cancer Lett.* 372 (2): 166–178. https://doi.org/10.1016/j.canlet.2016.01.001.

89 Wu, M.-R., Zhang, T., Gacerez, A.T. et al. (2015). B7H6-specific bispecific T cell engagers lead to tumor elimination and host antitumor immunity. *J. Immunol.* 194 (11): 5305–5311. https://doi.org/10.4049/jimmunol.1402517.

90 Fan, D., Li, Z., Zhang, X. et al. (2015). AntiCD3Fv fused to human interleukin-3 deletion variant redirected T cells against human acute myeloid leukemic stem cells. *J. Hematol. Oncol.* 8: 18. https://doi.org/10.1186/s13045-015-0109-5.

91 Satta, A., Mezzanzanica, D., Turatti, F. et al. (2013). Redirection of T-cell effector functions for cancer therapy: bispecific antibodies and chimeric antigen receptors. *Future Oncol.* 9 (4): 527–539. https://doi.org/10.2217/fon.12.203.

92 Zugmaier, G., Klinger, M., Schmidt, M., and Subklewe, M. (2015). Clinical overview of anti-CD19 BiTE® and ex vivo data from anti-CD33 BiTE® as examples for retargeting T cells in hematologic malignancies. *Mol. Immunol.* 67 (2 Pt A): 58–66. https://doi.org/10.1016/j.molimm.2015.02.033.

93 Fan, G., Wang, Z., Hao, M., and Li, J. (2015). Bispecific antibodies and their applications. *J. Hematol. Oncol.* 8: 130. https://doi.org/10.1186/s13045-015-0227-0.

94 Hirschhaeuser, F., Leidig, T., Rodday, B. et al. (2009). Test system for trifunctional antibodies in 3D MCTS culture. *J. Biomol. Screen.* 14 (8): 980–990. https://doi.org/10.1177/1087057109341766.

95 Breton, C.S., Nahimana, A., Aubry, D. et al. (2014). A novel anti-CD19 monoclonal antibody (GBR 401) with high killing activity against B cell malignancies. *J. Hematol. Oncol.* 7: 33. https://doi.org/10.1186/1756-8722-7-33.

96 Nagorsen, D., Kufer, P., Baeuerle, P.A., and Bargou, R. (2012). Blinatumomab: a historical perspective. *Pharmacol. Ther.* 136 (3): 334–342. https://doi.org/10.1016/j.pharmthera.2012.07.013.

97 Wu, J., Fu, J., Zhang, M., and Liu, D. (2015). Blinatumomab: a bispecific T cell engager (BiTE) antibody against CD19/CD3 for refractory acute lymphoid leukemia. *J. Hematol. Oncol.* 8: 104. https://doi.org/10.1186/s13045-015-0195-4.

98 Tabrizi, M., Bornstein, G.G., and Suria, H. (2010). Biodistribution mechanisms of therapeutic monoclonal antibodies in health and disease. *AAPS J.* 12 (1): 33–43. https://doi.org/10.1208/s12248-009-9157-5.

99 Braig, F., Brandt, A., Goebeler, M. et al. (2017). Resistance to anti-CD19/CD3 BiTE in acute lymphoblastic leukemia may be mediated by disrupted CD19 membrane trafficking. *Blood* 129 (1): 100–104. https://doi.org/10.1182/blood-2016-05-718395.

100 Linke, R., Klein, A., and Seimetz, D. (2010). Catumaxomab. *MAbs* 2 (2): 129–136. https://doi.org/10.4161/mabs.2.2.11221.

101 Haense, N., Atmaca, A., Pauligk, C. et al. (2016). A phase I trial of the trifunctional anti Her2 × anti CD3 antibody ertumaxomab in patients with advanced solid tumors. *BMC Cancer* 16 (1): 420. https://doi.org/10.1186/s12885-016-2449-0.

102 Löffler, A., Kufer, P., Lutterbüse, R. et al. (2000). A recombinant bispecific single-chain antibody, CD19 × CD3, induces rapid and high lymphoma-directed cytotoxicity by unstimulated T lymphocytes. *Blood* 95 (6): 2098–2103.

103 Newman, M.J. and Benani, D.J. (2016). A review of blinatumomab, a novel immunotherapy. *J. Oncol. Pharm. Pract.* 22 (4): 639–645. https://doi.org/10.1177/1078155215618770.

104 Topp, M.S., Gökbuget, N., Zugmaier, G. et al. (2014). Phase II trial of the anti-CD19 bispecific T cell–engager blinatumomab shows hematologic and molecular remissions in patients with relapsed or refractory B-precursor acute lymphoblastic leukemia. *J. Clin. Oncol.* 32 (36): 4134–4140. https://doi.org/10.1200/JCO.2014.56.3247.

105 Klein, J.S., Gnanapragasam, P.N.P., Galimidi, R.P. et al. (2009). Examination of the contributions of size and avidity to the neutralization mechanisms of the anti-HIV antibodies B12 and 4E10. *Proc. Natl. Acad. Sci.* 106 (18): 7385–7390. https://doi.org/10.1073/pnas.0811427106.

106 Brandl, C., Haas, C., D'Argouges, S. et al. (2007). The effect of dexamethasone on polyclonal T cell activation and redirected target cell lysis as induced by a CD19/CD3-bispecific single-chain antibody construct. *Cancer Immunol. Immunother.* 56 (10): 1551–1563. https://doi.org/10.1007/s00262-007-0298-z.

107 Leone, P., Shin, E.-C., Perosa, F. et al. (2013). MHC class I antigen processing and presenting machinery: organization, function, and defects in tumor cells. *JNCI J. Natl. Cancer Inst.* 105 (16): 1172–1187. https://doi.org/10.1093/jnci/djt184.

108 Löffler, A., Gruen, M., Wuchter, C. et al. (2003). Efficient elimination of chronic lymphocytic leukaemia B cells by autologous T cells with a bispecific anti-CD19/anti-CD3 single-chain antibody construct. *Leukemia* 17 (5): 900–909. https://doi.org/10.1038/sj.leu.2402890.

109 Köhnke, T., Krupka, C., Tischer, J. et al. (2015). Increase of PD-L1 expressing B-precursor ALL cells in a patient resistant to the CD19/CD3-bispecific T cell engager antibody blinatumomab. *J. Hematol. Oncol.* 8 (1): 111. https://doi.org/10.1186/s13045-015-0213-6.

110 Duell, J., Dittrich, M., Bedke, T. et al. (2017). Frequency of regulatory T cells determines the outcome of the T-cell-engaging antibody blinatumomab in patients with B-precursor ALL. *Leukemia* 31 (10): 2181–2190. https://doi.org/10.1038/leu.2017.41.

111 Lee, G.R. (2017). Phenotypic and functional properties of tumor-infiltrating regulatory T cells. *Mediators Inflamm.* 2017: 1–9. https://doi.org/10.1155/2017/5458178.

112 Horvath, C.J. and Milton, M.N. (2009). The TeGenero incident and the duff report conclusions: a series of unfortunate events or an avoidable event? *Toxicol. Pathol.* 37 (3): 372–383. https://doi.org/10.1177/0192623309332986.

113 Gökbuget, N., Dombret, H., Bonifacio, M. et al. (2018). Blinatumomab for minimal residual disease in adults with B-cell precursor acute lymphoblastic leukemia. *Blood* 131 (14): 1522–1531. https://doi.org/10.1182/blood-2017-08-798322.

14

Devices for Active Targeted Delivery: A Way to Control the Rate and Extent of Drug Administration

Jonathan Faro Barros[1,2], Phedra F. Sahraoui[1,2], Yogeshvar N. Kalia[1,2], and Maria Lapteva[1,2]

[1] *University of Geneva, School of Pharmaceutical Sciences, CMU, 1 rue Michel Servet, Geneva CH-1211, Switzerland*
[2] *University of Geneva, Institute of Pharmaceutical Sciences of Western, Switzerland CMU, 1 rue Michel Servet, Geneva CH-1211, Switzerland*

14.1 Introduction

Chronic disease can be defined as a health condition that continues or reoccurs over an extended period, this includes cardiovascular diseases, stroke, cancer, chronic respiratory diseases, diabetes, mental conditions, chronic pain, autoimmune diseases, genetic disorders, chronic infections as well as visual and hearing impairment [1, 2]. According to the World Health Organization (WHO), 60% of all deaths worldwide are due to chronic diseases [2]. Fortunately, efficient pharmacotherapies are available to treat these conditions; however, by definition, their chronicity necessitates a long-term administration of the therapeutic agents. The choice of the route of administration depends on a number of factors: the "ideal" delivery route should enable the delivery of the drug to its site of action (i) in therapeutically relevant amounts, (ii) in the appropriate timeframe, and (iii) in a targeted manner, i.e. avoiding distribution to therapeutically nonrelevant tissues.

Many chronically administered drugs utilize the traditional oral route [1]. In order to be effective, an orally administered drug needs to cross a number of barriers in order to reach its target site in sufficient amounts: it has first to be released from the dosage form, absorbed in the gastro-intestinal (GI) tract, escape first-pass hepatic metabolism, and be distributed to the therapeutic target site by the circulatory system. Unfortunately, in many cases, these drugs do not reach the target site either in therapeutically relevant amounts [3, 4] or in the right timeframe (poor pharmacokinetics) [5], or they are not sufficiently targeted (nonselective systemic distribution increases the risk of inducing off-target side effects) [6].

Moreover, among the elderly population, the accumulation of several chronic conditions leads to multiple treatments. Approximately 30% of adults aged 65 years and older in developed countries take five or more treatments [7] leading to numerous issues such as drug–drug interactions, cumulation of undesirable effects,

Targeted Drug Delivery, First Edition. Edited by Yogeshwar Bachhav.

and more importantly poor adherence. According to the WHO, only about 50% of patients with a chronic condition adhere to their treatments [8]. Lack of patient adherence to the medication is mainly due to the discomfort that conventional drug administration cause in the patient's life. Multiple dosages, narrow therapeutic windows, strict and complex dosing schedules, and systemic adverse side effects are some of the limitations of traditional oral administration. For a number of actives, oral delivery is simply impossible: for instance, proteic biopharmaceuticals degrade in the gastrointestinal (GI) tract due to their peptidic nature and need to be

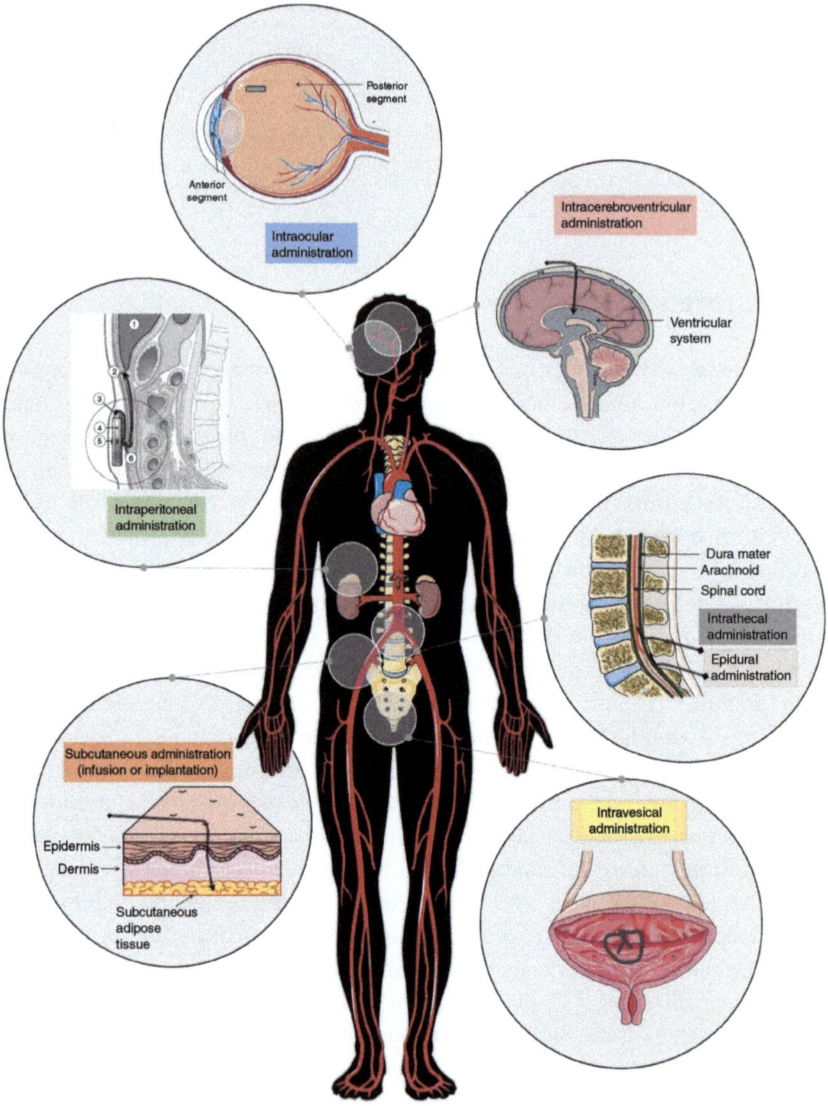

Figure 14.1 Main delivery routes for active targeted drug delivery. Source: Infographics adapted from Servier Medical Art https://smart.servier.com/.

administered parenterally. Drugs intended to treat central nervous disorders need to cross the additional blood–brain barrier in order to reach the target site. Hence, there is still room for improvement in drug delivery for the treatment of chronic conditions as a series of unmet needs remains to be addressed.

Therefore, alternative invasive or minimally invasive delivery routes have been investigated for the treatment of chronic conditions. As a counterweight to their invasiveness they bring several advantages: (i) the drug can be delivered in a targeted manner to poorly accessible sites – for example, to the hepatic venous system (intraperitoneal), to the central nervous system (intracerebroventricular [i.c.v.], intrathecal [IT], epidural), to the eye (intraocular), or to the bladder (intravesical), thus resulting in considerable dose sparing (Figure 14.1); (ii) the drug escapes GI tract and hepatic degradation (subcutaneous infusion); and (iii) the input kinetics can be tightly controlled by active drug delivery devices. Drug delivery can extend over several months, be initiated on patient demand, or follow complex input profiles including in response to physiological signals. Active drug delivery devices can deliver the drug at the target site exactly when it is needed, thus considerably improving the patient's quality of life.

This chapter focuses on active drug delivery devices, i.e. devices using an energy input or external driving force intended for long-term drug delivery and able to yield controlled drug delivery kinetics. The drug delivery systems covered include implantable and non-implantable devices requiring energy input for the long-term delivery of drugs: electrical, pneumatic, and osmotic pumps; micro-electromechanical systems (MEMS); nano-electromechanical systems (NEMS); iontophoresis and exclude bolus injections and minimally invasive injections (hollow microneedles, velocity-based devices, and microporation techniques); and drug delivery systems relying on passive drug diffusion/release (i.e. biodegradable implants, solid and dissolvable microneedles). However, the latter delivery systems are extensively reviewed in numerous publications [1, 9–11].

14.2 Macrofabricated Devices – Drug Infusion Pumps

The Food and Drug Administration (FDA) defines infusion pumps as "a medical device that delivers fluids, such as nutrients and medications, into a patient's body in controlled amounts" [12]. They often use the parenteral administration route and rely on energy input to deliver the active. While some are designed mainly for clinical use at the patient's bedside, ambulatory infusion pumps can be portable, wearable, or implantable [12], thus offering complex drug input kinetics while providing the patient autonomy and a higher quality of life. This section will focus on the latter. Ambulatory infusion pumps can be classified by their energy source and mechanism of fluid motion.

14.2.1 Peristaltic Pumps

One of the first patented infusion pumps was a peristaltic pump for blood transfusion [13]. Peristaltic pumps are electrically driven pumps where rollers are attached to a rotor compressing a flexible tubing, thus displacing the fluid inside the latter. A good example of such a pump are Medtronic's implantable SynchroMed™ pump systems (SynchroMed EL and SynchroMed II) for IT delivery of pain-management drugs. It is implanted in the anterolateral abdominal wall and a catheter delivers the drug specifically to the target site. The main advantage of this system is the possibility to program the pump remotely to deliver a wide range of flow rates, as well as enable a patient-triggered delivery in case of insufficient analgesia.

The FDA has approved systems for the IT delivery of ziconotide and IT/epidural delivery of morphine sulfate, both being indicated for the management of severe chronic pain. It is also recently approved for the implantable intravenous (IV) infusion of treprostinil for the treatment of a rare form of pulmonary arterial hypertension [14]. In 2015, the SynchroMed pump was used in a clinical trial to test an investigational anti-Parkinson's drug-using i.c.v. administration. However, the trial was terminated probably because of serious adverse events (including two linked to the pump implantation) and difficulties with the development and supply of infusion systems [15, 16]. Despite efficacious delivery, the SynchroMed pumps are still suffering from drawbacks such as the size of the pump itself while hosting a small drug reservoir that needs to be refilled every three to four months by a caregiver. Efforts toward the miniaturization of peristaltic pumps are underway (iPRECIO®; Alzet pumps, USA). Another setback with the current peristaltic pumps is their susceptibility to magnetic fields and thus magnetic resonance imaging (MRI)-linked disfunctions [1]. A similar pump: the CADD-Legacy Duodopa pump (Smiths Medical, USA) was successfully used to infuse a levodopa/carbidopa enteral gel intraduodenally in patients with Parkinson's disease with motor fluctuations and dyskinesias. It was concluded that this intervention is superior to conventional oral medications in preventing drug plasma concentrations fluctuations and thus fluctuations in motor performance [17]. A more recent study reports similar results after intrajejunal infusion of the same formulation using a CADD-Legacy pump [18].

14.2.2 Gas-Driven Pumps

Gas-driven pumps are often implantable and use a chlorofluorocarbon (CFC) gas as a driving force for drug delivery. The pump is composed of two chambers, respectively, containing the gas and the drug. The drug reservoir is compressed as the gas expands in its chamber as a result of body heat. Because body temperature is constant, the drug is delivered constantly but is tunable only by changing the drug concentration in the reservoir. Although the absence of an electrical energy source is an obvious advantage, the impossibility of drug input rate modulation is a drawback.

The Codman 3000 (Codman & Shurtleff, USA) can be cited as the first example of gas-driven implantable pumps. It was FDA approved for the IT delivery of morphine sulfate and hepatic arterial infusion of chemotherapy to the tumor site [14]. It also showed positive outcomes in the treatment of severe spasticity with baclofen [19]. Unfortunately, in 2018 the production of the Codman 3000 pump was terminated probably due to insufficient profitability, leaving the market niche to Flowonix

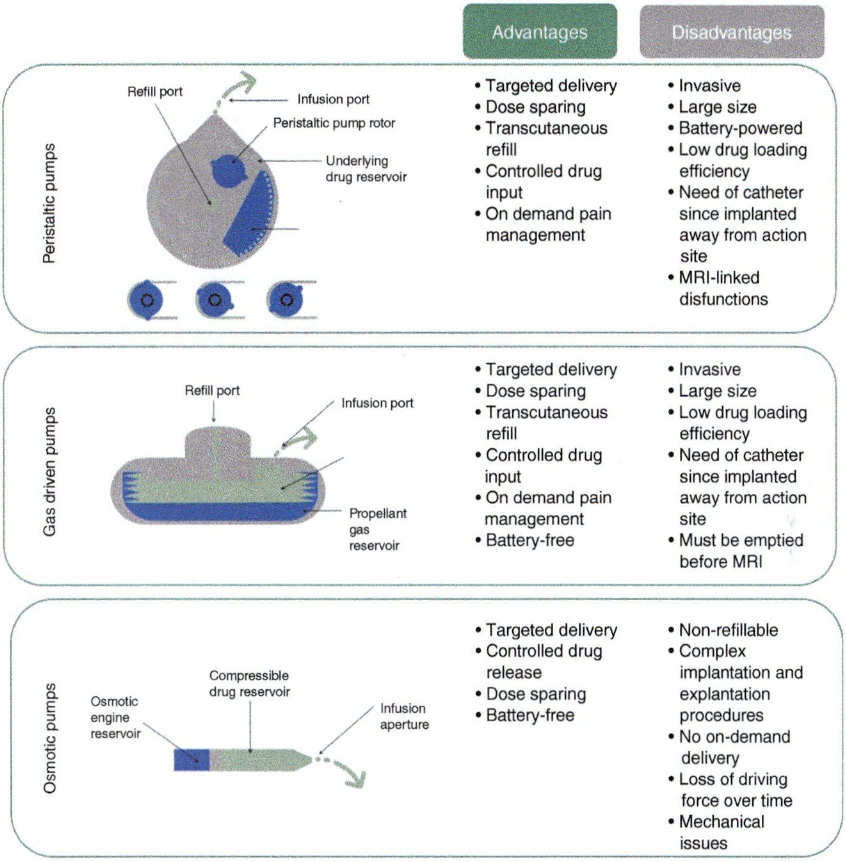

Figure 14.2 Drug delivery mechanisms, advantages, and disadvantages of different pumps.

Medical Inc. (USA) and its Prometra® pump. It has the same gas-expansion-based mechanism as the Codman 3000 pump; however, it is supplemented with electrically powered valves thus allowing remote control of drug delivery. Thus Prometra pumps could reach a mean accuracy of 97.1%, with a 90% confidence interval of 96.2–98.0% when administering morphine sulfate [20]. Nevertheless, one major drawback of these pumps is the need for reservoir emptying before MRI since the magnetic field can open the valves and induce an overdose [21]. Figure 14.2 summarizes the drug delivery mechanisms, advantages, and disadvantages of different pumps.

14.2.3 Osmotic Pumps

Osmotic pumps are small implantable devices that rely on osmotic energy. These pumps consist of a compressible reservoir containing the drug and a reservoir containing the "osmotic engine," composed of highly concentrated osmolytes (soluble salts) separated from the outer environment by a semipermeable membrane.

Once implanted, water from interstitial fluid permeates through the semipermeable membrane, attracted by the osmotic engine thus compressing the drug reservoir resulting in the drug solution release via an aperture [1]. Alza Corp. (USA; now acquired by Johnson & Johnson (USA)) developed Viadur® titanium implant for the 12-month delivery of leuprolide acetate indicated for the treatment of prostate cancer. The device was marketed by Bayer (Germany); however, it is no longer commercialized due to low cost-effectiveness, indeed the manufacturing of titanium osmotic pumps is expensive not only because of the raw material cost but also because the inner compartment requires a high smoothness. The Medici Drug Delivery System™ (Intarcia Therapeutics Inc., USA) is a subcutaneously implanted osmotic pump used for the long-term delivery of exenatide (three to six months) for the treatment of diabetes. Results from a 39-week, open-label, phase 3 trial ($n = 60$) point to good efficacy with a reduction in glycated hemoglobin in 90% of the patients [22]. Nevertheless, after a first FDA rejection in 2017 due to manufacturing issues, the system has just recently failed to obtain regulatory approval once more [23].

GemRIS™ and LiRIS™ are titanium-free osmotic pumps designed for the intravesical delivery of gemcitabine and lidocaine, respectively, to treat local bladder conditions. This innovative system developed by TARIS Biomedical LLC (now acquired by Johnson & Johnson, USA) consists of a double-lumen silicon tubing. The drug is contained in the bigger lumen in its solid form and acts as an osmotic engine itself attracting water from the urine across the permeable silicone, thus also leading to gradual drug dissolution and release via a 150 μm opening in the tubing. The second lumen contains nitinol (nickel titanium) wire, which coils upon bladder insertion and thus can reside inside the bladder [24]. The main advantages of this system reside in the fact that there are no moving parts, and more drug can be loaded as a solid than as a solution.

LiRIS is designed to release lidocaine drug over a two-week period. A clinical trial enrolling 20 patients with interstitial cystitis or bladder pain syndrome showed improvement in pain not only on day 14 (day of removal) but also after device removal. GemRIS is currently under clinical evaluation in several clinical trials (NCT02722538; NCT02720367, NCT03404791 [25]). TARIS Biomedical is also investigating the combination of gemcitabine with nivolumab (NCT03518320) [25], as part of a clinical collaboration with Bristol-Myers Squibb, and trospium for the treatment of overactive bladder (TAR-302; NCT03109379). TAR 302 could deliver the drug intravesically for 42 days, a 12-week study is ongoing [26].

14.2.4 Insulin Pumps

14.2.4.1 Diabetes and Insulin Product Development

Insulin is a proteic hormone secreted by the β-cells of the pancreas in response to elevated blood glucose levels, it is thus essential for normal carbohydrate, protein, and fat metabolism. Diabetes mellitus (DM) is a group of conditions characterized by an elevated glucose concentration [27, 28]. Type 1 DM results from autoimmune destruction of β-cells. Therefore, it is also termed juvenile-onset diabetes or insulin-dependent diabetes. Type 1 DM patients lack insulin production and are, hence, dependent on exogenous insulin for survival [28]. In contrast, type 2 DM is a metabolic disease where insulin is still secreted, only peripheral tissues suffer from insulin resistance: the tissue response to normal circulating insulin levels is decreased. Over time, β-cells need to secrete increasing amounts of insulin and become gradually unable to overcome insulin resistance [27, 28]. While early type 2 diabetes can be managed by dietary measures and oral pharmacotherapy [29], advanced type 2 DM and type 1 DM require daily insulin administration [29–31].

Insulin was first extracted from animal tissue and purified in 1921 by F. Banting and C. H. Best [32, 33]. Due to its proteic nature, insulin cannot be administered orally and must be injected. Since insulin is a highly conserved protein across species, the first insulins were isolated from bovine and porcine tissue. Unfortunately, the very short half-life of insulin [34] induced the need for very frequent injections. Moreover, endogenous insulin secretion follows a particular pattern: it consists of a basal secretion together with a postprandial pulsatile secretion in response to meal-related hyperglycemia [35]. Therefore, to increase patient comfort and to better mimic the endogenous insulin secretion, insulins with different pharmacokinetic properties (short-acting and long-acting) were gradually developed, first by formulation efforts and later by recombinant protein expression technology. Thus, the first synthetic human insulin was marketed in 1982 [36]. Subsequent efforts we directed at modifying the pharmacokinetic properties of the latter by altering the amino acid sequences (i.e. short-acting: insulin lispro, insulin glulisine, and insulin aspart or long-acting: insulin glargine, insulin detemir, and insulin degludec). Nevertheless, even with the availability of a wide range of insulin products, patients need to undergo multiple injections accompanied by as many finger punctures for blood sugar testing, not to mention the mental load induced by the need to manage their blood sugar.

Therefore, the overall "holy grail" of insulin delivery was to achieve a completely physiological insulin administration with the constant quest for normoglycemia [35, 37, 38], ideally in an automated way. The development of continuous insulin delivery systems was always accompanied by the development of methods for continuous glucose measurement [39]. Thus, drug infusion systems could be divided into either glucose-controlled feedback (closed-loop) devices or preprogrammed (open-loop) devices [40].

Figure 14.3 summarizes the parallel developments in insulin formulation, closed-loop infusion pumps, and efforts to produce automated closed-loop devices.

14.2.4.2 Open-Loop Insulin Delivery Systems

The first insulin pump, together with blood glucose monitoring, is attributed to A.H. Kadish. The pump was about the size of a backpack [39, 41]. First attempts of continuous insulin infusion were done using intravenous administration [42, 43]. In 1974, Slama et al. [42] used a portable peristaltic pump that weighed 1.4 kg and could be carried in a shoulder bag. In 1978, Pickup et al. described the clinical use of a miniature, battery-driven, syringe pump for the continuous subcutaneous infusion of insulin (CSII) [44]. The pump consisted of a prototype of the Mill Hill infuser,

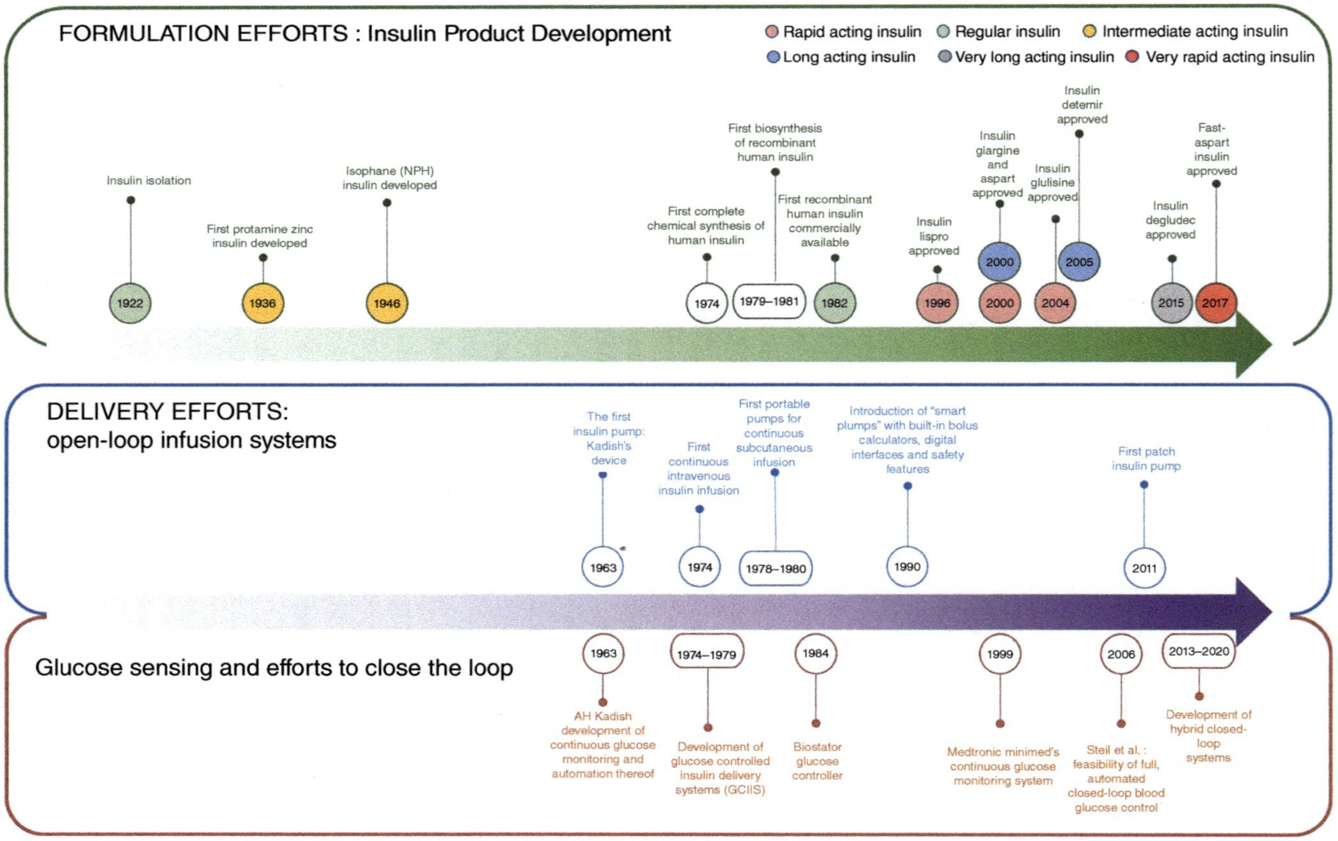

Figure 14.3 Timeline of efforts in formulation and infusion of insulin.

weighing 159 g, equipped with a 2 ml disposable syringe. Actrapid (animal monocomponent; Novo) [45] insulin was infused to 12 patients at a slow basal rate of 47 μl.h^{-1} and an eightfold higher rate of 375 μl.h^{-1} for 17 minutes before meals. The patients followed their usual subcutaneous insulin regimen on control days. It was concluded that CSII was feasible and provided good control of the blood glucose concentration in diabetics patients; however, the adjustment of the basal and preprandial insulin dose was challenging [44]. In the following years, similar studies were performed by Tamborlane et al. [46], Renner et al. [47], Irsigler and Kritz [48], and Hepp et al. [40].

The first pumps were cumbersome, heavy devices and the battery lasted for only a few hours. They had no safety alarms, offered limited flexibility in the basal insulin delivery rate, and were equipped with metallic needles. Clinical complications such as hyperglycemia and diabetic ketoacidosis raised from frequent needle displacement. Injection site infections were also reported [49].

Nevertheless, portable insulin infusion pumps have known an extremely quick development: within few years, devices became smaller, lighter, and user-friendly. As pumps started to receive global acceptance in the medical community, commercial devices appeared: the first was Dean Kamen's Auto-syringe (Figure 14.4a) [46, 49, 53]. A big improvement has been introduced by the device designed and manufactured by Siemens in 1979: the so-called portable insulin dosage-regulating apparatus (PIDRA; Siemens), later known as the Promedos E1. This pump, miniaturized to the size of a matchbox, could accommodate an increased volume of insulin, be easily refilled, and presented several safety features [40, 48]. Thus, the early 1980s saw the advent of numerous pumps: Promedos (Siemens, Germany), Betatron I and II (Eli Lilly, USA), SB-8000 (Sooil, South Korea), MiniMed 404 SP (Minimed Technologies, USA), and Nordisk Infuser (Novo, Denmark) [49] (Figure 14.4a). In numerous devices, the mechanism of delivery consisted of a syringe pump where the syringe piston is for instance actioned by a leadscrew [54]; numerous basal delivery rates could be selected together with bolus delivery modes [54]. The Nordisk Infuser could accommodate the highest volume of insulin (570 IU), while the Becton-Dickinson's B-D 1000, a spring-driven peristaltic pump, held the record of energy supply lifetime – 365 days. In 1985, the swiss company Disetronic produced the first 24-hour programmable insulin pump. Moreover, Disetronic devices were very light (100 g; MRS 3-infuser; MRS 1/H-Tron), shock- and water-proof, and presented numerous alarms and safety features including (i) a double control system that protected the patient against the possibility of a "pump running" i.e. an undesired delivery of insulin and (ii) the so-called "deadman switch" where the pump shuts off automatically if the patient does not react to the alarm (in case of a hypoglycemic event for instance) [54].

The modern insulin pump market is dominated by several manufacturers: Medtronic Mini-Med, Roche's Disetronic Medical Systems, Animas (Johnson & Johnson), Deltec (Smiths Group), Sooil, Nipro, and Insulet Cooperation Ltd. [49]. In addition to a further decrease in size, (Figure 14.4b,c) these pumps acquired the ability to account for actual glucose levels, carbohydrate intake, and thus adjust bolus delivery with the help of calculators [49].

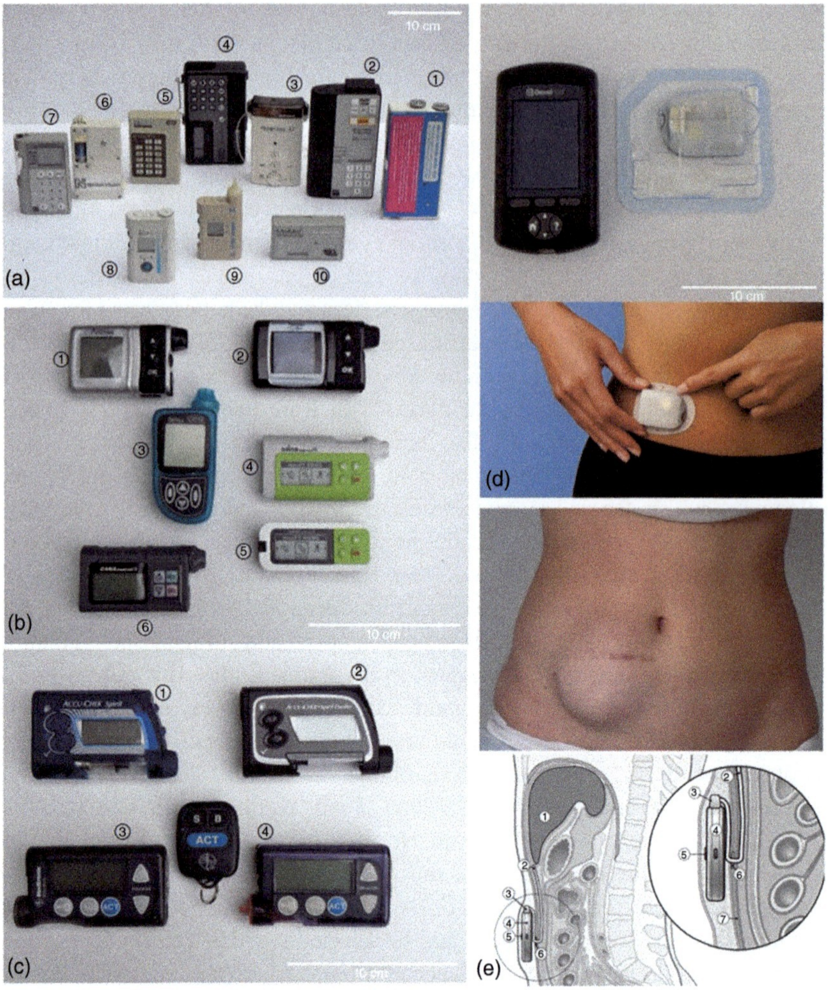

Figure 14.4 First generation of insulin pumps (a) (1978–1990s): 1. Mill-Hill Infuser; 2. Auto Syringe model AS6MP; 3. Promedos E1; 4. CPI 9100; 5. Betatron II; 6. Nordisk infuser; 7. Auto Syringe, model AS8MP, but it is known as "Travenol"; 8.H-Tron V 100; 9. H-Tron Hoechst; 10. Mini Med 504. Modern insulin pumps: (b) 1. Animas IR1200; 2. Animas vibe (Johnson & Johnson); 3. Cozmo 1700 (Deltec); 4. Dana Diabecare R; 5. The correspondent blood glucose meter; 6. Dana Diabecare. (c) 1 and 2. Accucheck Spirit and AccuChek Spirit Combo (Roche); 3 and 4. Paradigm (Medtronic Minimed) series 5 and 7: insulin reservoir of 1.8 and 3 ml, respectively. (d) OmniPod patch pump (pod and control unit). (e) Implantable pump for continuous intraperitoneal insulin infusion. 1: Liver; 2: catheter tip; 3: side port; 4: insulin pump; 5: refill port; 6: flange; 7: abdominal wall. Source: (a–d) [50, 51]/Adapted with permission Wiley. (e) Adapted from Dijk et al. [52].

In the UK, the Diabetes Control and Complications Trial and the UK Prospective Diabetes Study [55, 56] pointed to the fact that tight glycemic control is essential to prevent life-threatening DM-related complications such as diabetic retinopathy, nephropathy, and neuropathy. These studies have led to an increased interest in CSII as it offers accurate and physiological insulin delivery. In 2003, in an effort to elucidate the real medical and psychosocial impact of CSII, Weissberg-Benchell et al. [57] conducted a meta-analysis including 52 studies published between 1979 and 2001 encompassing a total of 1547 patients – only studies that compared CSII with other injection-based diabetes treatments either in the same individual or between treatment groups were included. The results indicated that glycated hemoglobin was lower in patients using CSII, with a greater benefit when used for at least one year. Blood glucose levels were also lower for patients using CSII. Complications such as pump malfunction, site infections, hypoglycemia, and diabetic ketoacidosis were also screened; despite the availability of only descriptive data, CSII therapy was associated with a decreased frequency of hypoglycemic episodes. Events of diabetic ketoacidosis were not evidenced in studies published after 1993 and the majority of studies reporting pump failures or catheter occlusions were published before 1988, suggesting the positive effect of technological progress on infusion devices. Overall patient satisfaction was also evaluated: a total of 19 studies investigated patient satisfaction and out of the 520 subjects approached, 325 chose to remain on the insulin pump after the completion of the study. Indeed a convenient and well-used insulin pump significantly improves the patient's quality of life, thus insulin pumps that initially were intended mainly for type 1 DM patients have now been widely accepted by type 2 DM patients as well [58].

However, Weissberg-Benchell et al. [57] also analyzed up to 16 studies that reported infections at the catheter site. And indeed, while the pump devices underwent continual improvement, the infusion sets remained a source of medical and practical problems [59]. In fact, infusion sets are responsible for many complications such as skin infections, contact dermatitis, and simple inconvenience [59], so issues with the infusion sets have been reported as the main reason for CSII therapy discontinuation. Thus, the suppression of complex infusion sets was the next evolution of insulin infusion systems: patch-like devices came to light. These devices are small enough to be attached directly to the skin and deliver insulin via a build-in cannula without the need of tubing. They can deliver both basal and bolus insulin [60]. Overall patch pumps can be classified into simple mechanical devices (V-Go by Valeritas and PAQ by CeQur) and full-featured electromechanical devices [60]. Among the latter, the OmniPod Insulin Management System (Insulet Corp., Bedford, MA) was the first patch pump marketed in the United States, it comprises a pump unit attached to an adhesive dressing (pod), which must be replaced every three days, and a wireless control unit with an integrated glucose meter. Many other devices exist but the OmniPod is one of the most popular devices (Figure 14.4d) [60]. Further developments consisted in further pump miniaturization and are described in further sections.

Implantable insulin pumps were developed for insulin delivery via the intraperitoneal route – the advantage of this method resides in the fact that it is more physiological – when insulin is directly infused intraperitoneally, it is absorbed into the portal vein. Indeed, in a healthy person, once secreted by the pancreas, the portal insulin concentration can be considerably higher than in the systemic circulation. Upon the first pass through the liver, between 20% and 80% of portal insulin is extracted by the latter, thus leading to low peripheral insulin levels. Whereas after subcutaneous (SC) injection, the systemic and portal vein insulin levels are equivalent, leading to both systemic hyperinsulinemia and hepatic hypoinsulinemia [61]. Consequently, upon intraperitoneal delivery, more predictable insulin profiles are obtained and glucagon secretion together with hepatic glucose production in response to hypoglycemia is improved. Fully implantable pumps (MiniMed MIP2007C, Medtronic) or external pumps with a percutaneous port such as the Accu-Chek Diaport system (Roche Diabetes Care) can be cited as infusion devices (Figure 14.4e). However, implantable pumps suffer from unpopularity [58] and complications such as cannula blockage, portal-vein thrombosis, and peritoneal infection could be observed, not to mention their higher cost and invasiveness. The MiniMed implantable pump was removed from the market on a worldwide scale in 2007 [52, 58].

14.2.4.3 Closed-Loop Insulin Delivery Systems

While patient-managed insulin pumps have significantly improved the patients' quality of life, an autonomous system able to secrete insulin at the right time and with the correct dose, without daily human intervention, thus completely replacing the endocrine role of the pancreas, has ever been the final aim of DM-related research. Such devices, often termed "artificial pancreas," are able to continuously monitor blood glucose, interpret the reading, and finally respond to glycemia by a controlled insulin delivery [62]. First reports on closed-loop insulin delivery systems were done almost simultaneously by several researchers in early 1970 [63–66] and paved the way for the first commercially available closed-loop device: the Biostator [67]. This large and complex apparatus consisted of a pump to ensure continuous intravenous blood withdrawal, a continuous glucose analyzer: a computer programmed to calculate the amount of insulin (or glucose, in case of hypoglycemia) to be infused and a computer-driven insulin intravenous infusion pump. Nevertheless, due to its size and use of IV route, the equipment was not suited for outpatient use [68].

In fact, for many years, the main barrier to the development of an "artificial pancreas" was the difficulties in the conception of implantable glucose sensors. Indeed long-term implantation of glucose sensors is hindered by poor biocompatibility: long-term implantation results in fibrin deposition on the electrodes or immune reaction [54]. Only in 1999, Medtronic's Minimed introduced the Continuous Glucose Monitoring (CGM) System: the glucose diffuses from interstitial fluid into the biosensor through a membrane and is enzymatically transformed to produce hydrogen peroxide by immobilized glucose oxidase enzyme. The hydrogen peroxide generated based on glucose interstitial concentration is then oxidized by electrode, creating an electric signal measurable by amperometry [53, 62, 69].

While the system successfully controlled overnight and interprandial glucose levels, the response time of the sensor was too slow to manage meal-related insulin requirements, and fragility of the sensor was also reported [53]. It was rapidly noticed that insulin dose calculation algorithms based on proportional-derivative control were suffering from major time lags, thus mathematical modeling, taking into account the metabolic system, was necessary [62]. In 2006, Steil et al. [70] were

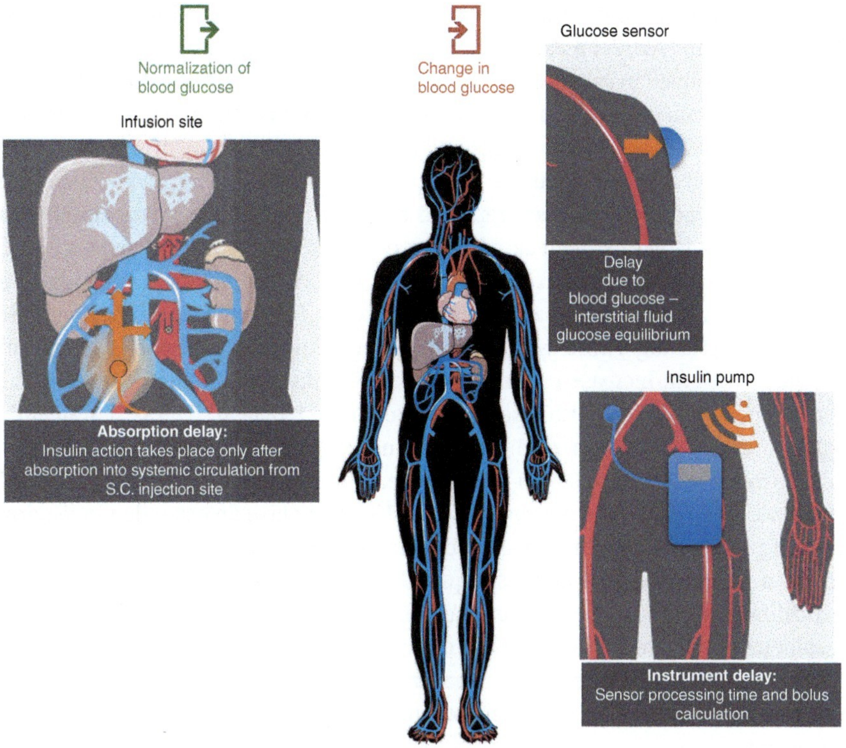

Figure 14.5 Delays between change and normalization in blood glucose in closed-loop systems. Source: Adapted from Servier Medical Art https://smart.servier.com/.

the first to demonstrate the feasibility of fully automated closed-loop blood glucose control and subcutaneous insulin delivery to manage glycemia [70]; however, despite significant progress in control algorithms, the system could not outperform an open-loop system and needed optimization.

Closed-loop systems experience major limitations from (i) glucose sensing and (ii) insulin delivery kinetics (Figure 14.5). Indeed the first hurdle consists in the fact that subcutaneous glucose sensors measure glucose concentration in the interstitial fluid. After a change in blood glucose levels, it takes 5 to 15 minutes for it to be detected in the interstitium [62]. While the pump interpretation of data and insulin dose calculation also requires some time, the main delay in insulin response is due to the absorption from the subcutaneous delivery site [62]. It can take up to one hour for regular insulin and around 10 minutes for rapid-acting insulins such as insulin analogs lispro, aspart, or glulisine [71].

As a consequence of these delays, a fully automated artificial pancreas has still not been developed. However, open-loop systems can be partially "closed" in the so-called "sensor-augmented therapy" of hybridization (use of the closed-loop only temporarily between meals). The patient must still intervene for meal-related insulin boluses.

In 2013, FDA approved the MiniMed 530G pump with an Enlite sensor and classified it as an "Artificial Pancreas Device System." Insulin infusion can be automatically suspended in case of low glucose detection. In hybrid closed-loop systems, the pump automatically controls basal insulin delivery based on the sensed values to control glucose levels between meals and overnight (MiniMed 670G insulin pump with a Guardian 3 sensor; Insulet's Omnipod Dash System). The "loop-closing" in the inter-meal periods has been reported to improve glycemic management and ensure better safety, treatment satisfaction, and sleep quality in DM1 patients [72]. The Diabeloop Generation 1 (DBLG1) hybrid closed-loop system consisting of a patch-pump (Cellnovo), a glucose sensor Dexcom G5, and a command module hosting a hybrid algorithm with customization settings (Diabeloop SA, France) associated with an increase in time-in-range (TIR; the proportion of time that the glucose concentration is within the target range 70–180 $mg.dl^{-1}$) was significantly higher in the DBLG1 group ($68.5 \pm 9.4\%$) than the open-loop, sensor-assisted pump group) ($59.4 \pm 10.2\%$; $p < 0.0001$) [73]. Indeed, in late 2016, Dexcom's G5 CGM system was the first sensor approved by the FDA to receive an expanded indication replacing fingerstick testing for adjusting treatment in diabetes patients [74], in fact for a long time, CGM devices required capillary blood glucose measurement as a calibration technique and were not used alone as a measurement technique in diabetes management [75]. The G5 mobile CGM system is composed of a small sensor inserted transcutaneously by an applicator that measures interstitial glucose. A transmitter is then attached to the sensor pod for wireless communication with the receiver.

In 2020, Medtronic Secured CE Mark for MiniMed™ 780G Advanced Hybrid Closed-Loop System, and preliminary clinical pivotal trial results indicate that an overall TIR of 75% was achieved, with an 82% TIR overnight. The built-in auto-correction feature contributed to 22% of all bolus insulin, and the SmartGuard™ feature (closed-loop mode) was utilized 95% of the time [76]. Closed-loop infusion systems for both insulin and glucagon also exist and are termed "Bionic pancreas." As an example, the iLet Bionic Pancreas was granted the "breakthrough device designation" by the FDA in 2019 [58].

Improvements to implantable glucose sensors are already on their way: while most of the sensors need to be changed every 7 to 14 days due to the enzymatic sensor lifetime, recently, a new CGM system has been approved using the first implantable long-term sensor. Developed by Senseonics, Inc. (Germantown, MD, USA), Eversense® is composed of a small sensor, 3.5×18.3 mm, that is subcutaneously implanted in the arm allowing continuous glucose measurement for up to 90 days. However, the placement of the sensor has to be done by a health care provider which makes it disadvantageous compared to other CGMs. A nonenzymatic detection of interstitial glucose allowed the extension of the duration of use from one week to three months: an optical sensor uses a glucose-reacting hydrogel that fluoresces through light-emitting diode (LED) stimulation (Figure 14.6). The intensity of the fluorescence is measured by photodiodes and communicates the signal to the transmitter, which calculates the concentration of interstitial glucose [78, 79]. The device was tested in a 90-day clinical trial to assess the accuracy of

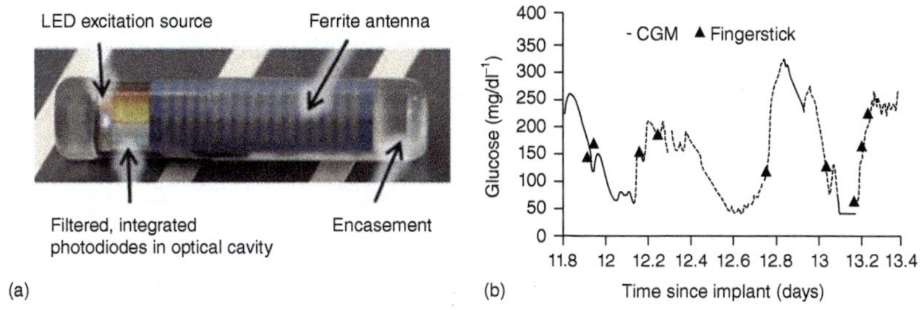

Figure 14.6 Next generation glucose sensing. (a) Implantable optical glucose sensor. (b) Optical glucose sensor data measurement overtime on a patient compared to traditional fingerstick measurement. Source: Adapted from Mortellaro and DeHennis [77].

glucose measurement compared to venous blood measurement in patients with type I diabetes [80]. Later, a 180-day clinical trial in type I and II diabetes patients was published in order to assess the safety and accuracy of the implantable CGM over an extended period. The outcomes of the study allow senseonic to receive CE Mark in 2016 [81]. Studies are currently ongoing to increase the use of CGM over 90 days.

While hybrid closed-loop systems constitute a major breakthrough in the management of diabetes, the research and development of those systems are expensive and time-consuming, resulting in often unaffordable devices. Exasperated by this aspect of pump therapy, patients, families, and caregivers gathered online under the hashtag "#WeAreNotWaiting," claiming more affordable devices. Hence started a do-it-yourself artificial pancreas system (DIY-APS) movement, aimed at sharing the knowledge about artificial pancreas devices and the provision of open-source hardware and software solutions, now known as the "Open Artificial Pancreas System" project [58, 82].

In conclusion, insulin pump therapy significantly improves the management of DM and prevents its complications: patients have lower glycated hemoglobin levels, fewer hypoglycemic events, and higher TIR values. The avoidance of multiple daily injections provides a better quality of life. However, many aspects still need to be optimized: infusion set associated issues and always increasing acquisition and functioning costs are the major issues [58], therefore there is still a lot of research in this domain. In parallel to technological advancements, novel insulins are being developed, for instance, Novo Nordisk's once-a-week insulin icodec has completed phase II trials [83]. In addition, biological therapies such as β-cell replacement therapies or stem cell therapies also seem promising and may offer a more physiological and intervention-free blood glucose management [84]. Furthermore, in the field of bioengineering, tissue engineering, and regenerative medicine, the development of the NanoGland, a novel silicon-based pancreatic islets of Langerhans encapsulation device, for the transplantation of human pancreatic islets is a step forward for the artificial replacement of body glands ("bioartificial pancreas"). Indeed, the advantages of transplanting an immunoisolation device containing the pancreatic islets over islet transplantation are numerous. Not only does it avoid the complications that the latter procedure has, such as graft failure, islet dispersion in the portal vein, lack of neovascularization, and life-long immunosuppressive treatments that are a real threat to a patient's life, but it also protects the islets from the patient's immune system, fosters revascularization, and extends islet survival. Therefore, toxic immunosuppressive drugs are no longer required. Moreover, better control of glycemia is provided compared to conventional injectable insulin. The islets function and viability of the NanoGland were demonstrated, *in vitro*, for over 30 days and up to 120 days, *in vivo*, posttransplantation in mice. Hence, this nanochannel-based device offers a novel and promising approach to the autotransplantation of human pancreatic islets [85–87].

14.3 Microfabricated and Nanofabricated Drug Delivery Devices

Most of the mechanical pump systems discussed previously have limitations such as large size and the need for a liquid dosage form [88]. This implies a decrease in the patient's quality of life due to the cumbersome size of these devices. In addition, the development of devices involving biopharmaceuticals has limited stability in liquid formulation.

Microfabrication technology of electromechanical systems enables most of these issues to be overcome. Moreover, the recent evolution of silicon-based computer chip manufacture allows a transfer of knowledge from the semiconductor industry to biomedical applications [89]. These new devices are thus composed of a semiconductor substrate integrating different structural elements such as drug reservoirs and programmable electronic circuits allowing a controlled release of therapeutics. MEMS and NEMS devices can be used to address medical needs not met by traditional drug administration devices. To be fully functional, a MEMS-based device should integrate different elements such as a power source, a drug reservoir, a control and programming circuitry that regulate drug delivery, and a wireless communication system [90].

MEMS and NEMS have significantly been developed and are becoming powerful tools in the biomedical field, offering state-of-the-art solutions for drug delivery devices related to precision and accuracy for either pulsatile or continuous administration [91].

14.3.1 Microelectromechanical Systems (MEMS)

14.3.1.1 Microchip-Based MEMS

The most rudimentary way of designing microsystems is to reduce the size of the reservoir containing an active drug. This reservoir containing a well-defined amount of drug must be capable of being triggered to release its contents. Santini et al. have extensively studied this reservoir-based system and were among the first to report a microchip that allowed a pulsatile release using a MEMS-based technology [92]. This silicon microchip was composed of an array of individual reservoirs capped with a gold thin layer. Each reservoir was connected to the electric circuit and could

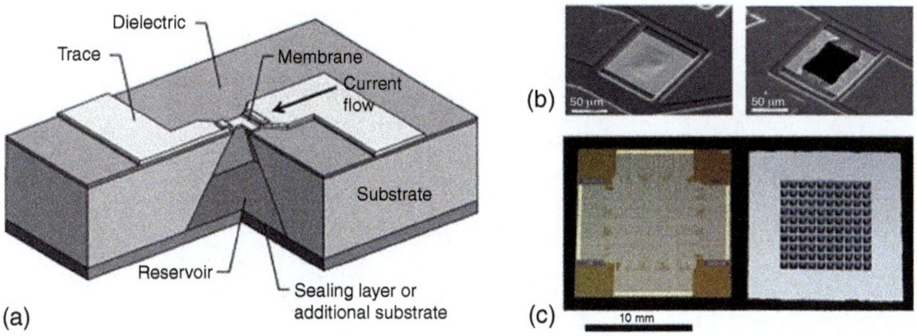

Figure 14.7 MEMS-based silicon microchip. (a) Representation of a single unit of a reservoir-based microchip. (b) Scanning electron micrographs image of platinum–titanium multiple layers membranes inactivated (closed; left) and electrothermally activated (opened; right). (c) Front- and back-side of a microchip composed of an array of 100-reservoir. Source: Adapted from Maloney et al. [88].

be selectively opened by applying an electric current. The gold membrane, located at the interface between the drug formulation and the interstitial fluid, was disrupted by either electrochemical dissolution (oxidation) [92] or electrothermal activation (fusion) of the metal membrane (Figure 14.7) [88].

The versatility of the device also relied on its ability to create a continuous release by the sequential and repeated opening of the reservoirs. In addition, a pulsed release in response to a sensor or the user's demand by telemetry could also be implemented in the device. Finally, different formulations could be accommodated in the reservoirs such as solids, liquids, or gels [92], but also sensors for example.

This multireservoir-based technology was then adapted by MicroCHIPS Biotech Inc., now part of Dare Bioscience (San Diego, CA, USA), to successfully achieve pulsatile delivery of leuprolide polypeptide, *in vivo,* to dogs over six months. The implantable device is composed of an array of 100 individual drug reservoirs. Each reservoir presents a 300 nl capacity and can be specifically triggered to deliver its content by electrothermal activation of the metal membrane. The microchip is then integrated into an implantable device composed of an electronic circuit with a telemetry-controllable power supply. The overall size of the implantable device is $4.5 \times 5.5 \times 1$ cm (Figure 14.8). An interesting aspect of the device is the ability to accommodate the leuprolide in a polyethylene glycol (PEG) matrix as a solid powder [93]. The use and control of the freeze-drying process allow biomolecules such as proteins and peptides to be less subject to chemical or physical degradation and increase their long-term stability [94]. With direct on-chip lyophilization, a stable formulation at 37 °C for six months was observed. After subcutaneous implantation in dogs, a controlled pulsatile release profile of leuprolide was obtained over 25 weeks. These results were consistent with data from other studies for subcutaneous injection of an extended-release formulation of leuprolide [93]. The microchip was also tested in a clinical trial for the delivery of human parathyroid hormone fragment (1–34) (hPTH(1–34)) for 20 days. The wirelessly programmable device was implanted subcutaneously in the abdomen of eight women with postmenopausal osteoporosis for four months. No device-related adverse events

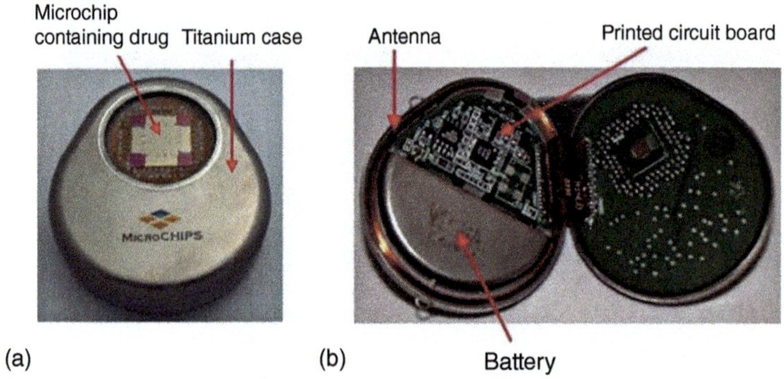

Figure 14.8 Microchip implantable device. (a) External view. (b) Electronic components composing the telemetric device. Source: Staples [90]/Adapted with permission John Wiley & Sons, Inc.

have been reported and the device was well tolerated by patients. Regardless of the fibrous capsule formation around the implant, similar pharmacokinetics of hPTH(1–34) were obtained for multiple subcutaneous injections. Biologic markers also showed an increase in bone formation after pulsatile administration. The one-month assessment of safety, tolerability, and bioequivalence showed promising results for the development of long-term implantable devices [95]. Since the acquisition of the technology by Daré Bioscience (San Diego, CA, USA) in 2019, the company is working on the application of a user-controlled long-acting and reversible contraceptive [96].

Other microchip-based devices have been investigated in order to extend the possibilities of incorporating the drug into the device in the form of polymeric films, in particular conducting polymers such as polypyrrole (PPy). In these devices, direct polymerization on the gold electrode incorporates the active compound into the polymeric film. The release of a single or multiple drugs is then controlled by the application of a negative electric potential (Figure 14.9) [97]. The drug loading of such devices is easier when compared to reservoir-based technology [90].

14.3.1.2 Pump-Based MEMS

MEMS-based technology enables miniaturization of the pump-based devices while preserving precision, accuracy, and reliability. Since the scale of the devices no longer allows the use of the electromechanical systems traditionally employed by these pumps, new fluid displacement technology is therefore being introduced.

An example of a miniaturized micropump is the Nanopump™ technology developed by Debiotech (Lausanne, VD, CH). The pump is based on the movement generated by an actuator to displace the drug formulation out of the pump chamber through the outlet valves. These actuators are made of piezo-electric material that possesses the ability to change their shape when a voltage is applied. The Nanopump is mounted on a ceramic plate containing the electrical circuit and is totally biocompatible (Figure 14.10). The pump allows a precise infusion at a rate of up to 2.5 $ml.h^{-1}$ with a size of only 10×6 mm [98]. Flow rate is controlled by adding

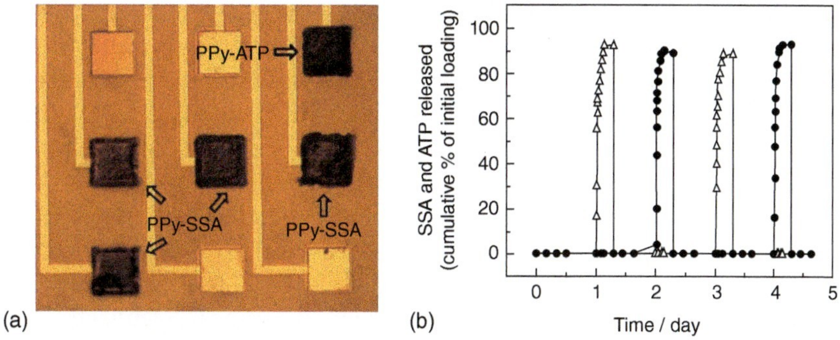

Figure 14.9 Conducting polymer-based device. (a) Metallurgical microscope image of the microchip containing nine gold microelectrodes – four doped by PPy/sulfonic acid (SSA) and one by PPy/adenosine triphosphate (ATP). (b) *In vitro* pulsatile release profile of SSA (Δ) and ATP (•) from a single microchip over days. Source: Adapted from Ge et al. [97].

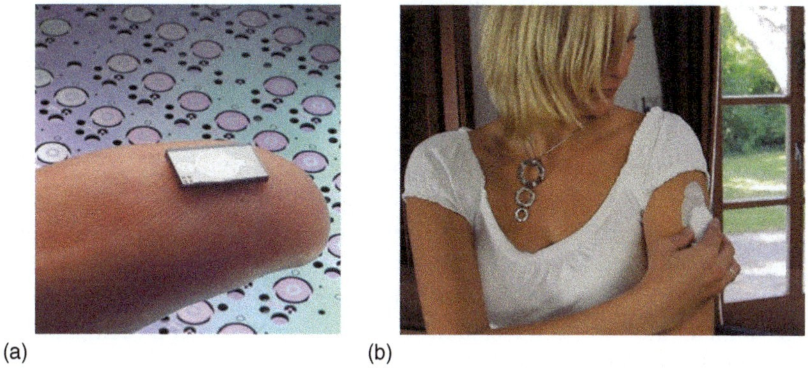

Figure 14.10 Debiotech devices (a) The Debiotech MEMS-based pump (b) The JewlPUMP™ containing MEMS-based pump applied on a patient's body. Source: Debiotech products page. Available at: https://www.debiotech.com/jewelpump/ (Accessed March 2, 2022 © 2021 Debiotech SA, All Rights Reserved).

sensors that send information and alerts to the user by connecting to an external device. The low manufacturing cost and high reliability of the disposable pump make it an interesting platform for many applications [99].

Debiotech integrates Nanopump into JewelPUMP™, a medical device for the management of type I and more recently type II diabetes. The pump patch is small, lightweight, and tubeless for a patient-friendly and continuous subcutaneous infusion of insulin [100]. The accuracy of the JewelPUMP was tested both *in vitro* and *in vivo* and compared to other commercially available pumps. The infusion accuracy over 24-hour in patients did not show a significant difference in flow rate error when compared to other conventional pumps. However, the JewelPUMP allowed earlier detection of catheter occlusion and was better accepted by patients, in particular, due to the absence of the tubing present in traditional pumps [101]. Despite some phase III clinical trial results in 2014 (NCT02097316), neither FDA nor

European Medical Agency (EMA) approvals have been granted [60]. Debiotech's technology is still under development and protected by many patents [102, 103].

Another application of the MEMS-based technology is the commercially available ophthalmic MicroPump™ system produced by Replenish Inc. (Pasadena, CA, USA) to treat patients with glaucoma and retinal-related diseases. The ocular pump was originally developed by the group of Humayun and consists of an electrically controlled mini drug pump [104]. The size of the device is 13 × 16 × 5 mm. It is composed of a refillable drug reservoir implanted in the subconjunctival space and a cannula that can be surgically inserted into the vitreous cavity [105]. Its driving force relies on the electrolysis of the water. Indeed, by applying an electrical potential to water, the latter is electrolyzed into oxygen and hydrogen. The gas generated increases the pressure inside the reservoir and pushes the drug formulation through the cannula and check valve. The drug delivery can be controlled by the applied current as it is directly correlated to the gas generated and therefore to the flow rate [104]. Different release profiles (continuous or pulsatile) can be obtained directly on target site over the duration of implantation, thus reducing the need for multiple ocular injections. Two versions of the MicroPump are currently available; Anterior MicroPump (AMP) for glaucoma and Posterior MicroPump (PMP) for retinal conditions. The implant is rechargeable with medication *in situ* via a 31-gauge needle of the Drug Refill System™ and can be controlled as well as charged by wireless telemetry via EyeLink™ external device. The implantation procedure as well as the safety assessment of the MicroPump was performed on a cohort of 11 patients with diabetic macular edema over 90 days. No surgical implantation complications or adverse events were reported. However, 4 of the 11 patients required additional ranibizumab to achieve the desired dose via intravitreal injection [106]. Finally, an evaluation of one-year implantation was conducted in 13 Beagle dogs showing biocompatibility as well as good safety profile [105].

14.3.1.3 MEMS – Efforts to Close the Loop

To be highly effective, a pharmaceutical treatment has to be adapted by the physician or patient himself. Ideally, the adaptation of treatment is based on the measurement of biomarkers and metabolites to monitor the evolution of the pathology through a physiological outcome. Biosensors are devices that can transpose a physiological signal into transmittable information such as an electrical current. The biosensors can be classed based on the signal transduction mechanism: (i) electrochemical that senses change in potential, impedance, charge accumulation, or current density, (ii) optical that relies on absorption/emission of an analyte, or (iii) mechanical transduction that senses the impact of an analyte on the shape or motion of a mechanical component [69]. The combination of these devices with the MEMS-based technology also allows the miniaturization of the sensors. Therefore, this advance allows the acquisition of patient signals that were previously inaccessible.

Furthermore, the main objective is to integrate a MEMS device in a closed-loop circuit in order to deliver the right amount of drug at the time, thanks to the sensor measurements. While biosensors have been in development for years and are presented above, other sensors have also been developed to measure various stimuli such as changes in pressure. Such sensors are interesting for hemodynamic monitoring in case of heart failure (HF) management. The St. Jude's cardioMEMS™ HF system recently acquired by Abbott is a pressure sensor approved by the FDA in 2014 [107]. Such a device is used to measure wirelessly pulmonary artery pressure in order to improve heart failure management and reduce heart failure-related hospitalizations. This has been tested through a randomized clinical trial that demonstrated 37% lower hospitalizations in subjects monitored by cardioMEMS [108]. Elman et al. developed an implantable MEMS for emergency situations – *in vitro* studies demonstrated the release of vasopressin in less than one minute. The implementation of the MEMS device with a pacemaker in a closed-loop circuit could improve the management of cardiovascular resuscitation [109]. Many other applications for the use of sensors in the pharmaceutical industry are being developed, such as tracking pills for compliance assessment [110, 111], acid reflux detection for gastroesophageal disease [112], sleep apnea [113], pressure monitoring stent [114, 115].

14.3.2 Nanofabricated Drug Delivery Devices

NEMS-based devices are the nanoscale level successor of MEMS-based devices, integrating mechanical and electrical functionality. Although both technologies overlap in numerous applications, by exploiting the unique properties at the nanometer scale such as surface forces (adhesion, friction, surface tension, etc.), NEMS present characteristics that might make them even more suitable than MEMS for controlled drug delivery [8]. For example, in the case of constant delivery, nanochannels or nanopores constitute a very accurate way to deliver therapeutics in a controllable manner (Figure 14.11). Fluids confined into nanometer structures display different physical behaviors from those observed at the micrometer scale or above because they approach the size corresponding to molecular scaling lengths [8, 117, 118]. Classical Fick's laws of diffusion are no longer applicable, which is not the case at the micrometer scale where the diffusion of the molecules is Fickian (release rate dictated by the gradient of concentration) and therefore the use of other devices such as pumps is required at that level.

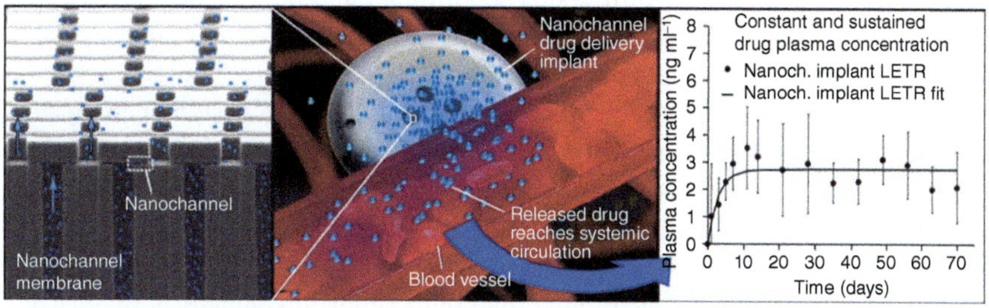

Figure 14.11 Implantable nanochannel device for constant drug delivery. Source: Adapted from Ferrati et al. [116].

The driving forces of molecular transport in nanochannels seem to be modulated through electrostatic interactions, molecule-to-surface interactions, and physical confinement, but our understanding of the diffusive transport in nanoconfinement still needs to be further investigated.

BioNEMS have been the subject of much research including their application for intratumoral drug delivery. Hood et al. have developed a multifunctional nanofluidic implant by combining a BioNEMS nanofluidic membrane with parallel nanochannels for the local controlled-drug release of chemo- and immune-therapeutics [119]. Successful proof-of-concept studies have been conducted *in vivo* in mice, showing the release of therapeutic drugs and contrast agents intratumorally, a cutting-edge technology for versatile applications with the potential to expand clinical utility for localized delivery [119]. In other studies, an implantable silicon nanochannel drug delivery system (nDS), constituted of up to 100 000 nanochannels.mm^{-2}, was developed for continuous sustained administration of cardioprotectants such as atorvastatin and resveratrol to manage cardiovascular diseases [120]. An nDS implant has also been developed for the long-term sustained and constant release of testosterone for the treatment of male hypogonadism [121]. In another study, a silicon micronanofluidic-based platform was used for the controlled administration of drugs and cell transplantation without the need for pumping mechanisms or actuation [122]. In addition, a proof-of-concept theranostic intraocular implant, based on nanofluidic technology, has been designed for sustained and controlled intraocular delivery of therapeutics such as bimatoprost and dexamethasone, through a silicon nanochannel membrane without the need for pumps, actuation or repeated clinical intervention (Figure 14.12) [123]).

Furthermore, Ferrati et al. developed another implantable drug delivery device with nanochannels as small as 2.5 nm. These nanochannels control the release of drugs by physical-electrostatic confinement, resulting in zero-order release kinetics of various molecules such as leuprolide, interferon α-2b, letrozole, and human growth factor at clinically-relevant doses, as shown in Figure 14.13. This device has shown to maintain sustained target doses for up to 70 days in dogs, rats, and mice after its implantation. This platform has the potential to improve long-term treatment, especially in chronic conditions [116].

The use of the nanochannels approach enables personalized patient treatment by tailoring the nanochannels' dimensions and number and hence offering great flexibility in drug dosage according to the patient's need. Such technology represents a significant advance in the field of drug delivery [124].

All these examples show the technological advances in the development of drug delivery systems for continuous and constant drug release. However, in some complex pathologies such as hypertension, rheumatoid arthritis, diabetes, cancers, and other diseases where pharmacokinetics is not constant within the day due to the circadian body clock, zero-order release kinetics can be inadequate and thus require a variable drug administration rather than a constant delivery dose. By using a low-intensity electric field, for instance, drug release can be modulated and thus customized according to patient needs [122, 125]. In line with this, tunable active delivery devices become more convenient for long-term therapeutic management

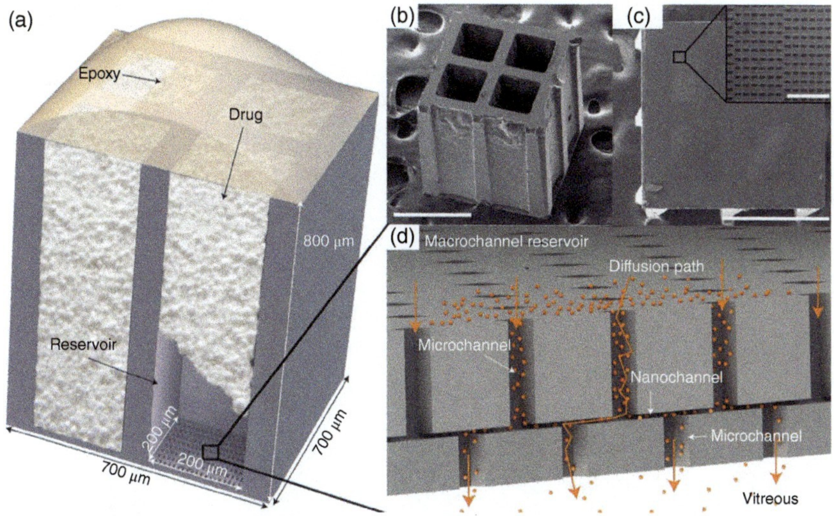

Figure 14.12 nViSTA device. (a) Device dimensions. After loading the drug in the macrochannel reservoirs, the system is then sealed with epoxy. (b) Scanning electron microscopy (SEM) image of the silicon membrane. (c) Microchannels SEM bottom view. (d) Drug diffusion path through the nanochannels (not to scale). Source: Di Trani et al. [123]/Adapted with permission Elsevier.

where the precise timing and accurate dosing are needed for synchronized drug release with the circadian behavior of the disease [118, 126, 127]. Such technologies could address unmet medical needs when tailored to individual physiology. To date, remotely active-controlled drug delivery systems have emerged showing that there is a growing interest in the development of such devices capable of tailoring drug delivery to individuals' needs in order to reach maximal treatment efficacy with minimal side effects. Thus, improving patient adherence and quality of life [128]. The ability to tightly regulate dosing with regard to timing, amount, and location of drug administration is a step forward in the area of precision medicine and chronotherapy, where the drug delivery is synchronized with the patient's body clock, mimicking the time-varying release of molecules in the body [8].

Despite advantages offered by implantable drug delivery systems, such devices also have some limitations such as biofouling, which is a phenomenon caused by the interactions with biological tissues and nonspecific adsorption of proteins on the surface of the biomedical implant. Thus, the device becomes inactive *in vivo* because it is recognized as a foreign body. A lot of researchers are trying to reduce such phenomena by functionalizing the surface of the implant, in order to preserve the full functionality of the device and improve its biocompatibility. One of the techniques that can be used to prevent biofouling is to use a technology of self-assembled monolayers (SAMs) of organic molecules [129–131]. SAM engineered surfaces are used in many studies in different fields and a wide variety of medical devices, for example, various biosensors are using SAM in order to attach bioreceptors and then detect the corresponding substrates (DNA, proteins, etc.). Therefore, SAM could either be used to

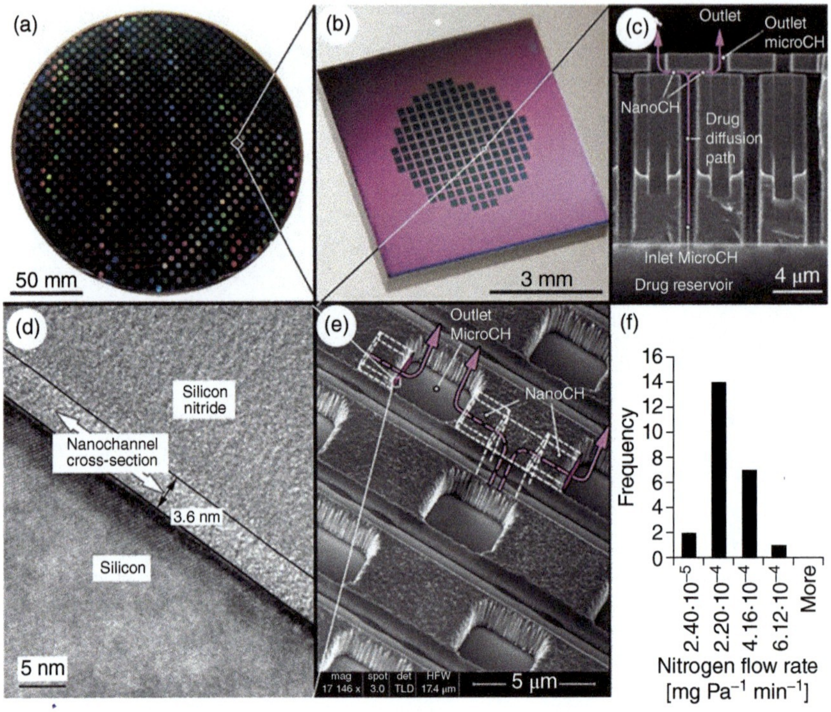

Figure 14.13 The implantable nanochannel device. (a) Image of a wafer containing ~700 nanochannel membranes; (b) the device with the nitride layer; (c) a cross-section showing the orientation of the micro- and nano-channels; (d) a transmission electron microscopy (TEM) image of a 3.6 nm nanochannel; (e) a SEM image of the outlet microchannels; (f) the results of quality-control gas testing realized on 24 membranes. The pink arrows indicate the drug delivery. Source: Ferrati et al. [116]/Adapted with permission Elsevier.

prevent the nonspecific adsorption of biomolecules on the surface of the device or in contrast, improve their adhesion by functionalizing with the appropriate receptors on biosensors in order to detect the corresponding substrate [129, 132, 133].

Despite some of the above-mentioned obstacles, the use of nanochannel technologies holds the promise to achieve higher patient compliance while enhancing therapeutic efficacy and minimizing systemic side effects, thereby improving clinical treatment of a wide range of conditions.

14.4 Noninvasive Active Drug Delivery Systems: Iontophoresis

Iontophoresis, a completely noninvasive physical enhancement technology, is a technique that uses a mild electric current (usually <0.5 mA.cm^{-2}) to enhance and facilitate the transfer of a drug with ionizable groups into and across biological membranes, enabling greater amounts to be delivered in a shorter time (Figure 14.14a,b) [134]. Electrotransport occurs through two principal mechanisms:

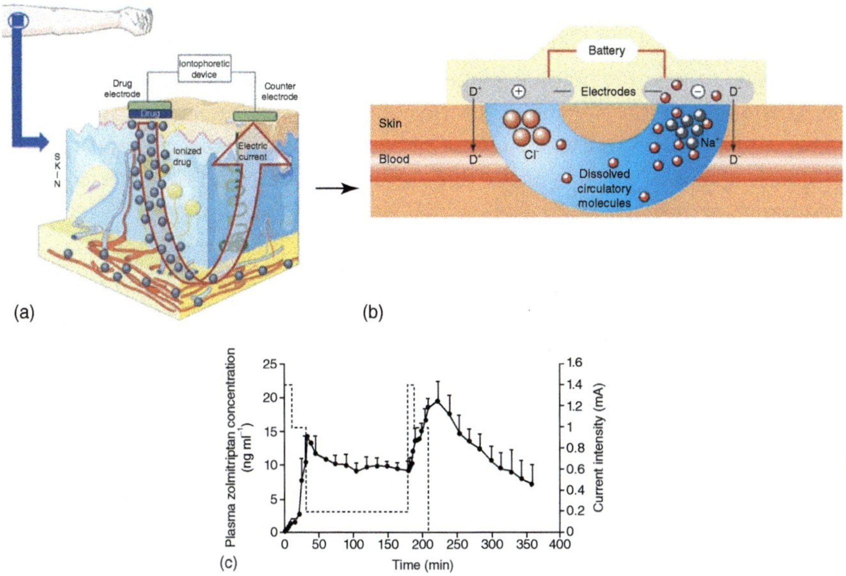

Figure 14.14 Iontophoresis for controlled drug delivery kinetics. (a) and (b) Principles of iontophoresis. (c) *In vivo* iontophoretic delivery of a multistep current profile of Zolmitriptan in Yorkshire swine. Source: Adapted from Kalia et al. and Patel et al. [134, 135].

electromigration (EM) that occurs when an electric field is applied to the charged species, and electroosmosis (EO), also known as convective solvent flow, from the anode to the cathode, again driven by the application of the electric field and resulting from the skin's negative charge under physiological conditions (isoelectric point [pI] of 4–4.5) [134, 136]. This also facilitates the electrotransport of neutral molecules from the anode but opposes cathodal electromigration of anions.

The transport of the drug can be expressed as follows, according to the Nernst–Planck theory (Eq. ((14.1))

$$J_{\mathrm{TOT}} = J_{\mathrm{P}} + J_{\mathrm{EO}} + J_{\mathrm{EM}} = \left[k_{\mathrm{p,DRUG}} + V_w + \left(\frac{i_d}{z_{\mathrm{DRUG}}F}\right) \times \frac{u_{\mathrm{DRUG}}}{\sum_1^i u_i c_i}\right] \times c_{DRUG}$$
$$= [k_{\mathrm{p,DRUG}} + V_w] \times c_{DRUG} + \frac{i_d t_{\mathrm{DRUG}}}{z_{\mathrm{DRUG}}F} \qquad (14.1)$$

where J_{TOT} is the total flux, J_{EM} and J_{EO} are the fluxes resulting, respectively, from EM and EO, and J_{P} is the passive flux; i_d is the applied current density; z_i, u_i, and c_i represent the valence, mobility, and concentration, respectively, of charge carriers in the system; V_w is the solvent permeability coefficient; and z_{DRUG}, u_{DRUG}, and c_{DRUG} are the corresponding values for the drug ($k_{\mathrm{p,DRUG}}$ and t_{DRUG}, represent the passive permeability coefficient and the transport number of the drug).

An iontophoretic assembly consists principally of two electrode compartments that are connected to a power source and microprocessor. One of the chambers contains the drug of similar polarity (i.e. cationic drug in the anodal chamber). When a constant small electrical current is applied, positively charged drugs are transported from the anode into the skin and negatively charged species from the cathode. Because the skin is a negatively charged membrane, its natural permselectivity (the preferential permeation of certain ionic species) favors the transport of cations, as well as neutral compounds. The intensity of the current applied usually determines the amount of compound delivered into and through the tissues, likewise, the area of the skin surface in contact with the electrode and the duration of the current application.

One of the main advantages of iontophoresis is the ability to control the drug input rate depending on the applied current intensity according to patient needs, allowing individualized therapy.

Migraine is a chronic disabling disease characterized by recurrent headaches associated with a broad panel of other symptoms, it manifests in attacks that last between 4 and 72 hours. Triptans are 5-HT(1B/1D) agonists that are very useful actives in the pharmaceutical arsenal to treat migraine. However, when the treatment is delayed, an incomplete response to triptans is observed. Therefore early treatment onset is critical in achieving a positive therapeutical outcome.

Depending on the administration route, numerous formulations of zolmitriptan (conventional and fast-dissolving tablet, nasal spray), a widely prescribed triptan to treat migraines, are marketed but there are still several unmet needs that should be addressed in order to reduce the onset time for therapeutic effect (fast-acting drug), improve the efficacy of the drug and thus improve patient compliance.

As discussed above, drug input kinetics can be tightly controlled using transdermal iontophoresis by adjusting parameters such as the current intensity and the application time. Therefore, basal and/or bolus administration can be achieved by modulating the current profile. Zolmitriptan is a good candidate for iontophoresis since it is a relatively hydrophilic molecule, charged positively at physiological pH ($pK_a \sim 9.52$), and has low molecular weight.

Patel et al. [135] conducted *in vitro* and *in vivo* studies using porcine skin and Yorkshire swine, respectively, in order to find the best formulation and iontophoresis conditions in the *in vitro* study and to assess the level of the drug reaching the blood flow and the feasibility of administering repeated bolus doses overlaid on a basal input rate in the *in vivo* study.

After an initial bolus, Zolmitriptan was detected in the blood after 2.5 minutes and the C_{max} was reached within 35 minutes. The results also showed a significant concentration of the drug persisted in a narrow range during a maintenance phase (t = 75–180 minutes) (low-level drug input rate). An increase of approximately 50% in blood drug levels has been reported after an application of a second pulse at t = 180 minutes within 10 minutes after increasing the current intensity from 0.2 to 1.4 mA (0.05–0.35 $mA.cm^{-2}$) as depicted in Figure 14.14c. *In vivo* data have shown that therapeutic amounts of Zolmitriptan can be achieved in a shorter period using transdermal iontophoresis compared to those from existing dosage forms, demonstrating the potential of this technique to provide a custom-made drug release profile.

The modulation of the current enables complex drug delivery profiles to be obtained for the delivery of higher molecular weight peptidic drugs. For example,

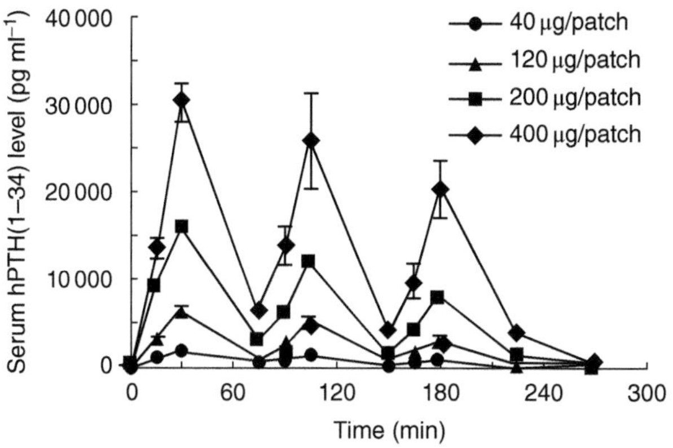

Figure 14.15 Serum hPTH(1–34) levels after three-pulse iontophoretic administration of hPTH(1–34) to 6M-OVX rats. Source: Adapted from Suzuki et al. [137].

three clear peaks were obtained in the serum of ovariectomized Sprague Dawley rats after administration of triple-pulse iontophoretic of hPTH(1–34) (doses: 40–400 µg/patch), performed by repeated 30-minute applications of a 0.1 mA.cm^{-2} current separated by 45-minute rest intervals, as shown in Figure 14.15.

It was originally thought that the relative contribution of electroosmosis became the predominant electrotransport mechanism over electromigration for compounds with high molecular weight (>3–4 kDa) [134, 136, 138, 139]. However, it turned out that the situation was more complex: electromigration has shown to be the predominant electrotransport mechanism in the successful iontophoretic delivery of cytochrome C (12.4 kDa) [140], RNase A (13.6 kDa) [140, 141], and human basic fibroblast growth factor (17.4 kDa), which were positively charged, and anionic RNase T1 (11.1 kDa) [142]. The experiments with the latter three proteins also showed that biological activity was retained post-iontophoresis. Moreover, a recent study has demonstrated that cetuximab (CTX) – a 152 kDa monoclonal antibody – could be delivered into the skin by iontophoresis [143], and this is predominantly based on electroosmosis (Figure 14.16a). Nevertheless, more studies are needed in order to elucidate the role of protein structure and the three-dimensional distribution of physicochemical properties on protein electrotransport.

Despite the potential applications and the commercial success of several iontophoretic devices, some of the FDA-approved products were withdrawn from the market due to some side effects, malfunction, and/or profitability reasons. LidoSite™, an epinephrine, lidocaine HCl Iontophoretic Patch was used to provide a fast (in 10 minutes) local anesthesia on intact skin but was withdrawn after only two years of being on the market due to limited commercial success. The Zecuity® sumatriptan iontophoretic transdermal system (Figure 14.16b) (Teva Pharmaceuticals, Israel) was suspected to cause burns in 2016 and discontinued in 2020 for safety reasons [145]. Likewise, IONSYS®, a fentanyl iontophoretic transdermal patch, used for acute postoperative pain (Figure 14.16c) has also been withdrawn from the market (twice) due to malfunction in the system (corrosion in the circuit system was identified in one of the lots of iontophoretic fentanyl patches)

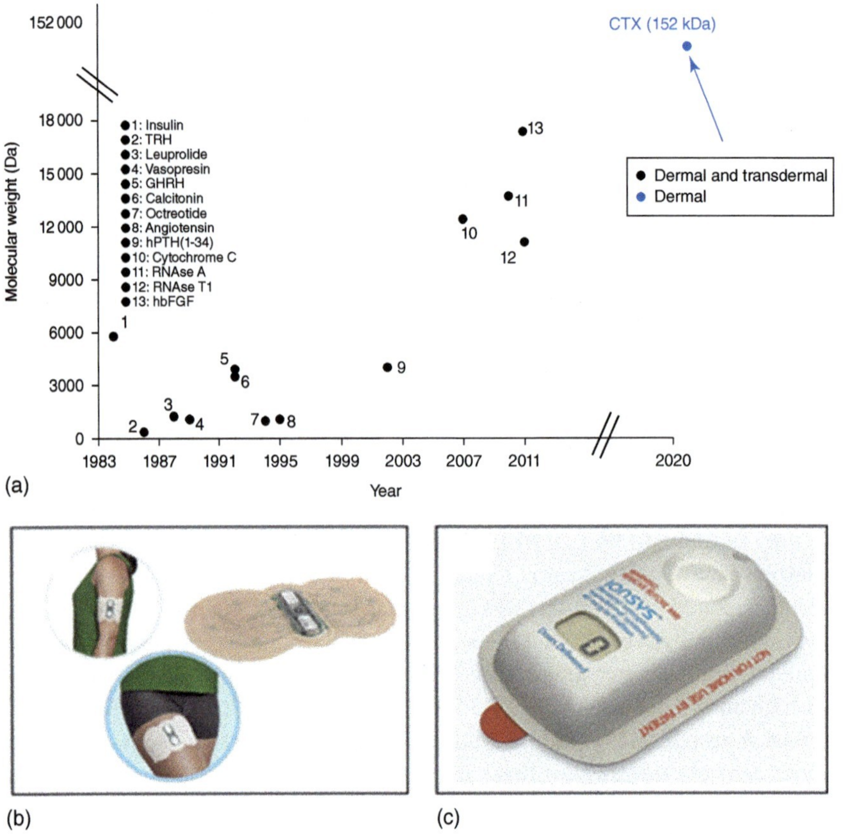

Figure 14.16 (a) A chronology of the studies on the iontophoretic delivery: evolution of the size as a function of time and iontophoretic devices. (b) Zecuity; (c) Ionsys®. Source: (a) Adapted from Dubey et al. [139]. (b) and (c) Adapted with permission from Bakshi et al. [144]/Adapted with permission Elsevier.

and business reasons. GlucoWatch®, an automatic noninvasive monitoring glucose device based on reverse iontophoresis to extract glucose through intact skin and provide glucose readings every 20 minutes for 12 hours, was discontinued from the market due to side effects such as mild burn and skin irritation and after complaints about its accuracy [144].

14.5 Conclusions

Chronic diseases require a long-term administration of drugs. Many efficient and safe drug molecules are available to treat a wide range of chronic diseases. However, despite a molecule possessing all the relevant efficacy and safety features, unless its administration can achieve the right concentration at the target at the right time, it will remain a suboptimal therapeutic agent. Extracorporeal or implantable active drug delivery devices of different sizes are able to deliver the drugs following complex infusion kinetics. The integration of sensors, closed-loop algorithms, and constant miniaturization significantly improves the patient's quality of life. A combination of drugs, smart materials, nanotechnology, and artificial intelligence has become a revolutionary alternative to address unmet medical needs, completely controlling drug release for on-demand administration (e.g., pain management) or sensor triggered administration (e.g., insulin infusion), not to mention the reduction of systemic drug-side effects by localized drug release. Thus, the use of these systems can drastically increase the efficacy of current treatments.

The specific implantation sites can yield targeted and localized drug delivery thus further reducing off-target side effects. From a technological point of view, it may not be utopic to say that in the near future chronically ill patients may entirely rely on an active drug delivery device to manage their condition; however, the low economic profitability of these devices, ever-increasing development, acquisition, and functioning costs still constitute an obstacle to their broad distribution.

Acknowledgments

The authors thank the University of Geneva for providing teaching assistantships for J.F.B., P.F.S., and M.L. M.L. also thanks the Neumann family from the Diabetesmuseum München, Veldener Str. 136, 81241 München, Germany, for providing illustrations of insulin pumps and enlightening discussions.

List of Abbreviations

AMP	anterior MicroPump
ATP	adenosine triphosphate
CFC	chlorofluorocarbon
CGM	continuous glucose monitoring
CSII	continuous subcutaneous infusion of insulin
DBLG1	Diabeloop Generation 1
DIY-APS	do-it-yourself artificial pancreas
DM	diabetes mellitus
EM	electromigration
EO	electroosmosis
FDA	Food and Drug Administration
GI	gastro-intestinal
hPTH(1–34)	human parathyroid hormone fragment (1–34)
i.c.v.	intracerebroventricular
IT	intrathecal
IV	intravenous
LED	light-emitting diode
MEMS	micro-electromechanically systems

MRI	magnetic resonance imaging
nDS	nanochannel delivery system
NEMS	nano-electromechanical systems
nViSTA	nanofluidic Vitreal System for Therapeutic Administration
PMP	posterior MicroPump
PPy	polypyrrole
SAM	self-assembled monolayers
SEM	scanning electron microscopy
SSA	sulfonic acid
TEM	transmission electron microscopy
TIR	time-in-range
WHO	World Health Organization

References

1 Pons-Faudoa, F.P., Ballerini, A., Sakamoto, J., and Grattoni, A. (2019). Advanced implantable drug delivery technologies: transforming the clinical landscape of therapeutics for chronic diseases. *Biomed. Microdevices* 21 (2): 47.

2 WHO (2005). Overview – preventing chronic diseases: a vital investment. https://www.who.int/chp/chronic_disease_report/part1/en/index1.html (accessed 27 April 2022).

3 Beg, S., Swain, S., Rizwan, M. et al. (2011). Bioavailability enhancement strategies: basics, formulation approaches and regulatory considerations. *Curr. Drug Delivery* 8 (6): 691–702.

4 Bolash, R.B., Niazi, T., Kumari, M. et al. (2018). Efficacy of a targeted drug delivery on-demand bolus option for chronic pain. *Pain Pract.* 18 (3): 305–313.

5 Yu, J., Zhang, Y., Yan, J. et al. (2018). Advances in bioresponsive closed-loop drug delivery systems. *Int. J. Pharm.* 544 (2): 350–357.

6 Chau, C.H., Steeg, P.S., and Figg, W.D. (2019). Antibody–drug conjugates for cancer. *Lancet* 394 (10200): 793–804.

7 Kim, J. and Parish, A.L. (2017). Polypharmacy and medication management in older adults. *Nurs. Clin. North Am.* 52 (3): 457–468.

8 Peeples, L. (2018). Medicine's secret ingredient – it's in the timing. *Nature* 556 (7701): 290–292.

9 Stewart, S.A., Domínguez-Robles, J., Donnelly, R.F., and Larrañeta, E. (2018). Implantable polymeric drug delivery devices: classification, manufacture, materials, and clinical applications. *Polymers (Basel)* 10 (12): 1379.

10 Guillot, A.J., Cordeiro, A.S., Donnelly, R.F. et al. (2020). Microneedle-based delivery: an overview of current applications and trends. *Pharmaceutics* 12 (6): 569.

11 Logomasini, M.A., Stout, R.R., and Marcinkoski, R. (2013). Jet injection devices for the needle-free administration of compounds, vaccines, and other agents. *Int. J. Pharm. Compd.* 17 (4): 270–280.

12 FDA (2018). Infusion pumps 2018. https://www.fda.gov/medical-devices/general-hospital-devices-and-supplies/infusion-pumps (accessed 27 April 2022).

13 Allen EE (1881). Instrument for transfusion of blood. Patent US249285A.

14 FDA. FDA-approved drugs. https://www.accessdata.fda.gov/scripts/cder/daf/index.cfm (accessed 27 April 2022).

15 ClinicalTrial.gov. (2015) Study on tolerability of repeat i.c.v. administration of sNN0031 infusion solution in patients with PD. https://clinicaltrials.gov/ct2/show/NCT02408562?term=medtronic+synchromed&draw=2&rank=1 (accessed 27 April 2022).

16 Paul, G., Zachrisson, O., Varrone, A. et al. (2015). Safety and tolerability of intracerebroventricular PDGF-BB in Parkinson's disease patients. *J. Clin. Invest.* 125 (3): 1339–1346.

17 Nyholm, D., Nilsson Remahl, A.I., Dizdar, N. et al. (2005). Duodenal levodopa infusion monotherapy vs oral polypharmacy in advanced Parkinson disease. *Neurology* 64 (2): 216–223.

18 Zulli, C., Sica, M., De Micco, R. et al. (2016). Continuous intra jejunal infusion of levodopa-carbidopa intestinal gel by jejunal extension tube placement through percutaneous endoscopic gastrostomy for patients with advanced Parkinson's disease: a preliminary study. *Eur. Rev. Med. Pharmacol. Sci.* 20 (11): 2413–2417.

19 Ethans, K.D., Schryvers, O.I., Nance, P.W., and Casey, A.R. (2005). Intrathecal drug therapy using the Codman Model 3000 Constant Flow Implantable Infusion Pumps: experience with 17 cases. *Spinal Cord* 43 (4): 214–218.

20 Rauck, R., Deer, T., Rosen, S. et al. (2010). Accuracy and efficacy of intrathecal administration of morphine sulfate for treatment of intractable pain using the Prometra® Programmable Pump. *Neuromodulation* 13 (2): 102–108.

21 Flowonix. (2016) Prometra® and Prometra® II programmable pumps magnetic resonance imaging (MRI) safety information 2016. https://flowonix.com/sites/default/files/pl-15200-02_-_prometra_and_prometra_ii_programmable_pumps_mri_scan_instructions.pdf (accessed 27 April 2022).

22 Henry, R.R., Rosenstock, J., Denham, D.S. et al. (2018). Clinical impact of ITCA 650, a novel drug-device GLP-1 receptor agonist, in uncontrolled type 2 diabetes and very high baseline HbA(1c): the FREEDOM-1 HBL (high baseline) study. *Diabetes Care* 41 (3): 613–619.

23 GlobalData GlobalData Healthcare Healthcare. (2020) Intarcia's diabetes drug implant ITCA 650 receives second FDA rejection. https://www.pharmaceutical-technology.com/comment/intarcia-itca650-fda-rejection (accessed 14 July 2020).

24 Nickel, J.C., Jain, P., Shore, N. et al. (2012). Continuous intravesical lidocaine treatment for interstitial cystitis/bladder pain syndrome: safety and efficacy of a new drug delivery device. *Sci. Transl. Med.* 4 (143): 143ra00.

25 ClinicalTrial.gov. https://clinicaltrials.gov.

26 Cutie C, Efros M, Sobol J, Gilleran J et al. (2019). Continuous intravesical delivery of trospium chloride significantly improves OAB symptoms: results of a phase 1b study. *49th ICS Annual Meeting*, Gothenburg, Sweden (3–6 September 2019).

27 DeFronzo, R.A. (2015). Pathogenesis of type 2 diabetes mellitus. In: *International Textbook of Diabetes Mellitus*, 371–400.

28 Pugliese, A. (2015). Immunopathogenesis of type 1 diabetes in Western society. In: *International Textbook of Diabetes Mellitus*, 442–453.

29 Chatterjee, S., Khunti, K., and Davies, M.J. (2017). Type 2 diabetes. *Lancet* 389 (10085): 2239–2251.

30 Scheen, A.J. (2017). Pharmacotherapy of "treatment resistant" type 2 diabetes. *Expert Opin. Pharmacother.* 18 (5): 503–515.

31 Daneman, D. (2006). Type 1 diabetes. *Lancet* 367 (9513): 847–858.

32 Tibaldi, J.M. (2014). Evolution of insulin: from human to analog. *Am. J. Med.* 127 (10 Suppl): S25–S38.

33 Rosenfeld, L. (2002). Insulin: discovery and controversy. *Clin. Chem.* 48 (12): 2270–2288.

34 Leal, M.C. and Morelli, L. (2013). Chapter 318 - Insulysin. In: *Handbook of Proteolytic Enzymes*, 3e (ed. N.D. Rawlings and G. Salvesen), 1415–1420. Academic Press.

35 Eaton, R.P., Allen, R.C., Schade, D.S., and Standefer, J.C. (1980). "Normal" insulin secretion: the goal of artificial insulin delivery systems? *Diabetes Care* 3 (2): 270–273.

36 Hilgenfeld, R., Seipke, G., Berchtold, H., and Owens, D.R. (2014). The evolution of insulin glargine and its continuing contribution to diabetes care. *Drugs* 74 (8): 911–927.

37 Skyler, J.S. (2010). Continuous subcutaneous insulin infusion – an historical perspective. *Diabetes Technol. Ther.* 12 (Suppl 1): S5–S9.

38 Tattersall, R.B. (1994). The quest for normoglycaemia: a historical perspective. *Diabet. Med.* 11 (7): 618–635.

39 Kadish, A.H. (1963). Physiologic monitoring of blood glucose. *Calif. Med.* 98 (6): 325–327.

40 Hepp, K.D., Renner, R., Piwernetz, K., and Mehnert, H. (1980). Control of insulin-dependent diabetes with portable miniaturized infusion systems. *Diabetes Care* 3 (2): 309–313.

41 Kadish, A.H. (1964). Automation control of blood sugar. I. A servomechanism for glucose monitoring and control. *Am. J. Med. Electron.* 3: 82–86.

42 Slama, G., Hautecouverture, M., Assan, R., and Tchobroutsky, G. (1974). One to five days of continuous intravenous insulin infusion on seven diabetic patients. *Diabetes* 23 (9): 732–738.

43 Deckert, T. and Lorup, B. (1976). Regulation of brittle diabetics by a pre-planned insulin infusion programme. *Diabetologia* 12 (6): 573–579.

44 Pickup, J.C., Keen, H., Parsons, J.A., and Alberti, K.G. (1978). Continuous subcutaneous insulin infusion: an approach to achieving normoglycaemia. *Br. Med. J.* 1 (6107): 204–207.

45 Teuscher, A. (2007). The history of insulin. In: *Insulin – A Voice for Choice* (ed. A. Teuscher), 10–13. Karger.

46 Tamborlane, W.V., Sherwin, R.S., Genel, M., and Felig, P. (1979). Reduction to normal of plasma glucose in juvenile diabetes by subcutaneous administration of insulin with a portable infusion pump. *N. Engl. J. Med.* 300 (11): 573–578.

47 Renner, R., Hepp, K.D., Mehnert, H., and Franetzki, M. (1979). Continuous intravenous insulin therapy with a miniaturized open-loop system. *Horm. Metab. Res. Suppl.* 8: 186–190.

48 Irsigler, K. and Kritz, H. (1979). Long-term continuous intravenous insulin therapy with a portable insulin dosage-regulating apparatus. *Diabetes* 28 (3): 196–203.

49 Alsaleh, F.M., Smith, F.J., Keady, S., and Taylor, K.M. (2010). Insulin pumps: from inception to the present and toward the future. *J. Clin. Pharm. Ther.* 35 (2): 127–138.

50 Diabetesmuseum München. Insulin pumps. https://diabetesmuseum.de/insulinpumpe (accessed 27 April 22).

51 Insulet Corporation. https://www.insulet.com.

52 van Dijk, P.R., Logtenberg, S.J.J., Gans, R.O.B. et al. (2014). Intraperitoneal insulin infusion: treatment option for type 1 diabetes resulting in beneficial endocrine effects beyond glycaemia. *Clin. Endocrinol.* 81 (4): 488–497.

53 Press, M. (2009). Glucose sensors and insulin pumps: prospects for an artificial pancreas. In: *Artificial Organs* (ed. N.S. Hakim), 77–91. London: Springer.

54 Elke, A., Tilman, S., Pia, M.H. et al. (2019). *Insulin Pump Therapy*. Berlin, Boston: De Gruyter.

55 Bailey, C.J. and Grant, P.J. (1998). The UK prospective diabetes study. *Lancet* 352 (9144): 1932.

56 Nathan, D.M., Genuth, S., Lachin, J. et al. (1993). The effect of intensive treatment of diabetes on the development and progression of long-term complications in insulin-dependent diabetes mellitus. *N. Engl. J. Med.* 329 (14): 977–986.

57 Weissberg-Benchell, J., Antisdel-Lomaglio, J., and Seshadri, R. (2003). Insulin pump therapy. A meta-analysis. *Diabetes Care* 26 (4): 1079–1087.

58 Kesavadev, J., Saboo, B., Krishna, M.B., and Krishnan, G. (2020). Evolution of insulin delivery devices: from syringes, pens, and pumps to DIY artificial pancreas. *Diabetes Ther.* 11 (6): 1251–1269.

59 Heinemann, L. and Krinelke, L. (2012). Insulin infusion set: the Achilles heel of continuous subcutaneous insulin infusion. *J. Diabetes Sci. Technol.* 6 (4): 954–964.

60 Ginsberg, B.H. (2019). Patch pumps for insulin. *J. Diabetes Sci. Technol.* 13 (1): 27–33.

61 Dirnena-Fusini, I., Åm, M.K., Fougner, A.L. et al. (2021). Physiological effects of intraperitoneal versus subcutaneous insulin infusion in patients with diabetes mellitus type 1: a systematic review and meta-analysis. *PLoS One* 16 (4): e0249611.

62 Cobelli, C., Renard, E., and Kovatchev, B. (2011). Artificial pancreas: past, present, future. *Diabetes* 60 (11): 2672–2682.

63 Albisser, A.M., Leibel, B.S., Ewart, T.G. et al. (1974). An artificial endocrine pancreas. *Diabetes* 23 (5): 389–396.
64 Pfeiffer, E.F., Thum, C., and Clemens, A.H. (1974). The artificial beta cell – a continuous control of blood sugar by external regulation of insulin infusion (glucose controlled insulin infusion system). *Horm. Metab. Res.* 6 (5): 339–342.
65 Kraegen, E.W., Campbell, L.V., Chia, Y.O. et al. (1977). Control of blood glucose in diabetics using an artificial pancreas. *Aust. N. Z. J. Med.* 7 (3): 280–286.
66 Mirouze, J., Selam, J.L., Pham, T.C., and Cavadore, D. (1977). Evaluation of exogenous insulin homoeostasis by the artificial pancreas in insulin-dependent diabetes. *Diabetologia* 13 (3): 273–278.
67 Young, A. and Herf, S. (1984). Biostator glucose controller: a building block of the future. *Diabetes Educ.* 10 (2): 11–12.
68 Clemens, A.H., Chang, P.H., and Myers, R.W. (1977). The development of Biostator, a glucose controlled insulin infusion system (GCIIS). *Horm. Metab. Res.* 7: 23–33.
69 Coffel, J. and Nuxoll, E. (2018). BioMEMS for biosensors and closed-loop drug delivery. *Int. J. Pharm.* 544 (2): 335–349.
70 Steil, G.M., Rebrin, K., Darwin, C. et al. (2006). Feasibility of automating insulin delivery for the treatment of type 1 diabetes. *Diabetes* 55 (12): 3344–3350.
71 Janež, A., Guja, C., Mitrakou, A. et al. (2020). Insulin therapy in adults with type 1 diabetes mellitus: a narrative review. *Diabetes Ther.* 11 (2): 387–409.
72 Sharifi, A., De Bock, M.I., Jayawardene, D. et al. (2016). Glycemia, treatment satisfaction, cognition, and sleep quality in adults and adolescents with type 1 diabetes when using a closed-loop system overnight versus sensor-augmented pump with low-glucose suspend function: a randomized crossover study. *Diabetes Technol. Ther.* 18 (12): 772–783.
73 Benhamou, P.-Y., Franc, S., Reznik, Y. et al. (2019). Closed-loop insulin delivery in adults with type 1 diabetes in real-life conditions: a 12-week multicentre, open-label randomised controlled crossover trial. *Lancet Digital Health* 1 (1): e17–e25.
74 FDA (2016). FDA expands indication for continuous glucose monitoring system, first to replace fingerstick testing for diabetes treatment decisions 2016. https://www.fda.gov/news-events/press-announcements/fda-expands-indication-continuous-glucose-monitoring-system-first-replace-fingerstick-testing (accessed 27 April 22).
75 Olczuk, D. and Priefer, R. (2018). A history of continuous glucose monitors (CGMs) in self-monitoring of diabetes mellitus. *Diabetes Metab. Syndr.* 12 (2): 181–187.
76 GlobeNewswire. (2020) Medtronic Presents U.S. Pivotal Trial Data for MiniMed™ 780G Advanced Hybrid Closed Loop System with Automated Correction Bolus Feature. https://www.globenewswire.com/news-release/2020/06/12/2047654/0/en/Medtronic-Presents-U-S-Pivotal-Trial-Data-for-MiniMed-780G-Advanced-Hybrid-Closed-Loop-System-with-Automated-Correction-Bolus-Feature.html (accessed 27 April 2022).

77 Mortellaro, M. and DeHennis, A. (2014). Performance characterization of an abiotic and fluorescent-based continuous glucose monitoring system in patients with type 1 diabetes. *Biosens. Bioelectron.* 61: 227–231.

78 Colvin, A.E. and Jiang, H. (2013). Increased in vivo stability and functional lifetime of an implantable glucose sensor through platinum catalysis. *J. Biomed. Mater. Res. Part A* 101 (5): 1274–1282.

79 FDA 2022. Eversense continuous glucose montioring system – P160048 2018. https://www.fda.gov/medical-devices/eversense-e3-continuous-glucose-monitoring-system-p160048s016 (accessed 27 April 2022).

80 Dehennis, A., Mortellaro, M.A., and Ioacara, S. (2015). Multisite study of an implanted continuous glucose sensor over 90 days in patients with diabetes mellitus. *J. Diabetes Sci. Technol.* 9 (5): 951–956.

81 Kropff, J., Choudhary, P., Neupane, S. et al. (2017). Accuracy and longevity of an implantable continuous glucose sensor in the PRECISE study: a 180-day, prospective, multicenter pivotal trial. *Diabetes Care* 40 (1): 63–68.

82 The Open Artificial Pancreas System project. https://openaps.org/what-is-openaps (accessed 14 July 2020).

83 Bajaj, H.S., Isendahl, J., Gowda, A., Stachlewska, K. et al. (2020). Efficacy and safety of switching to insulin icodec, a once-weekly basal insulin, vs insulin glargine U100 in patients with T2D inadequately controlled on OADs and basal insulin. 56e réunion annuelle 2020 de l'Association européenne pour l'étude du diabète (EASD) (22 September 2020).

84 Hussain, M.A. and Theise, N.D. (2004). Stem-cell therapy for diabetes mellitus. *Lancet* 364 (9429): 203–205.

85 Sabek, O.M., Ferrati, S., Fraga, D.W. et al. (2013). Characterization of a nanogland for the autotransplantation of human pancreatic islets. *Lab Chip* 13 (18): 3675–3688.

86 Sabek, O.M., Farina, M., Fraga, D.W. et al. (2016). Three-dimensional printed polymeric system to encapsulate human mesenchymal stem cells differentiated into islet-like insulin-producing aggregates for diabetes treatment. *J. Tissue Eng.* 7: 2041731416638198.

87 Vaithilingam, V., Bal, S., and Tuch, B.E. (2017). Encapsulated islet transplantation: where do we stand? *Rev. Diabet. Stud.* 14 (1): 51–78.

88 Maloney, J.M., Uhland, S.A., Polito, B.F. et al. (2005). Electrothermally activated microchips for implantable drug delivery and biosensing. *J. Controlled Release* 109 (1–3): 244–255.

89 Kumar, A. and Pillai, J. (2018). Implantable drug delivery systems: an overview. In: *Nanostructures for the Engineering of Cells, Tissues and Organs* (ed. A.M. Grumezescu), 473–511. William Andrew Publishing.

90 Staples, M. (2010). Microchips and controlled-release drug reservoirs. *Wiley Interdiscip. Rev. Nanomed. Nanobiotechnol.* 2 (4): 400–417.

91 Bhushan, B. (2007). Nanotribology and nanomechanics of MEMS/NEMS and BioMEMS/BioNEMS materials and devices. *Microelectron. Eng.* 84 (3): 387–412.

92 Santini, J.T. Jr.,, Cima, M.J., and Langer, R. (1999). A controlled-release microchip. *Nature* 397 (6717): 335–338.

93 Prescott, J.H., Lipka, S., Baldwin, S. et al. (2006). Chronic, programmed polypeptide delivery from an implanted, multireservoir microchip device. *Nat. Biotechnol.* 24 (4): 437–438.

94 Jain, D., Mahammad, S.S., Singh, P.P., and Kodipyaka, R. (2019). A review on parenteral delivery of peptides and proteins. *Drug Dev. Ind. Pharm.* 45 (9): 1403–1420.

95 Farra, R., Sheppard, N.F. Jr.,, McCabe, L. et al. (2012). First-in-human testing of a wirelessly controlled drug delivery microchip. *Sci. Transl. Med.* 4 (122): 122ra21.

96 Darebioscience. (2019) News-releases. https://darebioscience.com/microchips-biotech (accessed 27 April 2022).

97 Ge, D., Tian, X., Qi, R. et al. (2009). A polypyrrole-based microchip for controlled drug release. *Electrochim. Acta* 55: 271–275.

98 Piveteau, L.-D. (2013). Disposable patch pump for accurate delivery. *ONdrug-Delivery* (44): 16.

99 Debiotech S.A. (2020) Nanopump – Programmable Infusion Pump 2020. https://www.debiotech.com/page/index.php?page=product_01&id=2&id_prod=50 (19 August 2020).

100 Debiotech S.A. (2022) JewelPUMP - Continuous Subcutaneous Insulin Infusion. https://www.debiotech.com/page/index.php?page=product_01&id=1&id_prod=34#type=Brief (accessed 27 April 2022).

101 Borot, S., Franc, S., Cristante, J. et al. (2014). Accuracy of a new patch pump based on a microelectromechanical system (MEMS) compared to other commercially available insulin pumps: results of the first in vitro and in vivo studies. *J. Diabetes Sci. Technol.* 8 (6): 1133–1141.

102 Noth, A. and Chappel, E. (2016). Electronic control method and system for a piezo-electric pump patent. US Patent 9,316,220 B2, filed 19 December 2011 and issued 19 April 2016.

103 Bianchi, F. (2011). Container for storing a drug such as insulin patent. US Patent 2014/0166528 A1, filed 26 July 2012 and issued 19 June 2014.

104 Saati, S., Lo, R., Li, P.-Y. et al. (2009). Mini drug pump for ophthalmic use. *Trans. Am. Ophthalmol. Soc.* 107: 60–70.

105 Gutiérrez-Hernández, J.-C., Caffey, S., Abdallah, W. et al. (2014). One-year feasibility study of replenish MicroPump for intravitreal drug delivery: a pilot study. *Transl. Vision Sci. Technol.* 3 (4): 1.

106 Humayun, M., Santos, A., Altamirano, J.C. et al. (2014). Implantable MicroPump for drug delivery in patients with diabetic macular edema. *Transl. Vision Sci. Technol.* 3 (6): 5.

107 FDA (2020). CardioMEMS HF pressure measurement system. https://www.accessdata.fda.gov/scripts/cdrh/cfdocs/cfpma/pma.cfm?id=P100045 (accessed 27 April 2022).

108 Abraham, W.T., Adamson, P.B., Bourge, R.C. et al. (2011). Wireless pulmonary artery haemodynamic monitoring in chronic heart failure: a randomised controlled trial. *Lancet* 377 (9766): 658–666.

109 Elman, N.M., Ho Duc, H.L., and Cima, M.J. (2009). An implantable MEMS drug delivery device for rapid delivery in ambulatory emergency care. *Biomed. Microdevices* 11 (3): 625–631.

110 Hafezi, H., Robertson, T.L., Moon, G.D. et al. (2015). An ingestible sensor for measuring medication adherence. *IEEE Trans. Biomed. Eng.* 62 (1): 99–109.

111 FDA (2017). FDA approves pill with sensor that digitally tracks if patients have ingested their medication. https://www.fda.gov/news-events/press-announcements/fda-approves-pill-sensor-digitally-tracks-if-patients-have-ingested-their-medication (accessed 27 April 22).

112 Kwiatek, M.A. and Pandolfino, J.E. (2007). Prolonged reflux monitoring: capabilities of bravo pH and impedance-pH systems. *Curr. GERD Rep.* 1 (3): 165–170.

113 Jin, J. and Sánchez-Sinencio, E. (2015). A home sleep apnea screening device with time-domain signal processing and autonomous scoring capability. *IEEE Trans. Biomed. Circuits Syst.* 9 (1): 96–104.

114 Chow, E.Y., Chlebowski, A.L., Chakraborty, S. et al. (2010). Fully wireless implantable cardiovascular pressure monitor integrated with a medical stent. *IEEE Trans. Biomed. Eng.* 57 (6): 1487–1496.

115 Takahata, K., DeHennis, A., Wise, K.D., and Gianchandani, Y.B. (2003). Stentenna: a micromachined antenna stent for wireless monitoring of implantable microsensors. *Proceedings of the 25th Annual International Conference of the IEEE Engineering in Medicine and Biology Society (IEEE Cat No03CH37439)* (17–21 September 2003).

116 Ferrati, S., Fine, D., You, J. et al. (2013). Leveraging nanochannels for universal, zero-order drug delivery in vivo. *J. Controlled Release* 172 (3): 1011–1019.

117 Bruno, G., Di Trani, N., Hood, R.L. et al. (2018). Unexpected behaviors in molecular transport through size-controlled nanochannels down to the ultra-nanoscale. *Nat. Commun.* 9 (1): 1682.

118 Grattoni, A., Fine, D., Zabre, E. et al. (2011). Gated and near-surface diffusion of charged fullerenes in nanochannels. *ACS Nano* 5 (12): 9382–9391.

119 Hood, R.L., Bruno, G., Jain, P. et al. (2016). Nanochannel implants for minimally-invasive insertion and intratumoral delivery. *J. Biomed. Nanotechnol.* 12 (10): 1907–1915.

120 Sih, J., Bansal, S.S., Filippini, S. et al. (2013). Characterization of nanochannel delivery membrane systems for the sustained release of resveratrol and atorvastatin: new perspectives on promoting heart health. *Anal. Bioanal.Chem.* 405 (5): 1547–1557.

121 Ferrati, S., Nicolov, E., Zabre, E. et al. (2015). The nanochannel delivery system for constant testosterone replacement therapy. *J. Sex Med.* 12 (6): 1375–1380.

122 Trani, N.D., Grattoni, A., and Ferrari, M. (2017). Nanofluidics for cell and drug delivery. IEEE International Electron Devices Meeting (IEDM) (2–6 December 2017).

123 Di Trani, N., Jain, P., Chua, C.Y.X. et al. (2019). Nanofluidic microsystem for sustained intraocular delivery of therapeutics. *Nanomedicine* 16: 1–9.

124 Sanjay, S.T., Zhou, W., Dou, M. et al. (2018). Recent advances of controlled drug delivery using microfluidic platforms. *Adv. Drug Delivery Rev.* 128: 3–28.

125 Bruno, G., Geninatti, T., Hood, R.L. et al. (2015). Leveraging electrokinetics for the active control of dendritic fullerene-1 release across a nanochannel membrane. *Nanoscale* 7 (12): 5240–5248.

126 Fine, D., Grattoni, A., Zabre, E. et al. (2011). A low-voltage electrokinetic nanochannel drug delivery system. *Lab Chip* 11 (15): 2526–2534.

127 Bruno, G., Canavese, G., Liu, X. et al. (2016). The active modulation of drug release by an ionic field effect transistor for an ultra-low power implantable nanofluidic system. *Nanoscale* 8 (44): 18718–18725.

128 Di Trani, N., Silvestri, A., Bruno, G. et al. (2019). Remotely controlled nanofluidic implantable platform for tunable drug delivery. *Lab Chip* 19 (13): 2192–2204.

129 Barkam, S., Saraf, S., and Seal, S. (2013). Fabricated micro-nano devices for in vivo and in vitro biomedical applications. *Wiley Interdiscip. Rev. Nanomed. Nanobiotechnol.* 5 (6): 544–568.

130 Caldorera-Moore, M. and Peppas, N.A. (2009). Micro- and nanotechnologies for intelligent and responsive biomaterial-based medical systems. *Adv. Drug Delivery Rev.* 61 (15): 1391–1401.

131 Bixler, G.D. and Bhushan, B. (2012). Biofouling: lessons from nature. *Philos. Trans. R. Soc. London, Ser. A* 370 (1967): 2381–2417.

132 Gooding, J.J., Erokhin, P., Losic, D. et al. (2001). Parameters important in fabricating enzyme electrodes using self-assembled monolayers of alkanethiols. *Anal. Sci.* 17 (1): 3–9.

133 Schmaltz, T., Sforazzini, G., Reichert, T., and Frauenrath, H. (2017). Self-assembled monolayers as patterning tool for organic electronic devices. *Adv. Mater.* 29 (18): 1605286.

134 Kalia, Y.N., Naik, A., Garrison, J., and Guy, R.H. (2004). Iontophoretic drug delivery. *Adv. Drug Delivery Rev.* 56 (5): 619–658.

135 Patel, S.R., Zhong, H., Sharma, A., and Kalia, Y.N. (2009). Controlled non-invasive transdermal iontophoretic delivery of zolmitriptan hydrochloride in vitro and in vivo. *Eur. J. Pharm. Biopharm.* 72 (2): 304–309.

136 Marro, D., Guy, R.H., and Delgado-Charro, M.B. (2001). Characterization of the iontophoretic permselectivity properties of human and pig skin. *J. Controlled Release* 70 (1–2): 213–217.

137 Suzuki, Y., Nagase, Y., Iga, K. et al. (2002). Prevention of bone loss in ovariectomized rats by pulsatile transdermal iontophoretic administration of human PTH(1–34). *J. Pharm. Sci.* 91 (2): 350–361.

138 Lu, M.F., Lee, D., Carlson, R. et al. (1993). The effects of formulation variables on iontophoretic transdermal delivery of leuprolide to humans. *Drug Dev. Ind. Pharm.* 19 (13): 1557–1571.

139 Dubey, S., Perozzo, R., Scapozza, L., and Kalia, Y.N. (2011). Noninvasive transdermal iontophoretic delivery of biologically active human basic fibroblast growth factor. *Mol. Pharmaceutics* 8 (4): 1322–1331.

140 Cázares-Delgadillo, J., Naik, A., Ganem-Rondero, A. et al. (2007). Transdermal delivery of cytochrome C – a 12.4 kDa protein – across intact skin by constant-current iontophoresis. *Pharm. Res.* 24 (7): 1360–1368.

141 Dubey, S. and Kalia, Y.N. (2010). Non-invasive iontophoretic delivery of enzymatically active ribonuclease A (13.6 kDa) across intact porcine and human skins. *J. Controlled Release* 145 (3): 203–209.

142 Dubey, S. and Kalia, Y.N. (2011). Electrically-assisted delivery of an anionic protein across intact skin: cathodal iontophoresis of biologically active ribonuclease T1. *J. Controlled Release* 152 (3): 356–362.

143 Lapteva, M., Sallam, M.A., Goyon, A. et al. (2020). Non-invasive targeted iontophoretic delivery of cetuximab to skin. *Expert Opin Drug Delivery* 17 (4): 589–602.

144 Bakshi, P., Vora, D., Hemmady, K., and Banga, A.K. (2020). Iontophoretic skin delivery systems: success and failures. *Int. J. Pharm.* 586: 119584.

145 FDA (2020). FDA Drug Safety Communication: FDA evaluating the risk of burns and scars with Zecuity (sumatriptan) migraine patch. https://www.fda.gov/drugs/drug-safety-and-availability/fda-evaluating-risk-burns-and-scars-zecuity-sumatriptan-migraine-patch (accessed 27 April 2022).

15

Drug Delivery to the Brain: Targeting Technologies to Deliver Therapeutics to Brain Lesions

Nishit Pathak, Sunil K. Vimal, Cao Hongyi, and Sanjib Bhattacharya

Southwest University, Department of Pharmaceutical Science, 2-Tiansheng Road, Beibei, Chongqing 400715, PR China

15.1 Introduction

In 1959, the Nobel Prize winner physicist Richard Feynman delivered a lecture in which he illustrated the concept of "nanotechnology," a new technique as "adopting new tools to make smaller machines, at the atomic level." Feynman further discerned the possible use of nanoscience in the stipulation of medicine and its application. Feynman used a nanosized lancet by a nanomechanical surgeon to find out the heart problem and its cure [1]. Till now, nanomedicine have been used for plenty of applications, which deliver small particles. This application has gained the researcher interest and the proceeding of medicine to strengthen the nature of human activity and life. These interests combine exceptional efficiency, accuracy, reliability, adaptability, cost-effectiveness, control, and agility of the drug-delivery system [2, 3]. Nanotechnology methods have exclusively become a part of modern science and were being used in the field of medicine, such as tissue regeneration, infection prevention, cancer therapy, and subsequent cure as therapeutics. In the last century, cause of many neurodegenerative disorders of the central nervous system (CNS) and their pathological mechanisms was not well known, due to plenty of obstacles for diagnosis and subsequent cure [4, 5]. Advancement in nanotechnology brings micro- and macro-shape particles to handle complicated biological systems with higher selectivity and concurrently declines the objectionable oblique things. These advancements will have a dominant impingement in the field of neurology, exclusively prominent to the development of current therapeutic procedures [6]. Nanotechnology uses an advanced system to enhance the delivery of potential small-molecule medicine to pass over the blood–brain barrier (BBB) to cornerstone the neuron regeneration and act as multifunctional cells, neuroprotective cells, and cytoarchitecture brain, specifically by using different kinds of nanoparticles (NPs). Worldwide, central nervous system disorders (CNSDs), such as brain cancer and neurodegenerative diseases, affect millions of diverse range of people.

Targeted Drug Delivery, First Edition. Edited by Yogeshwar Bachhav.

15.2 Brain Tumor

Brain tumor is the most common CNS disease combined with initial metastases and cancer, harmful with tremendous fatality and the rate of suffering population with an approximate frequency of 6/100 000 [7]. As an example, glioblastoma multiforme [8], the maximum number described as dominant benign CNS tumor that exemplifies the most significant cause of cancer in person much below the age of 35 years [8]. World Health Organisation (WHO) officially classifies 120 types of central nervous system tumors based on molecular-level information generated from histology, neuropathology, and neuro-oncology [9]. The brain tumor is least aggressive (benign) to the maximum aggressive (malignant) according to its cell foundation and cell behavioral assets, as shown in Figure 15.1. Amongst brain tumors, gliomas are the most common with at least 80% of them being malignant and approximately 50% of them causing the *most* deaths. Despite the vital purpose of remedying gliomas, the general 5–12 months survival of patients with glioblastomas is much less than 5% [7, 10]. Solid tumors are linked to a variety of neurologic issues. Melanoma is the solid tumor with the greatest proclivity for multiple brain metastases, with up to 80% of patients developing multiple lesions [10]. Glioblastoma and gliomas originate from neuroglial ancestor cells that is inherently oncogenic, thus silently invades into the healthy tissues and makes it impossible for complete surgical resection [11].

US-FDA approved the most effective two medicines for glioma remedy that are particularly antiangiogenic medicine bevacizumab and temozolomide (TMZ), which got approval for clinical use [12]. Furthermore, the cure of malignant brain tumors is rather tricky. Most malignant brain tumors such as glioblastoma, glioma, were dealt in hospitals with the consolidation of surgical procedures, integral chemotherapy and radiotherapy.

Nevertheless, brain tumors were susceptible to persist early for theirs notably infiltrative, proliferative, and invasive nature [13]. Men are more likely to develop brain tumors compared to the women. They are most common in older adults, but can occur at any age. In children under the age of 14, brain tumors are the leading cause of cancer-related death [14]. Most cancer researcher face a huge problem to diagnose

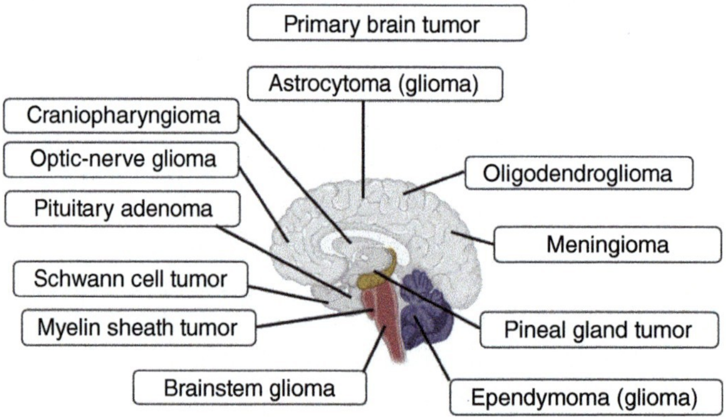

Figure 15.1 Primary common brain tumor types are common in children and adults.

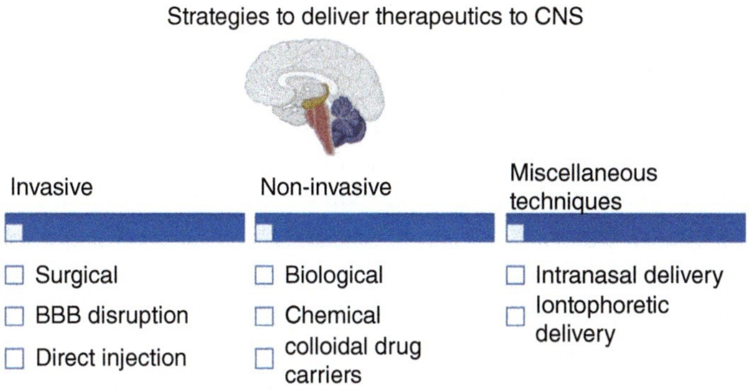

Figure 15.2 New strategies and approaches based on nano pharmaceutical carriers and deliver therapeutics to the central nervous system.

the metastatic disorder due to lack of molecular biomarker and subsequent targeted therapy against carcinogenesis and malignancies [15, 16]. In the majority of human populations, managing CNS disease is exceptionally challenging as the defensive hindrance of the CNS, the blood–cerebrum obstruction and the blood–cerebrospinal liquid boundary keeps up the brain microenvironment and neuronal movement, for legitimate working of the CNS. Transcriptomic and epigenetic studies for effective CNS conveyance have been examined [17]. These methodologies are summed up into three general classifications—noninvasive, obtrusive, and incidental procedures (Figure 15.2).

15.2.1 Obstacles to Brain Tumor-Targeted Delivery

Glioma oncogenesis is complex, with multiple barriers preventing drugs from reaching the tumor sites. The BBB, the blood–brain tumor barrier (BBTB), and a relatively small enhanced permeation and retention (EPR) effect are the three major obstacles to the treatment of the brain tumor. Specific stages of brain tumor growth necessitate barrier-targeting treatment strategies. BBB is the primary barrier to deal with tumor therapy [18, 19]. BBB is made up of microvascular endothelial cells and enclosing perivascular component consisting of the pericytes, basal lamina, astrocytes, and endothelial cells (ECs) that is collectively called as tight junction [20]. The transendothelial district grandstand many specific homes: low penetrability, exorbitant transendothelial electric obstruction, and low abundance of pinocytotic vesicles. The insulin and transferrin macromolecular polypeptides in the blood, with low efficiency and most of the particles remain sequestered intracellularly in the endothelial cells. The mechanism over the endogenous macromolecules that traverse the BBB coupled with nanomedicines and multifunctional transporters fit for targeted delivery conveyance and imaging for pharmaceutical applications [21, 22]. The tight intersections (TJs) of the endothelial cells were juxtaposed with large group transporters, chemicals, efflux and receptors that siphons off the multidrug release pathways distantly and constrain the particles inside the vascular cell to navigate via transcellular or paracellular fashion. Those physiologic, morphologic, and viable attributes to BBB to establish that the exogenous substrates and endogenous substrates do not move easily and are inhibited to reach at brain parenchyma [20, 23]. The BBB further competes with ion flow, and continues to balance for proper axonal and synaptic signalling by using neurotransmitters to assure the delivery of vitamins to the nerve cell. Various receptors and transporters shuttle the hydrophilic small molecule and pass on to the luminal endothelial plasma membranes, such as transferrin receptor low-density lipoprotein, endothelial boom factor, insulin, and glucose. The presence of varieties of receptors and transporters offer ample opportunities as a binding objective for drug-delivery components using active-targeting strategy [21]. BBTB serves as a natural guard for the brain, protecting it from toxic elements in the bloodstream while also providing it with the nutrients it needs to function properly [24, 25]. The difficulty of systemic medicines to efficiently traverse the BBTB is a key barrier and impediment in the development of medications for brain tumors. Drug delivery now focuses on transiently permeabilizing the BBTB or directly bypassing the BBTB via intratumoral or intraventricular injection [25]. The electrophrenic respiration effect in brain tumor tissues occurs as the tumor deteriorates, and nano-drug-delivery vesicles reach the tumor tissue through passive targeting [26]. Research on brain tumor-targeted drug-delivery systems has been explored totally based on these receptors [20–26].

The cerebrum has many ensuring shields, along with the head, meninges, cerebrospinal liquid BBB, and BBTB [27]. With the movement of cerebrum tumors, harmful cells begin to spread encompassing the stromal tissues. The human brain is the most complex phenomenon known in the universe, while the tumor tissue bunch attained of sufficiently immense volume, will be harmed further, and BBTB is undermined. BBTB exists among the cerebrum tumor tissues and fine vessels, forestalling the medication of most extreme hydrophilic particles and antitumor therapy for brain tumors [28]. Different sorts of BBB models have been built up, beginning from less upsetting monolayers of blood vascular endothelial cells (BECs) to higher prominent refined spheroid. In micro-chip style models, the handiest BBB designs can be mounted over segregation of BECs that were developed as a solitary monolayer at the abluminal part of transwell embeds [27, 28]. The precise remedy for brain tumors was based on drug transport capable of localizing, healing, and curing the brain tumor foci and in the meantime diminished the entanglement with peripheral healthy brain tissue [29]. That is why nano vehicles acquired a lot of attention in recent years for their tractable subsequent surface amendment and particle size and for better therapeutic efficacy. This section assesses the ongoing advances in nano-medicine for the cure of brain tumors: (i) BBB focusing on, that trans BBB to tranquilize conveyance and further tumor foci-focusing on sedate medication advantage, for the principal level of a cerebrum tumor while the BBB keeps on being unblemished; (ii) BBTB focusing cerebrum tumor using angiogenesis, and (iii) particle-size-driven control of EPR impact on brain tumor with irregular vascular endothelium.

Currently, several technologies have been developed to screen aptamers or peptides that own high specificity and binding efficiency for the BBB. The GMT-eight aptamer that collectively binds with U87 cells can further be used as a therapy for brain tumors, emblazoning nanocarrier developed for an anti-brain tumor effect [30]. Anticancer peptide nanoparticles (ACP-charged NPs) showed prominent low circulation of NPs in the liver, spleen, coronary heart, and lungs, whereas a lot of dissemination in cerebrum tumors. In this manner, paclitaxel-stacked ACP-charged NPs broadened the mean endurance time of brain tumor-bearing mice to 39% longer compared to low-molecular-weight protamine (LMWP)-charged NPs did. Rather, activatable ligands could be conjugated with PEG over pH-sensitive, esterase-labile, or stimuli-sensitive pharmaceutical nanocarriers [31]. Co-conveying chemotherapeutics with proteins can synergistically upgrade the antitumor effect. Guo et al. typified trail protein and doxorubicin (DOX) in liposomes for the cure of cerebrum tumors [32]. The consolidation of DOX and curcumin nanoformulations showed the highest cell lethality. Incubating U87 cells for 12 hours with 37 $ng\,ml^{-1}$ of reconsolidation protein or 1.0 $\mu g\,ml^{-1}$ of DOX alone did no longer significantly hinder cell blast (restraint sway <25%). The consolidation of these two medications viably hindered cell development (hindrance impact >50%). The middle endurance tumor time of mice with the liposomes-formulated medication becomes half and 23% prolonged than that of mice model with liposome-stacked or DOX-stacked liposomes, individually [33, 34].

15.2.2 Brain-Tumor-Focused Nano-Drug Delivery

Nanoparticles consistently form the pith of nano-bio material. It can be used as an advantageous surface for molecular clusters and functionalizing polymeric materials [35]. It can further be in the pattern of a nano-vesicle surrounded by a layer or a membrane. They are unparallel tools for targeting transporter that has promising applications for the finding and cure of perilous tissue (Figure 15.3). Various assortments of receptors exist on the BBB that can be used as the targeting motif for ligand-mediated trans-BBB transport [36]. Computer-aided drug design (CADD) is classified as structure-based virtual screening (SB–VS) and ligand-based virtual screening (LB–VS). Braim-targeting functional polypeptides are screened with CADD using loop II sequence and loop II tip sequence as the lead [37].

During *in vitro* cellular assay, all the three polypeptides conjugated with fluorescein isothiocyanate (FITC) were exclusively taken up by brain capillary endothelial cell (BCEC) cells, whereas control association for nAChR with high-affinity selectivity were bound to nicotinic acetylcholine receptors in HeLa cells. While during *in vivo* imaging, FITC-label characterized the peptide motif CDX, KC20.2S, and RVG29, all penetrated the cerebrum and kidney, indicating tremendous neuronal target capacity. Among three peptides, CDX was selected for modification with micellar materials poly(ethylene glycol)- poly(lactide) (PEG-PLA) because of its tissue selectivity. For instance, Li and coworkers used a 12-mer phage show peptide library to disconnect a peptide (implied as Pep TGN) for brain targeting. TGN were covalently conjugated with poly(ethylene-col)-poly(lactic-*co*-glycolic destructive) (PEG-PLGA) for drug-delivery application [38, 39]. The brain-targeting efficacy of coumarin-incorporated drugs with NPs becomes essentially better than that of those without coumarin incorporated into NPs.

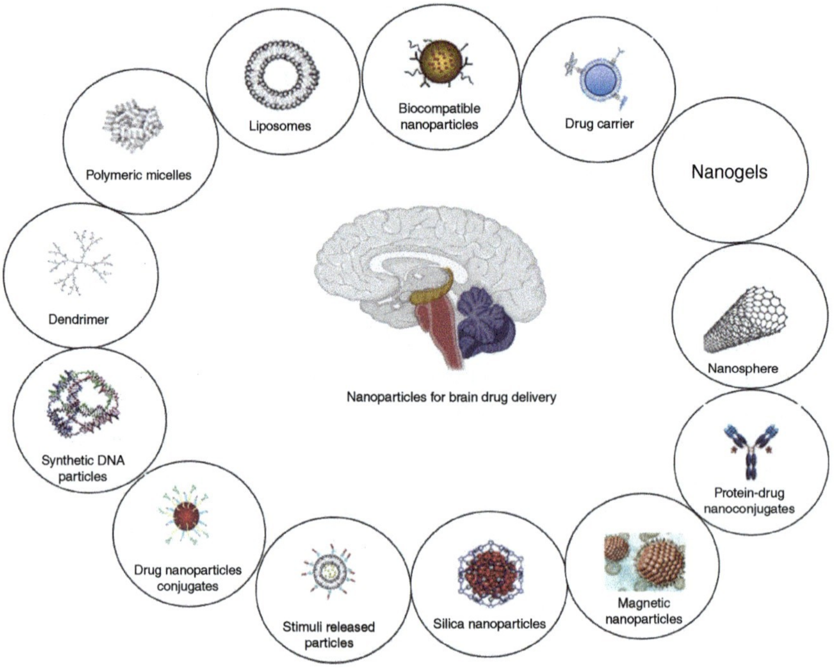

Figure 15.3 Different nano pharmaceuticals carriers for the drug delivery system for the CNS disorder and brain tumor.

Lu et al. first referred to a CBSA-altered PEG-PLA nanoparticle (CBSA-NP) for cerebrum-oriented delivery. They observed that increasing the floor CBSA thickness of the NP provided a better anti-brain tumor effect than that of the unmodified NPs. Nonetheless, poor selectivity is the pre-dominant problem of adsorptive-mediated targeting [40]. Polysorbate 80 coatings of nano vehicles can act with BBB and result in trans-BBB conveyance. Poly(n-butylcyano-acrylate) nanoparticles (p.c.- NPs) have been utilized for quite a while for brain-centric compatible structures [41]. Glucose transporter (GLUT-1) is found in high abundance on BBB and various tumors, presenting cerebrum tumor-directed medication facilitated by means of glucose digestion with the guidance of the glucose transporters. So, the connection of glucose particles to dendrimers is empowered by means of GLUT1 transporters as active targeting. Dhanikula et al., for example, investigated polyether-copolyester (PEPE) dendrimers with methotrexate carriers for diagnosis of gliomas and covalently linked them to D-glucosamine as that of the ligand targeted at GLUT-1 to improve BBB permeability and linked to the glucose metabolism. U87MG and U343 MGa cells were used to test the efficacy of MTX-loaded dendrimers [42]. Both U87MG and U343 MGa cells found that glucosylated PEPE dendrimers were endocytosed in significantly higher quantities than nonglucosylated PEPE dendrimers and had lower IC50 after MTX loading, implying that loading MTX in glucosylated PEPE dendrimers enhanced its effectiveness [42].

Cell-penetrating peptides (CPPs) are small cationic peptides and amphipathic peptides that can be speedily internalized across the cellular membranes. CPPs used to cargo the molecular shipment, including imaging agents (quantum dots and fluorescent dyes), capsules, liposomes, oligonucleotide/DNA/ RNA, peptide/protein, bacteriophages, and NPs into the cell. To enhance the brain transport of polyamidoamine (PAMAM) dendrimer, dendrimer and TAT peptides can be conjoined to magnetic NPs (MNPs) for the construction of an efficient and brain-targeted gene-delivery device (TAT-MNPs-PAMAM) [43].

Angiopep-2 has been used to assist in the treatment of brain tumors concentrated on ligands pointed toward the low-thickness lipoprotein-receptor-related protein (LRP) overexpressed on both glioma cells and BBB [44]. Xiang et al. utilized Chlorotoxin (ClTx), a 36-amino corrosive peptide, as targeting agent for the cerebrum glioma-focused medication with DOX-stacked liposomes. ClTx surprisingly intensified the take-up of liposomes by increasing the cytotoxicity into murine microvascular endothelium cell lines (BMECs) and glioma cell lines C6, U87MG, and U251. In BALB/c mice bearing U87 tumor xenografts, ClTx-charged liposomes had more accretion in the subcutaneous and intracranial tumor, higher tumor blast hindrance, and reduced hematotoxicity inside the armpit tumor model [45, 46]. Lactoferrin (Lf) is a remarkable brain-focused ligand that has been articulated to move over the BBB. More importantly, the receptor of Lf is a less-thickness LRP, which is overexpressed in glioma cells. Accordingly, a lactoferrin-conjugated PEG-PCL polymersome-assisted transport framework carrying multiple drug resistance (MDR) inhibitor tetrandrine (Lf-PO-DOX/Tet) and DOX boosted strong cytotoxicity contrary to C6 glioma cells. Lf-charged polymersome entered the cerebrum and gathered on the tumor region and inhibited tumor increment in glioma-bearing mice [47]. Guo et al. developed a PEG-PLGA NP drug-delivery system decorated with AS1411 (Ap) as the targeting ligand to facilitate anti-glioma delivery of paclitaxel (PTX). AS1411 acts as a DNA aptamer when tied up to nucleolin, which enters inside the plasma layer of both malignant cells and endothelial cells through angiogenic veins [48]. Bernardi et al. created indomethacin-stacked nanocapsules with poly(e-caprolactone), capric/caprylic triglyceride, and sorbitan monostearate and demonstrated its efficacy against glioma in an experimental glioma model. Pharmacokinetics of iron oxide NPs has been investigated for magnetic targeting as contrast agents and drug-delivery carriers in magnetic resonance imaging [49]. Iron oxide NPs have attracted much attention as contrast agents for magnetic resonance imaging, Yang and coworkers identified a PEG-modified, cross-linked starch MNP (PEG-MNP) that is ideal for magnetic targeting and has been circulating for a long time.

Besides accumulation in the brain, PEG-MNPs are also distributed inside the kidney, liver, lung, and spleen. The classic PEG-MNPs, which are functionalized with brain tumor targeting moieties, initiate the glioma tumor receptor-mediated transport across the BBB that could be put into the body for stable and long-term detection. Up to 1.0% injected dose/g tissue NP was found in the brain tumor. X-ray and histological investigations confirmed intense targeting and further implied a controlled contribution of passive mechanisms as well to the tumor tissue uptake by

NPs [50]. Cheng et al. described an EGF-altered AuNP promoted selectivity to the cerebrum tumor compared to untargeted conjugates. Nano-delivery system is optimistic for future delivery of a broad range of hydrophobic therapeutics medicine for the treatment of difficult-to-reach tumors and cancers [51]. In Figure 15.3, the distinctive category of NPs is shown for the cure of brain tumors and CNS disorders.

15.3 Neurodegenerative Diseases

CNS disease risk factors at the advanced age of the populace are steadily growing. However, nanoformulation efficacy toward the remedy of various neurological disorders offers better opportunity, especially in the neurodegeneration of the CNS [52, 53]. Neurodegenerative diseases that are commonly observed in society includes epilepsy, amyotrophic lateral sclerosis (ALS), Alzheimer's ailment (AD), a couple of cases of multiple sclerosis (MS), Parkinson's disorder (PD), brain cancer, schizophrenia, CNS infection (fungal and viral), cerebral ischemia, and brain malaria [54, 55].

Worldwide, millions of people suffer from neurodegenerative diseases, such as PD and AD. PD and AD are two illnesses that are growing and expanding in an alarming way among elderly people and putting on health care burden. Even though the research has improved our insight into these two disorders, the available medicines can only slightly alleviate their disease conditions [55]. *N*-methyl-D-aspartate (NMDA) and cholinesterase inhibitors (e.g. rivastigmine and donepezil), memantine are Food and Drug Administration (FDA)-approved drugs for AD and serve as receptor antagonists. However, the above-mentioned drugs acquire a limited benefit on harsh cognitive impairment and are thus incapable to pause the disease progression [56]. Xadago (safinamide) is a drug that improves motor symptoms while barely handling PD pathology [57]. Neurodegenerative disorders are incessant dynamic neuropathies portrayed by symmetrical loss of neurons with motor and psychological frameworks.

15.3.1 Alzheimer's Disease (AD)

AD is an ailment with the distinct signs of neurofibrillary tangles of intraneuronal hyperphosphorylated tau protein and extracellular amyloid-β (Aβ) plaques. Major motivation for focusing on dementia is that AD now impacts more than 24 million individuals globally and is anticipated to influence 115 million people by 2050 [58]. Both tau protein and Aβ are the significant cause of dementia and are neurotoxic to the brain [59]. To assess the role of oxidative stress in these processes, the death of many neuronal subdivision and enhanced levels of metal ions serve as pathological features that further contributes to Alzheimer's pathology [58, 60].

15.3.1.1 Alzheimer's Disease Focused on Drug Delivery

The most well-known cure for Alzheimer's disease is the hindrance of Aβ plaque development. Aβ plaques are noticeably disentangled after intracerebral infusion of an Aβ counteracting agent [61]. TfR-T12 peptide-modified PEG-PLA polymer was designed and transformed into melded with the Aβ neutralizer to deliver PTX for glioma therapy [62]. Later, with intravenous infusion of the combination protein, 3.5% of the infused per gram of brain tissue was attained. The consolidation of $A\beta_{1-42}$ in the CNS was decreased by 40% with the infused protein, without any ridge in the plasma $A\beta_{1-42}$ accumulation.

Nevertheless, the adequacy of the infusion protein was dominated by its immunogenicity and low strength of conjugation with Aβ neutralizer. Wheat germ agglutinin (WGA) may permit intranasal conveyance of the immune response to the cerebrum because WGA contains sugar particles and binds to glycosylated film segments of brain, thus expanding uptake through the olfactory mucosa [63]. Correlated with unmodified immune response (antibody), intranasal organization of the WGA-conjugated Aβ counteracting agent diminished the plaque length and the cerebral Aβ 40/42 substance of 5XFAD in transgenic mice. Exemplifying ability to combat Aβ, nanotherapeutics can prolong their blood circulation time and build up their bioavailability. Curcumin has the ability to cure AD since it represses the development of $A\beta_{1-42}$ oligomer and lessens the amyloid levels *in vivo*. Curcumin compound showed substantial bioavailability after infusion. However, due to the dispersion process, these NPs have some drawbacks, such as low-drug loading, drug expulsion during storage, and high polydispersity, as demonstrated by certain solid lipid nanoparticles (SLN) preparations. Curcumin to NPs or liposomes overcome this disadvantage and expand their affinity for $A\beta_{1-42}$ [64, 65]. Curcumin-formulated liposomes repressed the arrangement of oligomeric and fibrillar Aβ *in vitro*. When NPs tethered with apolipoprotein E (ApoE), a ligand for the low-thickness protein receptor present on the BBB, it boosts up the entrance to brain [66].

Mathew et al. combined curcumin-incorporating NPs with the Tet-1 peptide, a neuron-affine peptide with retrograde transport properties [67]. The curcumin nanoconjugate annihilated amyloid totals and diminished the oxidative stress [68]. Advantages of antioxidant is another ploy in Alzheimer's disease therapy because oxidative damage is an early incident of AD pathology [69]. Distinct antioxidants, including ferulic acid glutathione, nanoceria, and fullerenes, restrained the fibrillization of neuronal oxidative stress and Aβ peptide [70]. Chelating metallic particles were additionally valuable for turning off the Aβ plaque development. A copper chelator, D-penicillamine, breaks up Aβ totals *in vitro*. NPs covalently connected to D-penicillamine likewise facilitate the solubilization of Aβ-copper totals *in vitro* [71]. Despite the fact that NPs may further allow this hydrophilic medication to move across the BBB, there is no *in vivo* assessment of this technique.

A basic pathology factor for Alzheimer's disease progression is poor cholinergic neurotransmission, which acts as picking up information and memory impedances between neurons. Therefore, boosting cholinergic interaction with acetylcholine or cholinesterase (throb) inhibitors is a potential approach to tackle AD [72, 73]. The nanotubes crossed the BBB and were seized up by the synapses. However, polysorbate-80 NPs permit their infiltration of the BBB. The polysorbate 80 facilitates the transport of ApoE in the stream of blood [74]. Yu et al. bundled S14G-humanin in polymersomes to shield it from the peptidase cleavage over. The polymersomes decorated with lactoferrin are meant for BBB conveyance. Lactoferrin-charged polymersomes delivered 3.32-fold more load to the brain compared to unmodified polymersomes. Therefore, the modified polymersomes reversed the decrease of choline acetyltransferase (ChAT) activity caused by intracranial injection of $A\beta_{25-35}$ [75].

TGNYKALHPHNG (denoted as TGN) is a new 12-amino-acid ligand that was discovered in a prior work employing *in vivo* selection of a phage-displayed peptide library and has huge potential for brain transport. TGN was used as the first-order ligand for targeting and piercing the BBB in this way, which has been expressed to be a promising contender for the cure of Alzheimer's disease [76, 77]. TGN-functionalized NPs are applied to convey NP into the cerebrum of the mice made by intracerebroventricular infusion of $A\beta_{1-40}$ [78]. The conjugated NPs with TGN end up being a powerful cerebrum-concentrated-targeting framework, thereby delivering four times more payload to the brain compared to unmodified NPs. Despite the fact that oral and subcutaneous administration of nanoconjugate turned out to be groundbreaking, encapsulation in polymersomes enhanced its bioavailability [78, 79]. To embellish the BBB penetration, the polymersomes are coupled with OX26, a neutralizer coordinated with TfR. The OX26-modified polymersomes accumulated 1.26-fold more in the brain as compared to unmodified polymersomes [80].

The ramifications of the all-inclusive cerebrum medication were further demonstrated by general execution of scopolamine-treated rodents inside the Morris water-labyrinth test. The rodents model treated with NC-1900-stacked OX26-charged polymersomes showed effective development of memory and decline in aging compared to the rodents treated with polymersomes without modified OX26 [81].

Intranasal drug delivery is a technique to bypass the BBB and adequately deliver therapeutics to the cerebrum [82]. Vasoactive intestinal neuroprotective peptide was epitomized on NPs for intranasal delivery [83]. The addition of WGA improved brain delivery by around twofold. Functionalized WGA had a stronger ability to permeate the BBB, which propelled the spatial memory of dementia mice in a portion. The efficacy of the WGA-charged NPs was affirmed by the diminishing in acetylcholinesterase activity [84, 85]. Diminishing the size of the ligands minimized surrounding immunogenicity and toxicity because of lectin. The tiniest lectin, Odorranalectin might be utilized for brain-focused vehicles. Notwithstanding, the intranasal delivery is appropriate for higher efficacy because of the generally low portion administered over the nasal cavity [86].

Most scientific studies focused on AD, use methods to gracefully deliver therapeutics to the brain, including the systemic and local delivery to bypass the BBB, the brain-targeted ligand-modified systems, using intranasal and intracranial methods. In any case, barely much examinations were carried out about the distribution of therapeutics inside the cerebrum. The dissemination of such therapeutics in the solid cerebrum, such as neuroprotective, may not be as difficult in the light of their low harmfulness. Nevertheless, for toxic materials, selective distribution is required. Dual-targeting systems anchored to targeted cell-binding ligands and BBB-targeting ligands could be used to develop selective brain delivery.

15.3.2 Parkinson's Disease

Globally, PD is the second-largest neurodegenerative disease, which impacts 1–2% of the population beyond 65 years old. The selective loss of dopaminergic neurons in the substantia nigra and the accumulation of synuclein aggregates in the brainstem are hallmarks of Parkinson's disease, resulting in trouble-regulating movement [54, 55, 87]. Boosting the dopamine level inside the brain is the most commonly available method to cure Parkinson's Disease patients. Nevertheless, this cure no longer alters the state of the ailment or fixes the affected dopaminergic neurons. In spite of the fact that there were a few cure procedures exist for PD, including psychological, social electroconvulsive, undifferentiated organism, and practicing or real cure [54, 88, 89], this section will discuss the targeted drug delivery of therapeutic options in comparison to the therapeutic techniques.

15.3.2.1 Drug Delivery Focussed on Parkinson's Drug Disease

Glial-determined neurotrophic factor (GDNF) has a helpful neuroprotective effect on PD [90]. In any case, the utility of GDNF is constrained by its limited bioavailability to cross the BBB. Nonetheless, intravenously injected urocortin did not pass through the BBB. Whereas, NP modified with Lactoferrin as a targeting agent distributes 1.98-fold load of drug to brain compared to NPs unmodified ones [91]. The urocortin-stacked NPs lessened the striatal injuries in rodents caused by 6-hydroxydopamine (6-OHDA), as was evidenced by utilizing the impacts of social appraisals, a striatal transmitter test, and immunohistochemistry [92]. Hu et al. also demonstrated use of urocortin in lactoferrin-tagged NPs for PD cure [93]. For the targeted delivery of urocortin over odorranalectin-conjugated NPs, the internasal administration was used. Administration of odorranalectin-modified NPs significantly recovered the loss of dopaminergic cells caused by 6-OHDA and alleviated the reduction of neurotransmitters and reduced the rotational behavior [94, 95]. Gene therapy has distinct advantages over drug therapy because of its potential to reverse the progression of PD, restore dopaminergic neurons, or increase the level of dopamine synthetic enzymes [96]. For PD restoration, tyrosine hydroxylase (TH) activity inside the striatum is one of the available cures.

The OX26 TfR clone is available with a range of pre-conjugated fluorophores for histochemical and flow applications, as well as studies of dividing hematopoietic and tumor cell populations and metabolic activity. OX26 has been used to target TH-encoding plasmids encapsulated in liposomes [97]. Following their nasal administration, the striatal TH deviation was normalized and developed a level from 738 to 5486 pmol h, as indicated by milligram of protein. This cure further reversed the apomorphine-induced rotational behavior. Additionally, lactoferrin-tagged NPs have been utilized to deliver DNA to the brain and provided 4.2-fold higher gene articulation inside the cerebrum compared to unmodified NPs [92, 93]. Considering all these factors, PD is an advanced disease that is sometimes difficult to recreate independently in the 6-OHDA-treated rodent variant. The systemic rotenone model of PD has provided insights into the pathogenesis of PD by accurately replicating many aspects of human PD pathology. The rodent model is made with the guide of constant, nonstop exposure to rotenone model to mimic most extreme PD phenotype, along with Lewy body patches inside the nigral neurons, making this model helpful for demonstrating the efficacy of targeted drug delivery for PD cure [98, 99]. Gene therapy with lactoferrin-linked NPs is additionally evaluated against the rotenone-induced PD model as a drug conjugate [100]. Albeit many research studies utilize viral vectors for gene therapy of PD, safety is the principal concern in their application for medical cure [101]. Even though NP transfection proficiency is far less efficient and lower than that of the viral vectors, nonviral quality vehicle vectors were bearing the hope of predetermination [102, 103].

15.3.3 Cerebrovascular Disease

Cerebrovascular illness is a major cause of substantial disability, the second prevalent source of mortality, and the primary reason for the hospitalization of a large number of patients [78]. Stroke is a simple cerebrovascular disease; other causes consist of cerebral hemorrhage, cerebral embolism, and cerebral thrombosis. As the third leading cause of death in the industrialized world, stroke causes 15 million injuries and 5 million death per year [104]. Ischemic stroke cases are about 80% of all strokes prompted by hypoperfusion, apoplexy, or embolism. The remaining 20% of strokes are hemorrhagic in etiology and might be because of a fundamental vascular sore or hypertension as a result of endothelial dysfunction [105, 106].

15.3.3.1 Drug Delivery for Cerebrovascular Disease

The neuroprotective specialists attempt to cure stroke to forestall the oxidative strain caused by the recovery of blood discharge or reoxygenation injury [107]. To improve intravenous administration, BDNF is conjugated with polyethylene glycol (PEG), which diminishes its hepatic leeway and broadens its blood circulation time. The conjugate product was further modified with OX26 for brain targeting [108]. The OX26-BDNF-PEG conjugate turned out to be groundbreaking in bringing down the degree of infarction in rodents when exposed to 24 hours of center cerebral vein impediment. The cure impact boosted extensively better for OX26-BDNF-PEG conjugate compared to that of unmodified BDNF-PEG conjugate [108–110]. Daily intravenous administration normalized the neuronal thickness in the CA1 division of the hippocampus in rodents model of transitory forebrain ischemia [111]. This methodology could be utilized for the cure of reversible middle cerebral artery occlusion [107–111] Similarly, other neuroprotective factors, including VIP and bFGF, could be coupled with OX26, which could impact ischemia cure through intravenous infusion [112].

The neuroprotective effect is achieved through the calcium channel blockers or routes of free radical scavengers [113, 114]. Because of the better oxygen-conveying potential, hemoglobin (Hb) was considered to be a therapeutic candidate in the early stage of stroke cure [115]. The prototype molecule of this class of blood substitutes is diaspirin cross-linked hemoglobin (DCLHb). It is made up of cross-linking between the two alpha chains, which gives the molecule stability. To develop the oxygen-conveying potential and lessen the toxic effect of hemoglobin, various strategies were utilized, such as PEG-conjugation and diaspirin cross-linking, to provide useful outcomes for ischemia-driven cerebrum harm [76, 116]. SunBiol is a PEG-conjugated bovine-like hemoglobin with a diameter of 30–50 nm. Its plasma half-life is expanded to 9.6 hours in rodents, with slight nephrotoxicity. Cure with SunBiol fundamentally diminished the size of dead tissue and edema in rodents when blocked with a thrombotic blood cluster [117, 118].

Promoting tissue regeneration has the potential to help people in recovering their function after an injury [119]. Neuroregeneration can be stimulated by several factors, such as cell-based therapy, endogenous growth factors, and vascular regenerative therapy. Delivering the drugs to the subventricular zone, the neural stem-cell alcove, is ordinarily performed with intrathecal infusion [120]. To limit the invasiveness of the cure, EGF gets stacked directly into a hydrogel for epi-cortical vehicle-supported discharge, and EGF is conjugated with PEG to prolong its dispersion. The PEG-EGF-loaded hydrogel significantly increased the stimulation of neural stem cells/progenitor cells. The epicortical delivery methodology likewise turns out to be proficient in delivering erythropoietin, a glycoprotein with neuroprotective and neuro-regenerative potential for the damaged brain tissue [121]. Vascular endothelial growth factor (VEGF), an angiogenic increment component, advances neurogenesis and cerebral angiogenesis, limiting ischemic cerebrum injury. For intravenous infusion, VEGF-encoding plasmids were encapsulated in liposomes functionalized with transferrin ligand for intravenous injection, allowing them to penetrate the BBB. RT-PCR was used to quantify the amount of VEGF messenger RNA (mRNA) inside the ischemic cerebrum after 24 hours infusion of transferrin-coated liposomes compared to unmodified liposomes vehicle [122–124]. Reactive oxygen species (ROS) are involved in brain damage resulting in cerebral ischemia. The curative possibility of nerve growth factor (NGF) across neurological disorders may be realistically limited by its short half-life, based on the remarkable properties of carbon nanotubes (CNTs) for controlled drug target delivery [125]. Caspase-3 stimulation cause brain cell death after severe brain injury. Caspase-3 expression was downregulated by stereotactic administration of caspase-3 small interference RNA (siRNA)-bearing nanotubes, resulting in a neuroprotective effect in the ischemic brain [126, 127]. Linking the hemeoxygenase-1 (an antioxidant enzyme) gene with dexamethasone (an anti-inflammatory agent) further reduced the cramp caused by ischemia [128].

Intranasal drug delivery is one of the amazing noninvasive platforms to deliver therapeutics to the brain. Numerous therapeutics comprised of bFGF, recombinant human erythropoietin, deferoxamine, acidic fibroblast, BDNF, growth factors, HMGB1-restricting peptide, and ginsenoside Rb1 have been delivered to the cerebrum using intranasal route and resulted in substantial neuroprotective impacts in rodents with ischemia [30, 129, 130]. Despite the fact that intranasal delivery offers a short pathway for the brain to get access to the drugs, this strategy is restricted by the low-administered dose and low-residence time. Moreover, bioadhesive materials are needed to develop and increase the residence time and activation time, which may likewise cause extreme local immunogenicity and toxicity.

15.3.4 Inflammatory Diseases (ID)

CNS infection by viruses, bacteria, and fungal and parasitic pathogens can cause devastating deaths and neurological disability [131, 132]. The precise cure for these organisms crossing the BBB is not recognized. Neuroinvasive viruses typically gain access to the CNS through a variety of mechanisms, including (i) direct infection of endothelial cells and subsequent transcellular release of virus into the brain parenchyma, (ii) infection of peripheral immune cells that enter the CNS in a "Trojan Horse" mechanism, and (iii) paracellular entry following BBB breakdown, and retrograde virus transport from the peripheral nervous system [133–135].

15.3.4.1 Inflammatory Diseases (ID) Focused on Drug Delivery

Certain infectious agents have been shown to infect the brain tissue precisely by crossing the BBB without disturbing its permeability. Transcellular intrusion of BBB has been observed for bacterial and viral pathogens along with *Escherichia coli* [136], *Cit-Citrobacter freundii* [137], human *immunodeficiency infection kind 1* (HIV-1) [138, 139], and *Streptococcus pneumonia* [140]. In *C. freundii,* vesicular drug delivery has been proposed to elucidate the intracellular spot of several bacterial cells in a single film of vacuole-like frameworks, recommending that *C. freundii* attacks vacuoles, reproduces itself, and transcytoses by utilizing brain endothelial cells [136]. Explicit microbial ligand and BMEC receptor interaction have been now suggested as a systemic transcellular entrance for microbial agents. Consequently, the active infiltration of *E. coli* k-1 all over the BBB is interceded by utilizing a bacterial component that incorporates specific bacterial protein into the host cellular wall [137].

In a similar way, ongoing evince suggest that the HIV-1 infection enters the BBB with the guidance of gp120-assisted adsorptive endocytosis, a cellular mechanism that provides a gateway into BMEC [141]. Given that HIV-1-proteoglycan interactions depend absolutely on electrostatic contacts with natural components present in gp120 and sulfate organizations in proteoglycans, HIV-1 may likewise utilize those associations to smoothly enter and relocate over the BBB to attack the cerebrum. These findings were supported by the fact that heparinase and chondroitinase treatment of human brain microvascular endothelial cells (BMECs) reduced HIV-1 attachment, and gp120-deficient virus fails to bind and perform transcytosis through human BMEC [139]. Infected portions were entirely obstructed and were coated with heparin, indicating that HIV-1 binds to cell wall by means of proteoglycans. Additionally, roles of HBMEC have been revealed by transmission electron microscopy (TEM), showing the convergences of cytoplasmic vesicles of various sizes (from 150 nm to 5 μm) starting from a single basic virion of HIV-1 toward a thousand virions [138].

A very familiar aspect of disease is the higher permeability of BBB. Immune response is identified by the hasty production and release of inflammatory components, such as chemokines, cytokines, matrix metalloproteinases, and cellular adhesion molecules, at the site of disease. These components have been found to adjust the structure and integrity of the BBB. Perturbations of BBB and arbiters' intersections are responsible to trigger the various aspects of pathogenesis. The predominantly TNF-1 and IL-1, metalloproteinases, and cytokines play detrimental roles to open the BBB [142]. Therefore, it is been suggested that *S. pneumonia* [140] and HIV-1 [138, 139] take advantage to infiltrate the BBB. The trojan horse mechanism hypothesizes that the "bug" component of contaminated macrophages enables cerebrum endothelial cells to be taken up live inside the CNS by microglial cells. Such binding will be the initial phase across the passage of BBB over intact walls of capillaries, the section of invulnerable cells all over the BBB. This closeness could likewise trigger the part of infected immune cells, and the cerebrum endothelial cells are so close together, viral particles could spread infection to contaminant-resistant cells [132–135]. Macrophages stimulated by HIV-1 could cross the BBB allowing the infection for the upside passage into the CNS. Immunohistochemical assessment of the CNS tissue from the HIV-positive victims has demonstrated the presence of tight junction proteins and mononuclear cell-specific antigen CD-68 that changes the BBB and monocyte penetration. Interruption of tight intersection in veins has consistently been related to CD-68 marker that presents mononuclear cellular aggregates and microglial nodules [132].

15.3.4.2 Drug Delivery for the Treatment of Neuro-AIDS

Human immunodeficiency virus (HIV) illustrates some form of neurological dysfunction, most feasible due to passage of HIV-1 into the CNS that occurs earlier than the serological-diagnosis before the initiation of antiretroviral therapy for the HIV-positive patients, resulting in neurological complications in a number of the infected individuals [143, 144]. Microglia, macrophages, and astrocytes (in some extent) stimulate the cellular chemotaxis and the most adaptive immune response in the CNS [145]. Nanotechnology can improve antiretroviral drug delivery across the BBB, enhancing biodistribution and clinical advantage in case of neuroAIDS. HIV access to the CNS functions as a natural viral reservoir since most antiretroviral (ARV) drugs have inadequate or no delivery through the brain barriers. The BBB's structure, the presence of efflux pumps, and the production of metabolic enzymes all pose barriers to ARV drug-brain access. Clinical benefit of nanoART for HIV-1 disease is that it crosses the BBB and enhances the biodistribution into the brain [146, 147]. The magnetic NPs conjugated with the drugs vorinostat and tenofovir in a single dosage for the cure of neuroAIDS are also beneficial for activating latent viruses and killing them [148].

15.3.5 Drug Delivery for Multiple Sclerosis (MS)

Multiple sclerosis is a chronic disease of the CNS that affects the spinal cord, brain, and optic nerves, including a wide range of symptoms throughout the body. Neurotic features of MS are the demyelination by macrophages and T cells. Immune system encephalomyelitis is the most commonly used test model for human incendiary demyelinating ailment and various sclerosis (MS) [149]. Encephalomyelitis (EAE) is an unpredictable condition in which the association between immunopathological and neuropathological components prompts the key neurotic highlights of MS;

aggravation, axonal demyelination, and gliosis. The counter-administrative systems of irritation and remyelination that happen in EAE, can further fill in as a model for these procedures. Further, EAE is widely used as a model of malfunctioning neuro endothelial immune system disorders with a manifestation of complex neuropathology. Lot of the medications that were either approved or currently in use or under trial for future use in MS based are solely on testing on EAE model examinations. The use of encephalomyelitis as a model of MS is turned out to be chronic repulsing, due to the development of ceaseless relapse in animal models [150]. MS shows majority of signs of hyperactive immune system issues, including breakdown of the blood-cerebrum boundary. The primary aspects of the BBB in this disease can be related by two hypotheses: (i) endothelial cells of the brain serve as the tipping point of the inflammation, an early event that causes the induction of IFN-γ to process the MHC class II on the luminal of the endothelium cells. Then T cells can criss-cross the BBB upon acceptance of neural antigens cells conferred by MHC class II; (ii) T-cell activation in the periphery induces expression of adhesion molecules, such as integrins and selectins, to facilitate the interaction of endothelium with migratory T cells across the BBB [151]. Microglial cells act as antigen-presenting cells, which build the inflammable demyelination procedure. Upon incitement, astrocytes pump out various immunoregulatory particles, which incorporate prostaglandin E, IL-1, interferons, IL-3, and TNF. The second immunoregulatory highlight of astrocytes is their ability to mature MHC class II antigens explicitly after stimulation with IFN-1. Oligodendrocytes were the progenitor cells that combine and maintain the quality control of myelin inside the CNS. In case of MS, cytokines, such as TNF, can decay the cell by initiating the inflammatory response and starting the catastrophic demyelination [152]. Relapses of MS are frequently treated with repeated massive IV doses of glucocorticosteroids (GS), despite the established fact that oral and IV doses produce equal exposure. Glucocorticoid medication has been proven to speed up functional recovery in acute multiple sclerosis (MS) relapses, but no long-term functional benefit has been demonstrated. The target is to prevent the progressive tissue destruction with lack of axons, oligodendrocytes, and neurons, prompting permanent functional deficits [153].

15.4 Drug Delivery for CNS Disorders

There are several therapies for CNS disorders, such as radiotherapy (application of radiations), chemotherapy (application of chemicals), gene therapy (alterations in genes), immunotherapy (application of immunogenic agents), and surgical producers. All of these therapies have their own specific advantages and shortcomings [154–157]. Present-day, drug formulations target multiple sites and are capable to exhibit poly-pharmacological profiles, which is pivotal in the course of drug discovery. Neurological disorders are characterized by a gradual deterioration of neuronal cell structure and/or function in the brain and spinal cord. A number of unknown causes and factors, such as aging population, chronic brain disorders such as AD, PD, and stroke, may trigger neurological problems with a wide range of symptoms [158, 159].

Until recently, there is not an individual recombinant protein for the treatment of CNS disorder. FDA acknowledged that it is of prime importance that drugs must permeate BBB for CNS illness. It is important for drugs to cross BBB to reach brain tissue for effective treatment of brain disorders. Hence, it is one of the most important factors to be considered while developing a formulation for brain disorders. To penetrate the BBB from the bloodstream, a drug molecule must possess specific molecular weight, lipophilicity, and charge. On the other hand, the vast majority of small molecules (mw N 500 daltons, D), proteins, and peptides do not cross the BBB. The BBB also has additional enzymes (ecto- and endo-enzymes) that are present in large numbers inside the endothelial cells to protect the brain while solutes cross the cell membrane. Often, only some molecules with appropriate lipophilicity, molecular weight, and charge diffuse well from blood into the CNS. Approximately 98% of the small molecules and nearly all macromolecular medicines do not cross the BBB [30]. Therefore, drug-like compounds or molecules must cross the BBB through specific transporters and/or receptors located in the luminal region of the endothelial cells [158–161]. In the last 25 years, single epitope antibodies and ligands associated with the encephalon have been used to alter drug-delivery mechanisms. All previous clinical trials have failed because it makes easier for chemotherapeutic agents to enter the brain tissues. Even though, these drugs might not at once set off brain function restoration, but these therapeutics are probably crucial for minimizing the brain's gliotic reaction toward trauma and thus paving the way for neuronal and axonal growth. In contrast, assessment of localized transport of drug can be beneficial for tracking the effect of therapeutic healing procedures on brain function. However, the delivery of drugs to the brain has proved to be enormously complex and challenging [162, 163].

15.4.1 Tau Therapy

Tauopathy is identified with numerous different ailments, including AD, PD, Huntington's Disease (HD), and a lot of others neuronal disorders [164]. Alzheimer's disease, progressive supranuclear palsy, some frontotemporal dementias, corticobasal syndrome, and chronic traumatic encephalopathy are examples of tauopathies, which are progressive neurodegenerative disorders marked by tau-positive deposits throughout the brain. Comprehensive neuropsychological tests are necessary for detecting early signs of neurodegenerative disease. AD is an exemplary case of amyloidosis, the condition categorized as collection of amyloid fibrils originating from the misfolded proteins [165]. The autopsy procedure performed for deceased individuals have exhibited occurrence of Tau proteins as insoluble protein aggregates consisting of extensive phosphorylation. In addition, the distinction between aging development and tauopathy-directed Alzheimer's disease is yet to be understood, even AD accelerate the outbreak of aging [166]. Vimal et al. prepared the nanoconjugate (Au–PEG) that helped to reduce the amount of ambient tau burden, thereby minimizing the destructive impacts associated with migratory tau protein seeds [167]. Nanoconjugates serve as a plentiful opportunity as a resource pool to analyse the alterations for BBB for plethora of CNS disorder.

Although some successes are seen in case of *in vitro* implication of nanoconjugates, successfully effective targeting is still a challenge for *in vivo* implication of nanoconjugates. The impact of passive targeting is useful for its successful biomedical application. Gold nanoparticles (AuNPs) act as a kinetic stabilizer for the misfolded tau proteins, which are prone to form proteinous clusters. This helps in curtailing the oligomerization of phosphorylated Tau (Figure 15.4). Misfolded proteins accumulate as intracellular aggregates, such as neurofibrillary tangles and Lewy bodies, or extracellular plaques, such as senile and prion plaques, resulting in "proteopathies." Alzheimer's disease is diagnosed by the appearance of two distinct pathologies—senile plaques and neurofibrillary tangles (AD). A detailed understanding of the brain cells responsible for the clearance of protein aggregation caused by dysfunction is needed to fully comprehend the etiology of AD. AuNPs not only fulfill the purpose of nano companions but also alter the proteo-plasticity of tau microtubule interaction to restructure neuronal functioning associated with the cytoskeleton (Figure 15.4) [167]. Amyloid-beta's hypothesis keeps spinning in parallel to tau protein, BACE-1 protease inhibitor along with the GKS3β kinase inhibitor and numerous others are in the race for the improvement of active AD cure [167]. Extremely potent and successive drugs that are under trial for the treatment of AD incorporate the implication of activated tau vaccines and also the implication of antibodies designed for single epitopes. Other than the use of immunotherapeutic strategy, there are several other therapies for tau that are being investigated presently for neurodegeneration associated with Tau. These includes the modulation of tau hyperphosphorylation, checking the clustering of Tau, and checking the expression of its altered forms. The results of this clinical

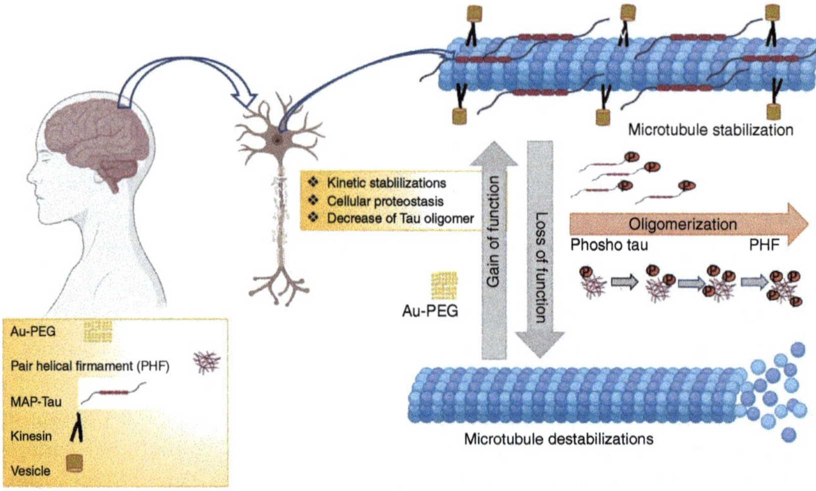

Figure 15.4 Tauopathy mechanism based on Au–PEG mediated amelioration. Hyperphosphorylated tau does not associate with microtubules as strongly as its unphosphorylated counterpart, resulting in destabilization of microtubule.

investigations are awaited. However, the complex Tau proteome and its distinct biological function deserve further exploration.

15.4.2 Immunotherapy

There are continuous obstacles to treating CNS-associated disorders due to complexity of the brain functioning. Therefore, only mildly effective remedies and no potent cure for CNS-related disorders are currently available. In case of passive immunotherapy, a patient is given immune cells or antibodies that can target the tumor cells. Especially, patient's own immune system is not stimulated; instead, immune cells are activated *in vitro* and injected into the patient. The use of external immunoglobulin for specific targets is implied for the treatment of CNS-associated disorders.

Currently, techniques to deal with CNS disorders consist of immunotherapy, radiotherapy, chemotherapy, surgical cure, and gene remedy, with every therapeutic strategy having its pros and cons [156, 157]. Active- or passive-targeting techniques using immunotherapeutic serve as important emerging approaches for anticancer therapy. Active immunotherapeutic techniques stimulate nonspecific immune response in case of tumor vaccines and nanomedicine. In the passive immunotherapeutic methods, application of the mAbs may also result in interaction with noxious counterparts or with the active immunological agents that might be effective component to counter neoplasm. In the disease, molecular surface markers (FOXP3$^+$, CD25$^+$, and CD4$^+$) are over expressed by activated effector T cells (Tregs) [168].

15.4.3 Gene Immunotherapy (GIT)

Gene therapy has been an effective measure to regulate the expression of the oncogenes [155, 169, 170]. The treatment of glioblastoma is also formulated using immunomodulatory gene therapies. Immunomodulatory and gene therapy is instrumental to access the tumor environment for further achieving the antitumor immune response. Speranza et al. formulated the adenovirus vector that can undergo the process of multiplication in host machinery to deliver the HSV TK gene (AdV-tk) for accentuation in the anti-PD-1 potency in mouse model for glioblastoma [171]. This strategy accentuated the AdV-tK and IFN signalling followed by increase in amount of PD-L1, where infiltration of CD8$^+$ T cells was permitted. For the patient of glioblastoma, this adjuvant therapy was successful apprehending the endurance from 30% to 80% [171]. Thus, it is important to use the synergistic approach to improve immune response through the immunotherapy. Although further scientific and clinical investigations are required for the optimization of adjuvant therapy due to the existing gap of distinct genes and antigens as treatment targets and the channel for the delivery is yet a barrier for glioma treatment. The use of GIT is beneficial for indirect antitumor effects and decrease in the resistance to drugs administered in cases of glioblastoma patients, but still there is a poor understanding of effective and distinct genes for glioblastoma. It is interesting that gene therapy is suggested to be an effective therapeutic treatment for glioblastoma.

15.4.4 Chemotherapy (CT)

Glioma and glioblastoma present a complicated immunosuppressive tumor microenvironment (TME) in the brain. Glioblastomas are low-immunogenic tumors located in a T-cell-depleted CNS region with an immune-suppressive microenvironment. The development of a repressive intratumor heterogeneity and TME is an enormous challenge because of the permeability challenge for the drugs across BBB. Chemotherapy and radiation therapy are usually used as treatment options for Glioblastoma multiforme (GBM) [172, 173].

Lollo et al. [174] have combined the immune-stimulating cytosine-phosphate-guanosine (CpG)-mediated immunotherapy with chemotherapy to obtain the high therapeutic index to reduce long-term relapse of glioblastoma. They designed a multifunctional lipid nanocapsule delivery system that consistently releases the immune-stimulating CpG agent and drug PTX to glioma tumor site. The orthotopic model of GL261 glioma mice showed higher survival with prolonged administration of coloaded lipid nanocapsules (PTX/CpG) compared to CpG and PTX alone. This concluded that for glioma tumor treatment, immune-chemo combined therapy approach may be more effective than only chemotherapy [174].

Kadiyala et al. developed nanodiscs modified with the high-density glioma tumor-specific antigens that also contained adjuvants and bioactive compounds [175]. The sHDL nanodiscs are designed as immune-chemo combined therapy for delivering CpG deoxynucleotides, a TLR9 agonist, with a chemotherapeutic docetaxel (DTX), to target gliomas. The nanodiscs are self-assembled synthetic apolipoprotein-I peptide that helps to target the tumor in the GL26 syngeneic glioma mouse. DTX-sHDL-CpG nanodiscs (discoidal shape) of 10–12 nm average size activate the antitumor immune responses by delivering drug to the tumor target site [175]. The immune-chemo combination therapy provides several benefits, such as low recurrence, reduced drug resistance, and reduced metastasis. Nevertheless, due to heterogeneity of the tumor cells, lack of unique antigens immunotherapy could be a challenge for investigating in the clinical trial.

15.4.5 Photoimmunotherapy (PIT)

Photoimmunotherapy is an oncological treatment that includes the photodynamic cure with immunotherapy cure [176]. It is an advanced therapy for glioblastoma that applies the use of photosensitizer's conjugation along a centred fragment of SCFv antibodies [177]. Its a type of immunotherapy that takes favour of a photosensitizer to consolidate along targeted SCFv antibodies and monoclonal antibodies. Due to the photoactivation effect in photodynamic therapy (PDT), intensive inflammatory factors (TNF-α and IFN-γ) and generation of ROS [178] are combined to enhance the antitumor immunity in glioblastoma [179, 180]. Glioblastomas can be fractionally induced by alterations (missense mutations, deletion mutations, and amplification mutations) in the EGFR, and gene mutations in glioblastoma can be accelerated by adjusting EGFR expression pattern on tumor surfaces [181].

Huang et al. [182] preferred the use of polylactic-*co*-glycolic acid (PLGA) that improves photo-immunoconjugates (PIC) delivery to cancer cells, removing a potential stumbling block to PIT efficacy. PLGA NPs were bound to the photo-immunoconjugate benzoporphyrin derivative and Mab cetuximab. EGFR-positive

U87 cells showed cytotoxicity when irradiated with activated PIC NPs [182]. Basically, the exertion of nanotechnology evolved common remedy outcomes compared with traditional PIC. Globally, PIT has evolved as promising tumor-targeted therapy compared to classic PDT because of the complementary antibodies attached to the NPs surface. PIT adds to the delivery of photosensitizers into tumor, thereby enhancing light-activated cytotoxicity and tumor contraction. Similar work has demonstrated that the imperceptible damage to non-cancerous cells also enhance the immunity of this approach [180]. Photoactivated nano-formulations illustrate higher targeting capacity and decrease tumor metastasis and relapse. Nevertheless, further investigation in this area of conjugation of photosensitizers to antibodies is desirable to overcome the pitfalls in drug-delivery practice. In the Table 15.1 we have summarized ongoing clinical trials using the different approaches to bypass the BBB.

Table 15.1 Summarizes ongoing clinical trials using the different approaches to bypass BBB in CNSDs.

Therapeutic strategy: CNS disorders	Therapeutic effects	Limitations
Brain tumor-targeted delivery: cytotoxic agents [183–186]	• TMZ has a 100% oral bioavailability. • Nitrosoureas and temozolomide have the ability to cross the BBB. • The clinical effectiveness of nitrosoureas chemotherapy in combination with radiotherapy has been studied. • Clinical effectiveness of etoposide and platinum salts in combination, as well as the availability of newer platinum derivatives (e.g. oxaliplatin) that are active against resistant tumors.	• Nausea, myelosuppression, hepatotoxicity, DOX cardiotoxicity, cisplatin nephrotoxicity, among other serious side effects. • In most cases, i.v. administration is used.
Brain tumor-targeted delivery: VEGFR inhibitors [187]	• Orally administered, biomarkers indicative of tumor responsiveness to VEGF inhibition may be available, vascular normalization may reduce interstitial fluid pressure and allow for better drug penetration. • Hypoxia in the tumor is reduced, and the treatment is well tolerated.	• Monotherapy leads to continued tumor development.
Brain tumor-targeted delivery: anti-VEGF antibody [188]	• In xenograft models, size reduction was observed, as well as good clinical efficacy in monotherapy.	• Severe toxicity, particularly when used in combination with irinotecan. • In most cases, i.v. administration is used.
Brain tumor-targeted delivery: EGFR inhibitors [189]	• Recognition of EGFR inhibition responsive phenotypes is a possibility.	• Restricted effectiveness in monotherapy including combination therapy, as well as radiotherapy, and serious side effects.

15.5 Future Prospects

Despite the availability of several existing therapeutics for the treatment of many neurodegenerative diseases, none of them are shown to reverse the disease with greater clinical outcomes and therefore only be used as palliative treatments [190]. Although nanotechnology has advanced quickly, and NPs are entities that have the potential to revolutionize treatment, imaging, and early detection of a variety of diseases, some inherent issues must be addressed. Toxicology should be considered when designing therapeutics. There are little data on the polymer's fate and long-term toxicity on neuronal cells, including the immunogenicity of polymers in the form of NPs, blood compatibility, RES capture, and brain deposition [191–193]. Nanoparticles comprise of various drug-delivery systems, such as liposomes, polymeric nanoparticles, micelles, and others, which provide a successful strategy because of their specific ability to target the BBB and help to deliver therapeutics to the brain more effectively than the drug alone. For example, nanocarriers have been shown to efficiently resolve the BBB and transmit curcumin to locations within the brain when appropriate ligand is anchored on surface to take full advantage of receptors overexpressed at the BBB. Besides the drug-delivery systems, the preferred brain drug delivery can be achieved by using the correct route of administration. The study of "systematically targeted drug delivery" as a treatment for brain tumors is promising. It will contribute to closing the gaps in our knowledge for the clinical and pathological conditions of brain tumors, laying the groundwork for various targeting options. Optimized-drug-delivery approaches for treating brain tumors could further that take into account the characteristics of brain tumor microenvironment to provide better therapeutic outcome.

15.6 Conclusions

Nano-assisted-targeted drug delivery is promising approach to cure and treat CNS disorders. The probable prosperity includes optimized drug distribution, noninvasive administration, reduced systemic toxicity, better treatment outcome, and revised compliance. In the future, researchers should find out the organelle and organ-targeted delivery to further improve the treatment of CNS disorders. Novel approaches using multidisciplinary sciences for glioblastoma treatment focus on nanobiotechnology, immunotherapy, gene therapy, and photoimmunotherapy. Nanomedicine provides new insights for the drug delivery to specific targets, as well as interaction of the drug molecules with the target receptors, which results in tumor tolerance to the specific drug. Recent advances of nanoscience improve the patient's adaptive and innate immunity to combat glioblastoma through the use of drug combinations. The efficacy of these approaches is further improved by using biodegradable (water-soluble hydrogels and matrices) and stimuli-sensitive nanomaterial (i.e. pH and temperature). The CNS challenges that must be conquered during immunotherapy incorporate differential mechanisms of action due to patients age, sex, cause of disease, and how severe the condition of patients.

Although the targeted delivery uses ligands to target pathological site, these biocompatible ligands are yet to evaluated further before clinical practice. Besides, metal-based NPs therapeutics are a potential approach to treat neurodegenerative disorders and brain injury. Additionally, one must monitor target sites, adjust dose, adjuvants, and follow clinical protocols to successfully administer drugs to bring the nano-based immune therapy into the clinical trials. Combination therapies and optimization of such targeted drug delivery need to be operated at the early stage of the disease to obtain superior therapeutic outcome, which could be more critical player in handling brain tumors, brain injury, and CNS disorders.

List of Abbreviations

6-ODHA	6-hydroxydopamine (6-OHDA)
ACP-charged NPs	amorphous calcium phosphate nanoparticles
AD	Alzheimer's disease
Aβ	β-amyloid
ALS	amyotrophic lateral sclerosis
ApoE	apolipoprotein E
ARV	antiretroviral
BBB	blood–brain barrier
BBTB	blood–brain tumor barrier
BCECs	brain capillary endothelial cells
BEC	blood vascular endothelial cells
BMEC	brain microvascular endothelial cells
CADD	computer Aided Drug Design
CBSA-NP	cationic bovine serum albumin-nanoparticle
ChAT	choline acetyltransferase
CITx	chlorotoxin
CNS	central nervous system
CNSD	centralnervous system disorders
CNT	carbon nanotube
CpG	cytosine-phosphate-guanosine
CPP	cell-penetrating peptide
DCLHb	diaspirin cross-linkedhemoglobin
DOX	doxorubicin
DTX	docetaxel
EAE	encephalomyelitis
EPR	enhanced permeation and retention
FDA	food and Drug Administration
FITC	fluoresce inisothiocyanate
GBM	glioblastoma multiforme
GDNF	glial-determined neurotrophic factor
GLUT	glutamine
GLUT-1	glucosetransporter type 1

GS	glucocorticosteroids
Hb	hemoglobin
HD	Huntington's disease
LRP	lipoprotein-receptor-related protein
MDR	multiple drug resistance
MNP	magnetic NPs
mRNA	messenger RNA
MS	multiple sclerosis
NGF	nerve growth factor
NMDA	*N*-methyl-D-aspartate
NP	nano-particles
p.c.-NP	poly(n-butylcyano-acrylate)
PAMAM	polyamidoamine
PCL	polycaprolactone
PD	Parkinson disease
PDT	photodynamic therapy
PEG	polyethylene glycol
PEG-PLA	poly(ethyleneglycol)- poly(lactide)
PEPE	polyether-copolyester
PIC	photo-immunoconjugates
PLGA	polylactic-co-glycolic acid
PTX	paclitaxel
ROS	reactive oxygen species
RT	radiotherapy
SB-VS	structure-based virtual screening
siRNA	small interference RNA
SLN	solid lipid nanoparticels
TEM	transmission electronmicroscopy
TH	tyrosine hydroxylase
TJ	tight junctions
TMZ	temozolomid
VEGF	vascular endothelial growth factor
WGA	wheat germ agglutinin
WHO	World Health Organisation

References

1 Feynman, R.P. (1960). There's plenty of room at the bottom. *Eng. Sci. (CalTech)* 594 (23): 22–36.

2 Jain, K.K. (2006). Role of nanotechnology in developing new therapies for diseases of the nervous system. *Nanomedicine (Lond.)* 1 (1): 9–12.

3 Modi, G., Pillay, V., and Choonara, Y.E. (2010). Advances in the tretment of neurodegenerative disorders employing nanotechnology. *Ann. N. Y. Sci.* 1184: 154–172.

4 Silva, G.A. (2006). Neuroscience nanotechnology: progress, opportunities and challenges. *Nat. Rev. Neurosci.* 7 (1): 65–74.
5 Bouffet, E., Tabori, U., Huang, A., and Bartels, U. (2010). Possibilities of new therapeutic strategies in brain tumors. *Cancer Treat. Rev.* 36 (4): 335–341.
6 Nicholas, M.K., Lukas, R.V., Chmura, S. et al. (2011). Molecular heterogeneity in glioblastoma: therapeutic opportunities and challenges. *Sem. Oncol.* 38 (2): 243–253.
7 Behin, A., Hoang-Xuan, K., Carpentier, A.F., and Delattre, J.Y. (2003). Primary brain tumours in adults. *Lancet* 361 (9354): 323–331.
8 Allard, E., Passirani, C., and Benoit, J.P. (2009). Convection-enhanced delivery of nanocarriers for the treatment of brain tumors. *Biomaterials* 30 (12): 2302–2318.
9 Louis, D.N., Perry, A., Reifenberger, G. et al. (2016). The 2016 World Health Organization classification of tumors of the central nervous system: a summary. *Acta Neuropathol.* 131 (6): 803–820.
10 www.cdc.gov/uscs/cancer (accessed 7 July 2020).
11 Nussbaum, E.S., Djalilian, H.R., Cho, K.H., and Hall, W.A. (1996). Brain metastases: histology, multiplicity, surgery, and survival. *Cancer* 78 (8): 1781–1788.
12 Laws, E.R., Parney, I.F., Huang, W. et al. (2003). Survival following surgery and prognostic factors for recently diagnosed malignant glioma: data from the Glioma Outcomes Project. *J. Neurosurg.* 99 (3): 467–473.
13 Liu, Y. and Lu, W. (2012). Recent advances in brain tumor-targeted nano-drug delivery systems. *Expert Opin. Drug Deliv.* 9 (6): 671–686.
14 Parrish, K.E., Sarkaria, J.N., and Elmquist, W.F. (2015). Improving drug delivery to primary and metastatic brain tumors: strategies to overcome the blood–brain barrier. *Clin. Pharmacol. Therapeutics* 97 (4): 336–346.
15 Fidler, I.J. (2011). The role of the organ microenvironment in brain metastasis. *Semin. Cancer Biol.* 21 (2): 107–112.
16 Arko, L., Katsyv, I., Park, G.E. et al. (2010). Experimental approaches for the treatment of malignant gliomas. *Pharmacol. Ther.* 128 (1): 1–36.
17 Pathan, S.A., Iqbal, Z., Zaidi, S. et al. (2009). CNS drug delivery systems: novel approaches. *Recent Pat. Drug Deliv. Formul.* 3 (1): 71–89.
18 Cardoso, F.L., Brites, D., and Brito, M.A. (2010). Looking at the blood–brain barrier: molecular anatomy and possible investigation approaches. *Brain Res. Rev.* 64 (2): 328–363.
19 Abbott, N.J., Patabendige, A.A., Dolman, D.E. et al. (2010). Structure and function of the blood–brain barrier. *Neurobiol. Dis.* 37 (1): 13–25.
20 Ge, S., Song, L., and Pachter, J.S. (2005). Where is the blood–brain barrier… really? *J. Neurosci. Res.* 79 (4): 421–427.
21 Preston, J.E., Abbott, N.J., and Begley, D.J. (2014). Transcytosis of macromolecules at the blood–brain barrier. *Adv. Pharmacol.* 71: 147–163.
22 Descamps, L., Dehouck, M.P., Torpier, G., and Cecchelli, R. (1996). Receptor-mediated transcytosis of transferrin through blood-brain barrier endothelial cells. *Am. J. Physiol.-Heart Circ. Physiol.* 270 (4): H1149–H1158.

23 Booth, R. and Kim, H. (2012). Characterization of a microfluidic in vitro model of the blood-brain barrier (μBBB). *Lab Chip* 12 (10): 1784–1792.

24 Khan, M., Sherwani, S., Khan, S. et al. (2021). Insights into multifunctional nanoparticle-based drug delivery systems for glioblastoma treatment. *Molecules* 26 (8): 2262.

25 Zhao, X., Ye, Y., Ge, S. et al. (2020). Cellular and molecular targeted drug delivery in central nervous system cancers: advances in targeting strategies. *Curr. Top. Med. Chem.* 20 (30): 2762–2776.

26 Zlokovic, B.V. (2008). The blood-brain barrier in health and chronic neurodegenerative disorders. *Neuron* 57 (2): 178–201.

27 Dyrna, F., Hanske, S., Krueger, M., and Bechmann, I. (2013). The blood-brain barrier. *J. Neuroimmune Pharmacol.* 8 (4): 763–773.

28 Hendricks, B.K., Cohen-Gadol, A.A., and Miller, J.C. (2015). Novel delivery methods bypassing the blood-brain and blood-tumor barriers. *Neurosurg. Focus* 38 (3): E10.

29 Pandit, R., Chen, L., and Götz, J. (2020). The blood-brain barrier: physiology and strategies for drug delivery. *Adv. Drug Deliv. Rev.* 165: 1–14.

30 Bukari, B., Samarasinghe, R.M., Noibanchong, J., and Shigdar, S.L. (2020). Non-invasive delivery of therapeutics into the brain: the potential of aptamers for targeted delivery. *Biomedicine* 8 (5): 120.

31 Cooley, M., Chhabra, S.R., and Williams, P. (2008). N-Acylhomoserine lactone-mediated quorum sensing: a twist in the tail and a blow for host immunity. *Chem. Biol.* 15 (11): 1141–1147.

32 Guo, L., Fan, L., Pang, Z. et al. (2011). TRAIL and doxorubicin combination enhances anti-glioblastoma effect based on passive tumor targeting of liposomes. *J. Control. Release* 154 (1): 93–102.

33 Pham, W., Zhao, B.Q., Lo, E.H. et al. (2005). Crossing the blood–brain barrier: a potential application of myristoylated polyarginine for in vivo neuroimaging. *Neuroimage* 28 (1): 287–292.

34 He, H., Li, Y., Jia, X.R. et al. (2011). PEGylated poly(amidoamine) dendrimer-based dual-targeting carrier for treating brain tumors. *Biomaterials* 32 (2): 478–487.

35 Salata, O.V. (2004). Applications of NPs in biology and medicine. *J. Nanobiotechnol.* 2 (1): 3.

36 Rauf, M.A., Rehman, F.U., Zheng, M., and Shi, B. (2019). The strategies of nanomaterials for traversing blood-brain barrier. In: *Nanomedicine in Brain Diseases* (ed. X. Xue), 29–57. Singapore: Springer.

37 Zhan, C., Yan, Z., Xie, C., and Lu, W. (2010). Loop 2 of *Ophiophagus hannah* toxin b binds with neuronal nicotinic acetylcholine receptors and enhances intracranial drug delivery. *Mol. Pharm.* 7 (6): 1940–1947.

38 Zhan, C., Li, B., Hu, L. et al. (2011). Micelle-based brain-targeted drug delivery enabled by a nicotine acetylcholine receptor ligand. *Angew. Chem.* 123 (24): 5596–5599.

39 Li, J., Feng, L., Fan, L. et al. (2011). Targeting the brain with PEG–PLGA NPs modified with phage-displayed peptides. *Biomaterials* 32 (21): 4943–4950.

40 Lu, W., Wan, J., She, Z., and Jiang, X. (2007). Brain delivery property and accelerated blood clearance of cationic albumin conjugated pegylated NP. *J. Control. Release* 118 (1): 38–53.

41 Tao, X., Li, Y., Hu, Q. et al. (2018). Preparation and drug release study of novel nanopharmaceuticals with polysorbate 80 surface adsorption. *J. Nanomater.* 2018: 11. https://doi.org/10.1155/2018/4718045.

42 Dhanikula, R.S., Argaw, A., Bouchard, J.F., and Hildgen, P. (2008). Methotrexate loaded polyether-copolyester dendrimers for the treatment of gliomas: enhanced efficacy and intratumoral transport capability. *Mol. Pharm.* 5 (1): 105–116.

43 Han, L., Zhang, A., Wang, H. et al. (2010). Tat-BMPs-PAMAM conjugates enhance therapeutic effect of small interference RNA on U251 glioma cells in vitro and in vivo. *Hum. Gene Ther.* 21 (4): 417–426. doi: 10.1089/hum.2009.087. PMID: 19899955.

44 Huang, S., Li, J., Han, L. et al. (2011). Dual targeting effect of Angiopep-2-modified, DNA-loaded NPs for glioma. *Biomaterials* 32 (28): 6832–6838.

45 Xiang, Y., Wu, Q., Liang, L. et al. (2012). Chlorotoxin-modified stealth liposomes encapsulating levodopa for the targeting delivery against Parkinson's disease in the MPTP-induced mice model. *J Drug Target.* 20 (1): 67–75. https://doi.org/10.3109/1061186X.2011.595490. PMID: 22149216.

46 Veiseh, O., Sun, C., Gunn, J. et al. (2005). Optical and MRI multifunctional nanoprobe for targeting gliomas. *Nano Lett.* 5 (6): 1003–1008.

47 Pang, Z., Feng, L., Hua, R. et al. (2010). Lactoferrin-conjugated biodegradable polymersome holding doxorubicin and tetrandrine for chemotherapy of glioma rats. *Mol. Pharm.* 7 (6): 1995–2005.

48 Guo, J., Gao, X., Su, L. et al. (2011). Aptamer-functionalized PEG–PLGA NPs for enhanced anti-glioma drug delivery. *Biomaterials* 32 (31): 8010–8020.

49 Bernardi, A., Bavaresco, L., Wink, M.R. et al. (2007). Indomethacin stimulates activity and expression of ecto-5′-nucleotidase/CD73 in glioma cell lines. *Eur. J. Pharmacol.* 569 (1–2): 8–15.

50 Cole, A.J., David, A.E., Wang, J. et al. (2011). Magnetic brain tumor targeting and biodistribution of long-circulating PEG-modified, cross-linked starch-coated iron oxide NPs. *Biomaterials* 32 (26): 6291–6301.

51 Cheng, Y., Meyers, J.D., Agnes, R.S. et al. (2011). Addressing brain tumors with targeted gold NPs: a new gold standard for hydrophobic drug delivery? *Small* 7 (16): 2301–2306.

52 Walker, J.E. (1983). Glutamate, GABA, and CNS disease: a review. *Neurochem. Res.* 8 (4): 521–550.

53 Teng, Y.D. (2014). For neural therapy and repair. *Cell Transplant.* 23 (6): 761–790.

54 Emard, J.F., Thouez, J.P., and Gauvreau, D. (1995). Neurodegenerative diseases and risk factors: a literature review. *Soc. Sci. Med.* 40 (6): 847–858.

55 https://www.who.int/mediacentre/news/releases/2007/pr04/en/ (accessed 9 November 2020).

56 Tan, C.C., Yu, J.T., Wang, H.F. et al. (2014). Efficacy and safety of donepezil, galantamine, rivastigmine, and memantine for the treatment of Alzheimer's disease: a systematic review and meta-analysis. *J. Alzheimers Dis.* 41 (2): 615–631.

57 Olanow, C.W. and Stocchi, F. (2016, 2016). Safinamide: a new therapeutic option to address motor symptoms and motor complications in mid-to late-stage Parkinson's disease. *Eur. Neurol. Rev.* 11 (Suppl. 2): 2–15.

58 Tanzi, R.E. and Bertram, L. (2001). New frontiers in Alzheimer's disease genetics. *Neuron* 32 (2): 181–184.

59 Younkin, S.G. (1998). The role of Aβ42 in Alzheimer's disease. *J. Physiol.-Paris* 92 (3–4): 289–292.

60 Bush, A.I. (2003). The metallobiology of Alzheimer's disease. *Trends Neurosci.* 26 (4): 207–214.

61 Brendza, R.P., Bacskai, B.J., Cirrito, J.R. et al. (2005). Anti-Aβ antibody treatment promotes the rapid recovery of amyloid-associated neuritic dystrophy in PDAPP transgenic mice. *J. Clin. Invest.* 115 (2): 428–433.

62 Sun, P., Xiao, Y., Di, Q. et al. (2020). Transferrin receptor-targeted PEG-PLA polymeric micelles for chemotherapy against glioblastoma multiforme. *Int. J. Nanomed.* 15: 6673.

63 Chauhan, N.B., Davis, F., and Xiao, C. (2011). Wheat germ agglutinin enhanced cerebral uptake of anti-Aβ antibody after intranasal administration in 5XFAD mice. *Vaccine* 29 (44): 7631–7637.

64 Ono, K., Hasegawa, K., Naiki, H., and Yamada, M. (2004). Curcumin has potent anti-amyloidogenic effects for Alzheimer's β-amyloid fibrils in vitro. *J. Neurosci. Res.* 75 (6): 742–750.

65 Yang, F., Lim, G.P., Begum, A.N. et al. (2005). Curcumin inhibits formation of amyloid β oligomers and fibrils, binds plaques, and reduces amyloid in vivo. *J. Biol. Chem.* 280 (7): 5892–5901.

66 Mulik, R.S., Mönkkönen, J., Juvonen, R.O. et al. (2010). ApoE3 mediated poly (butyl) cyanoacrylate NPs containing curcumin: study of enhanced activity of curcumin against beta amyloid induced cytotoxicity using in vitro cell culture model. *Mol. Pharm.* 7 (3): 815–825.

67 Mathew, A., Fukuda, T., Nagaoka, Y. et al. (2012). Curcumin loaded-PLGA NPs conjugated with Tet-1 peptide for potential use in Alzheimer's disease. *PLoS One* 7 (3): e32616.

68 Sharma, M., Dube, T., Chibh, S. et al. (2019). Nanotheranostics, a future remedy of neurological disorders. *Expert Opin. Drug Deliv.* 16 (2): 113–128.

69 Bonda, D.J., Wang, X., Perry, G. et al. (2010). Oxidative stress in Alzheimer disease: a possibility for prevention. *Neuropharmacology* 59 (4–5): 290–294.

70 Perry, G., Cash, A.D., and Smith, M.A. (2002). Alzheimer disease and oxidative stress. *J. Biomed. Biotechnol.* 2: 542340. https://doi.org/10.1155/S1110724302203010.

71 Cui, Z., Lockman, P.R., Atwood, C.S. et al. (2005). Novel D-penicillamine carrying NPs for metal chelation therapy in Alzheimer's and other CNS diseases. *Eur. J. Pharm. Biopharm.* 59 (2): 263–272.

72 Babic, T. (1999). The cholinergic hypothesis of Alzheimer's disease: a review of progress. *J. Neurol. Neurosurg. Psychiatry* 67 (4): 558–558.

73 Grossberg, S. (2017). Acetylcholine neuromodulation in normal and abnormal learning and memory: vigilance control in waking, sleep, autism, amnesia and Alzheimer's disease. *Front. Neural Circ.* 11: 82.

74 Bonda, D.J., Lee, H.G., Blair, J.A. et al. (2011). Role of metal dyshomeostasis in Alzheimer's disease. *Metallomics* 3 (3): 267–270.

75 Yu, Y., Pang, Z., Lu, W. et al. (2012). Self-assembled polymersomes conjugated with lactoferrin as novel drug carrier for brain delivery. *Pharm. Res.* 29 (1): 83–96.

76 Zhang, C., Wan, X., Zheng, X. et al. (2014). Dual-functional nanoparticles targeting amyloid plaques in the brains of Alzheimer's disease mice. *Biomaterials* 35 (1): 456–465.

77 Li, J., Feng, L., Fan, L. et al. (2011). Targeting the brain with PEG–PLGA nanoparticles modified with phage-displayed peptides. *Biomaterials* 32 (21): 4943–4950.

78 Gao, H., Pang, Z., and Jiang, X. (2013). Targeted delivery of nano-therapeutics for major disorders of the central nervous system. *Pharm. Res.* 30 (10): 2485–2498.

79 Tam, D.Y., Ho, J.W.T., Chan, M.S. et al. (2020). Penetrating the blood-brain barrier by self-assembled 3D DNA nanocages as drug delivery vehicles for brain cancer therapy. *ACS Appl. Mater. Interfaces* 12 (26): 28928–28940.

80 Pang, Z.Q., Fan, L., Hu, K.L. et al. (2010). Effect of lactoferrin- and transferrin-conjugated polymersomes in brain targeting: in vitro and in vivo evaluations. *Acta Pharmacol. Sin.* 31 (2): 237–243.

81 Mishima, K., Tsukikawa, H., Miura, I. et al. (2003). Ameliorative effect of NC-1900, a new AVP4–9 analog, through vasopressin V1A receptor on scopolamine-induced impairments of spatial memory in the eight-arm radial maze. *Neuropharmacology* 44 (4): 541–552.

82 Wu, H., Hu, K., and Jiang, X. (2008). From nose to brain: understanding transport capacity and transport rate of drugs. *Expert Opin. Drug Deliv.* 5 (10): 1159–1168.

83 Gao, X., Wu, B., Zhang, Q. et al. (2007). Brain delivery of vasoactive intestinal peptide enhanced with the NPsconjugated with wheat germ agglutinin following intranasal administration. *J. Control. Release* 121 (3): 156–167.

84 Gao, X., Tao, W., Lu, W. et al. (2006). Lectin-conjugated PEG–PLA NPs: preparation and brain delivery after intranasal administration. *Biomaterials* 27 (18): 3482–3490.

85 Li, J., Wu, H., Hong, J. et al. (2008). Odorranalectin is a small peptide lectin with potential for drug delivery and targeting. *PLoS One* 3 (6): e2381.

86 Wu, H., Li, J., Zhang, Q. et al. (2012). A novel small odorranalectin-bearing cubosomes: preparation, brain delivery and pharmacodynamic study on Aβ25–35-treated rats following intranasal administration. *Eur. J. Pharm. Biopharm.* 80 (2): 368–378.

87 Srikanth, M. and Kessler, J.A. (2012). Nanotechnology—novel therapeutics for CNS disorders. *Nat. Rev. Neurol.* 8 (6): 307–318.

88 Soler, R., Füllhase, C., Hanson, A. et al. (2012). Stem cell therapy ameliorates bladder dysfunction in an animal model of Parkinson disease. *J. Urol.* 187 (4): 1491–1497.

89 Alonso-Frech, F., Sanahuja, J.J., and Rodriguez, A.M. (2011). Exercise and physical therapy in early management of Parkinson disease. *Neurologist* 17: S47–S53.

90 Ron, D. and Janak, P.H. (2005). GDNF and addiction. *Rev. Neurosci.* 16 (4): 277–286.

91 Abuirmeileh, A., Lever, R., Kingsbury, A.E. et al. (2007). The corticotrophin-releasing factor-like peptide urocortin reverses key deficits in two rodent models of Parkinson's disease. *Eur. J. Neurosci.* 26 (2): 417–423.

92 Hu, K., Li, J., Shen, Y. et al. (2009). Lactoferrin-conjugated PEG–PLA NPs with improved brain delivery: in vitro and in vivo evaluations. *J. Control. Release* 134 (1): 55–61.

93 Hu, K., Shi, Y., Jiang, W. et al. (2011). Lactoferrin conjugated PEG-PLGA NPs for brain delivery: preparation, characterization and efficacy in Parkinson's disease. *Int. J. Pharm.* 415 (1–2): 273–283.

94 Huang, R., Han, L., Li, J. et al. (2009). Neuroprotection in a 6-hydroxydopamine-lesioned Parkinson model using lactoferrin-modified NPs. *J. Gene Med. A* 11 (9): 754–763.

95 Wen, Z., Yan, Z., Hu, K. et al. (2011). Odorranalectin-conjugated NPs: preparation, brain delivery and pharmacodynamic study on Parkinson's disease following intranasal administration. *J. Control. Release* 151 (2): 131–138.

96 Kordower, J.H. and Bjorklund, A. (2013). Trophic factor gene therapy for Parkinson's disease. *Mov. Disord.* 28 (1): 96–109.

97 Zhang, Y., Schlachetzki, F., Zhang, Y.F. et al. (2004). Normalization of striatal tyrosine hydroxylase and reversal of motor impairment in experimental parkinsonism with intravenous nonviral gene therapy and a brain-specific promoter. *Hum. Gene Ther.* 15 (4): 339–350.

98 Luk, K.C. and Lee, V.M.Y. (2014). Modeling Lewy pathology propagation in Parkinson's disease. *Parkinsonism Relat. Disord.* 20: S85–S87.

99 Gibb, W.R. and Lees, A. (1988). The relevance of the Lewy body to the pathogenesis of idiopathic Parkinson's disease. *J. Neurol. Neurosurg. Psychiatry* 51 (6): 745–752.

100 Kumari, S., Ahsan, S.M., Kumar, J.M. et al. (2017). Overcoming blood–brain barrier with a dual purpose Temozolomide loaded Lactoferrin NPs for combating glioma (SERP-17-12433). *Sci. Rep.* 7 (1): 1–13.

101 Patil, P.M., Chaudhari, P.D., Sahu, M., and Duragkar, N.J. (2012). Review article on gene therapy. *Res. J. Pharmacol. Pharmacodyn.* 4 (2): 77–83.

102 Mandel, R.J., Burger, C., and Snyder, R.O. (2008). Viral vectors for in vivo gene transfer in Parkinson's disease: properties and clinical grade production. *Exp. Neurol.* 209 (1): 58–71.

103 Davidson, B.L. and Breakefield, X.O. (2003). Viral vectors for gene delivery to the nervous system. *Nat. Rev. Neurosci.* 4 (5): 353–364.

104 Cooke, M.J., Wang, Y., Morshead, C.M., and Shoichet, M.S. (2011). Controlled epi-cortical delivery of epidermal growth factor for the stimulation of endogenous neural stem cell proliferation in stroke-injured brain. *Biomaterials* 32 (24): 5688–5697.

105 Gupta, A., Nair, S., Schweitzer, A.D. et al. (2012). Neuroimaging of cerebrovascular disease in the aging brain. *Aging Dis.* 3 (5): 414.

106 Tong, X., Yang, Q., Ritchey, M.D. et al. (2019). Peer reviewed: the burden of cerebrovascular disease in the United States. *Prevent. Chronic Dis.* 16: https://doi.org/10.5888/pcd16.180411.

107 Sun, Y., Feng, X., Ding, Y. et al. (2019). Phased treatment strategies for cerebral ischemia based on glutamate receptors. *Front. Cell. Neurosci.* 13: 168.

108 Bendix Johnsen, K., Burkhart, A., Bohn Thomsen, L., and Andresen, T. (2019). Targeting the transferrin receptor for brain drug delivery. *Prog. Neurobiol.* 181: 101665.

109 Zhang, Y. and Pardridge, W.M. (2001). Conjugation of brain-derived neurotrophic factor to a blood–brain barrier drug targeting system enables neuroprotection in regional brain ischemia following intravenous injection of the neurotrophin. *Brain Res.* 889 (1–2): 49–56.

110 Kamarudin, S.N., Iezhitsa, I., Tripathy, M. et al. (2020). Neuroprotective effect of poly(lactic-*co*-glycolic acid) NP-bound brain-derived neurotrophic factor in a permanent middle cerebral artery occlusion model of ischemia in rats. *Acta Neurobiol. Exp.* 80: 1–18.

111 Song, B.W., Vinters, H.V., Wu, D., and Pardridge, W.M. (2002). Enhanced neuroprotective effects of basic fibroblast growth factor in regional brain ischemia after conjugation to a blood-brain barrier delivery vector. *J. Pharmacol. Experimental Therapeutics* 301 (2): 605–610.

112 Jain, K.K. (2000). Neuroprotection in cerebrovascular disease. *Expert Opin. Investig. Drugs* 9 (4): 695–711.

113 Lewen, A., Matz, P., and Chan, P.H. (2000). Free radical pathways in CNS injury. *J. Neurotrauma* 17 (10): 871–890.

114 Siesjö, B.K., Zhao, Q., Pahlmark, K. et al. (1995). Glutamate, calcium, and free radicals as mediators of ischemic brain damage. *Ann. Thorac. Surg.* 59 (5): 1316–1320.

115 Gulati, A., Barve, A., and Sen, A.P. (1999). Pharmacology of hemoglobin therapeutics. *J. Lab. Clin. Med.* 133 (2): 112–119.

116 Vandegriff, K.D., Malavalli, A., and Olsen, S.D. (2018). Diaspirin Crosslinked Pegylated Hemoglobin. US Patent No. 10, 029, 001. Washington, DC: U.S. Patent and Trademark Office.

117 Lee, J., Lee, J., Yoon, S., and Nho, K. (2006). Pharmacokinetics of 125I-radiolabelled PEG-hemoglobin SB1. *Artif. Cells, Blood Substit. Biotechnol.* 34 (3): 277–292.

118 Ji, H.J., Chai, H.Y., Nahm, S.S. et al. (2007). Neuroprotective effects of the novel polyethylene glycol-hemoglobin conjugate SB1 on experimental cerebral thromboembolism in rats. *Eur. J. Pharmacol.* 566 (1–3): 83–87.

119 Erlandsson, A., Lin, C.H.A., Yu, F., and Morshead, C.M. (2011). Immunosuppression promotes endogenous neural stem and progenitor cell migration and tissue regeneration after ischemic injury. *Exp. Neurol.* 230 (1): 48–57.

120 Kolb, B., Morshead, C., Gonzalez, C. et al. (2007). Growth factor-stimulated generation of new cortical tissue and functional recovery after stroke damage to the motor cortex of rats. *J. Cereb. Blood Flow Metab.* 27 (5): 983–997.

121 Wang, Y., Cooke, M.J., Lapitsky, Y. et al. (2011). Transport of epidermal growth factor in the stroke-injured brain. *J. Control. Release* 149 (3): 225–235.

122 Wang, Y., Cooke, M.J., Morshead, C.M., and Shoichet, M.S. (2012). Hydrogel delivery of erythropoietin to the brain for endogenous stem cell stimulation after stroke injury. *Biomaterials* 33 (9): 2681–2692.

123 Sun, Y., Jin, K., Xie, L. et al. (2003). VEGF-induced neuroprotection, neurogenesis, and angiogenesis after focal cerebral ischemia. *J. Clin. Invest.* 111 (12): 1843–1851.

124 Zhao, H., Bao, X.J., Wang, R.Z. et al. (2011). Postacute ischemia vascular endothelial growth factor transfer by transferrin-targeted liposomes attenuates ischemic brain injury after experimental stroke in rats. *Hum. Gene Ther.* 22 (2): 207–215.

125 Hassanzadeh, P., Arbabi, E., Atyabi, F., and Dinarvand, R. (2017). Nerve growth factor-carbon nanotube complex exerts prolonged protective effects in an in vitro model of ischemic stroke. *Life Sci.* 179: 15–22.

126 Al-Jamal, K.T., Gherardini, L., Bardi, G. et al. (2011). Functional motor recovery from brain ischemic insult by carbon nanotube-mediated siRNA silencing. *Proc. Natl. Acad. Sci. U.S.A.* 108 (27): 10952–10957.

127 Budihardjo, I., Oliver, H., Lutter, M. et al. (1999). Biochemical pathways of caspase activation during apoptosis. *Annu. Rev. Cell Dev. Biol.* 15 (1): 269–290.

128 Patel, S.S. and Udayabanu, M. (2017). Effect of natural products on diabetes associated neurological disorders. *Rev. Neurosci.* 28 (3): 271–293.

129 Yenari, M.A. and Han, H.S. (2012). Neuroprotective mechanisms of hypothermia in brain ischaemia. *Nat. Rev. Neurosci.* 13 (4): 267–278.

130 Margulies, S. and Hicks, R. (2009). Combination therapies for traumatic brain injury: prospective considerations. *J. Neurotrauma* 26 (6): 925–939.

131 Wong, K.T. (2020). *Emerging CNS Infections. Infections of the Central Nervous System: Pathology & Genetics* (ed. K.T. Wong), 505–514. Kuala Lumpur, Malaysia: Department of Pathology, Faculty of Medicine, University of Malaya.

132 Huang, S.H. and Jong, A.Y. (2001). Cellular mechanisms of microbial proteins contributing to invasion of the blood–brain barrier: MicroReview. *Cell. Microbiol.* 3 (5): 277–287.

133 Bicker, J., Alves, G., Fortuna, A., and Falcão, A. (2014). Blood–brain barrier models and their relevance for a successful development of CNS drug delivery systems: a review. *Eur. J. Pharm. Biopharm.* 87 (3): 409–432.

134 Casadevall, A. (2010). Cryptococci at the brain gate: break and enter or use a Trojan horse? *J. Clin. Invest.* 120 (5): 1389–1392.

135 Santiago-Tirado, F.H. and Doering, T.L. (2017). False friends: phagocytes as Trojan horses in microbial brain infections. *PLoS Pathog.* 13 (12): e1006680.

136 Wang, Y. and Kim, K.S. (2002). Role of OmpA and IbeB in *Escherichia coli* K1 invasion of brain microvascular endothelial cells in vitro and in vivo. *Pediatr. Res.* 51 (5): 559–563.

137 Badger, J.L., Stins, M.F., and Kim, K.S. (1999). *Citrobacter freundii* invades and replicates in human brain microvascular endothelial cells. *Infect. Immun.* 67 (8): 4208–4215.

138 Liu, N.Q., Lossinsky, A.S., Popik, W. et al. (2002). Human immunodeficiency virus Type 1 enters brain microvascular endothelia by macropinocytosis dependent on lipid rafts and the mitogen-activated protein kinase signaling pathway. *J. Virol.* 76 (13): 6689–6700.

139 Bobardt, M.D., Salmon, P., Wang, L. et al. (2004). Contribution of proteoglycans to human immunodeficiency virus Type 1 brain invasion. *J. Virol.* 78 (12): 6567–6584.

140 Ring, A., Weiser, J.N., and Tuomanen, E.I. (1998). Pneumococcal trafficking across the blood-brain barrier. Molecular analysis of a novel bidirectional pathway. *J. Clin. Invest.* 102 (2): 347–360.

141 Banks, W.A., Freed, E.O., Wolf, K.M. et al. (2001). Transport of human immunodeficiency virus Type 1 pseudoviruses across the blood-brain barrier: role of envelope proteins and adsorptive endocytosis. *J. Virol.* 75 (10): 4681–4691.

142 Wolka, A.M., Huber, J.D., and Davis, T.P. (2003). Pain and the blood–brain barrier: obstacles to drug delivery. *Adv. Drug Deliv. Rev.* 55 (8): 987–1006.

143 Zhang, Y.L., Ouyang, Y.B., Liu, L.G., and Chen, D.X. (2015). Blood-brain barrier and neuro-AIDS. *Eur. Rev. Med. Pharmacol. Sci.* 19 (24): 4927–4939.

144 Saxena, S.K., Tiwari, S., and Nair, M.P. (2012). Nanotherapeutics: emerging competent technology in neuroAIDS and CNS drug delivery. *Nanomedicine* 7 (7): 941–944.

145 Pedemonte, E., Mancardi, G., Giunti, D. et al. (2006). Mechanisms of the adaptive immune response inside the central nervous system during inflammatory and autoimmune diseases. *Pharmacol. Ther.* 111 (3): 555–566.

146 Saxena, S.K., Maurya, V.K., Kumar, S., and Bhatt, M.L. (2020). Modern approaches in nanomedicine for NeuroAIDS and CNS drug delivery. In: *NanoBioMedicine* (ed. S.K. Saxena and S.M.P. Khurana), 199–211. Singapore: Springer.

147 Das, M.K., Sarma, A., and Chakraborty, T. (2016). Nano-ART and neuroAIDS. *Drug Deliv. Transl. Res.* 6 (5): 452–472.

148 Honeycutt, J.B. and Garcia, J.V. (2018). Humanized mice: models for evaluating NeuroHIV and cure strategies. *J. Neurovirol.* 24 (2): 185–191.

149 Jayant, R.D., Atluri, V.S., Agudelo, M. et al. (2015). Sustained-release nanoART formulation for the treatment of neuroAIDS. *Int. J. Nanomed.* 10: 1077.

150 Polman, C.H., Matthaei, I., De Groot, C.J.A. et al. (1988). Low-dose cyclosporin A induces relapsing remitting experimental allergic encephalomyelitis in the Lewis rat. *J. Neuroimmunol.* 17 (3): 209–216.

151 Lucchinetti, C.F., Parisi, J., and Bruck, W. (2005). The pathology of multiple sclerosis. *Neurol. Clin.* 23 (1): 77–105.

152 Owens, T. and Sriram, S. (1995). The immunology of multiple sclerosis and its animal model, experimental allergic encephalomyelitis. *Neurol. Clin.* 13 (1): 51–73.

153 Schmidt, J., Metselaar, J.M., Wauben, M.H. et al. (2003). Drug targeting by long-circulating liposomal glucocorticosteroids increases therapeutic efficacy in a model of multiple sclerosis. *Brain* 126 (8): 1895–1904.

154 Soussain, C., Ricard, D., Fike, J.R. et al. (2009). CNS complications of radiotherapy and chemotherapy. *Lancet* 374 (9701): 1639–1651.

155 Costantini, L.C., Bakowska, J.C., Breakefield, X.O., and Isacson, O. (2000). Gene therapy in the CNS. *Gene Ther.* 7 (2): 93–109.

156 Hanif, S., Muhammad, P., Chesworth, R. et al. (2020). Nanomedicine-based immunotherapy for central nervous system disorders. *Acta Pharmacol. Sin.* 41 (7): 1–18.

157 Freskgård, P.O. and Urich, E. (2017). Antibody therapies in CNS diseases. *Neuropharmacology* 120: 38–55.

158 Jeffrey, P. and Summerfield, S. (2010). Assessment of the blood–brain barrier in CNS drug discovery. *Neurobiol. Dis.* 37 (1): 33–37.

159 Wong, D.F., Tauscher, J., and Gründer, G. (2009). The role of imaging in proof of concept for CNS drug discovery and development. *Neuropsychopharmacology* 34 (1): 187–203.

160 Mansor, N.I., Nordin, N., Mohamed, F. et al. (2019). Crossing the blood-brain barrier: a review on drug delivery strategies for treatment of the central nervous system diseases. *Curr. Drug Deliv.* 16 (8): 698–711.

161 Jain, K.K. (2009). Current status and future prospects of nanoneurology. *Journal of Nanoneuroscience* 1 (1): 56–64.

162 Mukhtar, M., Bilal, M., Rahdar, A. et al. (2020). Nanomaterials for diagnosis and treatment of brain cancer: recent updates. *Chemosensors* 8 (4): 117.

163 Zeeshan, F., Mishra, D.K., and Kesharwani, P. From the nose to the brain, nanomedicine drug delivery. In: *Theory and Applications of Nonparenteral Nanomedicines* (ed. J. Siepmann), 153–180. Academic Press.

164 Kovacs, G.G. (2015). Invited review: neuropathology of tauopathies: principles and practice. *Neuropathol. Appl. Neurobiol.* 41 (1): 3–23.

165 DeLeo, A.M. and Ikezu, T. (2018). Extracellular vesicle biology in Alzheimer's disease and related tauopathy. *J. Neuroimmune Pharmacol.* 13 (3): 292–308.

166 Hernandez, F. and Avila, J. (2007). Tauopathies. *Cell. Mol. Life Sci.* 64 (17): 2219–2233.

167 Vimal, S.K., Zuo, H., Wang, Z. et al. (2020). Self-therapeutic NP that alters tau protein and ameliorates tauopathy toward a functional nanomedicine to tackle Alzheimer's. *Small* 16 (16): 1906861.

168 Grauer, O.M., Nierkens, S., Bennink, E. et al. (2007). CD^{4+} $FoxP^{3+}$ regulatory T cells gradually accumulate in gliomas during tumor growth and efficiently suppress antiglioma immune responses in vivo. *Int. J. Cancer* 121 (1): 95–105.

169 Kozielski, K.L., Ruiz-Valls, A., Tzeng, S.Y. et al. (2019). Cancer-selective NPs for combinatorial siRNA delivery to primary human GBM in vitro and in vivo. *Biomaterials* 209: 79–87.

170 Rosenberg, S.A. (1991). Immunotherapy and gene therapy of cancer. *Cancer Res.* 51 (18 Suppl): 5074s–5079s.

171 Speranza, M.C., Passaro, C., Ricklefs, F. et al. (2018). Preclinical investigation of combined gene-mediated cytotoxic immunotherapy and immune checkpoint blockade in glioblastoma. *Neuro Oncol.* 20 (2): 225–235.

172 Badie, B. and Berlin, J.M. (2013). The future of CpG immunotherapy in cancer. *Immunotherapy* 5 (1): 1–3.

173 Lim, M., Xia, Y., Bettegowda, C., and Weller, M. (2018). Current state of immunotherapy for glioblastoma. *Nat. Rev. Clin. Oncol.* 15 (7): 422–442.

174 Lollo, G., Vincent, M., Ullio-Gamboa, G. et al. (2015). Development of multifunctional lipid nanocapsules for the co-delivery of paclitaxel and CpG-ODN in the treatment of glioblastoma. *Int. J. Pharm.* 495 (2): 972–980.

175 Kadiyala, P., Li, D., Nuñez, F.M. et al. (2019). High-density lipoprotein-mimicking nanodiscs for chemo-immunotherapy against glioblastoma multiforme. *ACS Nano* 13 (2): 1365–1384.

176 Kobayashi, H. (2020). Theranostic near-infrared photoimmunotherapy. In: *Make Life Visible* (ed. Y. Toyama, M. Nakamura, A. Miyawaki, et al.), 219–225. Springer, Singapore.

177 Sandland, J. and Boyle, R.W. (2019). Photosensitizer antibody–drug conjugates: past, present, and future. *Bioconjug. Chem.* 30 (4): 975–993.

178 Qian, W., Qian, M., Wang, Y. et al. (2018). Combination glioma therapy mediated by a dual-targeted delivery system constructed using OMCN–PEG–Pep22/DOX. *Small* 14 (42): 1801905.

179 Li, F., Cheng, Y., Lu, J. et al. (2011). Photodynamic therapy boosts anti-glioma immunity in mice: a dependence on the activities of T cells and complement C3. *J. Cell. Biochem.* 112 (10): 3035–3043.

180 Burley, T.A., Mączyńska, J., Shah, A. et al. (2018). Near-infrared photoimmunotherapy targeting EGFR—shedding new light on glioblastoma treatment. *Int. J. Cancer* 142 (11): 2363–2374.

181 Mitsunaga, M., Ogawa, M., Kosaka, N. et al. (2011). Cancer cell–selective in vivo near infrared photoimmunotherapy targeting specific membrane molecules. *Nat. Med.* 17 (12): 1685–1691.

182 Huang, H.C., Pigula, M., Fang, Y., and Hasan, T. (2018). Immobilization of photo-immunoconjugates on NPs leads to enhanced light-activated biological effects. *Small* 14 (31): 1800236.

183 Kudarha, R.R. and Sawant, K.K. (2020). Chondroitin sulfate conjugation facilitates tumor cell internalization of albumin nanoparticles for brain-targeted delivery of temozolomide via CD44 receptor-mediated targeting. *Drug Deliv. Transl. Res.* 1–15.

184 Juillerat-Jeanneret, L. (2008). The targeted delivery of cancer drugs across the blood–brain barrier: chemical modifications of drugs or drug-nanoparticles? *Drug Discov. Today* 13 (23–24): 1099–1106.

185 Boulikas, T., Pantos, A., Bellis, E., and Christofis, P. (2007). Designing platinum compounds in cancer: structures and mechanisms. *Cancer Ther.* 5: 537–583.

186 Pugazhendhi, A., Edison, T.N.J.I., Velmurugan, B.K. et al. (2018). Toxicity of doxorubicin (DOX) to different experimental organ systems. *Life Sci.* 200: 26–30.

187 Ferraris, C., Cavalli, R., Panciani, P.P., and Battaglia, L. (2020). Overcoming the blood–brain barrier: successes and challenges in developing nanoparticle-mediated drug delivery systems for the treatment of brain tumours. *Int. J. Nanomed.* 15: 2999.

188 Chen, Y. and Huang, L. (2008). Tumor-targeted delivery of siRNA by non-viral vector: safe and effective cancer therapy. *Expert Opin. Drug Deliv.* 5 (12): 1301–1311.

189 Karyagina, T.S., Ulasov, A.V., Slastnikova, T.A. et al. (2020). Targeted delivery of 111In into the nuclei of EGFR overexpressing cells via modular nanotransporters with anti-EGFR affibody. *Front. Pharmacol.* 11: 176.

190 Kristjanson, L.J., Toye, C., and Dawson, S. (2003). New dimensions in palliative care: a palliative approach to neurodegenerative diseases and final illness in older people. *Med. J. Australia* 179: S41–S43.

191 Mushtaq, G., Khan, J.A., Joseph, E., and Kamal, M.A. (2015). Nanoparticles, neurotoxicity and neurodegenerative diseases. *Curr. Drug Metab.* 16 (8): 676–684.

192 Saraiva, C., Praça, C., Ferreira, R. et al. (2016). Nanoparticle-mediated brain drug delivery: overcoming blood–brain barrier to treat neurodegenerative diseases. *J. Control. Release* 235: 34–47.

193 Trippier, P.C., Jansen Labby, K., Hawker, D.D. et al. (2013). Target-and mechanism-based therapeutics for neurodegenerative diseases: strength in numbers. *J. Med. Chem.* 56 (8): 3121–3147.

Index

e

q

r